ECOLOGY OF ARABLE LAND – PERSPECTIVES AND CHALLENGES

Developments in Plant and Soil Sciences

VOLUME 39

Ecology of Arable Land – Perspectives and Challenges

Proceedings of an International Symposium, 9–12 June 1987
Swedish University of Agricultural Sciences, Uppsala, Sweden

Edited by
M. Clarholm
Department of Ecology and Environmental Research
Swedish University of Agricultural Sciences, Uppsala, Sweden

and
L. Bergström
Department of Soil Science
Swedish University of Agricultural Sciences, Uppsala, Sweden

Springer-Science+Business Media, B.V.

Library of Congress Cataloging in Publication Data

```
Ecology of arable land : perspectives and challenges / edited by M.
Clarholm and L. Bergström.
        p.   cm. -- (Developments in plant and soil sciences ; v. 39)
     "Proceedings of an international symposium 9-12 June, 1987, Swedish
University of Agricultural Sciences, Uppsala, Sweden."
   ISBN 978-94-010-6950-2
     1. Agricultural ecology--Congresses.  2. Soil ecology--Congresses.
3. Plant-soil relationships--Congresses.   I. Clarholm, Marianne.
II. Bergström, L.   III. Series.
S589.7.E26  1989
630'.2'745--dc20                                           89-15587
```

ISBN 978-94-010-6950-2   ISBN 978-94-009-1021-8 (eBook)
DOI 10.1007/978-94-009-1021-8

Published by Kluwer Academic Publishers,
P.O. Box 17, 3300 AA Dordrecht, The Netherlands.

Kluwer Academic Publishers incorporates
the publishing programmes of
Martinus Nijhoff, Dr W. Junk, D. Reidel and MTP Press.

Sold and distributed in the U.S.A. and Canada
by Kluwer Academic Publishers,
101 Philip Drive, Norwell, MA 02061, U.S.A.

In all other countries, sold and distributed
by Kluwer Academic Publishers Group,
P.O. Box 322, 3300 AH Dordrecht, The Netherlands.

Contents

vi

Preface

Agriculture in the industrial world has gone through dramatic changes over the past decades. Mechanization in combination with high inputs of fertilizers and pesticides has turned deficits of agricultural products into surplus. Over the same period we have experienced increased environmental problems in both the atmosphere and our water resources, which have been associated with the changes in management practices.

Concern about the potential pollution by nitrogen fertilizers as well as the low utilization efficiency of applied nitrogen by plants has created a need for a better understanding of nitrogen cycling in the plant-soil-water system. To achieve this, it is neccessary to study process interactions and process regulation in an ecosystem context. During the last decade many ecosystem studies have been initiated where more comprehensive sets of data have been gathered, mostly with a high resolution and synchronization in time.

A multidisciplinary research project emphasising the ecosystem approach was established in Sweden in 1979 entitled: "Ecology of Arable Land. The Role of Organisms in Nitrogen Cycling". The objective of the project was to investigate the functions of soil microogranisms and soil fauna, with particular attention to their importance for the circulation of nitrogen and carbon in four cropping systems. The experimental phase of the project was finished in 1985, while the termination date for the project was 1 July 1987.

A synthesis volume "Ecology of Arable Land —Organisms, Nitrogen and Carbon Cycling" edited by O Andrén, T Lindberg, K Paustian and T Rosswall will be published by Ecological Bulletins (Copenhagen) in 1989. A workshop was held in conjunction with the preparation of the synthesis volume in May 1987, where a number of leading scientists were invited to review the synthesis efforts to date. To take full advantage of their presence, an open symposium was arranged immediately after the workshop. The speakers were asked to cover "new and challenging findings" in their area of expertise and also to indicate future directions of research. This volume contains their contributions.

A common interest of the contributors is increasing the understanding of the turnover of carbon and inorganic nutrients in terestrial ecosystems. The authors approach this topic from different directions depending on their interests and expertise. Difficulties are identified in the quantification of below-ground production where death and re-growth, if incorporated into the calculations, can change production figures considerably as compared to values derived from "peak" estimates. The role of root-derived carbon is investigated in relation to nutrient competition between roots and microorganisms, the cost of N_2 fixation and the decompositon of organic nitrogen. Mycorrhizae use root-derived carbon and their roles in phosphorus conservation and in supplying nutrients to the host are exemplified. The contribution of different animal groups to soil fertility is reported for two native ecosystems, a forest and a grassland. Methodological as well as evolutionary discussions are given on nitrification and denitrification, two processes of increasing importance with generally increased inputs of nitrogen to ecosystems. With increased amounts of mobile nitrogen in ecosystems, the transfer of nitrogen between different ecosystems has also become apparent. Two papers describe nutrient movements across landscapes and ways to reduce contamination of surface- and ground-water.

Models are presented by many investigators for such different tasks as the simulation of long-term trends of nutrient status in separate soil types and for the prediction of crop outcome from remote sensing mesurements. Geostatistics and its possible use in investigations of animal populations and denitrificaiton rates are also presented. The book ends with a chapter containing general outlooks for future agriculture. All papers were revised in 1988 after referee treatment.

We thank the other members of the organizing committee, Olof Andrén, Ann-Charlotte Hansson, Leif Klemedtsson, Torbjörn Lindberg and Thomas Rosswall, for their generous participation in all possible ways, from the planning of the scientific programme to the practical parts involved in running the meeting. We are also grateful for the help from professor Eliel Steen in editing the papers.

The conference obtained financial support from the Swedish Council for Planning and Coordination of Research, Swedish Council for Forestry and Agricultural Research, Swedish Natural Science Research Council, Swedish Environmental Protection Board and Swedish University of Agricultural Sciences.

Uppsala, March 1989
Marianne Clarholm   Lars Bergström

M. Clarholm and L. Bergström (Eds.), Ecology of arable land, 1–7.

The application to agriculture of predictive plant production models based on regional experimental data

B.W.R TORSSELL
Department of Plant Husbandry, Swedish University of Agricultural Sciences, P.O. Box 7043, S-75007, Uppsala, Sweden

Key words: computer technique training, growth models, leys, parameter estimation, plant production.

Abstract

In Scandinavia complex and simple mechanistic models have both been applied to modelling above-ground plant production. Ignoring the need for detailed process understanding, it appears that for practical applications the less complex models provide as good predictions as the more complex. Basing the parameter estimates on representative agronomic experiments, the simple models offer a useful summary of existing agronomic data, and it is felt that in spite of their simplicity, realistic interactions may be calculated in many cases.

The combination of plant production models with pest models and with models for calculation of losses and economic returns, is now seen as a useful tool in management planning. Models are particularly useful when the planning concerns the current season and the calculations are based on current sampling of plant and weather and on weekly weather forecasts. However, the design and application of these techniques requires ordinary personnel to be specially trained, which presently is taken too lightly.

Introduction

It is now generally accepted that plant production models may have at least the following three applications to agricultural research and extension: i) a tool in data analyses, research planning and in education, ii) a means for predicting yield or harvest time and iii) a subroutine in larger models supporting calculations of nutrient flows, soil environment dynamics, pest control, crop management, and/or economic calculations.

A less common mode of expressing these applications is to regard the models: i) as a means of compiling existing data in large databanks and ii) as a means of extracting new information from our present knowledge of the agricultural system.

The first point is of particular interest in Scandinavian agricultural research because we possess huge volumes of plant production data from more than 50 years of Programmed Applied Research (PrAR).

The second point opens the possibility of using these data in combination with data from more detailed, recent measurements such as those from the Ecology of Arable Land project. Simulation of *e.g.* total biomass production of crops in relation to long-term soil dynamics and environmental effects then becomes possible, as convincingly shown by the Fort Collins Group at this conference (Cole *et al.*, Stewart and Cole, 1989).

In agricultural research there has often been some rivalry between the well established analysis of variance approach and the more daring systems analysis approach aiming at mechanistic, explanatory models. In Sweden we are now gradually easing this tension by realising that our Programmed Applied Research (PrAP) possesses important data for parameter estimation and model validation, whereas systems analysis provides a very rational way of analysing and understanding results from agricultural experimentation in general and from the PrAR in particular. The acceptance of this philosophy is documented in reports from research conferences in both horticulture (Torssell, 1986) and in agriculture (Torssell *et al.*, 1987). However, as will be shown in the discussion below, the implementation of these principles imposes real problems because of a lack of insight in the advanced techniques required.

Systems definition and level of complexity

Another reason for differences of opinions in the modelling approach concerns the level of complexity in relation to model definition and aim. From the early contribution of Duncan *et al.* (1967) and de Wit and his groups, mechanistic models of great complexity operating on short time scales have been developed at many research centers, cf. reviews by WMO (1983a, b), Hårsmar (1983) and Torssell *et al.* (1985). Certainly, these models have contributed significantly to our understanding of field crop physiology, soil dynamics and related subjects.

However, the author will claim that the very simple model concepts developed at an early stage for water balance (Slatyer, 1960), for growth-related weather indices (Fitzpatric and Nix, 1970) and growth accumulation expressed by a Gompertz curve (Thornley, 1976) have had at least as great an impact on the success of modelling plant growth as the complex models. Not always do the complex models agree with the simple models as found when comparing the precision of *e.g.* the complex grass production model of Hansen (1986) with the far simpler model of Torssell and Kornher (1983) which employs the integrating concepts of Slatyer (1960), Fitzpatric and Nix (1970) and Thornley (1976) referred to above.

Parameter estimation

The main point in this paper is to focus the attention on parameter estimation in plant production models. If such models are to be useful in an agricultural context, their parameters must reflect field situations.

Detailed measurement

The most common view on parameter estimation in growth models is to use data from controlled environments or detailed measurements in the field. The more detailed the measurements are, the more detailed mechanistic models may be developed and then used for further data analyses and parameter estimation. The data collection undertaken in the present project "Ecology of arable land" will most probably fall within this category. However, so far very little has been done in the project with regard to parameter estimation in plant production models, but potentially the data is very useful for such purposes. It is becoming more evident that such parameter estimations are required if the results are to be extrapolated to sites and other environments other than the rather specific experimental site in this particular project.

Programmed Applied Research (PrAR) data

A less common approach among biological systems analysts is to design the models so that most of their essential parameters can be estimated from results of ordinary field experiments. This will force the models towards simplicity and will tend to formulate mesurable parameters integrating broader processes.

We will illustrate this point with just one example, and return then to the simple mechanistic growth model of Torssell and Kornher (1983). The daily growth increment of biomass, W, may be expressed as

$$\Delta W/\Delta t = W \cdot R_s \cdot AGE \cdot GI \qquad (1)$$

where R_s is the initial (highest) value of the relative growth rate, AGE a function expressing the decline in R_s with growth and GI the weather index of Fitzpatric and Nix (1970).

The parameter R_s has thus a rather broad definition expressing the growth potential of the sward at the beginning of growth. By using PrAR data combined with corresponding weather and soil information, Kornher and Torssell (1983) were able to determine R_s in data sets which may be considered as huge in the context of parameter estimation in mechanistic models. Table 1 gives results from one of the data sets representing 300 site × year combinations. These R_s-values reflect by definition (Eq. 1) the nitrogen levels, but are independent of the weather. Accordingly, they could be used in Eq. (1) for predicting growth for any weath-

Table 1. Estimated values of the relative growth rate (R_s) in harvest time experiments with red clover — timothy mixtures in a two harvest system at different levels of nitrogen application (From Kornher and Torssell, 1983)

Ley year	Nitrogen kg/ha	Number of observations	Relative growth rate, R_s (g/g/week)	
			Spring growth	Regrowth
1	0	217	1.40	1.40
	90–100	334	1.55	1.50
	180–200	187	1.60	1.60
2	0	70	1.65	1.00
	90–100	166	1.75	1.20
	180–200	168	1.80	1.35

er situation within the range of the 20 years used for deriving the parameters.

Further, R_s-estimates using PrAR data are now emerging (Halling, 1986, 1987; Landström, unpubl.). Other well-known examples of using these sorts of large data sets are estimates of parameters expressing the change in the nutritional value of forage with time (Jönsson, 1981; Thorvaldsson and Andersson, 1986) or with weather (Thorvaldsson, 1987a, b; Thorvaldsson and Fagerberg, 1987).

The author would suggest that these trivial examples represent an important principle that obviously could be successfully applied to other models and in other large agronomic data sets, *e.g.* plant-water-nutrient relations (Jansson, pers. comm.).

The idea of using PrAR data for parameter estimates in various models has long been accepted in this program, as shown in a literature search covering the last 20 years (Torssell *et al.*, 1987; Table 2). There have been more papers concerned with regression models than with mechanistic models, and also the latter models have so far been confined to ley crops. However, growth models for grain and potato crops have been developed and a first order

estimate of model parameters on Swedish data has been made (Angus, unpubl.; Torssell, unpubl.; Yuen, unpubl.).

Practical model applications in Sweden

As a consequence of the distribution of papers shown in Table 2, examples of Swedish applications of growth models will be restricted to ley crops (Torssell and Kornher, 1983). The relevance of such applications depends very much on the precision and the generality of the model. The precision was proven by Torssell and Kornher (1983) and a strong indication of the generality of the model was given by Kornher and Petersen (1985) and recently by Kornher (unpubl.), who applied parameter estimates modified for north German conditions to British data (Fig. 1).

The various ways of model applications indicated in the Introduction are all exemplified by use of the Torssell and Kornher (1983) growth model, but then as a subroutine in larger models.

1) Data analyses, research planning: The model, in combination with regression models for nutri-

Table 2. Number of published Swedish papers on parameter estimates in plant-soil models in agriculture and horticulture (Torssell *et al.*, 1987)

Type of models	Subject				
	Cereals, pulses	Potatoes	Leys	Horticulture crops	Soil and water dynamics
Regression models					
Simple	17	8	9	–	2
Complex	1	–	5	–	–
Mechanistic models					
Simple	1	–	15	3	8
Complex	–	–	2	–	2

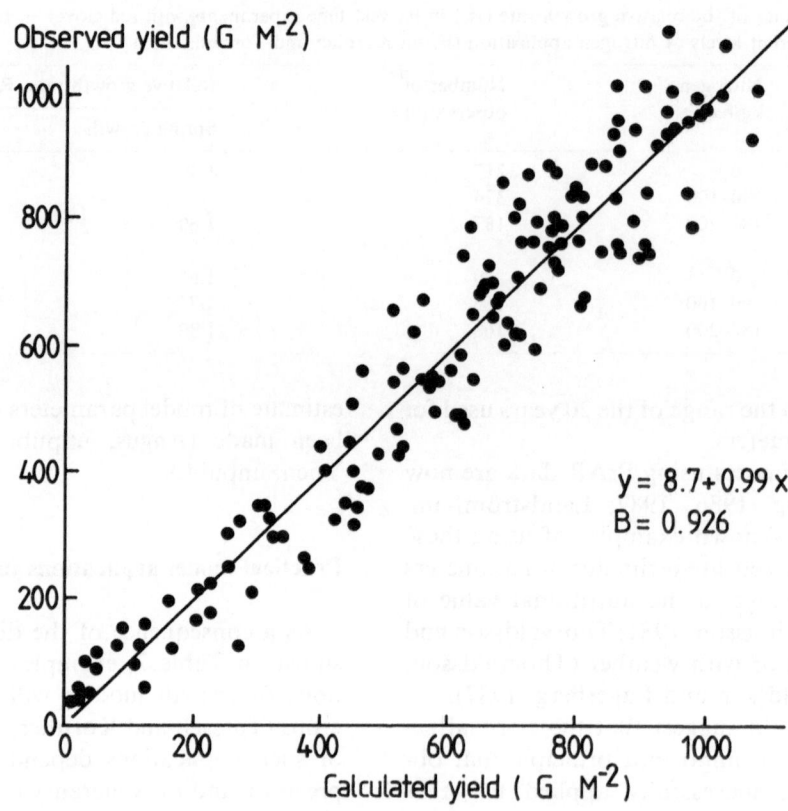

Fig. 1. Validation of the Torssell Kornher (1983) growth model using British data (Kornher, unpubl.).

tional value (Torssell *et al.*, 1983) forms the basis for planning and analyses of harvest time experiments in leys (Jönsson and Tuvesson, pers. comm.).

2) Teaching: Four years of experience in teaching with the model is available (Torssell *et al.*, 1985). To further increase its use and availability, a new PC-version has been produced (Lindström, 1987).

3) Yield and harvest time predictions: The first large scale, real time estimates of daily growth and accumulated ley production are now being produced by Dahlström (in manuscript; Fig. 2). Employing a nationwide weather grid information system in combination with the growth model, intriging, real time production patterns were revealed. These were validated by the Central Bureau of Statistics, but with a two months' time lapse.

The Torssell *et al.* (1983) extended model for estimating optimal harvest time in leys is used for calculating the harvest time, giving the optimum in the economic return of milk production (Fagerberg *et al.*, 1987; Fig. 3). This is achieved by combining

predictions of dry matter production with estimates of changes in protein content and digestibility. The predictions are based on current weather information and weekly weather forecasts (The Swedish Meteorological and Hydrological Institute) in a system that accordingly operates on an almost real time bases. In 1987 such predictions were attempted for six different Swedish regions.

As a final example of the model as a subroutine, reference is made to the nitrogen balance calculations on a farm scale performed by Kirchmann *et al.* (1987).

The present state of affairs

Considering the present situation in relation to modelling plant production in the agricultural context, two points emerge. 1) Mechanistic models for plant-water-nutrient relations in most crops are now available. 2) The idea of parameter estimation

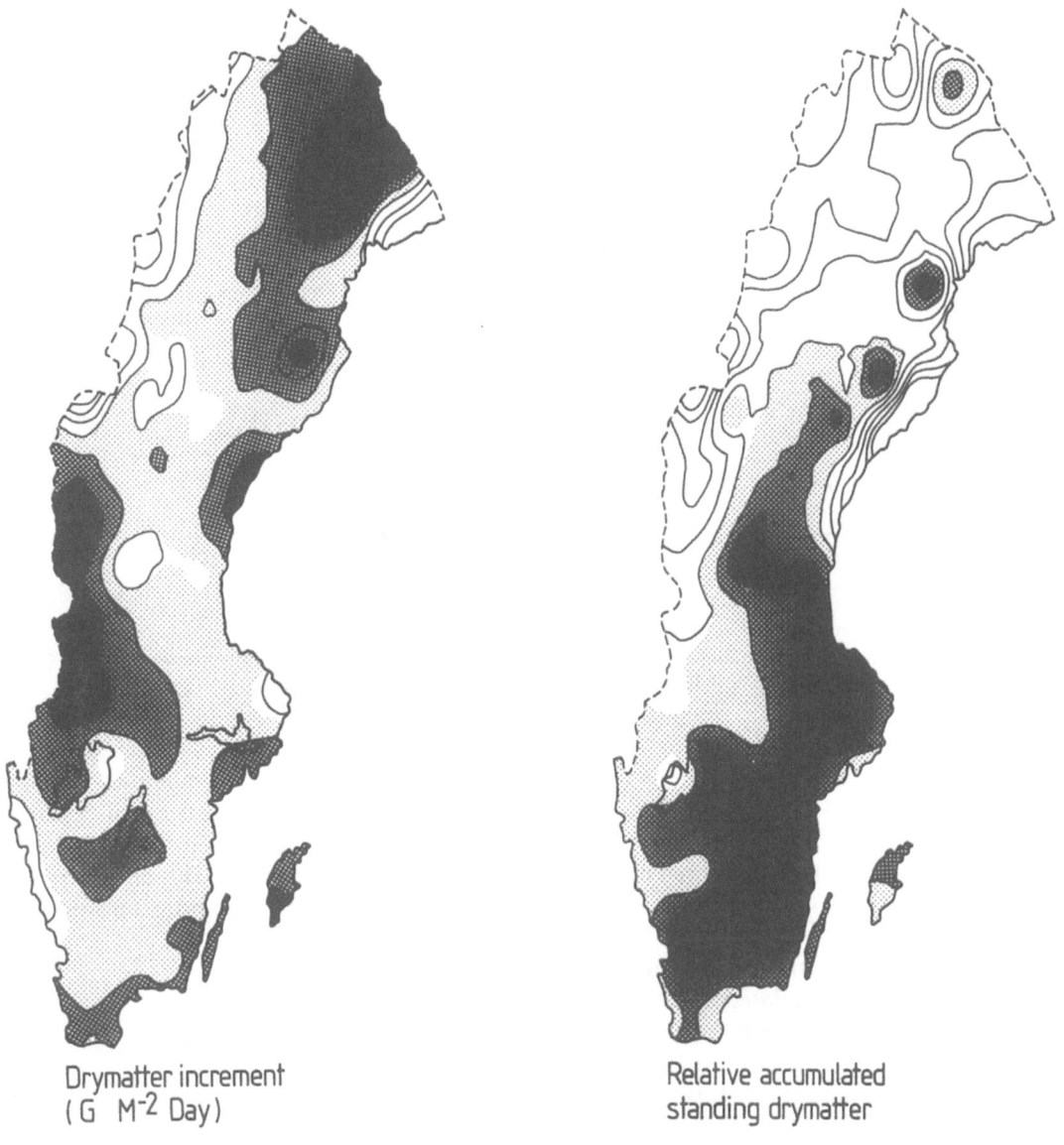

Drymatter increment
(G M⁻² Day)

Relative accumulated
standing drymatter

Fig. 2. Calculation of real time dry matter increment (left) and relative accumulated standing dry matter (right) on 1986.06.15 in Swedish leys (Dahlström, 1987). Darker areas indicate higher increments and dry matter, respectively.

in agronomic experiments is old and, in Sweden, employed in many simple regression models (Table 2).

However, the use of plant production models in applied agriculture and horticulture research is rare (Table 2), and one might speculate on the reasons why. The author believes that the main practical obstacle is just that presently there are too few scientists and extension workers who dare to apply

modelling-simulation as a method in plant-soil science. The technique required is still very demanding, where the main bottlenecks are: 1) Methods of optimizing several parameters simultaneously in large data sets. 2) Capability of solving the rather intricate problems of producing simulation programs for practical applications. 3) Uncertainty of how to apply simulation models in practical predictions.

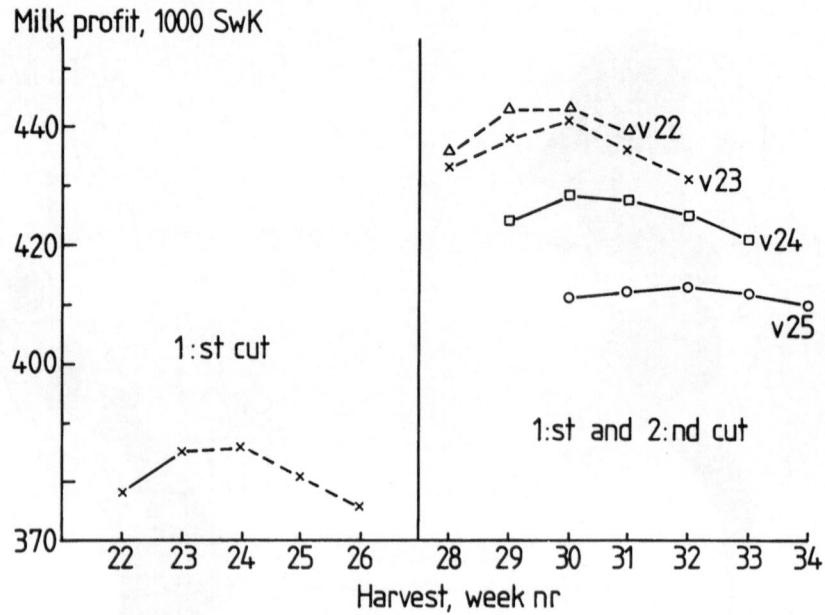

Fig. 3. Example of harvest time prediction in timothy based on calculation of the profit of milk production; with one cut (left) and with two cuts (right). Numerals in the right figure indicate harvest week for the first cut (Fagerberg *et al.*, 1987).

Perspectives and challenges

It appears that there is a potential for use of plant production models in agricultural research and extension, but that this potential is far from fully utilised. The most common explanation of this state of affairs among potential users is that the models are not adequately validated and thus unreliable. The author believes that the real reason is the shortage of scientists who are willing to commit themselves to this technique and that this hesitation is related to resource allocation. Practical application of the new modelling technique is presently still very demanding in terms of skilled personnel. In times of diminishing funds we tend to allocate our resources to safe, well-known methods rather than to systems analysis adventures. One countermeasure to this short-sighted view applicable within available resources is stronger leadership from responsible group members aiming at training and education of scientists and extension officers in -advanced techniques. Obviously, this input must in turn be based on intelligent identification of research areas where the modelling technique is profitable and preferably also where supporting datasets are to be found. However, broader in-

troduction of such new techniques requires more than training and identification of research tasks: new funds are needed to employ and keep skilled personnel and to acquire the necessary hardware for practical application and extension.

References

Cole C V, Stewart J W B, Parton W J, Ojima D and Schimel D S 1989 Modeling land use effects on soil organic matter dynamics in the North American great plains. *In* Ecology of Arable Land — Perspectives and Challenges. pp 89–98. Eds. M Clarholm and L Bergström. Kluwer Academic Publishers, Dordrecht, The Netherlands.

Dahlström B 1987 Spatial estimation of biomass in realtime based on a growth model for temporary grasslands. (*Manuscript*).

Duncan W G, Loomis R S, Williams W A and Hanau R 1967 A model for simulating photosynthesis in plant communities. Hilgardia 38, 181–205.

Fagerberg B, Karlsson S and Torssell B W R 1987 Harvest time predictions in leys based on weather forecasts and milk gain calculations. (*In Swedish.*) (Skördetidsprognoser i slåttervall grundade på väderleksprognoser och vinstberäkning.) Sveriges lantbruksuniversitet, Konsulentavd. rapporter Allmänt 105, 3:1–3:10.

Fitzpatric E A and Nix H A 1970 The climatic factor in Australian grassland ecology. *In* Australian Grasslands, pp 3–26. Ed. R M Moore, ANU Press, Canberra.

Halling M 1986 Growth of timothy and red clover in relation to weather and time of autumn cutting. 1. Yield and initial relative growth rate subsequent spring. Sw. J. Agric. Res. 16, 153–160.

Halling M 1988 Growth of timothy and red clover in relation to weather and time of autumn cutting. 2. Storage carbohydrates in autumn and spring. Sw. J. Agric. Res. 18, 161–170.

Hansen G K 1986 Heimdal — analyses and simulation of crop growth, water and carbon balance. Diss. The Royal Veterinary and Agricultural University, Copenhagen.

Hårsmar P O 1983 The effect of climatic factors on growth and development of plants: Potential and actual. (*In Norwegian.*) (Klimafaktorers innvirkning pa planters vekst og produksjon, potensiell og aktuell, i Norden. En litteraturstudie.) Norges landbrukshøgskole, Fysisk Institutt, Ås.

Jönsson N 1981 Quality changes in ley crops. (*In Swedish.*) (Kvalitetsförändringar hos vallväxter.) Sveriges lantbruksuniversitet, Inst. för växtodling, Rapport 93.

Kirchmann H, Torssell B W R and Roslon E 1987 A simple model for ntrogen balance calculations of temporary grassland-ruminant systems. Sw. J. Agric. Res. 18, 3–8.

Kornher A and Torssell B W R 1983 Estimation of parameters in a yield prediction model for temporary grassland using regional experimental data. Sw. J. Agric. Res. 13, 145–155.

Kornher A and Petersen E C 1985 Erfahrungen mit einem Simulations modell über die Ertragsbildung auf Grüland. Jahrestagung der Arbeitsgemeinschaft Grüland und Futterbau. Gesellschaft für Pflanzenbauwissenschaften 1985.

Lindström L 1987 Using personal computers to improve the education in plant husbandry at the Swedish University of Agricultural Sciences. Progress Report. Dept. of Plant Husbandry, Uppsala.

Slatyer R O 1960 Agricultural climatology of the Katherine Area, N.T. Tech. Pap. Div. Ld. Res. reg. Surv., C.S.I.R.O. Aust. no. 6.

Stewart J W B and Cole C V 1989 Influence of elemental interactions and pedogenic processes in soil organic matter dynamics. Plant and Soil 115, 199–209.

Thornley J H M 1976 Mathematical Models in Plant Physiology. Academic Press, London.

Thorvaldsson G 1987 The effects of weather on nutritional value of timothy in northern Sweden. Acta Agriculturae Scandinavica 37, 305–319.

Thorvaldsson G 1988 The morphological and phenological development of timothy as affected by weather, and its relation to nutritional value. Acta Agriculturae Scandinavica 38, 33–48.

Thorvaldsson G and Andersson S 1986 Variation in timothy dry matter yield and nutritional value as affected by harvest date, nitrogen, fertilization, years and local in Northern Sweden. Acta Agriculturae Scandinavica 36, 367–385.

Thorvaldsson G and Fagerberg B 1987 The effects of weather on nutritional value and phenological development of timothy. Sw. J. Agric. res. 18, 51–59.

Torssell B W R 1986 The systems ecology of crop production — a research tool in regional experiment programs. (*In Swedish.*) (Växtproduktionens systemekologi — ett hjälpmedel i försöksverksamheten.) Försöksledarmötet i Alnarp 1986. Sveriges lantbruksuniversitet, Konsulentavd. rapporter Trädgård 310, 9:1–9:8.

Torssell B W R and Kornher A 1983 Validation of a yield prediction model for temporary grasslands. Sw. J. Agric. Res. 13, 125–135.

Torssell B W R, Jönsson N and Kornher A 1983 A systems approach to planning research in temporary grassland production. Sveriges lantbruksuniversitet, Inst. För växtodling, Rapport 123.

Torssell B W R, Jönsson N and Roslon E 1987 Systems analyses as a complement in field experiments. (*In Swedish.*) (Systemanalys som komplement till försök och rådgivning.) Sveriges lantbruksuniversitet, Konsulentavd. rapporter Allmänt 107, 14:1–14:10.

Torssell B W R, Svensson B and Ohlander L 1985 Plant production models — research tools with broad application. (*In Swedish.*) (Växtodlingsmodeller — forskningsinstrument med bred tillämpning.) Sveriges lantbruksuniversitet, Konsulentavd. rapporter Allmänt 81, 19–38.

WMO 1983a Guidelines on Crop-Weather Models. Ed. G W Robertson. World Climatic Applications Programme, WCP-50, Geneva.

WMO 1983b Weather-based Mathematical Models for Estimating Development of Ripening Crops. Ed. G W Robertson. WMO, Agricultural Meteorology, CAgM Report no. 15, Geneva.

M. Clarholm and L. Bergström (Eds.), Ecology of arable land, 9–20.
© 1989 Kluwer Academic Publishers.

Primary productivity of natural grass ecosystems of the tropics: A reappraisal

S.P. LONG[1], E. GARCIA MOYA[2], S.K. IMBAMBA[3], A. KAMNALRUT[4], M.T.F. PIEDADE[5], J.M.O. SCURLOCK[6], Y.K. SHEN[7] and D.O. HALL[6]

[1]*Dept. of Biology, University of Essex, Colchester CO4 3SQ, UK;* [2]*Centro de Botanica, Colegio de Postgraduados, Chapingo, Mex 56230, Mexico;* [3]*Dept. of Botany, University of Nairobi, PO Box 30197, Nairobi, Kenya;* [4]*Faculty of Natural Resources, Prince of Songkhla University, PO Box 123, Hat Yai 90110, Thailand;* [5]*INPA, C.P. 478, Manaus 9000, Brazil;* [6]*Dept. of Biology, King's College, London W8 7AH, UK and* [7]*Shanghai Institute of Plant Physiology, 300 Fenglin Road, Shanghai 200032, China.*

Key words: biomass, decomposition, *Distichlis spicata*, *Echinochloa polystachya*, *Eulalia trispicata*, *Lophopogon intermedius*, *Pennisetum mezianum*, primary production, primary productivity, *Themeda triandra*

Abstract

Studies of net primary production in four contrasting tropical grasslands show that when full account is taken of losses of plant organs above- and below-ground these ecosystems are far more productive than earlier suggested. Previous values have mainly been provided by the International Biological Programme (IBP), where estimates of production were based on a change in vegetation mass alone and would not necessarily have taken full account of organ losses and turnover. Calculation at three of our sites based on established methodology using changes in plant mass alone (*i.e.* that used by the International Biological Programme, IBP) proved to be serious underestimates of when account was taken of losses simultaneously with measurement of change in plant mass. Accounting for the turnover of material at these three sites resulted in productivities up to five times higher than were obtained using the standard IBP procedure. An emergent C_4 grass stand at a fourth site in the Amazon achieved a productivity which approached the maximum recorded for agricultural crops. In this case, productivity values, when organ losses were taken into account, only slightly exceeded that obtained with IBP methods. The findings reported here have wider implications, in prediction of global carbon cycling, remote sensing of plant productivity and impact assessment of conversion to arable cropping systems.

Introduction

The needs of many of the developing countries of the world to increase food production both for home consumption and for export cash crops are, to varying extents and in the short term prospect, satisfied by destruction of natural ecosystems and replacement by agroecosystems. Although destruction of rainforest for agricultural expansion is well known, other communities of the tropics are also being lost. Indeed the areas with the most pressing needs for increasing arable food production are not the tropical rainforest zones, but the semi-arid tropics where grasslands form the natural vegeta-

tion. Tropical grasslands, excluding wetlands, occupy 15 million square kilometres and, in terms of both land area and biomass, are second only to tropical forests in importance (Lieth, 1978). In addition, periodically inundated grasslands, a form of wetland ecosystem, occupy 0.7–1.0 million square kilometres in South America alone, of which one-third is in the Amazon basin (W. J. Junk, pers. comm.).

Cursory examination of productivity figures for arable and natural grassland systems of the tropics might suggest that conversion of land to arable would greatly increase total dry-matter productivity (Buringh, 1980; Lieth, 1978). Knowledge of these natural ecosystems is based mainly on the International Biological Programme (IBP) studies. As we outline below, the methodology used in these studies may have lead to a serious and variable underestimation of production and turnover of plant biomass in these communities.

A comment, on previous productivity studies

Net primary production (P_n) is the total photosynthetic gain, less respiratory losses, of plant matter by vegetation occupying a unit area of ground (Cooper, 1975; Coupland, 1979; Jordan, 1981; Lieth, 1978; Linhurst and Reimold, 1978; Milner and Hughes, 1968; Roberts *et al.*, 1985). Over any one period, this must equal the change in plant mass (ΔW) plus any losses through death (L), both above- and below-ground (Roberts *et al.*, 1985):

$$P_n = \Delta W + L \qquad (1)$$

Essentially, P_n is the measure of the amount of plant matter available to consumer organisms.

Previously, Bourliére and Hadley (1970) reported net above ground production for 22 tropical grasslands. Their values were based on peak standing dry matter alone, whilst below-ground production was not considered. The most extensive data on the productivity of tropical grasslands is provided by the IBP studies, reviewed by Singh and Joshi (1979). They considered 21 studies (8 published) of net primary production in tropical grasslands of India and Africa. The few estimates of below ground net primary production (BP_n) provided by these studies were obtained by 'trough-peak analysis' of the changes in underground biomass with time (Singh and Joshi, 1979). Essentially, this is a method proposed by Milner and Hughes (1968) for the IBP, and will be referred to as the 'IBP standard method'. In this method, when biomass (W) per unit area was found to increase between harvest intervals, BP_n over the interval was considered to equal the increase, but where W decreases or remains unchanged over an interval between harvests, BP_n is assumed to be 0. The annual BP_n is then obtained by summing the estimates for each harvest interval, *i.e.* the positive increments are summed. For this method to yield a value equal

to the actual production, formation of new plant material and death must be separated in time – any overlap of the two processes will lead to underestimation of BP_n. Above-ground net primary production (AP_n) of these tropical grasslands was estimated mainly by 'trough-peak analysis' of live and standing dead material (Singh and Joshi, 1979). This is an elaboration of the 'IBP standard method', where AP_n is determined via a decision matrix (Singh *et al.*, 1975). Positive increments in biomass are summed, as in the 'IBP standard method', but where positive increments in the amount of standing dead coincide with positive increments in biomass, these are also added to the total. This procedure would correct for amounts of material lost through death during periods of biomass increase, if no decomposition occurs – however, since this is unlikely the method will again lead to underestimation. Three assumptions which could seriously influence the estimate of production therefore underlie the procedures used in the IBP studies of tropical grasslands:

1) That during any interval between harvests death of material does not occur if production is occurring and vice versa, *i.e. formation of new organs or parts of organs must not overlap with loss through death of organs or parts of organs.* In mixed grasslands this assumption will fail to take account of the different cycles of growth and death in different species, thus a species which is most productive at the time when total biomass is declining would be ignored. Even in monotypic stands this is obviously a gross over-simplification, since sequential senescence of leaves and tillering patterns mean that old leaves and stems will be dying as new ones form.

2) That P_n cannot be negative, either above- or below-ground, since only positive increments are considered. Since these grasslands are dominated by perennials with underground or surface storage organs significant transfers of assimilate between above- and below-ground components will be expected. For example, as dry or cold seasons approach translocation of matter from the shoots to below-ground storage organs can occur. During this period the above-ground biomass will decline and the below-ground biomass rise. The computational procedures used in the IBP studies would give a positive BP_n and zero AP_n, and therefore a positive total P_n, even though in reality no new material is formed, but existing material reallocated.

3) That all increases in biomass represent P_n. Variability in biomass over the study sites will mean that the average biomass recorded on different dates may vary simply because of random fluctuations in the samples obtained. An apparent positive increment in biomass between two dates could result simply by random chance and although over the year these random fluctuations may be self-cancelling, selection of only positive increments in calculating P_n, will mean that the greater the variability between the samples the greater the overestimation of production (Singh, *et al*, 1984). Singh *et al*. (1975) in their analyses of the US-IBP grassland productivity studies suggest that this error may be compensated for by only including positive increments where the biomass on one sampling date is significantly greater ($p < 0.1$) than on the previous sampling date. However, this will mean that small, but real increases in biomass may go undetected. In a relatively uniform natural coastal grassland, Hussey and Long (1982) found that harvests of 40 quadrats, of the optimum sampling size, placed by a randomised design were necessary to provide a mean estimate of biomass with a standard error within 10% of the mean. Thus, even at this level of sampling a 25% increase in biomass between two dates would not be statistically significant. In the IBP tropical grassland studies sample numbers were considerably lower, *e.g.* 10 quadrats per harvest (Singh and Yadava, 1974).

Assumption 1 would therefore lead to underestimation of P_n, assumption 2 to overestimation and assumption 3 without statistical restraints would also lead to overestimation, but with statistical restraints could lead to substantial underestimation. However, despite these theoretical objections Singh and Joshi (1975) in an analysis of some 30 methods of computing P_n as applied to the data gathered for U.S. grasslands in the IBP studies, concluded that the 'trough-peak' analyses used in calculating P_n in the tropical grassland studies, and many other IBP grassland sites, were among the best procedures. Determination of actual net primary production would require correction of change in biomass for losses due to death, as defined in Equation 1. Losses (L) may be measured by correcting the change in dead vegetation mass between sampling intervals by the amount lost due to decomposition measured by use of litter bags or labelling of dead material (Roberts *et al.*, 1985) or by direct observation of organ death (Jackson *et al.*, 1986). Methods based on simultaneous measurement of ΔW and L of Eqn. 1 make none of the assumptions made in the IBP standard method and other 'trough-peak' analyses. No such method was included in the comparisons made by Singh *et al.* (1975) who only considered methods in which it was assumed that production could be estimated from changes in vegetation mass alone. Methods were evaluated on their ability to discriminate between different grasslands and although 'trough-peak' analysis methods were favoured, it was noted that all methods were closely correlated and that a constant correction factor could be used to relate different methods. For example, multiplication of P_n estimated from the peak biomass method by 1.41 would give P_n as measured by the 'IBP standard method'. Linthurst and Reimold (1978) compared P_n estimated for three U.S. coastal grasslands obtained from peak biomass, as used by Bourlière and Hadley (1971), and from the 'peak-trough analysis', as used by Singh and Joshi (1975), with a method in which P_n was calculated from change in biomass corrected for simultaneous losses through death. Two important conclusions can be reached from this work. 1) That P_n estimated by both use of peak biomass and 'trough-peak' analysis seriously underestimated P_n, by 50%–85% and by 10%–70%, respectively. 2) That the degree of underestimation varied markedly with location and with species composition and that there was no constant conversion factor. This conclusion is in sharp contrast to that of Singh *et al.* (1975), but unsurprising in view of the fact that the errors accruing from the three assumptions underlying the IBP methods will vary independently. Similar conclusions may be drawn from other comparisons of methods made in temperate grasslands (Hofstra and Bradbury, 1976; Long and Mason, 1983). These findings bring the IBP estimates of primary production in tropical grasslands into question, not only may these be serious underestimates, but they may even fail to rank grasslands correctly in terms of their actual productivity. Although the potential errors in the methods used in most of the IBP studies of grassland P_n were well recognised (Cooper, 1975; Coupland, 1979) the degree of underestimation was not quantified; this would require application of the IBP standard methods and methods correcting for losses on the same vegetation. Overestimation of P_n further affects understanding of ecosystem functioning in tropical grasslands, since the estimates of P_n provided the basis for the extrapolation to rates of vegetation death and decomposition in the IBP syntheses of tropical grasslands (Singh *et al.*, 1979).

This study determines productivity using methods which allow correction for simultaneous losses of biomass through death, both above- and below-ground (Roberts *et al*, 1985) in four tropical grasslands and provides a basis for examining the extent to which previous methodology may have underestimated production.

Study sites

Details of the four study sites are summarised in Table 1.

Site 1: Kenya

This site was typical dry savannah grassland within the Nairobi National Park. Rainfall is bimodal with the major rains falling in May and a secondary peak of rainfall in October. Growing seasons are normally May–August and October–December. However, in 1984 (immediately preceeding this study) the May rains failed. The site was protected by an exclosure, but would normally be lightly grazed by wildebeest and other game animals. Similar land in Kenya is currently being ploughed for sorghum, cotton and pineapple production.

Site 2: Thailand

Measurements were made at the Prince of Songkhla University field site at Klong Hoi Khong, about 40 km from Hat Yai. The climate here is wet monsoonal, with maximum rainfall in October–December and a brief dry season in February/

March. The site is a semi-natural humid grassland with occasional stands of dipterocarp trees, typical of grassland found throughout southern Thailand. A grass community is maintained by periodic burning. To avoid complications in determining productivity, the site was protected by a fire-break for the duration of this study. This site is now a relict of the original grassland, the surrounding land having been cleared for production of rice, sorghum and a variety of other arable crops.

Site 3: Mexico

The site was a saline grassland at the Colegio de Postgraduados field site, Montecillos (about 50 km from Mexico City). The climate is dry sub-humid, with maximum rainfall in July/August and a dry season from December to February. The study area contained an almost pure stand of a halophytic grass. The grass community is maintained by burning, every 2–3 years.

Site 4: Brazil

The study site was a stand of emergent macrophyte vegetation on the island of Marchantaria, about 40 km from Manaus. This has a tropical rainforest climate, uniformly warm, with heavy rainfall except during a short drier season from August to September. Seasonality of vegetation growth is determined predominantly by the river level rather than by climate, the growing season coinciding with the period of inundation. Many of these sites are now burnt in the dry season and planted to arable crops which can be harvested before the sites are flooded during the next wet season.

Methods

Sites 1, 2 and 3

To determine P_n, following the definition of Equation 1, changes in biomass were measured at monthly intervals. Dry weight of both live and dead vegetation present at each site was determined monthly by clipping to ground level, sets of 20

Table 1. Site descriptions

Terrestrial grasslands	1. Montecillos Chapingo Mexico	2. National Park Nairobi Kenya	3. Klong Hoi Kong Hatya Thailand	4. Marchantaria Manaus Brazil
Location	19°28′N 98 55′W	1°0.5′S 36 49′E	6°0′N 100 56′E	03°20′S 60°00′W
Elevation (m)	2220	1500	100	20
Precipitation (mm yr^{-1})	700	950	2100	3130
Average monthly temperatures (°C):				
Minimum	12 (January)	12 (July)	27 (December)	26 (February)
Maximum	19 (May)	19 (Feb.)	29 (June)	28 (August)
Solar radiation (MJ m^{-2} yr^{-1})	6810	6500	5970	6410
Soil type	Solonet	Black clay vertisol	Humic gley	Alluvial deposits
Grassland type	Saline	Dry Savanna	Humid 'Savanna'	Emergent macrophyte
Dominant species	*Distichlis spicata* (L.) Greene	*Pennisetum mezianum* Leeke *Themeda triandra* Forsk	*Eulalia trispicata* Henr. *Lophopogon intermedius* A. cam.	*Echinochloa polystachya* Hitche
Common crops on conversion to agricultural use	Pasture	Sorghum Pineapple	Sorghum Rice	Maize

quadrats (0.25 × 1.0 m) located by a randomised block design. Soil cores were removed from the centre of 5–20 of these quadrats, to a depth of 15 cm, and organic material extracted by washing over a sieve of 2 mm mesh. Use of 1 mm and 0.5 mm mesh sizes did not recover significantly greater (t, $p > 0.05$) quantities of live root, but the increased capture of dead matter greatly increased the time required for the sorting of samples. For dead material, ability to pass through a 2 mm mesh sieve was used throughout as the arbitrary division between recognisable dead vegetation and particulate organic matter. Preliminary studies suggested that extraction of soil cores to 15 cm depth was adequate to remove more than 90% of the root system by weight. Above-ground material was sub-sampled to approximately 100 g fresh weight before sorting. Below-ground material was divided into fine roots (ca. < 1 mm dia.) and coarser material. Fine root matter was sub-sampled to 1.0 g and the coarser material sorted entirely.

Live leaves were separated from dead on the basis of tissue necrosis, dead portions being removed from otherwise green leaves. Stems were sorted likewise, taking care to remove dead sheaths from living stems. Roots were divided on a similar basis, using vital staining (tetrazolium salts) where visual discrimination was not otherwise possible.

Tetrazolium staining was cross-checked by microscopic examination of cut root tissues where the results of surface staining were ambiguous. The sorted plant material was then thoroughly washed and dried to constant weight at *ca.* 90°C (Roberts *et al.*, 1985).

Losses of live and dead plant material, L in Eqn. (1), were assumed to be by death and decomposition *in situ*, respectively. Large herbivores were excluded from each site by a fence, where necessary. Thus total loss of material (L) equals the change in dead matter (ΔD) plus the loss due to decomposition (A):

$$L = \Delta D + A \qquad (2)$$

The quantity ΔD was estimated every month from the harvested samples. Decomposition losses (A) were determined monthly by litter bags. Decomposition of dead shoot material was measured at the ground surface, whilst for roots and rhizomes the litter bags were inserted 5 cm below the ground, with the soil carefully replaced above the bag to minimise disturbance. A portion of dead material obtained at random from each harvest (approx 2 g) was placed in each of 20 litter bags of 2 mm nylon mesh, 8.0 cm × 6.0 cm, and recovered after one month from the field. Contents were washed over a sieve of 2 mm mesh and dried to constant weight.

Table 2. Quantities of vegetation and rates of its decomposition recorded at monthly intervals

Terrestrial grasslands	Montecillos Chapingo Mexico		National Park Nairobi Kenya		Klong Hoi Kong Hatyai Thailand		Marchantaria Manaus Brazil	
	Shoots	Roots/ rhizomes	Shoots	Roots/ rhizomes	Shoots	Roots/ rhizomes	Shoots	Roots
Biomass (g m^{-2})								
Max	430	568	337	195	442	520	6840	460
Mean	279	389	222	109	338	340	4450	218
Min	161	229	32	61	244	205	540	40
Dead vegetation (g m^{-2})								
Max	821	896	525	218	1048	188	1140	–
Mean	586	573	352	167	782	109	575	–
Min	273	306	66	128	634	70	40	–
Decomposition rates (g g^{-1} mo^{-1})								
Max	0.28	0.16	0.18	0.32	0.21	0.46		–
Mean	0.09	0.10	0.12	0.21	0.14	0.27		–
Min	0.01	0.03	0.05	0.08	0.07	0.14		–
Total decomposition (g m^{-2} yr^{-1})	606	686	454	433	1340	414		–

The loss of material from the bags is a measure of the rate at which a random sample of the dead vegetation at the site at the start of each interval between harvests would decompose over the month. Controls in which the litter bags were filled, taken to the field and then immediately returned and processed showed no significant loss of material.

Relative rates of decomposition were expressed as the proportion of initial dry weight lost within the month. In the context of this study decomposition represents loss of dead material due to leaching, breakdown to small particles (< 2 mm dia.) and consumption by invertebrates. Total decomposition (A) was estimated as the sum of the monthly products of relative decomposition rate and mean quantity of dead material (Table 2).

Site 4

The cyclical inundation of the study site in Brazil is described by Fig. 3. Due to the large size of the plants (up to 12.0 m long), and since the study area is covered by water except during the three-month low-water season, a different approach to measurement of P_n than that used at the first three sites was necessary.

The vegetation of this site was a monospecific stand of *Echinochloa polystachya*. This species forms unbranched annual stems of up to 60 or more nodes. Monthly harvests consisted of 15 individual plants uprooted for dry weight determination. Harvested material was separated into five categories before drying; live stems, leaves (lamina and sheaths), adventitious roots, dead leaves and dead stems. In addition, a further 20 stems were marked and the monthly increase in length and node number recorded. Uprooted plants and quadrats were chosen at random points along a 100 m transect. Production was then calculated as the product of the number of new nodes and the sum of the mean weights of internodes, leaf lamina and leaf sheaths determined from destructive harvests for that month. Direct estimation of new root production was not possible in plants which grew up through several metres of water. Similarly, estimation of root losses to flowing water would be equally difficult. For this community root production could therefore only be estimated from the sum of the positive increments in root biomass between months. However, since roots constituted only a very small (< 10%) fraction of the plant biomass, we assume that this would not have a large effect on the estimate of total production. Production per plant was then multiplied by the mean number of plants counted in 20 quadrats of 1.0 m square to obtain dry weight per unit area.

Results and discussion

Changes in biomass at the first three sites are illustrated in Fig. 1. At all sites living shoot material could be found in every month of the year. On average 50% of the biomass was constituted by underground roots and rhizomes. This proportion

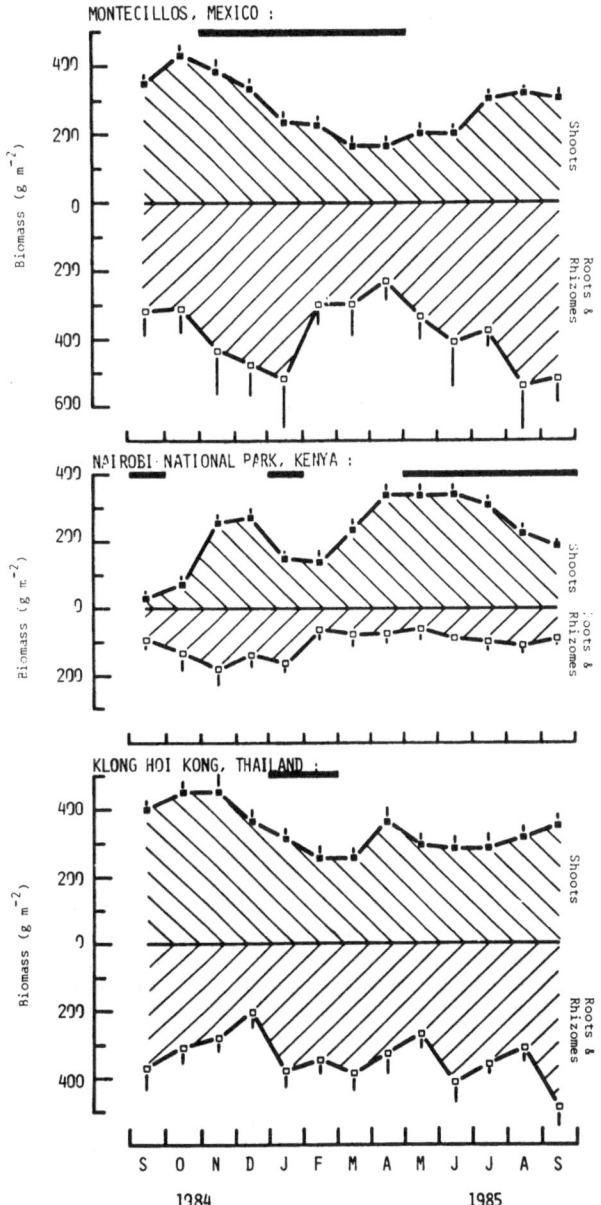

Fig. 1. Changes in shoot and in root and rhizome biomass. *i.e.* living material. Points represent the mean and the vertical bars ±1 s.e. Horizontal bars indicate periods of low rainfall. *i.e.* months in which precipitation was less than 50 mm.

rose substantially during periods of drought, *e.g.* 70% at Nairobi in September 1984. Amounts of dead material above-ground rose at all three sites over the year, on average by $350\,g\,m^{-2}$, probably reflecting the exclusion of fire and large grazers from the sites during the year.

At these grassland sites, relative rates of decomposition of dead vegetation varied from 3% to 32% per month, below-ground material decomposing more rapidly than that above-ground (Table 2). The higher rates correlated with periods of high rainfall, the lowest with periods of drought. The total amounts of material decomposed at each site are given in Table 2. Some losses through death were recorded on each interval between harvests at each site emphasising the importance of considering this factor throughout the year. The total amount of material that died during the year (L) was determined (Eqn. 2), and the value substituted into Equation 1, to give the total net primary production (P_n). Figure 2 summarizes the productivities obtained by this procedure. Estimates at all three grassland sites are substantially higher than the P_n of $800\,g\,m^{-2}$ which has been suggested as an average of annual P_n for tropical grasslands obtained in previous studies (Jordan, 1981; Lieth, 1978). Since both biomass changes and losses through death were measured at regular intervals (Fig. 1), it was possible to use our data to determine the values that would be obtained with 1) the 'standard IBP method' (Milner and Hughes, 1968; Singh *et al.*, 1975) and the peak biomass method (Bourliére and Hadley, 1970) (See Table 3). Both methods underestimate primary production by two- to five-fold in the first three grassland sites. Although substantial changes in amounts of dead vegetation occurred at each site within a year (Table 2) differences between consecutive harvests were not significant (t, $p > 0.1$), and following the recommendations of Singh *et al.*, (1975) should not be included in estimating P_n.

The underestimation of P_n by the IBP standard methods results from failure to include losses through mortality. Our estimate of mortality is heavily dependent on the estimate of decomposition. The mesh size of the litter bags would exclude larger invertebrates, *e.g.* termites, leading to an underestimation of decomposition. Enclosure of material into litter bags will alter the microenvironment for decomposition. Wiegart and

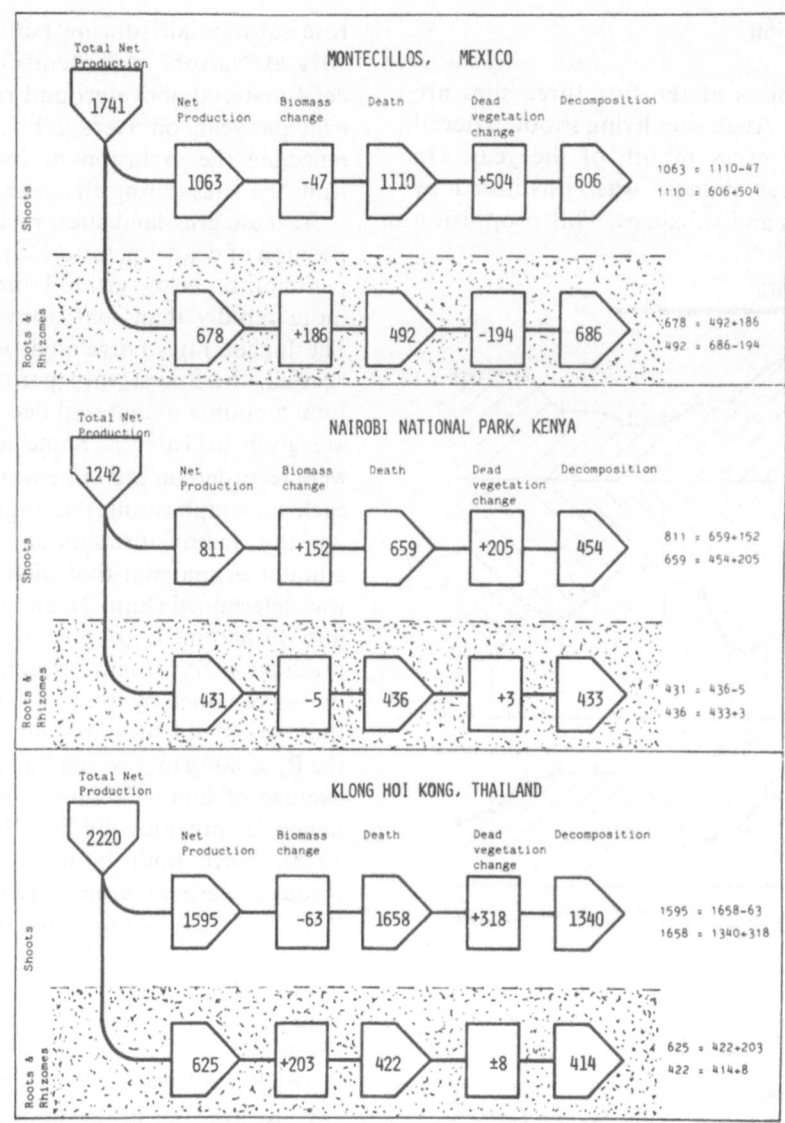

Fig. 2. Net primary production, its partitioning and fate over the 12 months from September 1984. Fluxes are illustrated by the arrowed boxes and the net changes in conserved quantities by rectangles; units g m^{-2}. From left to right: **Total Net Production**, the sum of above- and below-ground production; **Net Production** (P_n), the sum of the change in biomass and the amount of material lost through death; **Biomass Change**, (ΔB) the change in the dry-weight of living material; **Death** (L), the total amount lost through death which equals the sum of the change in dead vegetation and the total amount of dead material decomposed; **Dead Vegetation Change** (ΔD), the change in the dry-weight of dead material; and **Decomposition** (A), the total loss of dead vegetation through decomposition and breakdown. For example, at Montecillos site total net primary production (P_n) was 1741 g m^{-2} for the 12 months, this was obtained by summing P_n for shoots (1063 g m^{-2}) and for roots and rhizomes (678 g m^{-2}). Net primary production for shoots was the sum of the changes in shoot biomass over the year (-47 g m^{-2}) and the loss of shoot biomass through dead shoots (504 g m^{-2}), which in turn was the sum of the changes in the weight of dead shoots (504 g m^{-2}) and the amount lost through decomposition (606 g m^{-2}). The same procedure was applied to the below-ground components to obtain the P_n of the roots and rhizomes.

Table 3. Net Primary Production; a comparison of the estimates obtained by taking account of losses through death and below-ground production, with estimates from biomass change alone

Net Primary Production ($g m^{-2} yr^{-1}$)	Montecillos, Mexico	Nairobi National Park, Kenya	Klong Hoi Kong, Thailand	Marchantaria Manaus Brazil
1) Accounting for mortality (including below-ground organs)	1741	1242	2220	9925
2) Accounting for mortality (above-ground only)	1063 (39%)*	811 (35%)	1595 (28%)	9425 (5%)
3)'IBP standard method' (including below-ground organs)	740 (56%)	663 (47%)	570 (74%)	8680 (18%)
4) Maximum biomass (above-ground only)	430 (75%)	337 (75%)	442 (80%)	6300 (37%)

*Figures in parenthesis indicate underestimation of productivity as a percentage of net primary production over the 12 months calculated as: 1) the sum of all changes in biomass and losses due to mortality (eqn. 1), for both above- and below-ground); 2) as for 1), but for above-ground vegetation only; 3) the sum of all positive changes in above- and below-ground biomass; and 4) the difference between the peak biomass above-ground during the year.

Evans (1964), in comparing decomposition of dead vegetation in open quadrats and in litter bags, found the latter to under-estimate decomposition losses by about 10%. No account could be taken of the losses of organic matter as root exudates or to mycorrhizal associations, which from the very limited data on this topic may account for as much as 40% of plant productivity (Bowen, 1980). Thus, although our estimates are very much higher than those obtained on the same data by the IBP methods, they are still likely to be conservative relative to the true biological productivity of these plant communities. In considering the first three grassland sites, the 'standard IBP method' would classify Klong Hoi Khong as the least productive and Nairobi National Park as the most productive. When account is taken of losses through death and below-ground production, the ranking is reversed (Table 3). Thus the implicit assumption that the 'IBP' and other harvest methods, even though they underestimate true P_n, would identify the more productive sites must now be questioned.

Changes in biomass at site 4 in Brazil are shown in Fig. 3. Net primary production over one annual cycle is estimated at 9925 $g m^{-2}$ (Fig. 4). Root biomass formed a much smaller part of total standing biomass in this ecosystem, at most 8% (Table 2). The plants have an almost continuous supply of water and do not possess below-ground perennating organs, apparently obtaining nutrients via adventitious roots attached to the nodes of the long stems. A low root biomass and productivity would therefore be expected. Root production esti-

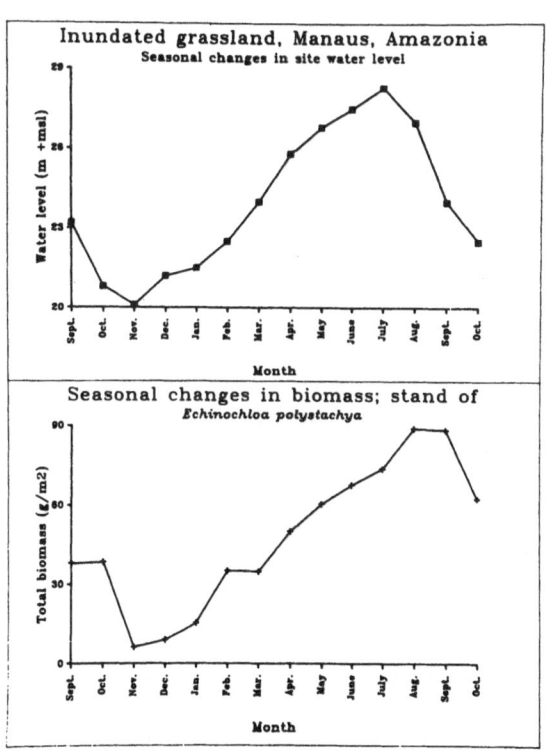

Fig. 3 a. Change in water level at the study site near Manaus, Brazil. The site was at ca. 21 m above mean sea level. b. Change in total dry matter of an *Echinochloa polystachya* stand over the same period in 1985/6.

mated as the positive increments in biomass was 505 $g m^{-2}$ (Fig. 4).

Since monthly biomass changes were recorded, it is also possible to estimate production for site 4

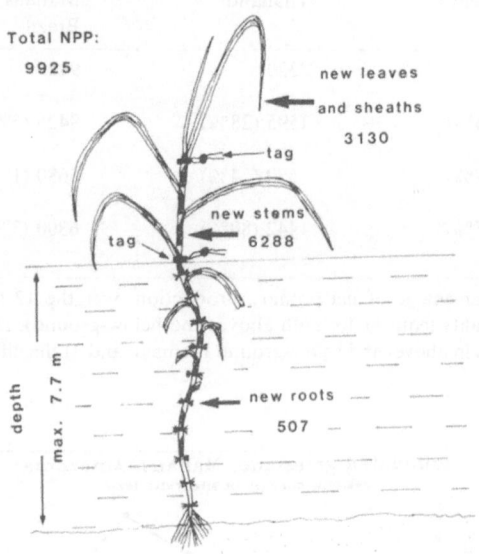

Inundated grassland near Manaus, Amazonia
(pure stand of Echinochloa polystachya)
Nov./85-Nov./86

Total NPP:
9925

new leaves
and sheaths
3130

tag

new stems
6288

tag

depth max. 7.7 m

new roots
507

Fig. 4. Gains in stem, leaf and root material of an *Echinochloa polystachya* stand determined by monthly growth measurements and demographic data, to determine net primary production (NPP). Figures are dry matter per unit ground area (g m⁻²) over the period Nov. 1985–Nov. 1986.

using both the 'IBP standard method' and from peak biomass (Table 3). Both these methods result in slightly lower estimates of productivity than when mortality is taken into account, although the differences are not so marked as for the terrestrial grassland sites. The relationship of P_n estimated by different methods in tropical grasslands is highly variable in contrast to the conclusions of Singh *et al.*, (1975), but in agreement with the variability noted in coastal grasslands by Linthurst and Reimold (1978).

The seasonal growth exhibited by *E. polystachya*, with a large increment in biomass during the year and highly seasonal phenology (a feature more in common with temperate than with tropical grass-lands), means that most of the turnover of material recorded here took place towards the end of the growing season. Production of *E. polystachya*, is notable in approaching and exceeding the maximum annual productivity of agricultural crops; for example, 8800 g m⁻² (Beadle *et al*, 1985).

Mean below-ground biomass as a proportion of

mean total biomass was 58%, 33%, 50% and 5%, respectively, for the sites in Mexico, Kenya, Thailand and Brazil. The reported values for root biomass are of total dry weight. Although vigo-rously washed samples could still be contaminated by soil mineral particles leading to overestimation of mass, microscopic examination of samples during sorting did not reveal obvious contamina-tion. Samples from the sites in Thailand and Mexico were combusted in a muffle furnace. The ash contents of these samples were 10%–20%, again suggesting that mineral particle contamina-tion was not leading to serious overestimation. Nevertheless the possibility remains that the values given for root and rhizome masses are overesti-mated because of some degree of soil particle contamination. However, since we have calculated production and decomposition from changes in mass rather than from maximum amounts, the errors in mass estimates due to soil contamination will be self-cancelling in calculating both decom-position and production (Fig. 1 and Table 2). Be-low-ground dead matter similarly formed a sub-stantial part of total dead matter at the terrestrial grassland sites (Table 2). Annual below-ground biomass turnovers, estimated as production divided by mean biomass were 1.7, 4.0 and 1.8 for the sites in Mexico, Kenya and Thailand respective-ly. Annual turnovers of dead matter below-ground measured as decomposition divided by mean dead matter, were 1.2, 2.6 and 3 8, respectively, for Mexico, Kenya and Thailand (Fig. 2). This em-phasises the need to consider not only the quantity of below-ground biomass, but also its rate of turnover in any study attempting to determine the productivity of a tropical grassland.

The findings of this study have important im-plications to ecological understanding of tropical grasslands in three areas: 1) The true input of organic matter to the ecosystem, including that within the soil, will have been undervalued, par-ticularly in the case of the terrestrial grasslands. Thus the potential environmental impact of vegeta-tion removal will not have been fully appreciated in the event of clearance for agricultural use. 2) Assi-milation of carbon into plant matter and the input to roots and rhizomes below-ground will have been underestimated. In view of the extent of tropical grasslands, this would have a significant effect on estimates of global carbon cycling and rates of atmospheric CO_2 rise. Previous studies suggested that productivity of terrestrial grasslands alone in the tropics is $13.5\,Gt\,yr^{-1}$ (Lieth, 1978), about 9% of total terrestrial production. The proportion attributable to grasslands would rise to more than 25% of total terrestrial production if the minimum 3-fold underestimate found here for three grassland sites is generally applicable. If the more modest underestimation of production in other grass ecosystems is also taken into account, the considerable significance of natural grass ecosystems to total terrestrial production becomes apparent. An accurate estimate of productivity of grasses is essential for the establishment of a baseline against which the effects of the global rise of CO_2 levels may be assessed. 3) Recently, techniques for the remote sensing of vegetation biomass have been widely used. Both satellite and aircraft remote sensing give estimates of standing biomass, after calibration of the techniques against 'ground truth' measurements obtained by conventional methods. Productivity is then estimated by, for example, the maximum biomass. However, on the basis of this study such methodology cannot be used alone to provide a reliable estimate of biological production because it ignores death and below-ground production (Table 3) leading to an unpredictable degree of underestimation of actual production. The solution to this problem lies in the development of techniques with adequate ground verification which simultaneously estimate biomass and its death, or which estimate the ability of the vegetation to intercept light for photosynthesis and relate this to biomass production (Tucker *et al.*, 1986; Warrick, 1986).

Acknowledgements

This work was carried out with the support of the United Nations Environment Programme (UNEP) under project FP/4102-83-06(2405) 'Primary productivity and photosynthesis'. The authors are grateful to the many people who have advised on the development of the project or assisted in the field, particularly Jose Nunes de Mello and Wolfgang Junk in Brazil; Jenesio Kinyamario in Kenya; Alberto Escalente in Mexico; and John Evenson and Kritsanapong Laksanapokin in Thailand. We thank Dave Heath, Tim Gray and Mike Roberts of the University of Essex for their comments on draft versions of the manuscript; and Doreen Radford for typing the manuscript.

References

Beadle C L, Long S P, Imbamba S K, Hall D O and Olembo R J 1985 Photosynthesis in relation to plant production in terrestrial environments. UNEP/Tycooly, Oxford.

Bowen G D 1980 Misconceptions, concepts and approaches in rhizosphere biology. *In* Contemporary Microbial Ecology. Ed. D C Ellwood. pp 283–304. Academic Press, London.

Bourlière F and Hadley M 1970 The ecology of tropical savannas. Ann. Rev. Ecol. Syst. 1, 125–152.

Bradbury I K and Hofstra G 1976 Vegetation death and its importance in primary production measurements. Ecology 57, 209–211.

Buringh P 1980 Limits to the productive capacity of the biosphere. *In* Future Sources of Organic Raw Materials. Eds L E St. Piere and G R Brown. pp 325–383. Pergamon Press, Oxford.

Cooper J P (Ed.) 1975 Photosynthesis and Productivity in Different Environments. IBP Vol. 3. Cambridge University Press, Cambridge.

Coupland R T (Ed.) 1979 Grassland Ecosystems of the World. IBP Vol. 18. Cambridge University Press.

Hussey A and Long S P 1982 Seasonal changes in weight of above- and below-ground vegetation in a salt marsh at Colne Point, Essex. Journal of Ecology 70, 757–771.

Jackon D, Long S P and Mason C F 1986 Net primary production, decomposition and export of *Spartina anglica* on a Suffolk Salt marsh. J. Ecology 74, 647–662.

Jordan C F (Ed.) 1981 Tropical Ecology. Hutchinson Ross, Stroudsberg.

Lieth H F H (Ed.) 1978 Patterns of Primary Productivity in the Biosphere. Hutchinson Ross, Stroudsberg.

Linthurst R and Reimold R J 1978 An evaluation of methods for estimating the net primary production of estuarine angiosperms. J. Appl. Ecol. 15, 919–932.

Long S P and Mason C F 1983 Saltmarsh ecology. Blackie, Glasgow.

Milner C and Hughes R E 1968 Primary Production of Grassland. IBP Handbook 6. Blackwell, Oxford.

Roberts M J, Long S P, Tieszen L L and Beadle C L 1985 Measurement of plant biomass and net primary production. *In* Techniques in Bioproductivity and Photosynthesis, 2nd edition. Eds J Coombs, D O Hall, S P Long and J M O Scurlock. pp 1–19. Pergamon Press, Oxford.

Singh J S and Joshi M C 1979 Tropical grasslands primary production. *In* Grassland Ecosystems of the World. Ed. R T Coupland. pp 197–218. Cambridge University Press, Cambridge.

Singh J S, Lauenroth W K and Sernhorst R K 1975 Review and assessment of various techniques for estimating net aerial primary production in grasslands from harvest data. Bot. Rev. 41, 181–232.

Singh J S, Singh K P and Yadava P S 1979 Tropical grasslands ecosystems synthesis. *In* Grassland Ecosystems of the World.

Ed. R T Coupland. pp 231–240. Cambridge University Press, Cambridge.

Singh J S and Yadava P S 1974 Seasonal variation in composition, plant biomass and net primary productivity of a tropical grassland at Kurukshetra, India. Ecol. Monogr. 44, 351–376.

Singh J S, Lavenroth W K and Steinhurst R K 1975 Review and assessment of various techniques for estimating net aerial primary production in grasslands from harvest data. Bot. Rev. 41, 181–232.

Singh J S, Lavenroth W K, Hunt H W and Smith D M 1984

Bias and random errors in estimators of net root production: A simulation approach. Ecology 65, 1760–1764.

Tucker C J, Fung I Y, Keeling C D and Gammon R H 1986 Relationship between atmospheric CO_2 variations and a satellite-derived vegetation index. Nature 319, 195–199.

Warrick R A 1986 Photosynthesis seen from above. Nature 319, 181.

Wiegert R G and Evans F C 1964 Primary production and the disappearance of dead vegetation on an oldfield in south eastern Michigan. Ecology 45, 49–62.

M. Clarholm and L. Bergström (Eds.), Ecology of arable land, 21–29.
© 1989 Kluwer Academic Publishers.

A reassessment of shoot/root and root/organic matter interactions

DAVID C. COLEMAN[1], DIANA W. FRECKMAN[2], GAMIN M. WANG[3] and JOHN D. GOESCHL[4]

[1] *Department of Entomology and Institute of Ecology, University of Georgia, Athens, GA 30602, USA*
[2] *Department of Nematology, University of California-Riverside, Riverside, CA 92521, USA. Present address: National Science Foundation, 1800 G Street, NW, Washington, DC 20550, USA*
[3] *SRI International, 333 Ravenswood Ave., Menlo Park, CA 94025, USA*
[4] *Duke University Phytotron, Department of Botany, Duke University, Durham, NC 27706, USA*

Key words: shoot/root interactions, root/organic matter interactions, phloem transport, ^{14}C, ^{11}C

Abstract

Recent studies of root/organic matter (OM) interactions are reviewed, which lead to some interesting results on OM dynamics, and root/microbial interactions. We also examine shoot/root interactions in grasses. Our recent studies of intra-generic differences in plant function with respect to labile carbon dynamics show marked differences between mycorrhizal-infected and uninfected plants. We attempt to link and compare shoot/root/OM dynamics, using radiotracer techniques.

Introduction

Over the last century, scientists have noted the close correlations and strong associations roots have with organic matter. The famous Danish pedologist/soil scientist P.E. Müller (1887) had excellent pictures of roots and forest floor leaves, which showed their close associations. These relationships are reviewed by St. John et al (1983). For both root distribution in tropical forests (Table 1), and mycorrhizal hyphal growth pattern (Table 2), there was increased growth in the presence of organic matter, compared to bulk soil, with lower organic matter content.

More recent laboratory and greenhouse studies have extended the research from presence or absence of roots/mycorrhizae to measurement of principal processes, including: decomposition, respiration, and directions of carbon flow, in root-soil-microbial systems.

We review two approaches to root-organic matter (O.M.) dynamics in some detail. The first approach, exemplified by the studies of Helal and Sauerbeck (1984, 1986), has measured the "priming effect" of root–soil–organic matter interactions.

Table 1. Root growth experiments in two tropical terra firme forests near Manaus, Brazil. (From St. John et al., 1983)

	Control	OM
Site 1		
Mean root lengths ± SE	9.3 ± 2.4	135 ± 44.9
Sum of ranks	138.5	239.5
Number of observations	14	13
Site 2		
Mean root lengths ± SE	14.3 ± 3.7	125 ± 45.8
Sum of ranks	179.5	285.5
Number of observations	15	15

Table 2. Hyphal lengths in decimeters (± SE) in the VAM hyphal growth experiment. (From St. John et al., 1983)

Pot 1	
Organic matter	Control
17.9 ± 4.8	7.5 ± 3.1
Pot 2	
Organic matter	Control
10.7 ± 2.2	3.5 ± 0.8

Using a replicated set of divided root chambers (Fig. 1), where roots were confined to the middle part with a fine mesh screen, Helal and Sauerbeck

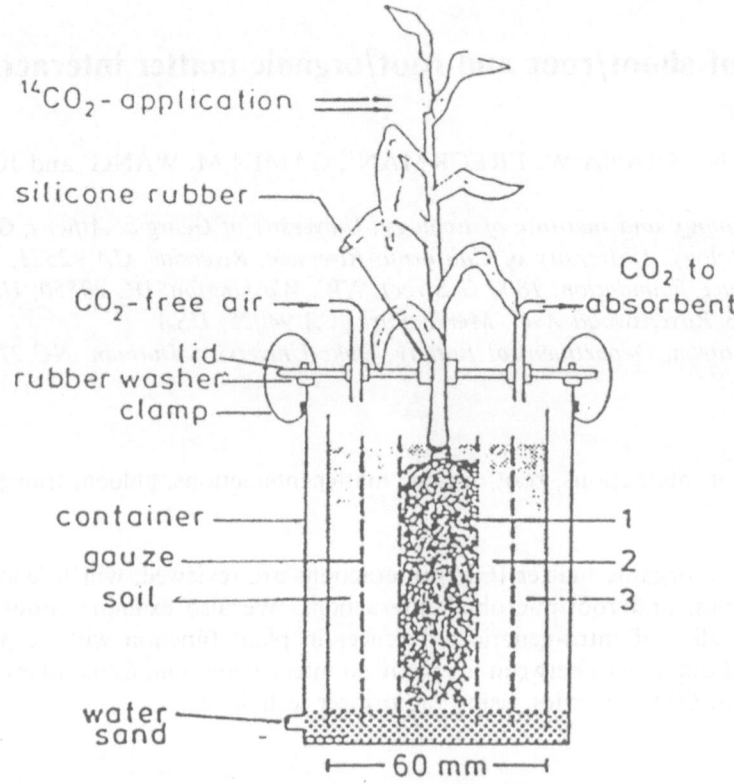

Fig. 1. Arrangement for studying turnover processes in soil zones of different proximity to roots: a) root zone, 2) intermediate zone, 3) outer zone. (From Helal and Sauerbeck, 1986)

(1986) followed changes in soil biomass C, and the resident organic C in a Parabraunerde (Luvisol, Argillic brown earth (sand/silt/clay = 2.7/83/14.3%)). Using photosynthetically-fixed ^{14}C to trace plant-derived carbon over a 30 day period, Helal and Sauerbeck (1986) estimated a mean of 68% plant-derived microbial biomass C measured by the chloroform fumigation and incubation method (CFIM) (Jenkinson and Powlson 1976), versus some 32% which was soil-derived (Table 3). Helal and Sauerbeck (1986) measured decreases in soil organic-C in different zones of their divided root chambers. The decreases were small, but significant, with the largest amount (7.4%) decreasing in the most proximal zone 1 cm to either side of the growing roots (Table 4).

The contribution of the rhizosphere to O.M. dynamics has been studied extensively over the last decade or more, with the pioneering studies of Barber and Martin (1976) and Barber and Lynch (1977) noting the amount and extent of rhizosphere activity. Coleman *et al.* (1977) extended these ideas to include a "priming", of sorts, of subsequent nutrient mineralization by microbivorous fauna active in the root/rhizosphere region. This was developed further in a conceptual model (Trofymow and Coleman, 1982), and tested with some success, by Elliott *et al.* (1979), Ingham *et al.* (1985) and Clarholm (1985).

The other approach measures the suppression and competition effect of growing roots on soil organic matter dynamics. Reid *et al.* (1982) addressed problems of aggregate stability. Decomposition rates of plant roots in the presence of living roots of maize and perennial ryegrass were studied by Reid and Goss (1982). Following up on the observations of earlier workers (Fuhr and Sauerbeck, 1968; Shields and Paul, 1973; Jenkinson 1977) that ^{14}C-labelled residues in soil decomposed more slowly in the presence of living cereal crop roots, than in fallow, Reid and Goss (1982) performed experiments to test how much of the effects

Table 3. Increase of microbial biomass in planted soil and contribution of plant and soil carbon to biomass growth as a function of proximity to roots. (Helal and Sauerbeck, 1986).
Biomass-C content on day 0 (= initial content); 12 mg C/zone, 36 mg C/pot
Biomass-C content of unplanted soil on day 30 (= control): 11.8 mg C/zone, 35.4 mg C/pot

| | Biomass-C after planting | | Biomass increase during planting | | | | | |
	mg C	% of initial	Total mg C	Plant derived mg ^{14}C	% of total	Soil derived mg C	% of total
Root zone	59	492	47	38.3	82	8.7	18
Intermediate zone	26	217	14	6.2	44	7.8	56
Outer zone	22	183	10	3.5	35	6.5	65
Sum	107	297	71	48	68	23	32

Table 4. Decrease in soil-organic carbon in soil zones differing in proximity to roots after 30 days planting with maize (initial content = 1610 mg C/zone). (Helal and Sauerbeck, 1986)

	Organic carbon mg/zone	Decrease % of initial content
Control (unplanted)	1590	1.2
Root zone	1491	7.4
Intermediate zone	1532	4.9
Outer zone	1546	4.0
Mean	1523	5.4

were due to indirect inhibition of microbial activity by reduced aeration, or water supply.

Reid and Goss (1982) grew seedlings of barley (*Hordeum vulgare* L.) in soil, pulse-labelling them several times with $^{14}CO_2$. At the flowering stage, shoots were harvested, and the roots left to decompose for another 67 days. Soils were then sieved (2 mm. mesh) into a series of glass pots, for the main experiment.

Seedlings of either *Lolium perenne* or *Zea mays* were placed in the experimental jars, and grown to maturity. $^{14}CO_2$ evolved from the jars was measured with and without plants, and marked decreases in evolved CO_2 were found in the presence of maize roots or ryegrass roots. Reid and Goss (1982) postulated that the roots took up labelled organic carbon, which reduced available C for microbial activity. As all standard abiotic variables (aeration, H_2O, inorganic nutrients, *etc.*) were kept optimal in both fallow and experimental containers, root competition should not have been a factor.

There are certain areas of uncertainty, in the above mentioned experiment, which deserve clarification: First, what happened to the microbial biomass C through the 24–42 days of the experiments? Is there a chance that microbivorous fauna, or other biotic agents, were more active in the growing-plant treatments which would decrease microbial biomass? If the organic C was taken up by the roots, why didn't it move to the shoots? Were there any transport costs? Were there any mycorrhizae present in either of the sets of experiments and what effect did they have?

Additional nutrient elements beyond C, such as N and P, might have been influential in the aforementioned interactions. There are several possibilities to follow, both in terms of organic P and N production (Helal and Sauerbeck, 1984; McLaughlin *et al.*, 1987; Parton *et al.*, 1983) and in the area of enzyme activities. There are many unknowns involved in root *versus* mycorrhizal enzyme activities, or whether these activities are shared (Barea and Brown, 1975; Coleman *et al.*, 1983; Hedley and Stewart, 1982; St. John and Coleman, 1983).

Shoot/root allocation studies

We decided to extend our studies of source/sink phenomena from root and soil O.M. to changes in sources and sinks inside the plant itself. Our research on physiological aspects of sources and sinks of carbon in plant/soil systems has been under way for about one year. The following are some preliminary results.

We used two ecotypes of the African savannah grass, *Panicum coloratum*, which have been in use by Dr. S. J. McNaughton and associates over the last ten years. The ecotypes are termed "high" and "low" responders to grazing pressure. The high

^{11}C REAL TIME CARBON FLOW DETECTION

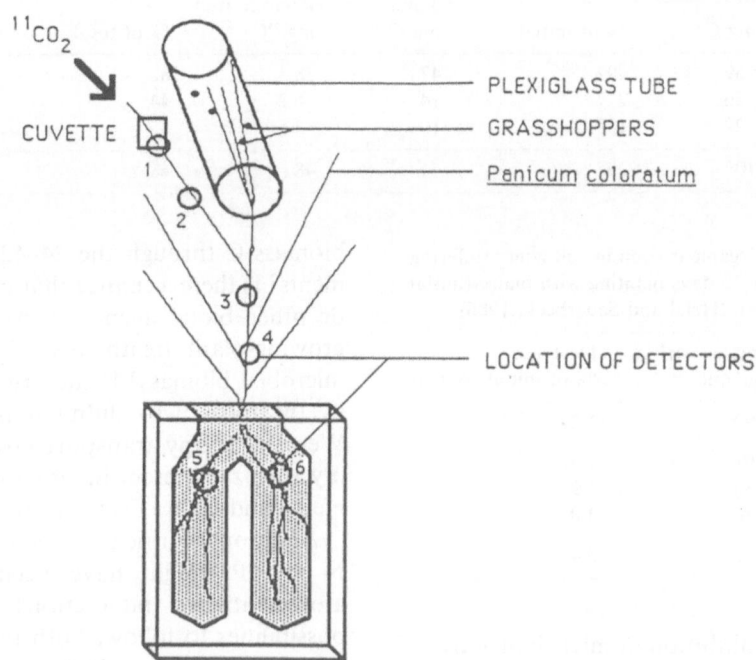

$^{11}CO_2$

CUVETTE

PLEXIGLASS TUBE

GRASSHOPPERS

Panicum coloratum

LOCATION OF DETECTORS

Fig. 2. Arrangement of plants and equipment used to measure $^{11}CO_2$ assimilation, and translocation of ^{11}C labelled photosynthates in *Panicum coloratum* L. This diagram shows typical detector locations on one tiller of *Panicum* in a split root pot which allows one side of the root system to be infected with mycorrhizae, and the other side not infected.

responder grows more leaves and stems after grazing by large ungulates, while the response of the low responder is much less (McNaughton 1983).

Our experimental regime was devised to grow test plants in special split-root growth chambers (Fig. 2), and follow the patterns of translocation of Carbon-11 labelled photosynthate, at detectors located at various sites down the stem from the labelling cuvette (Fig. 3). The sensitivity of the apparatus (Magnuson *et al.* 1982) is a few pico-curies, and the amounts of labelled translocate are as low as a few nanomoles.

Translocation rates for the two ecotypes were markedly different. The low response translocation was fast, about 2.5 cm/minute, and the concentration of ^{11}C in the photosynthate was low, ca. 5.7 nCi at point 3 (the leaf sheath), compared to

12.7 nCi at point 3 for the high responder (Table 5). The high responder translocation rates averaged 1.3 cm/minute between points 2 (leaf) and 4 (stem), only half of the low-responder ecotype. We believe that the first type of plant would quickly resupply the roots with carbon products during daylight hours, while the second type of plants would retain more of the carbon in the upper part of the plant.

Additional studies were made with *Panicum* plants with no vesicular-arbuscular mycorrhizal (VAM) infection (control), or in a split-root apparatus with VAM infection on one side of the split-root chamber. The control plant (Fig. 4) had moderate transport speeds, and low concentration of labelled solutes in the phloem. Most importantly, a much greater amount of ^{11}C photosynthate was being transported through the roots (*i.e.,* detector #5) on the same side of the split-root pot

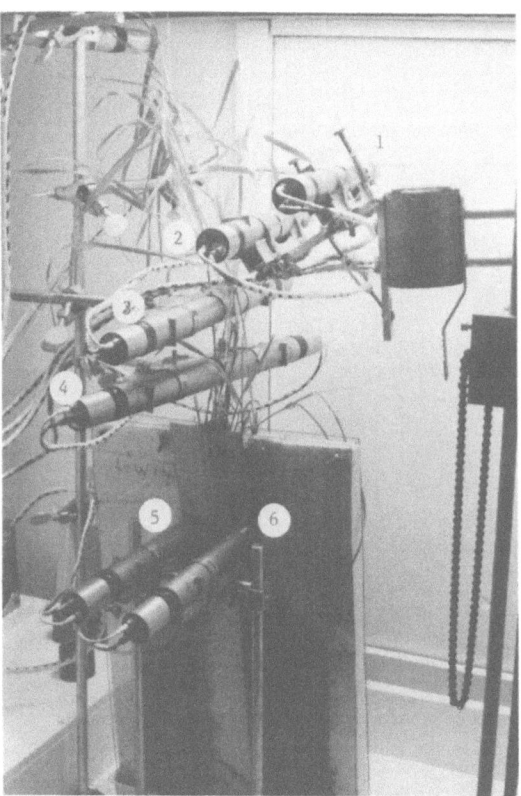

Fig. 3. Photograph of a specimen of *Panicum* showing the location of a photosynthetic cuvette (1) covering a 2.5 cm region of one leaf. Air with 350 ppm CO_2 and dew point = 18°C, flows through the cuvette then to gas analysis instruments to continuously measure net carbon exchange rate and transpiration. At the appropriate time a valve is switched to a similar flow of air with trace amounts of $^{11}CO_2$. A pair of detectors in location (1) measures the assimilation of $^{11}CO_2$, and the storage and export of ^{11}C photosynthates from the labelled leaf. To measure ^{11}C activity translocated through the phloem, (3) and (4) monitor activity in the leaf sheath and stem respectively, and (5) and (6) monitor activity transported through the roots in the left and right sides respectively of the split root pot. Note that this specimen (and all others) have several tillers, but only the labelled one is shown in the diagrams.

as the tiller than through roots on the opposite side (*i.e.* #6). In contrast, the mycorrhizal plant (Fig. 5) had high transport speed and high concentration of labelled solutes in the phloem (Wang *et al.* (1989). These results are consistent with the 30% higher carbon exchange rate and phloem loading rate in the plants with mycorrhizae.

This compares favorably with recent studies of Douds *et al.* (1988). They found that from 3–3.9% more photosynthate was transferred to the VAM side as compared to VAM-free side of a split-root pot, using *Carrizo citrange* seedlings. Mycorrhizal *Allium porrum* translocated 7% more C to roots than did non-mycorrhizal plants of similar mass

(Snellgrove *et al.* (1982). More details of our experimental results are presented in Wang *et al.* (1989).

Concluding remarks

Although work must be carried out expeditiously, due to the 20.4 minute half-life of the C-11 isotope, the research has some strong features. A complete experiment can be set up and run in six hours, and the same plants can be used repeatedly, on more than one day.

Our work augments and amplifies the results of

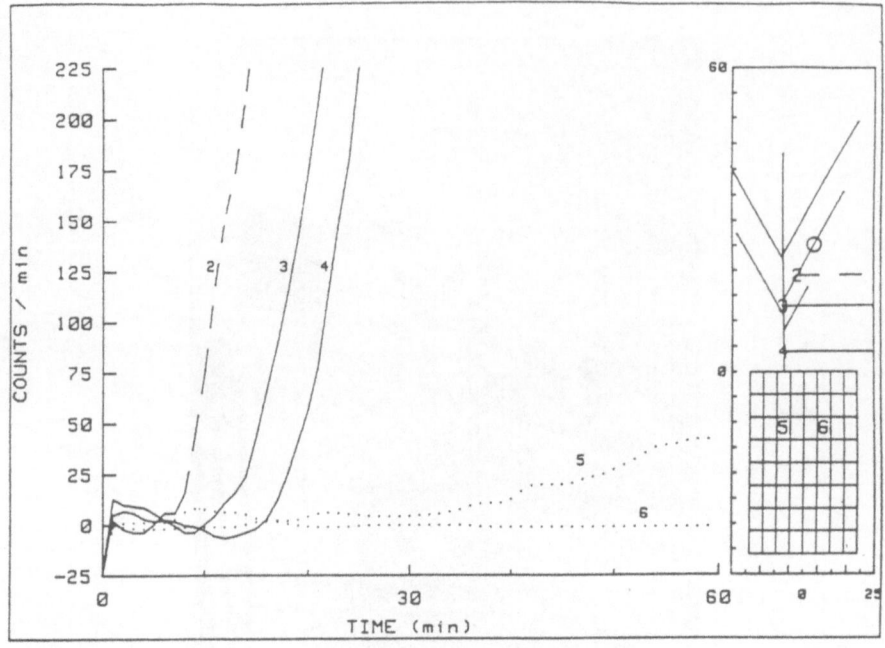

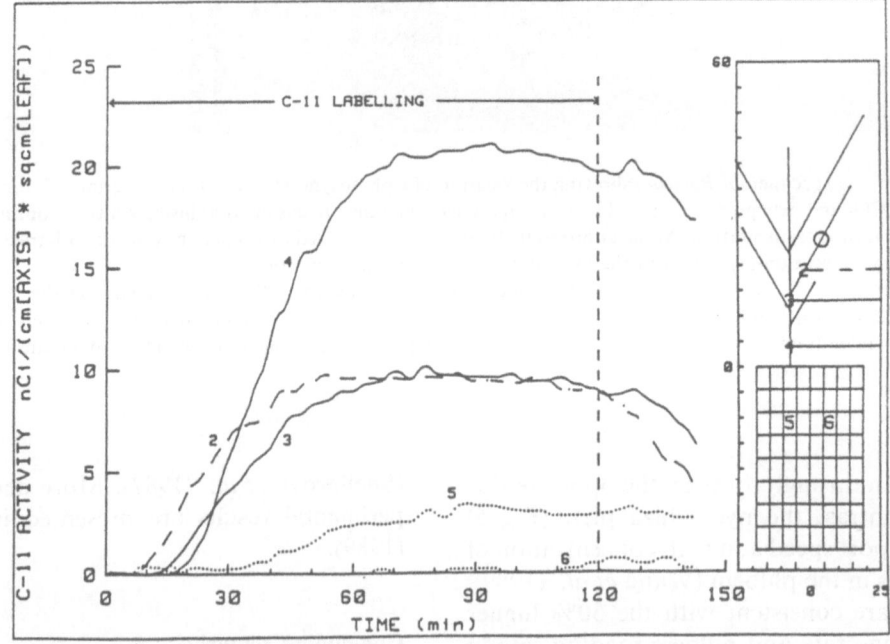

Fig. 4. **Top**. Arrival of the ^{11}C tracer front at various detector locations on a tiller of a control (i.e. non-mycorrhizal) specimen of *Panicum coloratum* L. Continuous (steady-state) labelling with ^{11}CO$_2$ in the circled region of one leaf (see plant diagram at right) began at time = 0. A line was fitted by computer to each of the rising ^{11}C activity curves, and extrapolated to zero counts/min to determine the arrival time. For example the arrival time at detector #2 (dashed curve) was 7.6 min, and at #4 it was 16.8 min, giving an elapsed time of 9.2 min. The distance between detectors # 2 and #4 was 16.2 cm, thus the mean speed of translocation in this specimen was 1.75 cm min^{-1}.

Bottom. Levels of ^{11}C activity at the same detector locations as above, during and following the 120 min period of ^{11}CO$_2$ labelling, *i.e.* Extended Square Wave input pattern (ESW). These data are corrected for background, detector sensitivity, specific activity, and the mean steady-state transit time from the point of application (circled area of leaf) to each detector location. Thus the final 20 min of steady-state activity at each location is proportional to the amount of transported carbon in the phloem (detector #2, #3, #5, & #6), or the phloem plus sink assimilation (#4). For example the steady-state level of activity in the phloem of this specimen (detector #2) was calculated to be 9.0 nCi per cm of leaf axis per cm^2 of labelled leaf area.

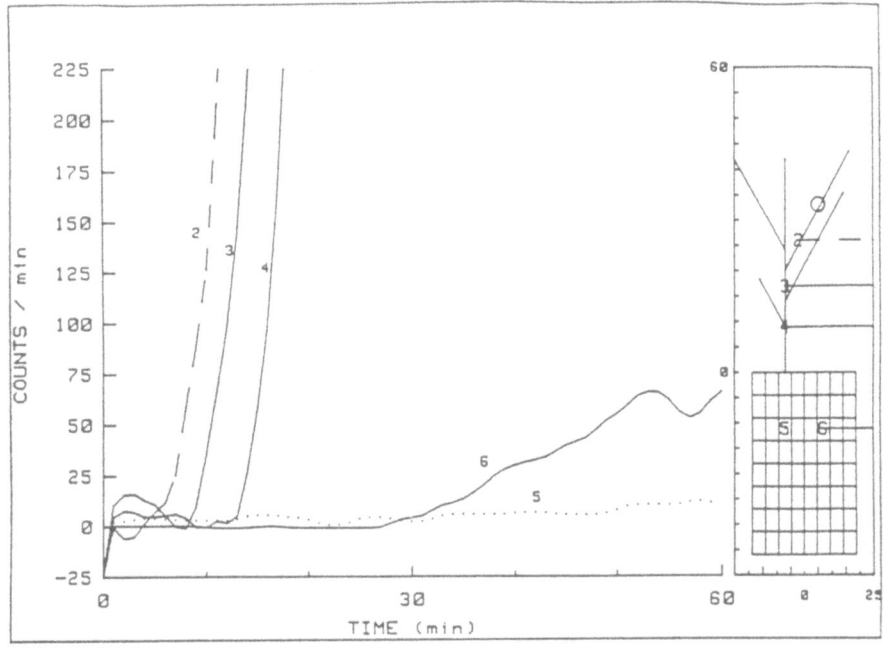

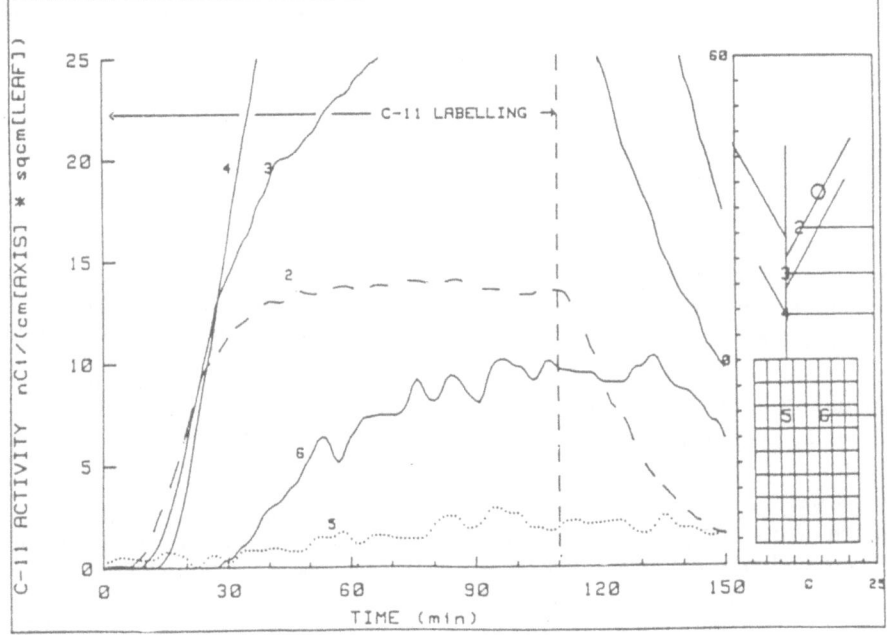

Fig. 5. **Top**. Arrival of the ^{11}C tracer front at the numbered detector locations on a tiller of a specimen of *Panicum coloratum* L. infected with mycorrhizae on one side of a split root pot (*i.e.* detector location #6 on right hand side of grid) The mean speed of translocation from detector #2 to #4 was 2.4 cm min[-1].

Bottom. The higher steady-state level of ^{11}C activity at detector #2 (also #3 & #4 off scale) is consistent with the higher rates of photosynthesis and export (*i.e.* phloem loading rate, see Goeschl and Magnuson, 1986) from the labelled leaf of this mycorrhizal infected plant.

Table 5. Comparisons of the speed of transport, and the ^{11}C activity level (proportional to the concentration of photosynthetic products) in the phloem of the two "ecotypes" of *Panicum, i.e..*, one which can sustain low levels of grazing and another which can sustain high levels of grazing.

The significance of the speed and concentration of phloem transport is that at a given loading rate (and thus at a given transport rate in mol/s) plants with a high speed and low concentration have high values of "unloading conductance", and have low speeds and high concentrations in the phloem.

Panicum "ecotype"	Replicates	1a	2a	3a	Mean
Low Grazing Response	Speed (2–4) cm/min	1.8	2.4	2.6	2.3
	Activity #2 f (conc.) in phloem (nCi)	2.5	7.5	7	5.7
		1b	2b	3b	
High Grazing Response	Speed (2–5) cm/min	1.2	1.3	1.4	1.3
	Activity #2 f (conc.) in phloem (nCi)	10	17	11	12.7

other workers such as Warembourg (1989), who has effectively studied whole-plant dynamics using carbon-14.

The root/shoot studies presented extend the concept of source and sink strengths across the entire range of shoot/root/rhizosphere (including mycorrhizae) and soil OM intractions. When these interactions are considered as they occur in space and in time, including activities of heterotrophic microbes and fauna, then one has a powerful scheme for holistic studies, which have considerable sensitivity, as well.

Future research work on biota/soil organic matter interactions will pay big dividends as we attempt to unravel some of the hidden aspects of biotic interactions in plant/soil systems.

Acknowledgements

We thank Drs. Y. Fares and C.E. Magnuson for their assistance in the development and operation of the ^{11}C apparatus. This work was conducted in the Duke University Phytotron under funding provided by NSF Grant 86-00605 to the University of Georgia.

References

Barber D A and Martin J K 1976 The release of organic substances by cereal roots into soil. New Phytol. 76, 69–80.

Barber D A and Lynch J M 1977 Microbial growth in the rhizosphere. Soil Biol. Biochem. 9, 305–308.

Barea J M and Brown M E 1974 Effects on plant growth produced by *Azotobacter paspali* related to synthesis of plant growth regulating substances. J. Appl. Bacteriol. 37, 583–593.

Clarholm M 1985 Possible roles for roots, bacteria, protozoa, and fungi in supplying nitrogen to plants. pp 355–365. *In* Ecological Interactions in Soil. Eds. A H Fitter, D Atkinson, D J Read and M B Usher. Special Publishing Service of British Ecological Society No. 4. Blackwell, Oxford.

Coleman D C, Reid C P P and Cole C V 1983 Biological strategies of nutrient cycling in soil systems. Adv. Ecol. Res. 13, 1–55.

Coleman D C, Cole C V, Anderson R V, Blaha M, Campion M K, Clarholm M, Elliott E T, Hunt H W, Schaefer B and Sinclair J 1977 Analysis of rhizosphere-saprophage interactions in terrestrial ecosystems. pp 299–309. *In* Soil Organisms as Components of Ecosystems. Eds. U Lohm and T Persson. Ecological Bulletins (Stockholm) No. 25.

Douds Jr., D D, Johnson C R and Koch K E 1988 Carbon cost of the fungal symbiont relative to net leaf P accumulation in a split-root VA mycorrhizal symbiosis. Plant Physiol. 86, 491–496.

Elliott E T, Coleman D C and Cole C V 1979 The influence of amoebae on the uptake of nitrogen by plants in gnotobiotic soil. pp 221–229. *In* The Soil–Root Interface. Eds. J L Harley and R S Russell. Academic Press, London.

Fuhr R and Sauerbeck D 1968 Decomposition of wheat straw in the field as influenced by cropping and rotation. pp. 241–250. *In* Isotopes and Radiation in Soil Organic Matter Studies, Proceedings of the IAEA/FAO Symposium. Vienna.

Goeschl J D and Magnuson C E 1986 Physiological implications of the Münch-Horwitz theory of phloem transport: Effects of loading rates. Plant, Cell Environm. 9, 95–102.

Hedley M J and Stewart J W B 1982 Method to measure microbial phosphate in soils. Soil Biol. Biochem. 14, 377–385.

Helal H M and Sauerbeck D 1984 Influence of plant roots on C and P metabolism in soil. Plant and Soil 76, 175–182.

Helal H M and Sauerbeck D 1986 Effect of plant roots on carbon metabolism of soil microbial biomass. Z. Pflanzenernaehr. Bodenkd. 149, 181–188.

Ingham R E, Trofymow J A, Ingham E R, and Coleman D C 1985 Interactions of bacteria, fungi, and their nematode grazers: Effects on nutrient cycling and plant growth. Ecol. Monogr. 55, 119–140.

Jenkinson D S 1977 Studies on the decomposition of plant material in soil. V. The effects of plant cover and soil type on the loss of carbon from ^{14}C labelled ryegrass decomposing under field conditions. J. Soil Sci. 28, 424–434.

Jenkinson D S and Powlson D S 1976 The effects of biocidal

treatments on metabolism in soil. V. A method for measuring soil biomass. Soil Biol. Biochem. 8, 209–213.

Magnuson C E, Fares Y, Goeschl J D, Nelson C E, Strain B R, Jaeger C H and Bilpuch E C 1982 An integrated tracer kinetics system for studying carbon uptake and allocation in plants using continuously produced $^{11}CO_2$. Rad. Environ. Biophys. 21, 51–65.

McLaughlin M J, Alston A M and Martin J K 1987 Transformations and movement of P in the rhizosphere. Plant and Soil 97, 391–399.

McNaughton S J 1983 Serengeti grassland ecology: The role of composite environmental factors and contingency in community organization. Ecol. Monogr. 53, 291–320.

Müller P E 1887 Studien über die natürlichen Humusformen und deren Einwirkung auf Vegetation und Boden. J. Springer, Berlin.

Parton W J, Persson J and Anderson D W 1983 Simulation of organic matter changes in Swedish soil. pp 511–516. *In* Analysis of Ecological Systems: State-of-the-Art in Ecological Modelling. Eds. W K Lauenroth, G V Skogerboe and M Flug. Elsevier Scientific Publishing, New York.

Reid J B and Goss M J 1982 Suppression of decomposition of ^{14}C-labelled plant roots in the presence of living roots of maize and perennial ryegrass. J Soil Sci. 33, 387–395.

Reid J B, Goss M J and Robertson P D 1982 Relationship between the decreases in soil stability effected by the growth of maize roots and changes in organically bound iron and aluminum. J. Soil Sci. 33, 379–410.

St. John T V and Coleman D C 1983 The role of mycorrhizae in plant ecology. Can. J. Bot. 61, 1005–1014.

St. John T V, Coleman D C and Reid C P P 1983 Association of vesicular-arbuscular mycorrhizal hyphae with soil organic particles. Ecology 64, 957–959.

Shields J A and Paul E A 1973 Decomposition of ^{14}C-labelled plant material under field conditions. Can. J. Soil Sci. 53, 297–306.

Snellgrove R C, Splittstoesser W E, Stribley D P and Tinker B P 1982 The distribution of carbon and the demand of the fungal symbiont in leek plants with vesicular-arbuscular mycorrhizae. New Phytol. 92, 75–87.

Trofymow J A and Coleman D C 1982 The role of bacterivorous and fungivorous nematodes in cellulose and chitin decomposition in the context of a root/rhizosphere/soil conceptual model. pp 117–137. *In* Nematodes in Soil Ecosystems. Ed. D W Freckman. Univ. Texas Press, Austin.

Wang G M, Coleman D C, Freckman D W, Dyer M I, McNaughton S J, Acra M A and Goeschl J D 1989 Carbon partitioning patterns of mycorrhizal versus non-mycorrhizal plants. Real-time dynamic measurements using $^{11}CO_2$. New Phytol. (submitted).

Warembourg F R and Roumet C 1989 Why and how to estimate the cost of symbiotic N_2 fixation? A progressive approach based on the use of ^{14}C and ^{15}N isotopes. Plant and Soil 115, 167–177.

M. Clarholm and L. Bergström (Eds.), Ecology of arable land, 31–41.
© 1989 Kluwer Academic Publishers.

Why and how to estimate the cost of symbiotic N_2 fixation? A progressive approach based on the use of ^{14}C and ^{15}N isotopes

F.R. WAREMBOURG and C. ROUMET
Centre L. Emberger, CNRS-BP 5051, F-34033 Montpellier Cedex, France

Key words: carbon cost, lupin, N_2 fixation, PEP carboxylase, red clover, respiration, soybean

Abstract

The process of symbiotic nitrogen fixation, though of obvious advantage to legumes in situations in which nitrogen is limiting, results in substantial penalty to the host plant in terms of cost of maintenance, synthesis and nitrogen reduction. Accurate estimates of costs are difficult to obtain because of the lack of simple methods to measure N_2 fixation and associated energy consumption. In relation to these difficulties, a multiple-step approach involving isotopes ($^{14}CO_2$–$^{15}N_2$) methodologies is described.

The estimation of net respiratory cost associated with the N_2 reduction activity in near-natural conditions was achieved using simultaneous $^{14}CO_2$ and $^{15}N_2$ labelling. It gives a minimum value of 2.5 mg C/mg N fixed. This value was corrected by the estimation of the amount of carbon saved through the process of CO_2 fixation by the PEP carboxylase of the nodules, using $^{14}CO_2$ in the soil atmosphere. This gives a real respiratory cost of 4 mg C/mg N fixed.

Abbreviations. IRGA = Infrared Gas Analyser; PEP = Phosphoenolypyruvate

Introduction

Although nitrogen is very abundant in nature, it often limits plant productivity because the only unlimited form, atmospheric N_2, is only available to a very limited range of organisms: free living N_2-fixing microorganisms, or those symbiotically associated with higher plants. Despite the selective advantage of freeing themselves from the nutrient constraints of the environment, these N_2-fixing organisms are far from being competitive in many situations. This is partly due to the high cost of nitrogen reduction activity. In agriculture, the imperative need to enhance biological N_2-fixation as a means of avoiding the increasing cost of nitrogen fertilizers has prompted scientists to estimate the energy involved in the process.

The capacity for coupling photosynthesis with the reduction of atmospheric N_2 has given N_2-fixing associations, such as the legume-Rhizobium symbiosis, an outstanding ability to colonize and to survive in nitrogen-deficient environments. In addition, recent studies (see Schubert, 1982; Gadal, 1983) have also shown that in the long history of coevolution between symbionts and hosts, carbon- and energy-saving processes (CO_2- and H_2-recycling activities) which can improve the efficiency of symbiotic N_2 fixation have emerged.

To further understand the adaptative strategies of legumes in the wild and to improve the important role that leguminous crop play in agriculture (both for man and animal nutrition as well as for contributing to the N balance of agrosystems), there is an urgent need to estimate efficiency of N_2 fixation and its variation between symbioses. The effectiveness with which carbohydrates are used in the process of reducing atmospheric N and the consequences on the whole carbon economy of the plant have to be determined. A fair amount of information has already been summarized on the cellular (Bergersen, 1971; Gutschick, 1982; Schubert, 1982) and whole plant (Atkins *et al.*, 1978; Minchin *et al.*, 1981; Schubert and Ryle, 1980) levels. Judging from the range of estimates,

2.2 to 19.4 mg C/mg N_2 fixed (Phillips, 1980), the most appropriate method of measuring efficiency at any level of organization is not completely clear. This is mainly due to methodological difficulties in measuring N_2 reduction and energy utilization.

We have developed new methods for simultaneous measurements of N_2 fixation and carbohydrate consumption in order to meet the requirements for a correct estimate of the costs involved at the whole plant level. The objectives of this paper are to present and discuss this approach which involved the use of $^{14}CO_2$- and $^{15}N_2$-labelling during short periods of time.

How to estimate the whole plant energy cost of N_2 fixation?

Prerequisites for a proper estimate are the following:

1. The rate of N_2 fixation is not constant. It changes with environmental factors (including those which affect photosynthesis), with plant and nodule age, as well as with plant phenology (Sinclair and de Wit, 1975). Furthermore, due to concomitant H_2 evolution (Dixon, 1967) and CO_2 fixation by nodule carboxylases (Christeller *et al.*, 1977), processes that are also influenced by the same factors, the efficiency of N_2 fixation may greatly vary with time. Hence, an accurate estimation of the costs of N_2 fixation can only be made over short periods of time.

2. Estimation of energy costs can be based on measurements of carbohydrate consumption as indicated by CO_2 production. It is thus necessary to distinguish the CO_2 produced for growth and maintenance activities of roots and nodules from the CO_2 produced in the nitrogen reduction and assimilation activities.

3. Measurement of respiration, associated with the process of N_2 fixation during a given period, has to be coupled with an exact measurement of the amount of N_2 fixed during the same period. However, most of the respiration data in previous studies were based on acetylene reducing activity, which is not a true measurement of N_2 fixation (Knowles, 1981). The only absolute estimate is through $^{15}N_2$ exposure of the nodulated system and subsequent measurement of the ^{15}N content of the plant.

4. Plant integrity and environmental conditions must be preserved in order to allow extrapolation of the results to field and community situations.

Previous studies of the respiratory costs of N_2 fixation suffer from the lack of a direct method for separating root and nodule activity for growth and maintenance, from that associated with the process of N_2-fixation. Methods based on linear regression of respiration on nitrogenase activity, measured either with acetylene (Witty *et al.*, 1983) or H_2 evolution (Rainbird *et al.*, 1984), have the great advantage of being non-destructive. They still require a fair amount of artificial conditions for the plants and the results need to be converted in order to provide true estimates of fixed N_2.

CO_2 respiratory loss

Experimental approach

In our studies of the carbon and nitrogen economy of legumes (Fernandez and Warembourg, 1987; Warembourg *et al.*, 1984), including the cost of N_2 fixation, the principle was as follows: since N_2 fixation is known to be directly dependent on current photosynthesis (Hardy and Havelka, 1975; Kouchi *et al.*, 1986), it is assumed that by exposing the plants to $^{14}CO_2$ during one photoperiod, the N_2-fixing activity during that period and the following night will be supported solely by ^{14}C assimilates. The $^{14}CO_2$ efflux from roots and nodules will, therefore, include all of the $^{14}CO_2$ produced in the N_2 reduction process which occurs during those 24 hours. The procedure was (1) to expose the aerial part of the plant to $^{14}CO_2$ during one day and the root system to $^{15}N_2$ during the same day and the following night, (2) to measure the $^{14}CO_2$ evolved from the soil during and after this exposure period. This was repeated several times during the growing period in order to detect differences in carbon use during the life cycle of the plants.

The plants were grown outside in soil, in PVC containers 30 cm high and 16 cm wide (see Warembourg *et al.*, 1982 for details). For some labelling experiments, sets of containers held either nodulated or non-nodulated plants; in other experiments, only nodulated plants were used. Soil and aerial atmospheres were carefully separated by a silicone sealant placed around the base of the plants stem and around the cover that maintained the containers air tight. The aerial parts of the plants were enclosed in a plastic chamber and exposed to an atmosphere containing $^{14}CO_2$. The CO_2 concentration was maintained at 340 ppm and temperature was continuously regulated to match the ambient conditions outside the chamber. $^{15}N_2$ was injected into the soil atmosphere of half of the containers, in order to raise the concentration of ^{15}N to approximately 2.5%. During the time of exposure this atmosphere was circulated in a closed circuit with a pump and CO_2 was trapped in a NaOH solution. The level of O_2 was maintained automatically at a concentration of 19%. After the labelling period (24 h) the containers were opened and thoroughly flushed with outside air for one hour, in order to free the system of all remaining ^{15}N. The other half of the containers were continuously flushed with CO_2-free air and the effluent CO_2 was trapped in NaOH solutions contained in 'snap tubes'. An automatic system allowed a change of tubes every hour thus making it possible to follow the time-course of CO_2 and $^{14}CO_2$ evolution during the labelling period and the following days. Measurements were made by titration and by scintillation counting of one aliquot. At the end of the measurement period, the plants were harvested and C and N contents estimated, using classical dry combustion and Kjeldahl methods respectively. ^{14}C was measured by scintillation counting and ^{15}N by mass spectrometry. The ^{14}C and ^{15}N contents of the plant parts were used to calculate C incorporation and nitrogen fixation following the period of labelling.

Analysis of root respiration curves

As illustrated by the results obtained for red clover (Fig. 1A), the hourly rate of $^{14}CO_2$ evolution after exposure of the plants to $^{14}CO_2$ showed two distinctive phases: a period of intensive activity which ended within two days, followed by a period of lower activity which extended to one week. During this second phase, the amount of $^{14}CO_2$ lost

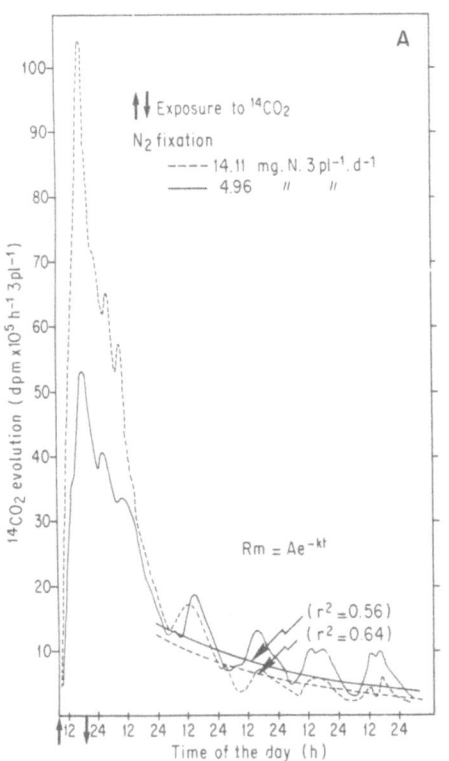

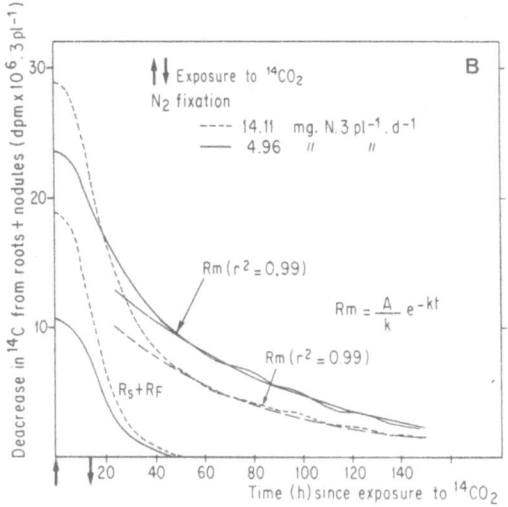

Fig. 1. Time course of hourly rates of $^{14}CO_2$ production (A) and ^{14}C decrease (B) from roots plus nodules by respiration of red clover (*Trifolium pratense* L.) fixing different amounts of N_2. The shoots were exposed to $^{14}CO_2$ during one day (results are for 3 plants). r^2 = regression coefficients, R_m, R_s and R_f = maintenance, synthesis and N_2-fixation respiratory components.

per unit of time showed an exponential decrease significantly described ($P < 0.05$) by an expression of the type:

$$y = Ae^{-kt} \tag{1}$$

where y is the hourly rate of $^{14}CO_2$ production, A the initial rate, k the average rate constant, and t the time in hours.

This later phase of $^{14}CO_2$ production was attributed to the turnover of labelled structures in roots and nodules (*e.g.* enzymes) and associated with maintenance activities (Ryle *et al.*, 1976). It was called the *Rm* component of the $^{14}CO_2$ efflux. Daily fluctuations were significantly correlated with soil temperature changes (Warembourg, 1983).

Integration of equation (1) makes it possible to account for the total amount of labelled carbon lost by roots and nodules in maintenance. The new exponential expression is of the type

$$Y = \frac{A}{k} e^{-kt} \tag{2}$$

where A/k is the total amount of $^{14}CO_2$ lost in the *Rm* component. Y is the fraction still remaining in the root plus nodules, k the average rate constant.

As indicated in Fig. 1B, this new expression (2) describes the data very satisfactory ($P < 0.01$). To account for the period of $^{14}CO_2$ assimilation, the delay in translocation from the leaves to the roots, the synthesis of compounds, and an average turnover time for these compounds, the *Rm* process has been calculated from 24 h after the beginning of exposure to $^{14}CO_2$. The difference between this RM component and the total $^{14}CO_2$ efflux accounted for the intense respiratory activity and was associated with the synthesis and N_2 fixing and activities in roots and nodules. It was called Rs + Rf component of the $^{14}CO_2$ efflux.

A similar pattern has been shown for the respiratory activity measured at other periods of the life cycle of red clover and with soybean plants labelled with $^{14}CO_2$ (Warembourg, 1983).

Estimation of the respiratory cost of N_2 fixation

Because the soil atmosphere must be kept in a closed system during exposure to $^{15}N_2$, the plants used for detailed respiration measurements were not those exposed to $^{15}N_2$. To avoid errors due to differences between plants, we used a double check control: 1) In plants exposed to $^{15}N_2$ the total $^{14}CO_2$ efflux was measured during the same period. 2) One day before labelling, the nitrogenase activity of all plants was measured, using the C_2H_2 reduction technique. We then calculated two ratios:

a) (difference between plants in N_2 fixation)/difference in $^{14}CO_2$ efflux and

b) ($^{15}N_2$ fixed during 24 h)/(C_2H_2 reduced during 2 h).

We found a very significant correlation ($r = 0.95$) between N_2 fixation activity and the $^{14}CO_2$ efflux. We also found that fixation activity as measured by $^{15}N_2$ was significantly correlated with the activity as measured by C_2H_2 reduction. Thus we had a mean of precisely estimating the amount of N_2 fixed by the plants used for detailed respiration measurements.

From the analysis of the $^{14}CO_2$ effluxes, total $^{14}CO_2$ respiration of the root system can be described as the sum of three additive components: maintenance R_m, synthesis R_s and nitrogenase activity R_f. It is similar to what is described for total CO_2 (Mahon, 1977) and can be expressed as follows:

$$R = R_m + R_s + R_f \tag{3}$$

$$R_s + R_f = s_{r+n}W_{r+n} + fQ_N \tag{4}$$

where $R_s + R_f$, the respiratory component for synthesis and nitrogenase activities, is calculated from the curves as the difference between total and maintenance (R_m) respiratory effluxes, and W_{r+n} is the ^{14}C content of roots + nodules measured at the end of the respiration measurement. These ^{14}C values are converted to mg C using the specific activity of the $^{14}CO_2$ supplied to the plants. Q_N is the amount of N_2 fixed during the labelling period measured from the ^{15}N content of the plant. s_{r+n}, the coefficient for root + nodule synthesis and f, the coefficient for N_2 reduction (or N_2 reduction efficiency, the amount of C per g of fixed N_2) are the unknowns. They can be calculated by solving two equations if data are available for two plants or two sets of plants fixing different amounts of N_2.

For red clover, the size and the number of nodules did not permit separation from the roots. It was assumed that the synthesis coefficient, s_{r+n}, did not differ for plants of the same age, regardless of a possible difference in the proportion of ^{14}C contained in roots vs. nodules.

Table 1. Estimated costs of nitrogen fixation in soybean (*Glycine max* L. Merr c.v. Hodgson) and red clover (*Trifolium repens* L.) at various times during the growing period (in mg C/mg N₂ fixed)

Time of the year (date)	18/05	31/05	14/06	5/07
Time after sowing (days)	54	69	82	97
Cost for soybean (mg C/mg N₂)	2.5	2.5	7.6	7.6
Time after sowing (days)	93	106	120	141
Cost for red clover (mg C/mg N₂)	2.8	3.0	3.2	4.6

In soybean, the nodules were separated from the roots thus allowing measurement of their ^{14}C content (Warembourg, 1983). In addition, respiratory effluxes were also measured on non-nodulated roots, thus allowing estimation of the respiratory component for root synthesis and therefore the coefficient s_r. Assuming that this coefficient was identical for non-nodulated and nodulated roots of the same age, we used it in the $R_s + R_f$ equation (4) modified as follows:

$$R_s + R_f = s_r W_r + s_n W_n + f Q_N \qquad (5)$$

with s_r and s_n the coefficients for roots and nodules synthesis respectively, W_r and W_n being the ^{14}C structural carbon in roots and nodules. As before, the two unknowns s_n and f were calculated by solving equations.

Estimates of f, the carbon cost associated with the process of N₂ fixation, ranged from 2.5 to 7.6 mg C/mg N₂ fixed for soybean, with the maximum occurring just prior to the decrease of N₂ fixation (Table 1). It varied from 2.8 to 4.6 in red clover also showing an increase with time and therefore with nodule age. These values are well within the range of literature values estimated by other means and the lower values are below the most recent theoretical estimation of carbon costs: 3.4 mg C respired/mg N₂ fixed (Schubert, 1982).

Discussion

Our calculations do not account for the costs of nodule growth and maintenance. They do include the cost of N₂ reduction, H₂ production, N assimilation and transport. The decrease of efficiency caused by H₂ evolution is partly corrected for by the activity of an hydrogenase which can act as an energy saving process. A comparison can be made with the estimates reported by Rainbird *et al.* (1984), indicating 2.5 mg C for the process of reducing 1 mg of N₂, or 2.9 mg C if the cost of the H₂ evolution is taken into account. On the basis of theoretical assumptions another 0.7 mg of C was associated with the transport processes. This was slightly higher than our minimum estimate. The value reported by Witty *et al.* (1983) using acetylene and oxygen ranged from 2.6 to 6.4 mg C/mg N₂ fixed, assuming a $C_2H_2:N_2$ ratio of 3:1. In our studies, it can be suggested that part of the increase in carbon cost recorded for both soybean and clover with nodule age may be explained by an increasing production of H₂ (Bethlenfalvay and Phillips, 1977) or by a decrease in CO₂ fixation by PEP carboxylase activity, as reported by Coker and Schubert (1981).

Indeed, the rate of CO₂ fixation in nodules may account for recycling 30% or more of the respiratory efflux of CO₂ (see Schubert, 1982). As suggested by some authors (Atkins *et al.*, 1978; Salsac *et al.*, 1984), the discrepancy between values of C/N energy cost and the theoretical values would be reduced by accurate data on the amount of respiratory CO₂ recaptured by PEP carboxylase activity in the nodules.

We have recently began to improve our estimate of the costs of N₂ fixation by determining the quantitative relationships between CO₂ and N₂ fixation.

CO₂ recycling activity

Whole plant qualitative studies

A method has been developed in order to investigate the role of the PEP carboxylase in the overall carbon economy of leguminous plants in relation to N₂ fixation, and its variation with internal (plant phenology and age) and external factors. It has been used for comparative studies of fixed carbon incorporation in different species, such as ureide- and amide-transporting plants, in different cultivars, or in plants associated with different strains of Rhizobium.

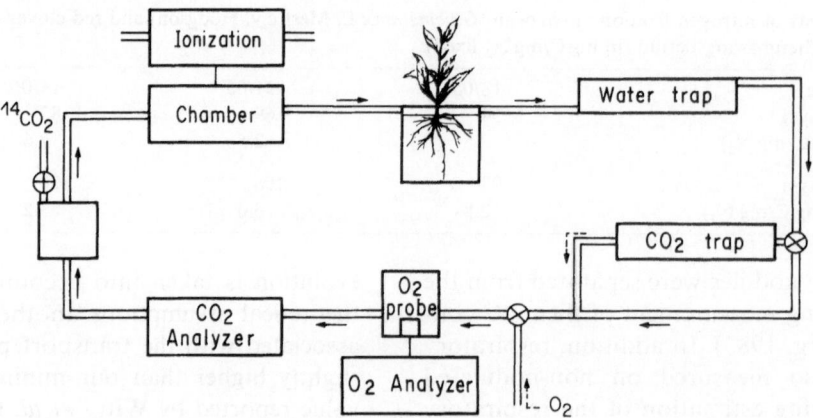

Fig. 2. Equipment used to expose the root system of legumes to $^{14}CO_2$ at constant specific activity and CO_2 concentration during long periods of time for measuring PEP carboxylase activity.

The schematic diagram in Fig. 2 illustrates the equipment used. The nodulated system is closed and exposed to $^{14}CO_2$ over long periods of time. To prevent dilution of the $^{14}CO_2$ with the respiratory efflux of the nodules, a device has been introduced to regulate both CO_2 and specific activity using an IRGA and an ionization chamber respectively. Oxygen concentration was also regulated to avoid any change in the various respiratory processes due to the confined atmosphere during the exposure period, which lasted from a few hours to several days.

Figure 3 represents diurnal fluctuations of both processes for *Lupinus albus* L. The nodulated roots were exposed for 3 hours in an atmosphere contain-

ing $^{14}CO_2$ at a CO_2 concentration of 0.3%, similar to that occurring in the rhizosphere of plants. N_2 fixation activity was measured using the acetylene reduction method. Similar results have been obtained for soybean. Both species showed a significant correlation between carboxylase and nitrogenase activities (r = 0.78).

Next step will be to use simultaneously $^{15}N_2$ and $^{14}CO_2$ in the root atmosphere in order to investigate the role of fixed carbon in the nitrogen transport and metabolism of the whole plant.

The principal advantage of this method is to integrate variations that may occur on an hourly or daily basis, both in the total amount of CO_2 fixed and in the fate of incorporated carbon in relation to

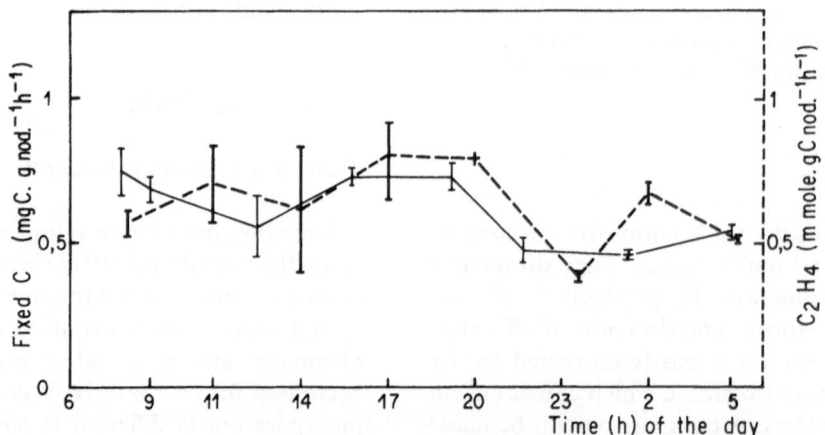

Fig. 3. Diurnal fluctuations of nodule CO_2- and N_2-fixation activities in white lupin (*Lupinus albus* L.), 38 days old. Light was maintained from 6 to 21 OC.

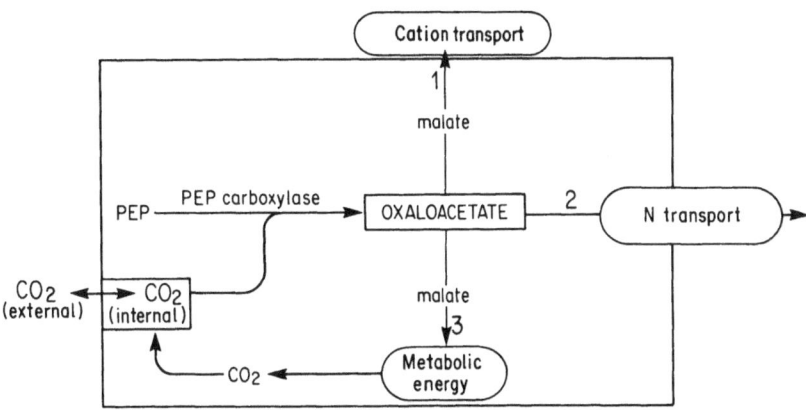

Fig. 4. Proposed roles of CO_2 fixation in legume nodules (adapted from Coker and Schubert, 1981).

N_2 fixation. It allows many qualitative studies but is not amenable to use for the quantitative determination of the CO_2 fixed by carboxylation.

Quantitative estimation

Measuring the importance of CO_2 fixation by the legume nodules on a whole plant level is difficult for many reasons which can be explained by the various roles of this CO_2 in the overall energetics of the N_2 fixation process.

As indicated by the schematic diagram in Fig. 4 there are three possible uses for the CO_2 fixed: 1. It provides organic acids, mainly malate, that may be utilized as counter ions for transport of cations in place of NO_3^- in N_2-fixing plants. 2. The oxaloacetate produced can provide carbon skeletons for asparagine and aspartate biosynthesis, the main forms of N transport in amide-exporting plants. 3. In addition, organic acids may end up in energy-yielding substrates for plant or bacterial metabolism, and mainly for nitrogen reduction. This portion will therefore be dissipated as CO_2.

So far, the only '*in vivo*' assay to measure CO_2 fixation has been to expose the nodules or the nodulated root system to an atmosphere enriched in $^{14}CO_2$, and then to measure the amount of ^{14}C incorporated into the plant. This approach supposes that external and internal specific activities are identical, but that condition can only be achieved by a high external CO_2 concentration. This will also decrease the dilution effect of respired CO_2 on specific activity. In addition, the fact that

part of the CO_2 fixed will end up in respired CO_2 requires measurements of incorporated carbon before respiration occurs. For these reasons, the actual estimates have been done on a very short term basis: a few minutes exposure at high CO_2 concentration with high specific activities.

Nodulated roots of soybean plants, 50 days old, were exposed to $^{14}CO_2$ for 5 minutes at a CO_2 concentration of 5%. The plants were then harvested, starting immediately after the end of the exposure period. As indicated by Fig. 5, the decrease in ^{14}C attributed to loss by respiration reached almost 72% in 4 hours. According to Coker and Schubert (1981), the rate of CO_2 fixation measured immediately after exposure to $^{14}CO_2$ may provide the most reliable estimate of the true amount of CO_2 fixed. When compared with the rate of C_2H_2 reduction, the molar ratios of CO_2 fixed per C_2H_2 reduced were around 1.16. Assuming a $C_2H_2 : N_2$ ratio of 3; this corresponds to 3.5 mol CO_2 fixed/mol N_2 fixed (or 1.5 mg C/mg N_2 fixed). Our results coroborate the estimates of Coker and Schubert's (1981) which ranged between 1 and 3.4 mol CO_2 fixed/mol N_2 fixed. However, concerns have been raised about the effect of high CO_2 pressure on nitrogenase activity (Mulder and Van Veen, 1960) and on the diffusion of inorganic forms of carbon (CO_2 and HCO_3^-) in the plant tissues as suggested by Atkins *et al.* (1978). This carbon may be released as CO_2 after pulse labelling, thus explaining a decrease in the plant ^{14}C. Therefore, further investigations are necessary in order to confirm the proportion of fixed carbon used to supply metabolic energy to the N_2 fixation process.

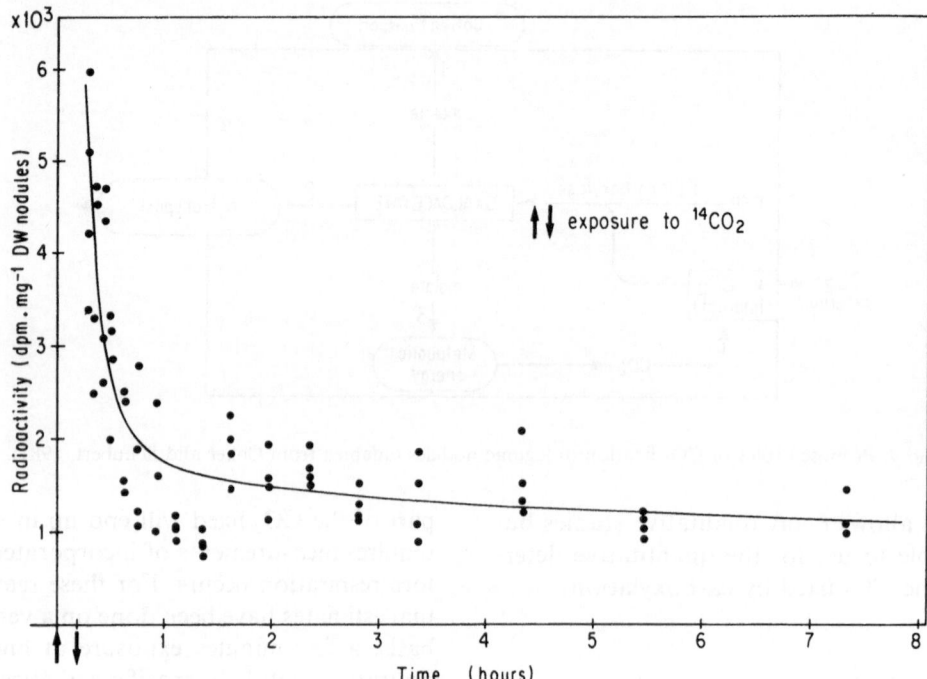

Fig. 5. Time distribution ^{14}C in soybean plants after pulse labelling of the root systems with $^{14}CO_2$ during 5 minutes at 5% CO_2.

Approaching a real estimation of the cost of N_2 fixation

The methodologies presented in this paper provide useful information for an approach to measuring the real cost of N_2 fixation. This can be illustrated by the results obtained for respiration and CO_2 fixation of soybean plants at the beginning of the flowering phase.

– The first methodology described, based on the measure of CO_2 efflux, does not take into account the CO_2 used for nodule growth and maintenance. So, the value of 2.5 mg C/mg N_2 fixed (see above) reflects for the whole plant the minimum net carbon cost associated with N_2 fixation.

– The second approach indicates that an important part of the CO_2 respired inside the nodule is fixed by carboxylase activity: 1.5 mg C/mg N_2 assuming a $C_2H_2:N_2$ ratio of 3:1. Twenty eight per cent (0.4 mg C/mg N_2) remained in the plant (Fig. 5), as organic carbon and/or in nitrogen compounds and 72% (1.1 mg C/mg N_2) were released outside the nodule as CO_2. This later production of CO_2 is accounted for in the 2.5 mg C/mg N_2 measured in respiration.

From Fig. 6 it is obvious that:
– In terms of C balance for the whole plant, the cost of N_2 fixation is 2.5 mg C/mg N_2.
– Concerning the amount of photoassimilates that are diverted to the N_2 fixation process ending into CO_2, the reduction of 1 mg N_2 requires 2.9 mg of C.
– Estimation of the real cost in terms of C substrates degraded (total production of CO_2 inside the nodule) will account for 1.4 + 1.5 mg C: the primary CO_2 production, + 1.1 mg C, the CO_2 recycled and dissipated as CO_2, a total of 4 mg C/mg N_2 which is close to the theoretical value of 3.4 (Schubert, 1982). It is now clear that the cost of N_2 fixation determined by external CO_2 efflux could not be compared with the theoretical estimation of N_2 reduction. The outside CO_2 efflux reflects the net C cost for the whole plant. The inside CO_2 efflux including the CO_2 recycled (50%) reflects more precisely the real C cost for the process of N_2 reduction.

This estimate of 4 mg C/mg N_2 may have to be corrected if (as it is certainly the case) the oxydative metabolism of sucrose (from the phloem) and dicarboxylic acids (from CO_2 fixation) does not generate the same amount of ATP equivalents.

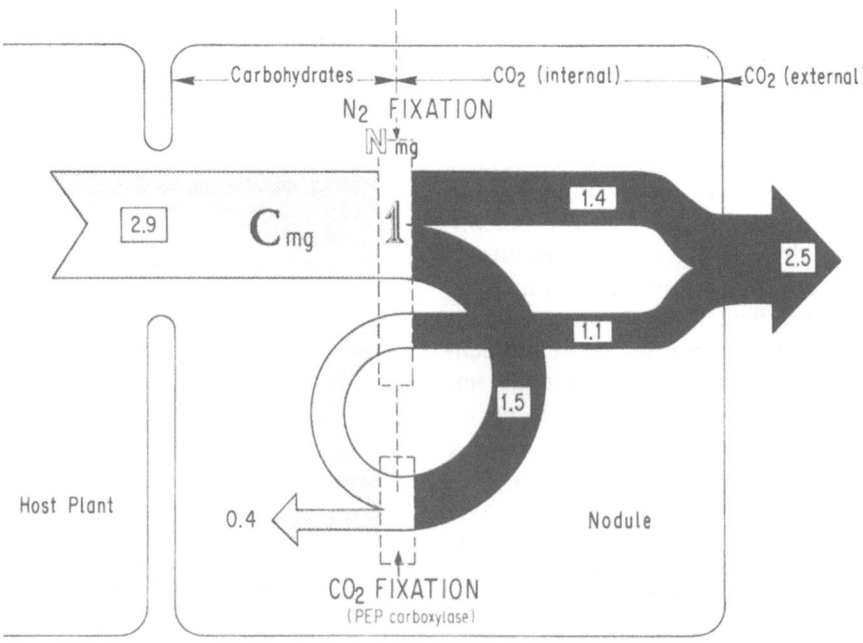

Fig. 6. Schematic diagram illustrating the carbon economy of soybean nodules in the process of N₂ fixation. Values are based on whole plant measurements using $^{14}CO_2$ and $^{15}N_2$.

The same kind of approach can be used for the highest measured values of N₂ fixation costs, which in soybean were estimated in the range of 7.6 mg C/ mg of N₂ fixed. In that case, the low efficiency may have been partly related to a decrease in carboxylase activity with nodule age, as already reported by Coker and Schubert (1981).

Perspective of future research

In 1980, Phillips declared that 'the field of symbiotic fixation still suffers some considerable uncertainties regarding energetic costs and efficiencies'. Since then, experimental techniques of measuring energy demand for fixation have improved considerably, particularly in the analysis of the various component processes (Pate *et al.*, 1981), but also the energy cost of the whole plant (Rainbird *et al.*, 1984; Ryle *et al.*, 1983; Witty *et al.*, 1983). In the latter, determination of the actual rate of N₂ fixation was lacking, and it is commonly accepted that use of $^{15}N_2$ represents the most accurate approach. It appears from our results that the simultaneous use of $^{14}CO_2$ and $^{15}N_2$ allows an estimation of the true cost of N₂ fixation in terms of carbon for the whole plant. Besides, even though it requires a fair amount of experimental care and the access to a mass or emission spectrometer, this method can be applied in complex systems in near-natural conditions. Its use can considerably improve our understanding of the energy budget of different symbioses, different host species, or strains of Rhizobium at different phenological stages and under various environmental conditions. It can also be extended to non-legume symbioses such as *Frankia* associations.

Nevertheless, it can only assess the net efficiency of N₂ fixation without distinction of the processes which can affect it. Indeed, it does not take into account hydrogen production, which decreases efficiency, nor hydrogenase and PEP carboxylase activities which improve it. As shown in this paper, complementary studies are possible in order to estimate such processes. Their understanding are of great importance for comparisons between symbioses or between different strains of Rhizobium within the same host species. The results of such studies can have implications in selection, and genetic programs in order to produce more efficient symbioses.

The energy costs of symbiotic N₂ fixation are

now definable. Quantitative or semi-quantitative estimates are available from measurements and extrapolations of known biochemistry. Considerable knowledge can now be obtained through well-designed experiments.

Appreciation of the quantitative ecological consequences of energy costs, their variability, and the role of environmental factors can now be considered. There are also many potential agricultural and biotechnological applications such as increasing biofixation in the field. Free-living organisms are of greater inherent interest for most biotechnologists, but symbiotic fixers can be greatly improved for agriculture purposes. The recent research in efficiency of N_2 fixation has paved the way for new selection programmes based on energetics and better allocation of photosynthates for fixation, and during longer periods of time during the life cycles of host plants. Genetics may also contribute in producing more efficient Rhizobium strains through manipulation of PEP carboxylase and hydrogenase activities.

In conclusion, even though symbiotic N_2 fixation appears as an energy-intensive process, several-fold greater than the industrial Harber process, the fact that it can be improved implies a challenging task for future research.

Acknowledgements

The authors wish to thank Mrs. F. Lafont for technical assistance, N. Combes for typing and L. Miller for help in correcting the manuscript.

References

Atkins C A, Herridge D F and Pate J S 1978 The economy of carbon and nitrogen in nitrogen-fixing annual legumes. *In* Isotopes in Biological Dinitrogen Fixation. pp 211–242. Vienna: Int. At. Energy Agency.

Bergersen F J 1971 Biochemistry of symbiotic nitrogen fixation in legumes. Annu. Rev. Plant Physiol. 22, 121–140.

Bethlenfalvay G J and Phillips D A 1977 Ontogenic interactions between photosynthesis and symbiotic nitrogen fixation in legumes. Plant Physiol. 60, 419–421.

Christeller J T, Laing W A and Sutton W D 1977 Carbon dioxide fixation by lupin root nodules. I. Characterization, association with phosphoenolpyruvate carboxylase, and correlation with nitrogen fixation during nodule development. Plant Physiol. 60, 47–50.

Coker G T and Schubert K R 1981 Carbon dioxide fixation in soybean roots and nodules. I. Characterization and comparison with N_2 fixation and composition of xylen exudate during early nodule development. Plant Physiol. 67, 691–696.

Dixon R O D 1967 Hydrogen uptake and exchange by pea root nodules. Ann. Bot. 31, 179–188.

Fernandez M P and Warembourg F R 1987 Distribution and utilization of assimilated carbon in red clover during the first year of vegetation. Plant and Soil 97, 131–143.

Gadal P 1983 Phosphoenolpyruvate carboxylase and nitrogen fixation. Physiol. Veg. 21, 1069–1074.

Hardy R W F and Havelka U D 1975 Photosynthate as a major factor limiting N_2 fixation by field-grown legumes with emphasis on soybeans. *In* Symbiotic Nitrogen Fixation in Plants. Ed. P S Nutman. pp 421–439. Cambridge University Press, London.

Gutschick V P 1982 Energetics of microbial fixation of dinitrogen. *In* Advances in Biochemical Engineering 21, 109–167.

Knowles R 1981 The measurement of nitrogen fixation. *In* Current Perspectives in Nitrogen Fixation. Eds. A H Gibson and W E Newton. pp 327–333. Australian Acad. Sci., Canberra, Australia.

Kouchi H, Akao S and Yoneyama T 1986 Respiratory utilization of ^{13}C labelled photosynthate in nodulated root systems of soybean plants. J. Exp. Bot. 37, 985–993.

Mahon J D 1977 Respiration and the energy requirement for nitrogen fixation in nodulated pea roots. Plant Physiol. 60, 817–821.

Minchin F R, Summerfield R J, Hadley P, Roberts E H and Rawsthorne S 1981 Carbon and nitrogen nutrition of nodulated roots of grain legumes. Plant Cell Environ. 4, 5–26.

Mulder E G and Van Veen W L 1960 The influence of carbon dioxide on symbiotic nitrogen fixation. Plant and Soil 13, 265–278.

Pate J S, Atkins C A and Rainbird R M 1981 Theoretical and experimental costing of nitrogen fixation and related processes in nodules of legumes. *In* Current Perspectives in Nitrogen Fixation. Eds. A H Gibson and W E Newton. pp 105–116. Australian Acad. Sci. Canberra.

Phillips D A 1980 Efficiency of symbiotic nitrogen fixation in legumes. Ann. Rev. Plant Physiol. 31, 29–49.

Rainbird R M, Hitz W D and Hardy W F 1984 Experimental determination of the respiration associated with soybean/Rhizobium nitrogenase function, nodule maintenance and total nodule nitrogen fixation. Plant Physiol. 75, 49–53.

Ryle G J A, Arnott R A, Powell C E and Gordon A J 1983 Comparison of the respiratory effluxes of nodules and roots in six temperate legumes. Ann. Bot. 52, 469–477.

Ryle G J A, Cobby J M and Powell C E 1976 Synthetic and maintenance respiratory losses of [14]CO_2 in uniculum barley and maize. Ann. Bot. 40, 571–586.

Salsac L, Drevon J J, Zengbé M, Cleyet-Marel J C and Obaton M 1984 Energy requirement of symbiotic nitrogen fixation. Physiol. Vég. 22, 509–521.

Schubert K R 1982 The energetics of biological nitrogen fixation. Workshop Summaries. I. Am. Soc. Plant Physiol. 1–30.

Schubert K R and Ryle G J A 1980 The energy requirements for nitrogen fixation in nodulated legumes. *In* Advances in Legume Science, Eds. R J Summerfield and A H Bunting. pp. 85–96. Royal Botanic Gardens, Kew, England.

Sinclair T R and de Wit C T 1975 Photosynthate and nitrogen requirements for seed production by various crops. Science 189, 565–567.

Warembourg F R 1983 Estimating the true cost of dinitrogen fixation by nodulated plants in undisturbed conditions. Can. J. Microbiol. 29, 930–937.

Warembourg F R, Haegel B, Fernandez M P and Montange D 1984 Distribution et utilisation des assimilats carbonés en relation avec la fixation symbiotique d'azote chez le soja (*Glycine max* L. Merril). Plant and Soil 82, 163–178.

Warembourg F R, Montange D and Bardin R 1982 The simultaneous use of ¹⁴CO₂ and ¹⁵N₂ labelling techniques to study the carbon and nitrogen economy of legumes grown under natural conditions. Physiol. Plant. 56, 46–55.

Witty J F, Minchin F R and Sheehy J E 1983 Carbon costs of nitrogenase activity in legume root nodule determined using acetylene and oxygen. J. Exp. Bot. 34, 951–963.

M. Clarholm and L. Bergström (Eds.), Ecology of arable land, 43–52.

Plant- and soil-related controls of the flow of carbon from roots through the soil microbial biomass

J.A. VAN VEEN, R. MERCKX[1] and S.C. VAN DE GEIJN[2]
Research Institute Ital, P.O. Box 48, 6700 AA Wageningen, The Netherlands. Present address:
[1]*Laboratory of Soil Fertility and Soil Biology, Catholic University of Leuven, B-3030 Leuven, Belgium and*
[2]*Centre for Agrobiological Research, P.O. Box 14, 6700 AA Wageningen, The Netherlands*

Key words: microbial biomass, organic matter decomposition, rhizosphere, root material production, soil nutrient status, soil structure and texture

Abstract

The flow of carbon from plant roots through the microbial biomass is one of the key processes in terrestrial ecosystems. Roots release considerable amounts of organic materials which are utilized by microbes as substrate for biosynthesis and energy supply. The fate of photosynthates and other organic material in the soil-root environment under different conditions was studied using ^{14}C-tracers. Soil structure and texture had a large effect on the turnover of the ^{14}C-labelled materials through the microbial biomass. Finer, clayey soils tended to be more 'preservative' than coarser, sandy soils, *i.e.* larger amounts of ^{14}C were incorporated in microbial biomass and soil organic matter fractions in clayey soils than in sandy soils.

The soil nutrient status also appeared to affect organic matter turnover. At limiting plant-nutrient concentrations the utilization of ^{14}C-labelled photosynthates seem to be hampered. Plant roots influenced the transformation of glucose and crop residues and the effect was attributed to plant-induced changes in mineral nutrient status. The mechanisms of this process and the consequences are discussed.

A number of areas for future research are identified, including the potentials for manipulating rhizodeposition.

Introduction

Below ground net primary production is the main source of organic matter for a wide variety of soil ecosystems (Milchunas *et al.*, 1985). Photosynthetically fixed carbon is translocated from the above-ground parts of the plant to the roots and from the roots into the surrounding soil. The carbon compounds are utilized by microbes for biosynthesis and energy production. Since microbial activity is the 'motor' driving soil nutrient cycles, the production and utilization of root-derived carbon is a key issue of the functioning of soil ecosystems. Since Hiltner (1904) proposed the term 'rhizosphere' for the zone of increased microbial activity in and around roots, numerous studies have been made mainly on the quality of soluble root exudates in artificial growth systems (*e.g.*

Rovira, 1969; Vancura *et al.*, 1977). More recently studies using ^{14}C-CO_2 as tracer were reported on the quantity of carbon delivered from growing roots under more realistic conditions. Most ^{14}C-CO_2-labelling studies have been made in growth cabinets of different degree of sophistication. It is clear that best estimates of the total production of carbon by roots can be made when plants are exposed throughout their entire life cycle to an atmosphere containing ^{14}C-CO_2 at a constant specific activity (Newman, 1985). However, pulse labelling is often more feasible for field measurements, and when interpreted cautiously it may give additional information (Keith *et al.*, 1986).

Since the early studies of Martin (1971; 1975) and Warembourg and Paul (1973), utilizing the ^{14}C-CO_2 labelling techniques, knowledge on several aspects of the production of root derived carbon has considerably increased. However, many questions remain, in particular on the mechanisms of the utilization of the root-material by microbes and the consequences for the cycling of C and N. In this paper we will deal with plant-related factors such as species differences and growth stage, which determine the production of root derived carbon, on the basis of own research and data from literature. Soil-related factors such as soil-type and the mineral nutrient status, which might control the turnover of the root-derived carbon compounds by the microbial biomass, will also be discussed.

Plant-related aspects of the production of root-derived carbon

Estimates of total input of root derived carbon from arable crops range from approx. 900 to 3000 kg C per ha per year (Table 1). The proportion of the total assimilated carbon by plants, that is released into the soil ranges from 16 to 33% (Keith *et al.*, 1986; Sauerbeck and Johnen, 1976; Warembourg and Paul, 1973). Several factors have been recognized to influence the distribution of photosynthates within a plant and thus the input to soil. On the basis of their data from pulse labelling experiments in the field, Keith *et al.* (1986) showed that the proportion of the photosynthates that is translocated to the roots decreased substantially in the course of the growing season. They labelled plants in the field by a 1 day exposure to ^{14}C-CO_2 and harvested 3 weeks after labelling for analysis. After 6 weeks of growth of wheat 50% of the recovered ^{14}C was found in shoots, 13% in roots, 8% in soil and 29% was respired as ^{14}C-CO_2 from the soil-root compartment. At the end of the growing season, after 22 weeks of growth, the distribution of ^{14}C over shoots, roots, soil, and ^{14}C-CO_2 respired, was 97%, 0.4%, 0.7% and 1.5% respectively. Since most growth cabinet studies have been

Table 1. Input of C ($kg\,C\,ha^{-1}\,y^{-1}$) to soil by growing plants during a growing season

Input kg C/ha	% of crop yield	Reference
1300	30	Keith *et al.*, 1986
1500		Martin and Puckridge, 1982
1200	81	Jenkinson and Rayner, 1977
2950		Johnen and Sauerbeck, 1977
941	31	Lucas *et al.*, 1977

made with relatively young plants, of up to 6–8 weeks, extrapolations on the basis of these data to total inputs of carbon for an entire growing cycle of a plant are disputable.

The distribution pattern of photosynthates may differ according to plant species, phenological stage and/or growth conditions. Similar to the observations by Helal and Sauerbeck (1986) we found, that over the same growth period relatively more ^{14}C remains in the shoots of maize as compared to wheat (Table 2). The change in the ^{14}C-allocation of young Douglas fir during a growing season is opposite to the aforementioned change with wheat plants. In the later growth stages much more ^{14}C is translocated to belowground parts of the trees than in early stages (Gorissen and Van Veen, 1988).

It has often been observed that the fraction of total ^{14}C-fixed, that is evolved from the soil-root compartment as ^{14}C-CO_2 is smaller than the fraction remaining in the roots (Table 3). This is remarkable, since the opposite seems to be more logical. In his review on the fate of carbon translocated to roots, Lambers (1987) showed on the basis of literature data the C-utilization for root respiration as a proportion of the net amount of carbon translocated to the roots, to average 0.6 ranging between 0.35 and 0.7. So, when also turnover of carbon through the rhizosphere microflora resulting in even more ^{14}C-CO_2, is included, values of 0.2 for the ratio between C-CO_2 respired and total C translocated below-ground (roots + soil + - rhizosphere as calculated from the data given in Table 3) seem to be rather questionable.

Soil related factors controlling the utilization of root-derived organic materials by micro-organisms

It is widely accepted, that the production of carbon compounds by roots either by exudation or otherwise, is sufficiently large to support considerable growth of microbes in the rhizosphere. It is, therefore, surprising, that Whipps and Lynch (1983) calculated, that the root material production during the initial 3 weeks of growth of barley was not sufficient to sustain the observed growth of the microbial biomass in the rhizosphere. Only 10% or less of the acquired carbon was estimated to be produced by the roots. The apparent discrepancy was explained by assuming overestimation of mi-

Table 2. ^{14}C distribution over shoot, root, soil respiration and soil for different plants after 28 days of growth in a ^{14}C-CO_2 atmosphere in sandy soils

	Wheat	Maize	Douglas fir*	
			A	B
Shoots	49	69	65	37
Roots	27	28	25	55
Soil respiration	19		9	8
Soil	5	3	1	1

*Three years old trees: A = early stage of growth; B — later stage of growth

crobial biomass or underestimation of root material production. They concluded that the amount of carbon leaving the roots limits luxuriant microbial growth and may therefore create heavy competition for substrates on the root surface, as well as in the rhizosphere.

Microbial biomass measurements in rooted soils using the chloroform fumigation methods to determine the total utilization of root derived compounds by the microbial biomass must be considered with caution, since the decomposition of root material during the incubation period following chloroform fumigation might contribute considerably to the (^{14}C)-CO_2 evolution. New approaches solving the problem of decomposition of non-microbial biomass material during incubation after fumigation have been suggested, whereby biomass is determined on the basis of extractions of material directly after fumigation (Brookes *et al.*, 1985; Merckx and Martin, 1987). Nevertheless comparisons of data from the conventional measurements provide insight into the relative utilization of root-derived material by the soil microbial biomass.

Only a small fraction of soil microbial biomass profits from the supply of organic compounds from the roots. After 6 weeks of wheat growth in an atmosphere, constantly labelled with ^{14}C-CO_2, less than 6% of the microbial biomass was ^{14}C-labelled (Merckx *et al.*, 1985). The amount of ^{14}C in the microbial biomass represented approximately 20% of the total amount of ^{14}C present in the soil. When growing maize in the same soil and under the same conditions also approximately 20% of the residual ^{14}C in the soil was found in the microbial biomass. However, in the maize-cropped soil ^{14}C-labelled biomass was 16–17% of the total microbial biomass. The larger ^{14}C-labelled fraction of the total microbial biomass in the soil cropped with maize as compared to the wheat-cropped soil indicates a better availability of the maize-derived material. Whipps and Lynch (1985) reviewed data on the distribution of root-derived ^{14}C in soil over water soluble and insoluble fractions for different plants showing that materials which originated from maize were relatively more water soluble than from wheat.

Soil structure and texture have been indicated as

Table 3. Distribution of ^{14}C (%) amongst plant and soil components after growth in a $^{14}CO_2$ atmosphere

Plant	Age	Labelling	Author	Shoots	Roots	Soil	Rhizosphere-CO_2
Maize	25 days	continuous	Helal and Sauerbeck, 1983	56.0	31.0	2.0	11.0
Wheat	21 days	continuous	Whipps and Lynch, 1983	52.5	18.6	11.5	17.3
Wheat	24 days	continuous	Martin and Kemp, 1986	51.6	31.0	5.9	11.6
Wheat	28 days	continuous	Merckx *et al.*, 1985	48.8	27.3	4.7	19.2
Wheat	112 days	1 pulse at 49 days	Martin and Kemp, 1986	59.2		14.1	26.7
Wheat	112 days	1 pulse at 70 days	Martin and Kemp, 1986	90.5		5.2	4.3

crucial factors in the control of the turnover of organic matter through the soil microbial biomass (Van Veen *et al.*, 1984; 1985). Differences in the decomposition rates of organic material, *i.e.* glucose, microbial tissue and crop residues, in different soils could be explained by assuming that organic matter is physically better protected against decomposition in clayey soils than in sandy soils and that clayey soils have a larger capacity to preserve microbes (Van Veen *et al.*, 1985) than have sandy soils. Similar effects of soil type on the turnover of root derived material were observed by Merckx *et al.* (1985). When comparing sandy with clay-rich soils, we found that in the clay-rich soil (1) a higher proportion of the label was retained at harvest, (2) the proportion of the label lost by below-ground respiration decreased with time and (3) the proportion of soil ^{14}C incorporated in microbial biomass decreased with time.

In an attempt to sort out the clay mineral effect from other effects *e.g.* pH, microflora and organic matter quality, we carried out an additional experiment in which the sandy soil was amended with pure clay (5% K-illite). The results of the ^{14}C distribution were very similar to those measured in the natural clay-rich soil. The pH of the soil was unaffected by the clay addition. Microbial biomass ^{14}C-dynamics, however, followed a different pattern (Fig. 1). Again the incorporation percentage was higher in the clay-amended than in the unamended

soil but there was no decrease with time. The incorporation percentage seemed to attain a constant level around $\pm 60\%$. The results show that even with relatively small clay additions, effects similar to those observed in the silty-clay loam could be provoked. Physical protection of organic matter and of the newly formed micro-organisms by the presence of clay minerals could account for the observed effects.

The importance of plant induced changes in energy and nutrient status of soil for the turnover of organic matter

Qualitative changes in root exudation, due to changes in mineral nutrition, have been reported. (Kraffczyk *et al.*, 1984; Trolldenier and von Rheinbaben, 1981a, b). Changes in microbial population in the rhizosphere, as a result of a change in mineral nutrition of the plant, usually were interpreted as a consequence of this altered exudation pattern (Turner *et al.*, 1985).

Direct effects of the mineral nutrient supply on the microbial growth in the rhizosphere have always been neglected as the carbon supply for living roots has always been considered to be the key factor controlling microbial growth (Newman, 1985; Whipps and Lynch, 1983). However, more recent data indicate that the mineral nutrient status of a soil might directly influence microbial dynamics. On the basis of the low utilization efficiency of 15% of root-derived materials by the microbial biomass, Helal and Sauerbeck (1986) suggested that the carbon supply is not the limiting factor for microbial growth in the rhizosphere. We investigated the influence of soil nutrient status on the production of root-derived material and associated microbial growth in the rhizosphere of maize by growing plants for 42 days in an atmosphere containing $^{14}CO_2$ (Merckx *et al.*, 1987). The soil was fertilized to two levels and microbial growth was measured by a fumigation technique and dilution plating.

When comparing the results from the soils with the two mineral nutrient amendments we observed that in the more nutrient-poor soil (1) a higher proportion of ^{14}C accumulated after 6 weeks, (2) total microbial biomass remained unchanged, (3) ^{14}C-labelling of the biomass levelled off whereas

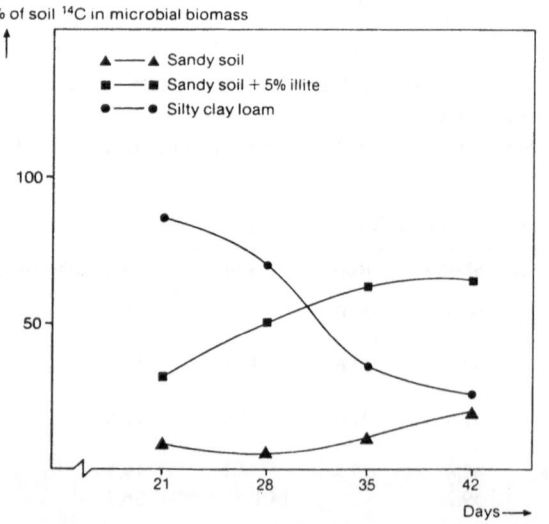

Fig. 1. Biomass ^{14}C formation in different soils during 6 weeks of growth of wheat in a ^{14}C-CO_2 labelled atmosphere:

actual inputs of ^{14}C still increased and (4) a higher percentage of the soil ^{14}C residue remained in an easily extractable or available form. In particular the last observation was of interest as after 5 and 6 weeks of growth of maize 52–35% of the residual ^{14}C in the soil receiving the low nutrient-amendment was in extractable form as opposed to 14–16% in the higher fertilized soil. On the basis of those data, we concluded that the mineral nutrient supply may severely limit microbial utilization of root-released materials. Helal and Sauerbeck (1986) suggested a very low efficiency of substrate utilization as the explanation for the decreased rate of turnover of organic matter in nutrient-poor soils. However, Elliott *et al.* (1981) supposed that a decreased uptake rate due to mineral nutrient stress rather than the utilization efficiency may be decisive.

The implications of these observations are important. If microbial growth is not only determined by the availability of carbon, but also of mineral nutrients, plants might have dual and counteracting effects on microbial activity in the soil. On the one hand plants stimulate microbial activity through the supply of organic substrate, but on the other hand they might limit microbially mediated processes, through depletion of mineral nutrients.

To investigate this we designed two experiments in which we studied the effect of the soil nutrient status on the turnover of ^{14}C labelled glucose and crop residues. The soil nutrient status was modified by additions of mineral fertilizers and/or the presence of plants.

In the first experiment, field moist, sandy soil was amended with 0, 0.5 and 1.0 g Sporumix kg^{-1}. Sporumix is a commercial fertilizer mix containing 14% N (60/40 NH_4^+/NO_3^--N), 16% P_2O_5, 18% K_2O and trace elements (Merckx *et al.*, 1987). Maize was grown in pots containing 1 kg soil, kept at constant moisture (additions to constant weight) and at constant temperature. Part of the soil was kept unplanted. After 3, 4 and 5 weeks of incubation planted soil was made free of roots by handpicking and homogenised. Glucose (^{14}C-labelled, 4.6 mg C. 50 g^{-1} moist soil) was added to samples of both previously cropped and unplanted soil. Subsequent incubation was carried out for four days.

Fertilized soil had a slightly higher CO_2 evolution than unfertilized samples (Table 4). For the unplanted treatment the incubation period after

Table 4. ^{14}C-CO_2-evolution over 4 days (% of input) after addition of ^{14}C-glucose to a sandy soil, previously incubated for 3, 4 and 5 weeks either with or without growth of maize and amended with 0, 0.5 and 1.0 g Sporumix. kg^{-1} soil

Previous incubation		^{14}C-CO_2		
		3 weeks	4 weeks	5 weeks
Fertilization level				
− plant	0	41	–	40
	0.5	43	–	41
	1.0	44	–	44
+ plant	0	37	33	34
	0.5	43	35	34
	1.0	46	34	34

fertilization had no effect. However, for the soils planted with maize the proportion of ^{14}C-glucose respired as [14]CO_2 in 4 days was significantly reduced: for unfertilized soils already after 3 weeks, but after 5 weeks for all fertilizer levels to the same extent. In all cases glucose had disappeared within the 4-day incubation period. Plants did thus markedly influence the efficiency of ([14]C-) glucose utilization, probably by inducing changes in the mineral nutrient status of the soil.

In another experiment we studied the effect of fertilizer addition on decomposition of crop residues. The design of the experiment was similar, except that only two fertilizer levels were studied (0 and 1.0 g Sporumix kg^{-1}). Prior to incubation, and immediately after fertilization 1.3 g of ^{14}C-labelled young wheat plant residues were added per kg moist soil. After 2, 3, 4, 5 and 6 weeks of incubation either water (2 ml), glucose (10 mg C) or N (7.5 mg N-NH_4^+) were added per 50 g moist soil to samples of the previously planted or unplanted soil.

In the unplanted soil the ^{14}C-CO_2 evolution from the crop residues was decreased by fertilization, two weeks after the start of incubation (Fig. 2). In unplanted low fertility soils the addition of N decreased the decomposition rate (^{14}C-CO_2 evolution) as compared to the control treatment. In the fertilized soils N-addition slightly increased the ^{14}C-CO_2 evolution. Another difference between the two fertilizer treatments was the gradual change in the response of the ^{14}C-CO_2 respiration to the addition of unlabelled glucose. In the fertilized situation glucose always stimulated, whereas the unfertilized samples showed this only after more than 4 weeks (Fig. 2).

Similar trends were observed in planted soils, as shown in Fig. 3 for the samples after 6 weeks. Glucose addition stimulated, while N-addition restricted ^{14}C-CO_2 evolution both in planted and unplanted soils. Fertilized soils showed higher ^{14}C-CO_2 respiration from decomposed crop residues in planted but not in unplanted soils (after 6 weeks: Fig. 3).

As to the mechanism involved in these changes one might at present only speculate. A working hypothesis might be that the available mineral N is a prerequisite for biosythesis processes. In the absence of N, available glucose or other readily available substrate will be incorporated into cells as reserve material without much loss for energy-requiring conversions. Addition of glucose to fertilized soils will lead to a higher proportion respired, as compared to unfertilized soils.

Plants will compete for N, mineralized by the microbial biomass, and reduce chances for effective growth of the biomass. This will happen primarily in unfertilized soils, but after prolonged plant growth N will also be depleted in fertilized soils. This would explain the disappearance of glucose

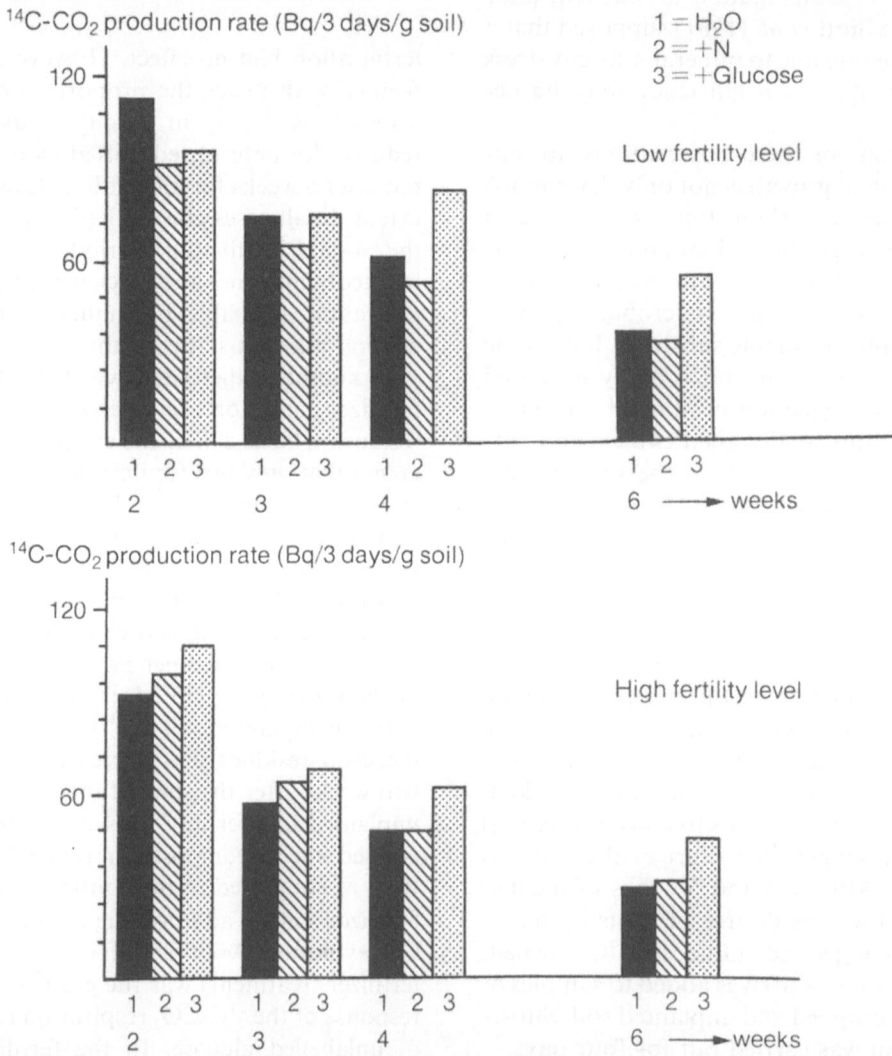

Fig. 2. Effect of water (1), N- (2) and glucose (3) addition on the decomposition rate of ^{14}C-labelled wheat residues in an unplanted sandy soil, initially amended with 0 (low fertility) or 1 (high fertility) g Sporumix kg^{-1} soil. Decomposition rates were measured after 2, 3, 4 and 6 weeks of incubation.

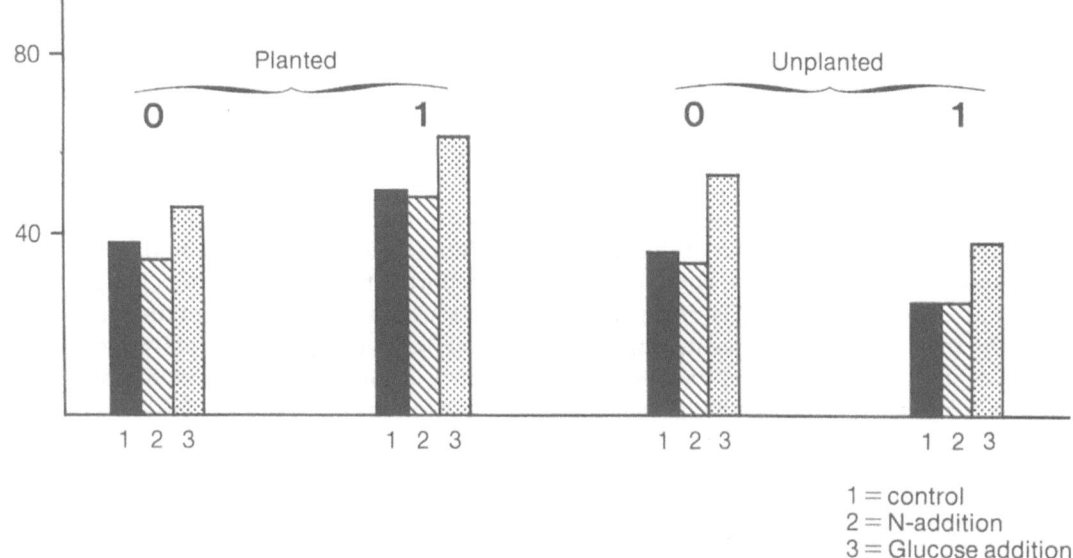

14C-CO2 production rate (Bq/3 days/g soil)

1 = control
2 = N-addition
3 = Glucose addition

Fig. 3. Effect of water (1), N- (2) and glucose (3) addition on the decomposition rate of 14C-labelled wheat residues in a sandy soil initially amended with 0 or 1.0 g Sporumix kg⁻¹ soil after 6 weeks incubation with and without growth of maize.

with a relatively low loss through respiration after prolonged plant growth (Table 4).

Incorporation of young wheat plant residue will initially enhance availability of easily decomposable C-compounds, but after two weeks this source might be exhausted. N-addition did decrease ^{14}C-CO_2 evolution in unfertilized soils whereas it slightly stimulated in the fertilized treatment (Fig. 2). One might suppose that stimulation of ^{14}C-CO_2 evolution in the fertilized soil might result from a stimulated break-down of more recalcitrant material with a low N-content. Addition of glucose to fertilized soil also stimulated ^{14}C-CO_2 respiration. In unfertilized soils this started only after 4 weeks, possibly as a result of the gradually enhanced availability of mineralized N. Substrate competition might also have interfered.

The data sets from these experiments are not yet complete, but present results clearly show the complex nature of the effects of plants on the turnover of carbon in the soil. Much more research is needed to improve our understanding of the basic mechanisms of these processes.

A direct or indirect effect of the prevailing balance between mineral nutrients and accessibility of the C-substrate on the composition of the microbial population might also be involved. Such a

dependence has been reported by Söderström *et al.* (1983) who found that the microbial activity in coniferous forest podzols decreased after application of NH_4NO_3. The effect was observed 3 months after addition and was still evident after 3–5 years. As the microbial activity in forest soils is generally fungal dominated, a shift in the composition of the microbial population or a direct inhibitory effect on essential enzymatic processes such as lignolytic decomposition which are associated mainly with fungi, could explain these effects.

Perspectives for future research

The utilization of ^{14}C-CO_2 as a tracer to follow the fate of photosynthates has considerably improved the understanding of the mechanisms of root material production by growing plants (Helal and Sauerbeck, 1983; 1986; Keith *et al.*, 1986; Martin, 1977; Martin and Kemp, 1986; Merckx *et al.*, 1985; 1987; Whipps and Lynch, 1983; 1985). As yet most data are from growth cabinet studies with relatively young plants. However, extrapolations to field conditions for estimating the total input of organic matter in soil during a growing season or, in case of perennial plants and trees, for a longer, fixed period should be done with great care. Better estimates need *in situ* measurements, whereby pulse labelling techniques might be adequate.

Recently, we observed a temporary, significant decrease of ^{14}C-CO_2 evolution rate from the soil-root system immediately following exposure of trees to moderate exposure to O_3 and SO_2, *i.e.* 180 and $52 \mu g\,m^{-3}$ during 3 and 28 days respectively (Gorissen and Van Veen, 1988). This might be the result of a delay in the translocation of ^{14}C-labelled photosynthates to below-ground compartments. The potential consequences of this effect for the vitality of the trees are great. Frequent, temporary, inhibitions of the translocation of carbon to the roots due to increased levels of air-pollutants could well affect the energy supply and consequently the vitality of roots and associated microbes, such as mycorrhiza. In the long run this may then affect uptake of water and nutrients from soil. To put this in a proper perspective with regard to the assessment of tolerable levels of atmospheric pollutants, both *in situ* measurements and more laboratory studies are necessary. Here the use of ^{11}C to trace the fate of photosynthates directly after exposure to pollutants could be helpful (Coleman and co-workers, this issue). A large problem for *in situ* measurements are older, larger trees, which, of course, cannot be labelled entirely using a carbon tracer. Attempts are now made to study the effects of atmospheric pollutants on trees *in situ* using gas-chambers which allow for the exposure of a single branch to pollutants.

Proper understanding of the effect of several important soil-related control factors of production of root material and its utilization by the rhizosphere microbial population is hampered by technical problems. Assessment of the soil-temperature effect needs separate control of above- and below-ground compartments (Martin and Kemp, 1986), whereas studies on the effect of soil moisture are hindered by the large demand for water by growing plants which makes it extremely difficult to maintain a constant soil moisture level. The essential impact of soil structure and texture on the transformation of root-derived products and other organic materials in soil has been discussed before (Van Veen *et al.*, 1984; Van Veen and Van Elsas, 1988). Here, studies are needed to identify the key elements of the soil system which mainly control the microbial processes and to characterize them quantitatively.

The results from investigations on the effect of the nutrient status of the soil on the utilization of root-derived products and other organic materials in soil have changed our views on the relative importance of the levels of substrates for microbial growth and activity. It seems that the balance between C-substrates and mineral nutrients is rather delicate with regard to its effect on the composition and activity of the soil microbial population. The consequences could be larger than what is reported here. By inducing changes in the nutrient status of soil, plants might directly influence microbial growth and microbially mediated processes, such as mineralization and immobilization or denitrification. For instance, Haider *et al.* (1985, 1987) reported that the mineralization of soil organic N was higher in planted than in unplanted soil. Merckx *et al.* (1987) observed that the nutrient status of the soil also affected the growth of specific microbial populations. In recent years numerous studies have been conducted to explore the possibilities of manipulating the soil ecosystem through the introduction of beneficial micro-organisms (*e.g.* see Lynch, 1983). However, the establishment of introduced, new, organisms in the complex soil system has been shown to be difficult (Van Veen and Van Elsas, 1988). Even newly introduced specific rhizosphere bacteria, such as associative symbiotic N-fixers, has often been reported to be unsuccessful in establishing an effective plant-growth stimulating microflora. Another approach which might be more effective in establishing a beneficial rhizosphere population, is the breeding of cultivars with a specific rhizodeposition. Neal *et al.* (1970, 1973) showed that by chromosome-substitution wheat cultivars were obtained with markedly different rhizosphere populations. As compared to manipulation of soil life by introduction of new strains, the use of plant cultivars inducing a favourable rhizosphere population might create a more stable situation. In a long-term trial of up to 7-years of monocultures of water melon most cultivars resistant to *Fusarium oxysporum* showed severe wilt after 4–5 years (Hopkins *et al.*, 1987). Only one cultivar, moderately resistant in greenhouse tests had a unique resistance which was effective throughout the 7 years trial. The suppressive factor was sensitive to anti-biological treatments and it was concluded that this particular cultivar promoted a biological control factor in soil. A combination of both biotechnological approaches could also be feasible. Rennie and Larson (1979) showed that the growth and thus, the establishment of two introduced N_2-fixing bacteria in the rhizosphere was different for different cultivars. An exciting possibility is therefore to tailor root exudation to the establishment of particular micro-organisms. Modern genetic technologies are available to plant breeding. However, successful utilization of the vast potential of the genetic techniques in this area should be preceded by a proper understanding of the ecology of the rhizosphere.

References

Brookes P C, Landman A, Pinden G and Jenkinson D S 1985 Chloroform fumigation and the release of soil nitrogen: Rapid direct extraction method to measure microbial biomass nitrogen in soil. Soil Biol. Biochem. 17, 837–842.

Elliott E T, Cole C V, Fairbanks B C, Woods L E, Bryant R J and Coleman D C 1983 Short term bacterial growth, nutrient uptake and ATP turnover in sterilized, inoculated and C-amended soil: The influence of N-availability. Soil Biol. Biochem. 15, 85–91.

Gorissen A G and Van Veen J A 1988 Temporary disturbance of translocation of assimilates in Douglas firs caused by low levels of ozone and sulphur dioxide. Plant Physiol. 88, 559–563.

Haider K, Mosier A and Heinemeyer O 1985 Phytotron experiments to evaluate the effect of growing plants on denitrification. Soil Sci. Soc. Am. J. 49, 636–641.

Haider K, Mosier A and Heinemeyer O 1987 Effect of growing plants on denitrification at high soil nitrate concentrations. Soil Sci. Soc. Am. J. 51, 91–102.

Helal H M and Sauerbeck D R 1983 Method for studying turnover processes in soil layers at different proximity to roots. Soil Biol. Biochem. 15, 223–235.

Helal H M and Sauerbeck D R 1986 Effect of plant roots on carbon metabolism of soil microbiol biomass. Z. Pflanzenernähr. Bodenkd. 149, 181–188.

Hiltner L 1904 Über neuere Erfahrungen und Probleme auf dem Gebiet der Bodenmikrobiologie und unter besonderer Berücksichtigung der Gründüngung und Bracke. Arb. Dtsch. Landwirtsch. Ber. 98, 59–78.

Hopkins D L, Larkin R P and Elmstrom G W 1987 Cultivar-specific induction of soil suppressiveness to *Fusarium* wilt of water melon. Phytopathology 77, 607–611.

Jenkinson D S and Rayner J H 1977 The turnover of soil organic matter in some of the Rothamsted classical experiments. Soil Sci. 123, 298–305.

Johnen B G and Sauerbeck D R 1977 A tracer technique for measuring growth, mass and microbial breakdown of plant roots during vegetation. *In* Soil Organisms as Components of Ecosystems. Eds. U Lohm and T Persson. Ecol. Bull. (Stockholm) 25. 366–373.

Keith H, Oades J M and Martin J K 1986 Input of carbon to soil from wheat plants. Soil Biol. Biochem. 18, 445–449.

Kraffczyk I, Trolldenier G and Beringer H 1984 Soluble root exudates of maize: Influence of potassium supply and rhizosphere micro-organisms. Soil Biol. Biochem. 16, 315–322.

Lambers H 1987 Growth, respiration, exudation and symbiotic associations: The fate of carbon translocated to the roots. *In* Root Development and Function. Eds. P J Gregory, J V Lake and D A Rose. pp 125–145. Soc. Exp. Bot., Seminar Series 30, 1987.

Lucas R E, Holtman J B and Connor L J 1977 Soil carbon dynamics and cropping practices. *In* Agriculture and Energy. Ed W Lockertz. pp 333–351. Academic Press, London.

Lynch J M 1983 Soil Biotechnology. Blackwell Scientific Publications, Oxford. 191 p.

Martin J K 1971 ^{14}C-labelled material leached from the rhizosphere of plants supplied with $^{14}CO_2$. Aust. J. Biol. Sci. 24, 1131–1142.

Martin J K 1975 ^{14}C-labelled material leached from the rhizosphere of plants supplied continuously with $^{14}CO_2$. Soil Biol. Biochem. 7, 395–399.

Martin J K 1977 Factors influencing the loss of organic carbon from wheat roots. Soil Biol. Biochem. 9, 1–7.

Martin J K and Kemp J R 1986 The measurement of C transfers within the rhizosphere of wheat grown in field plots. Soil Biol. Biochem. 18, 103–108.

Martin J K and Puchridge D W 1982 Carbon flow through the rhizosphere of wheat crops in South Australia. *In* The Cycling of Carbon, Nitrogen, Sulphur and Phosphorus in Terrestrial and Aquatic Ecosystems. Eds. I E Galbally and J R Freney. pp 77–81. Australian Academy of Science, Canberra.

Merckx R, Den Hartog A and Van Veen J A 1985 Turnover of root-derived material and related microbial biomass formation in soils of different texture. Soil Biol. Biochem. 17, 565–569.

Merckx R and Martin J K 1987 Extraction of microbial biomass components from rhizosphere soils. Soil Biol. Biochem. 19, 371–376.

Milchunas D G, Lauenroth W K, Singh J S, Cole C V and Hunt H W 1985 Root turnover and production by ^{14}C dilution: implications of carbon partitioning in plants. Plant and Soil 88, 353–365.

Neal J L, Atkinson T G and Larson R I 1970 Changes in the rhizosphere population of selected physiological groups of bacteria related to substitution of specific pairs of chromosomes in spring wheat. Plant and Soil 39, 209–212.

Newman E I 1985 The rhizosphere: Carbon sources and microbial populations. *In* Ecological Interactions in Soil. Ed. A H Fitter. pp 107–121. Spec. Publ. No 4 of the British Ecological Society, Blackwell Scientific Publ. Oxford.

Rennie R J and Larson R I 1979 Dinitrogen fixation associated with disomic chromosome substitute lines of spring wheat. Can. J. Bot. 57, 2771–2775.

Rovira A D 1969 Plant root exudates. Bot. Rev. 35, 35–56.

Sauerbeck D R and Johnen B G 1976 Root formation and decomposition during plant growth. *In* Soil Organic Matter Studies. pp 141–148. Proceedings of Symposium IAEA/FAO/GSF, Braunschweig.

Söderström B, Bååth E and Lundgren B 1983 Decrease in soil microbial activity and biomasses owing to nitrogen amendments. Can. J. Microbiol. 29, 1500–1506.

Trolldenier G and von Rheinbaben W 1981a Root respiration and bacterial population of roots. I. Effect of nitrogen source, potassium nutrition and aeration of roots. Z. Pflanzenernähr. Bodenkd. 144, 366–377.

Trolldenier G and von Rheinbaben W 1981b Root respiration and bacterial population of roots. II. Effect of nutrient deficiency. Z. Pflanzenernähr. Bodenkd. 144, 378–384.

Turner S M, Newman E I and Campbell R 1985 Microbial population of ryegrass root surfaces: Influence of nitrogen and phosphorus supply. Soil Biol. Biochem. 17, 711–715.

Vancura V, Prikryl Z, Kalachová L and Wurst M 1977 Some quantitative aspects of root exudation. Ecol. Bull. (Stockholm) 25, 381–386.

Van Veen J A, Ladd J N and Frissel M J 1984 Modelling C and N turnover through the microbial biomass in soil. Plant and Soil 76, 257–274.

Van Veen J A, Ladd J N and Amato M 1985 Turnover of carbon and nitrogen through the microbial biomass in a sandy loam and a clay soil incubated with $[^{14}C(U)]$ glucose and $[^{15}N](NH_4)_2SO_4$ under different moisture regimes. Soil Biol. Biochem. 17, 747–756.

Van Veen J A and Van Elsas J D 1988 Impact of soil structure and texture on the activity and dynamics of the soil microbial population. Proc. 4th Intern. Symp. on Microbial Ecol. pp 481–488.

Warembourg F R and Paul E A 1973 The use of $^{14}CO_2$ canopy techniques for measuring carbon transfers through the plant-soil system. Plant and Soil 38, 331–345.

Whipps J M and Lynch J M 1983 Substrate flow and utilization in the rhizosphere of cereals. New Phytol. 95, 605–623.

Whipps J M and Lynch J M 1985 Energy losses by the plant in rhizo-deposition. Ann. Proc. Phytochem. Soc. Eur. 26, 59–71.

M. Clarholm and L. Bergström (Eds.), Ecology of arable land, 53–62.

Carbon and nitrogen dynamics along the decay continuum: Plant litter to soil organic matter

JERRY M. MELILLO[1], JOHN D. ABER[2], ARTHUR E. LINKINS[3], ANDREA RICCA[1], BRIAN FRY[1] and KNUTE J. NADELHOFFER[1]

[1]The Ecosystems Center, Marine Biological Laboratory, Woods Hole MA 02543, USA; [2]Institute for the Study of Earth, Oceans and Space, University of New Hampshire, Durham, NH 03824, USA; [3]Biology Department, Clarkson University, Potsdam, NY 13676, USA

Key words: carbon, cellulose, decomposition, $\delta^{13}C$, $\delta^{15}N$, lignin, nitrogen, stable isotopes

Abstract

Decay processes in an ecosystem can be thought of as a continuum beginning with the input of plant litter and leading to the formation of soil organic matter. As an example of this continuum, we review a 77-month study of the decay of red pine (*Pinus resinosa* Ait.) needle litter. We tracked the changes in C chemistry and the N pool in red pine (*Pinus resinosa* Ait.) needle litter during the 77-month period using standard chemical techniques and stable isotope analyses of C and N.

Mass loss is best described by a two-phase model: an initial phase of constant mass loss and a phase of very slow loss dominated by degradation of 'lignocellulose' (acid soluble sugars plus acid insoluble C compounds). As the decaying litter enters the second phase, the ratio of lignin to lignin and cellulose (the lignocellulose index, LCI) approaches 0.7. Thereafter, the LCI increases only slightly throughout the decay continuum indicating that acid insoluble materials ('lignin') dominate decay in the latter part of the continuum.

Nitrogen dynamics are also best described by a two-phase model: a phase of N net immobilization followed by a phase of N net mineralization. Small changes in C and N isotopic composition were observed during litter decay. Larger changes were observed with depth in the soil profile.

An understanding of factors that control 'lignin' degradation is key to predicting the patterns of mass loss and N dynamics late in decay. The hypothesis that labile C is needed for 'lignin' degradation must be evaluated and the sources of this C must be identified. Also, the hypothesis that the availability of inorganic N slows 'lignin' decay must be evaluated in soil systems.

Introduction

Decay processes in an ecosystem can be thought of as a continuum beginning with fresh plant litter and leading to the formation of refractory soil organic matter. The early stages of this continuum have been intensively researched over the past two decades in laboratory studies lasting weeks to months and in field studies lasting one to two years. The results of many of these studies have been reviewed by Swift *et al.* (1979) and Melillo *et al.* (1984). The general pattern early in decay has been recognized (Olson 1963, Howard and Howard 1974) and the rates of mass loss have been correlated with both initial litter quality (Fogel and Cromack, 1977; Melillo *et al.* 1982; Minderman, 1968) and environmental variables such as temperature and moisture (Bunnell and Tait, 1977; Bunnell *et al.*, 1977; Jansson and Berg, 1985; Meentemeyer, 1978). Patterns of N accumulation and release, and factors controlling them have been identified for the early stages of decomposition (Aber and Melillo, 1980; 1982; Berg and Staaf, 1981).

Our knowledge of the later stages of decay along the continuum is much poorer. We know little about the patterns of change of the various C fractions and nutrient pools in litter material during the later stages of decay and we have no detailed knowledge of factors that control them.

As an example of this continuum, we review the most important results of a 77-month field study of decomposition of red pine (*Pinus resinosa* Ait.) needle litter. We tracked the changes in a number of C fractions and in the N pool in the litter along the decay continuum using wet chemistry methods and stable C isotope analyses. Based on our results and previous studies of decomposition, we suggest directions for future research.

Methods

Site

This ongoing study is being conducted in a 1.3 ha red pine plantation at the Harvard Forest, Petersham, Massachusetts, USA. The plantation is situated on an Entic Haplorthrod (Spodosol) of the Gloucester series. The soil is of glacial origin and is very stony (Lyford, 1963; 1964). Low and poorly drained areas were excluded from the study site. The plantation has a deep forest floor and an A2 albic (E) horizon. When the soil was characterized in 1978 the O2 horizon had an ash-free mass of 41.5 Mg/ha and a forest floor depth of 8.0 cm.

The plantation was established in 1925 following harvest of mixed hardwoods that had developed on an old field abandoned in the mid-1800's. Hardwood weeding was carried out several times during the first five years after establishment. Red pine accounts for 89% of the basal area in the plantation, with the remainder composed of white spruce (*Picea glauca* (Moench)), red maple (*Acer rubrum* L.), and red oak (*Quercus rubra* L.). There is no shrub layer or understory.

Field methods

Decay of pine needle litter was studied using litterbags (Bocock *et al.*, 1960; Melillo *et al.*, 1984). Freshly fallen red pine needles were collected in the fall of 1978 and air dried to a constant weight. A known amount (usually 10 g) of air-dried needles was placed in fiberglass mesh bags (20 × 20 cm, 2 mm mesh). On 26 October 1978, the bags of needles were anchored in the litter layer of the forest floor in the pine plantation. At intervals of 2, 5, 7, 9, 11, 13, 15, 17, 22, 24, 29, 36, 59, 66 and 77 mo, five bags were retrieved. After removing fine roots growing into the bags and the large arthropods, we conducted a series of chemical analyses on the needles.

Soil samples of the forest floor and three depths (0–15 cm, 15–20 cm, 30–45 cm) in the mineral soil were taken in the fall of 1986. These samples were analyzed as described below.

Standard chemical analyses

Original red pine litter and litter in various states of decay was air dried, ground to 20 mesh in a Wiley mill, and oven dried at 50°C for 48 h. Subsamples were used to determine moisture content (105°C for 48 h), ash content (500°C for 12 h), total N and proximate C fraction composition. The total N analyses were performed on litter samples digested in a mixture of concentrated sulfuric acid and mercury catalyst using a block digestor (Anonymous, 1977).

The proximate C fractions were isolated by a series of extractions. Fats, oils, waxes, and other non-polar soluble materials were removed using methylene chloride (Anonymous, 1976) in a modified block digestor tube for 5 h at 50°C. A subsample of the residue was extracted with hot water (100°C for 3 h) and filtered to remove the polar extractables.

The water-insoluble residue was then digested with 72% H_2SO_4 for 1 h. Secondary hydrolysis was carried out in an autoclave at 120°C for one h (Effland, 1977). The resulting solution was filtered and the filtrate analyzed for acid soluble carbohydrates using the phenol sulfuric acid assay (Dubois *et al.*, 1956) using D-glucose standards and measuring the absorbance at 490 nm. The acid soluble carbohydrates will also be referred to as 'cellulose.' What remained was considered the acid insoluble fraction, and will be referred to as 'lignin.'

Stable isotope analyses

Red pine litter in various states of decay, and forest floor and mineral soil samples were analyzed for stable C and N isotopic compositions (Minagawa *et al.*, 1984). Samples were ground with wireform CuO in a mortar and pestle, then sealed with Cu metal into Vycor tubes under vacuum and combusted at 900°C for 1 h. Carbon dioxide and N_2 were separated and purified via cryogenic distillation on a high vacuum manifold, and isotopic compositions measured with a Finnigan 251 isotope ratio mass spectrometer.

The results from the mass spectrometer analyses are expressed in terms of $\delta(X)$ values, which are parts per thousand differences from a standard:

$$\delta(X) = [(R_{sample}/R_{standard}) - 1] \times 10^3$$

where X is ^{13}C or ^{15}N and R is the corresponding ratio $^{13}C/^{12}C$ or $^{15}N/^{14}N$. The δ values are measures of the relative amounts of heavy and light isotopes in a sample. Increases in these values signify relative increases (*e.g.*, more ^{13}C relative to ^{12}C) in the amount of the heavy isotope components. Standard reference materials are C in the PeeDee limestone and N in the atmosphere (Peterson and Fry, 1987). The precision of the C and N measurements is ±0.1‰.

Results

Pattern of mass loss and changes in carbon fraction dynamics

At the end of the 77-month study the red pine needle litter had lost 83% of its initial mass. Although a classical decay curve fits the pattern of mass loss, we choose to divide the decay sequence into two distinct phases: Phase I – an initial period of constant mass loss, dominated by the loss of acid soluble carbohydrates (0–44 mo) and Phase II – a

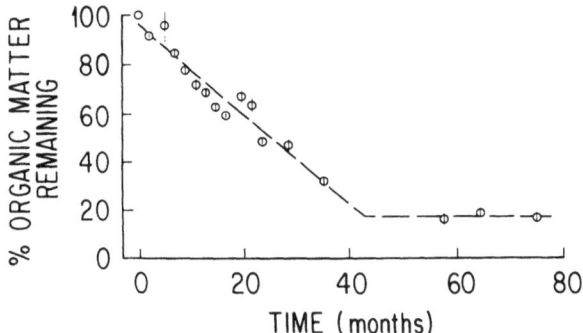

Fig. 1. Percentage of original mass of red pine litter (organic matter) remaining *versus* time in months, dashed line follows two-phase mass loss, bars indicate ± S.E.

period of very slow, or, negligible mass loss (44–77 mo) (Fig. 1). Different rate-determining mechanisms control mass loss during these two phases.

We have used the ratio of lignin concentration in plant litter to the concentrations of lignin plus acid soluble carbohydrates in the litter as an index of the plant material's susceptibility to microbial attack. We refer to this index as the lignocellulose index (LCI). As the LCI increases, we expect that plant litter becomes increasingly resistant to microbial attack. During the first 59 months of decay, the LCI in red pine needle litter increased from 0.45 to 0.67 and it remained in the mid-0.6 range during Phase II of decay (Fig. 2). The LCIs in the forest floor and the upper 15 cm of mineral soil in the pine plantation were 0.72 and 0.73 respectively.

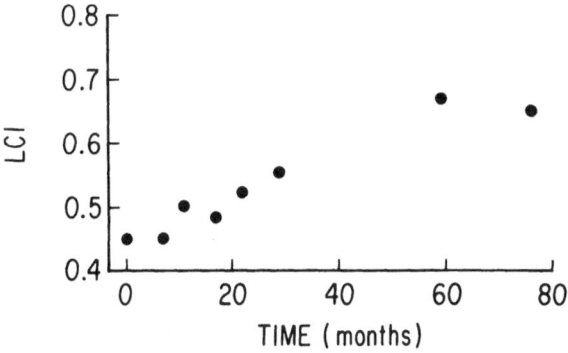

Fig. 2. The lignocellulose index (LCI) or the fraction of 'lignin' in the lignocellulose component of the decaying litter *versus* time in months.

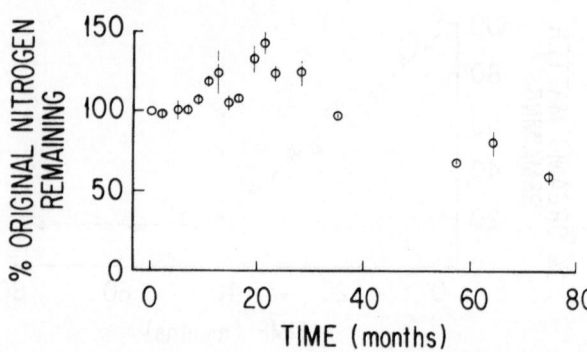

Fig. 3. Percentage of original nitrogen remaining in the decaying litter *versus* time in months, bars indicate ± S.E.

Nitrogen dynamics

Nitrogen dynamics in the decaying red pine needles (Fig. 3) followed roughly the two-phase pattern; first a net accumulation of N in the decomposing litter complex (immobilization phase) followed by a net release of N (mineralization phase).

The maximum amount of N accumulated per unit of initial litter mass (Nmax) during the immobilization phase was $1.8 \, mg \, N \, g^{-1}$ initial litter. We observed Nmax 22 months into the decay after which the net mineralization phase began. After 77 months of decay, when only 17 percent of the initial litter mass remained, the litter complex still contained N corresponding to 59 percent of its original content.

The C:N ratio declined throughout the 77-month period. The initial material had a C:N ratio of 140. At Nmax it had dropped to 86, and it continued to decline during the net mineralization phase. It also declined with depth in the soil profile of the pine plantation. The lowest C:N ratio (16.6) occurred in the 30–45 cm depth of the mineral soil.

Stable isotope analyses

We observed small changes in the C and N isotopic composition during the 77-month decay sequence (Figs. 4 and 5). The initial litter material had a $\delta^{13}C$ value of −26.7‰. Over the first 29 months of decay, the litter became slightly heavier (less negative) suggesting a loss of C fractions that are depleted in ^{13}C such as tannins or non-polar

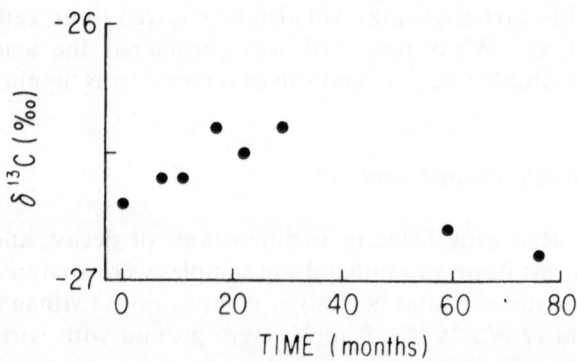

Fig. 4. The $\delta^{13}C$ of the decaying litter *versus* time in months.

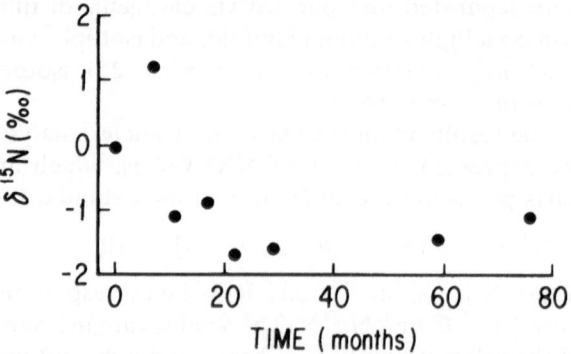

Fig. 5. The $\delta^{15}N$ of the decaying litter *versus* time in months.

components. We did observe a rapid loss of non-polar compounds and water-soluble components early in decay. Lignin, which is also depleted in ^{13}C, was conserved during this time period.

Between the twenty-ninth and fifty-ninth months, the material became slightly depleted in ^{13}C, a $\delta^{13}C$ decrease of 0.4. During this time a C fraction enriched in ^{13}C, namely cellulose (Benner *et al.*, 1987; Nadelhoffer and Fry, 1988), was being lost from the litter material, while lignin was being conserved relative to cellulose; that is, the absolute amount of cellulose declined more rapidly than the absolute amount of lignin.

We saw a change in C isotopic composition along the depth gradient from the forest floor to 45 cm into the mineral soil. The $\delta^{13}C$ of the organic matter increased from −26.2 to −24.6. This suggests more rapid lignin loss, or influx of soluble C or both.

The changes in the N isotopic composition were also small during the litter decay sequence (Fig. 5). During the immobilization phase the litter became

Table 1. Carbon and nitrogen isotopic composition for layers of soil in red pine plantation.

Soil layer	Depth below forest floor (cm)	$\delta^{13}C‰$	%C	$\delta^{15}N‰$	%N
Forest floor		− 26.2	45.38	0.2	1.10
Mineral soil	0–15	− 25.6	5.33	4.8	0.23
Mineral soil	15–30	− 24.9	2.07	7.6	0.11
Mineral soil	30–45	− 24.6	1.66	8.1	0.10

enriched in ^{14}N (a $\delta^{15}N$ decrease of between 2 and 3‰). Over the remainder of the 77-month decay sequence, the mineralization period, the litter became only slightly enriched in ^{15}N, even though the loss of N was large during the mineralization phase. In the samples taken along the depth gradient extending from the forest floor into the mineral soil we saw an increase in $\delta^{15}N$ of 8‰, with the deepest soils being the most enriched in ^{15}N (Table 1).

Discussion

Patterns of mass loss and changes in carbon fraction dynamics

The dynamics of the lignocellulose fraction of decaying plant litter become increasingly important as decomposition proceeds. This is apparent for the decay of red pine litter and for the decay of other litter types. We have recently gained some insight into the factors controlling lignocellulose degradation in a completed long-term laboratory study of the relationship between extracellular enzyme activity and C fraction dynamics during the decomposition of three deciduous leaf litters (Linkins, pers. comm.).

The early stage of the decay process is characterized by the relatively rapid disappearance of cellulose from the litter and maximum rates of cellulase activity. Eventually cellulase activity declines precipitously and finally the decomposition process is characterized by very low rates of mass loss and enzyme activity. This last stage of decay is reached as the litter composition approaches an LCI of 0.7.

Low cellulase activities combined with LCI values approaching 0.7 suggest that the litter remaining is largely lignin-encrusted cellulose. As litter approaches an LCI of 0.7 the major limitation to decomposition will be lignin degradation since lignified cellulose must be delignified before it can be hydrolyzed (Crawford, 1981; Eriksson, 1978; Kirk et al., 1978).

Although the factors that control lignin degradation are not well understood, there is evidence that the availability of labile C compounds like simple sugars can enhance the rate of lignin breakdown (Crawford, 1981; Drew and Kadam, 1979; Reddy, 1984). The labile compounds facilitate lignin degradation by functioning as cometabolites. In forest ecosystems soluble C leached from fresh litter and root exudates are probably important sources of these labile C compounds (Smith, 1969). Research needs to be done to further our understanding of the relationship between root exudates and lignocellulose degradation.

There is some evidence from laboratory studies that N can slow lignin decay rate. Research done on the lignolytic enzyme system of the fungus *Phanerochaete chrysosporium* led to the conclusion that at least for this fungus even low levels of either ammonium or some amino acids could reduce both the formation of lignin-degrading enzymes and the rate of lignin degradation (Fenn et al., 1981; Keyser et al., 1978). It must be noted that research done on the lignolytic enzyme systems of other fungi have not shown N suppression to be an important controller of lignin degradation (Kirk, 1987; Reddy, 1984). Carefully designed field experiments need to be done to test the hypothesis that N compounds slow the rate of lignin degradation.

Lignocellulose index

During Phase II, the LCI undergoes only a slight change. It increases with depth in the forest floor and the surface layer of mineral soil. Interestingly,

Table 2. Lignocellulose indexes (LCI) for forest and agricultural soils in North America

	Horizon	LCI	Reference
Forest soils			
Keene, New Hampshire:			
white pine stand	L	0.50	Waksman *et al.*, 1928
	F	0.66	Waksman *et al.*, 1928
	H	0.73	Waksman *et al.*, 1928
	A–1	0.78	Waksman *et al.*, 1928
	B–1	0.76	Waksman *et al.*, 1928
Surry, Maine:			
spruce stand	L	0.52	Waksman *et al.*, 1928
	H	0.73	Waksman *et al.*, 1928
Mt. Desert Island, Maine:			
hardwood/spruce	L	0.45	Waksman and Reuszer, 1932
	F	0.63	Waksman and Reuszer, 1932
	H	0.78	Waksman and Reuszer, 1932
Agricultural soils			
Missouri:			
'summit' soil	A	0.83	Waksman and Stevens, 1930
Hays, Kansas:			
Chernozem	A	0.75	Waksman and Stevens, 1930
Edmonton, Alberta:			
Chernozem	0–25 cm	0.81	Waksman and Stevens, 1930
Indianhead, Saskatchewan:			
brown soil	1–20 cm	0.80	Waksman and Stevens, 1930
Brandon, Manitoba:			
Chernozem	1–20 cm	0.79	Waksman and Stevens, 1930
	25–30 cm	0.77	Waksman and Stevens, 1930
Prairie, Great Plains:			
Carrington loam	A	0.81	Waksman and Stevens, 1930

the biologically active upper mineral soil layers from a number of forest and agricultural sites in North America have LCI values in the range of 0.7–0.8 (Table 2).

From these observations, we have begun to develop a model of the transformation of plant litter to soil humus (Fig. 6). The process begins with litter materials of different initial chemical composition, including different initial LCIs, falling to the soil. There, through the activities of the microbial community, the chemistries of the different litter materials are reduced to a least common denominator in terms of chemical quality; that is, regardless of their initial chemistries, all litter materials eventually reach a common chemistry with an LCI in the 0.7 to 0.8 range.

This simple model has some interesting implications for our understanding of the factors that control decay rate. Up to the point when litter materials reach an LCI of 0.7 to 0.8, decay rate is a function of initial litter quality and environ-

mental conditions such as temperature, moisture and availability of 'exogenous' labile C and N to the decomposing litter complex. For below-ground litter, soil texture may also be an important controller. It can influence the degree to which organic matter is physically protected from microbial attack and it can influence the redox potential of the soil. Once litter materials pass through the 'decay filter' where all litter materials are reduced to a least common denominator in terms of chemical quality, then environmental conditions alone may control decay rate.

Nitrogen dynamics

That such a large amount of N remained after 77 months of decay (when less than 20% of initial litter material remained) raises important questions about the factors controlling N mineralization in the litter late in the decay sequence. The association

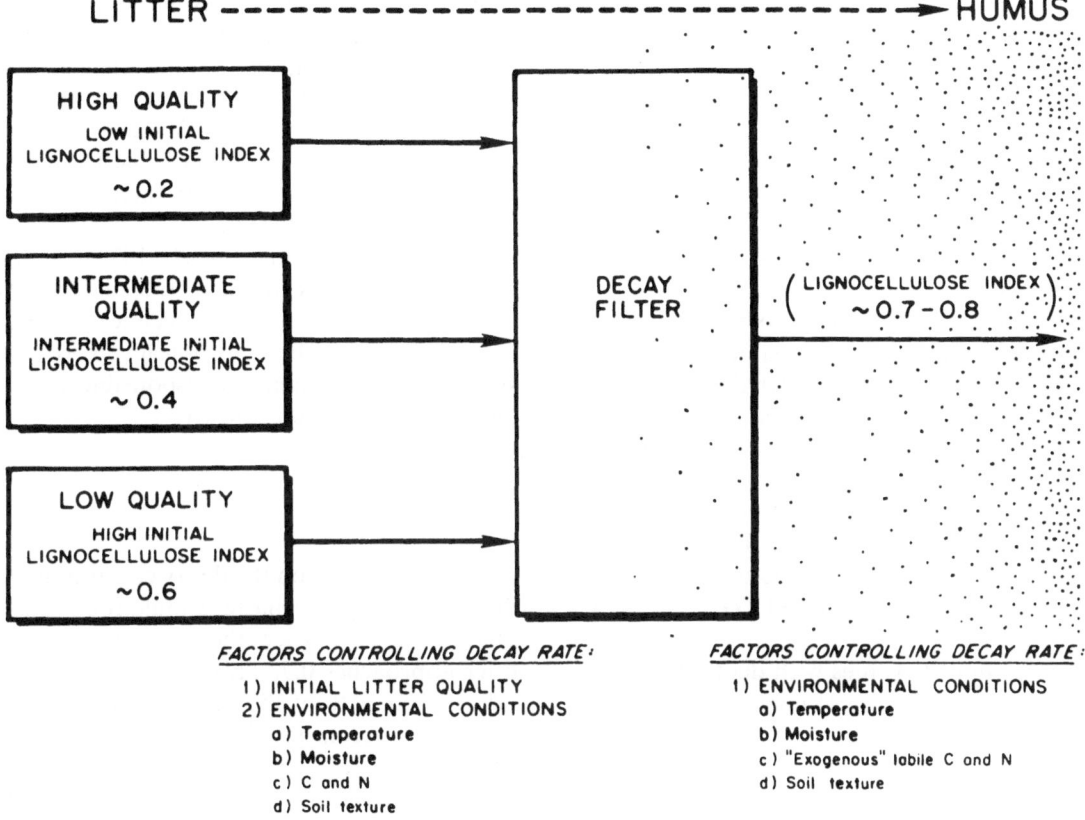

Fig. 6. Simple model of the transformation of plant litter to soil humus.

between N release from the litter late in decay and the breakdown of the lignocellulose complex, the dominant C component of well decomposed litter, is not clearly understood and deserves study.

As the red pine litter decomposed over the 77 months, its N concentration increased proportionally to accumulated mass loss (Aber and Melillo, 1980; Melillo *et al.*, 1984). The equation describing this relationship is:

$$Y = -58.48X + 114.50 \quad (n = 17, \ r^2 = 0.93)$$

where Y is the percentage of the initial mass remaining at a specific time in the decay sequence, and X is the N concentration at that time. Rearranging equation (1) to predict N concentration:

$$[N] = \frac{\%OM - 114.50}{-58.48}$$

If the relationship holds for the entire sequence, then with 0.01 percent of the initial litter mass

remaining, the predicted N concentration of the remaining material would be 1.96%. This is the same as the N concentration of the organic matter in the 15–30 cm layer of the mineral soil at the red pine site.

Stable isotope analyses

The generality of the pattern of changes observed in C isotope abundances during decay is not known because the literature contains no reports of these changes in decaying forest litter. Salt marsh grass, *Spartina alterniflora*, exposed to anaerobic decomposition (buried in the marsh sediment) for eighteen months was found to be about 2‰ more negative than fresh litter (Ember *et al.*, 1987). A 'selective preservation' of ^{13}C depleted lignin was used to explain the change in isotopic composition. In a companion study, where Spartina litter decomposed in an aerobic environment (at the marsh surface), no fractionation was found. Large changes in $\delta^{13}C$ values in decomposing litter may only occur when decay takes place in an anaerobic environment and this may be related to the fact that lignin degradation is minimal in anaerobic environments.

The changes in C isotopic composition with soil profile depth have been reported for many terrestrial ecosystems including forests (Dzurec *et al.*, 1985; O'Brien and Stout, 1978; Schleser and Pohling, 1980; Schleser and Bertram, 1981; Stout *et al.*, 1975). The $\delta^{13}C$ values of soil organic matter often increase between 1 and 2‰ between the surface and 45 cm in the profile. This increase is most often explained as the consequence of microbial decomposition processes. Alternatively, inputs of C to the deep soil layers could have a different isotopic composition than inputs to the surface. This alternative explanation was carefully considered in a shrubland ecosystem and was rejected (Dzurec *et al.*, 1985).

If increases in $\delta^{13}C$ values with depth in the soil profile are the consequence of microbial processes, interesting implications appear for the origins of refractory soil organic matter. It has long been held (Waksman, 1926) that refractory soil organic matter has at least two sources: 1) lignins and some of the fats and waxes of the original plant material; and 2) products of microbial synthesis where the C sources are the more labile compounds of the original plant litter. The increases in $\delta^{13}C$ with decay suggest that compounds depleted in ^{13}C such as the lignin are ultimately less important as sources of C for refractory soil organic matter than are some of the more labile C compounds that are enriched in ^{13}C and used for maintenance and growth by microbes. We are not saying that microbes do not break down lignin or use it to build microbial biomass. What we are saying is that they use lignin less efficiently than they use simple carbon compounds in the building of microbial biomass.

While changes in N isotopic composition of decaying forest litter have not been reported in the literature, a number of forest soil profiles have been analyzed for ^{15}N content. A study of a soil from a spruce (*Picea* sp.) forest in France (Mariotti *et al.*, 1980), demonstrated a $\delta^{15}N$ increase with depth in the profile. A large 8.7‰ increase was observed in the surface soil layers (the forest floor and upper 10 cm of mineral soil) and only small increases

deeper in the profile. Soil from two Wisconsin oak forests showed increasing $\delta^{15}N$ values with depth and ^{15}N enrichment of organic matter when incubated aerobically in laboratory microcosms (Nadelhoffer and Fry, 1988). We observed a similar pattern in the red pine stand (Table 1). However, it is not a universal pattern in forest soils. In two studies, one in Maine (Shearer *et al.*, 1978) and the other in Belgium (Riga *et al.*, 1971), a rapid increase in $\delta^{15}N$ was found in the surface layers to about 40 cm followed by a decline in the deepest layers. This could be light N leached from the soil surface. The mechanisms responsible for the different patterns is not known.

Future research

Future research on the decay continuum of plant litter to soil organic matter should focus on the late phase of decay, when the LCI of the remaining litter approaches 0.7. This occurs for many litters when 15 to 30% of the original material remains. This 'well decomposed' material is apparently a combination of lignin-encrusted cellulose and products of microbial activity, and it is rich in N.

A more complete understanding of the C and N budgets of ecosystems requires that we learn more about the factors that control decay of litter material late in the continuum. Of particular interest are the relationships between rhizosphere activities and decay of the residual litter. Roots and associated mycorrhizae are sources to the soil of labile C compounds, amino acids and enzymes, all of which may play roles in the decomposition of refractory litter components that remain late in the decay continuum.

References

Aber J D and Melillo J M 1980 Litter decomposition: Measuring relative contribution of organic matter and nitrogen to forest soils. Can. J. Bot. 58, 416–421.

Aber J D and Melillo J M 1982 Nitrogen immobilization in decaying hardwood leaf litter as a function of initial nitrogen and lignin content. Can. J. Bot. 60, 2263–2269.

Anonymous 1976 Alcohol-benzene and dichloromethane solubles in wood and pulp. T204-OS-76. Technical Association of the Pulp and Paper Industry, Atlanta, Georgia, USA.

Anonymous 1977 Individual/simultaneous determinations of nitrogen and/or phosphorus in BD acid digests. Industrial

Method Number 329-74 W/B. Technicon Industrial Systems, Tarrytown, New York, USA.

Benner R, Fogel M L, Sprague E K and Hodson R E 1987 Depletion of ^{13}C in lignin and its implications for stable carbon isotope studies. Nature 329, 708–710.

Berg B and Staaf H 1981 Leaching, accumulation and release of nitrogen in decomposing forest litter. *In* Terrestrial Nitrogen Cycles Eds. F E Clark and T Rosswall. pp 163–178. Ecological Bulletin, Volume 33, Stockholm, Sweden.

Bocock K L, Gilbert O, Capstick C K, Twinn D C, Ward J S and Woodman M J 1960 Change in leaf litter when placed on the surface of soils with contrasting humus types. I. Losses in dry weight of oak and ash leaf litter. J. Soil Sci. 11, 1–9.

Bunnell F L, Tait D E N, Flanagan P W and Van Cleve 1977 Microbial respiration and substrate weight loss. I. A general model of the influences of abiotic variables. Soil Biol. Biochem. 9, 33–40.

Bunnell F L and Tait D E N 1977 Microbial respiration and substrate weight loss. II. A model of the influences of chemical composition. Soil Biol. Biochem. 9, 41–47.

Crawford R L 1981 Lignin Biodegradation and Transformation. John Wiley and Sons, New York 154 p.

Drew S W and Kadam K L 1979 Lignin metabolism by *Aspergillus fumigatus* and white rot fungi. Deo. Ind. Microbiol. 125, 227–332.

Dubois M K, Gilles K A, Hamilton J K, Rebers P A and Smith F 1956 Colorimetric method for determination of sugars and related substances. Anal. Chem. 28, 350–356.

Dzurec R S, Boutton T W, Caldwell M M and Smith B N 1985 Carbon isotope ratios of soil organic matter and their use in assessing community composition changes in Curlew Valley, Utah. Oecologia 66, 17–24.

Effland M J 1977 Modified procedure to determine acid soluble lignin in wood and pulp. TAPPI 60, 143–144.

Ember L M, Williams D F and Morris J T 1987 Processes that influence carbon isotope variations in salt marsh sediments. Mar. Ecol. Prog. Ser. 36, 33–42.

Eriksson K E 1978 Enzyme mechanisms involved in cellulose hydrolysis by the white rot fungus *Sporotrichum pulverulentum*. Biotech. Bioeng. 20, 317–332.

Fenn P, Chois S and Kirk T K 1981 Lignolytic activity of *Phanerochaete chrysosporium*: Physiology and suppression by NH_4^+ and L-glutamate. Arch. Microbiol. 130, 66–91.

Fogel R and Cromack K Jr 1977 Effect of habitat and substrate quality on Douglas fir litter decomposition in western Oregon. Can. J. Bot. 55, 1632–1640.

Howard P J A and Howard D M 1974 Microbial decomposition of tree and shrub leaf litter. I. Weight loss and chemical composition of decomposing litter. Oikos 25, 341–352.

Jansson P E and Berg B 1985 Temporal variation of litter decomposition in relation to simulated soil climate: Long-term decomposition in a Scots pine forest. V. Can. J. Bot. 63, 1008–1016.

Keyser P, Kirk T K and Zeikus J G 1978 Ligninolytic enzyme system of *Phanerochaete chrysosporium*: Synthesized in the absence of lignin in response to nitrogen starvation. J. Bacteriol. 135, 790–797.

Kirk T K, Yang H H and Keyser P 1978 The chemistry and physiology of the fungal degradation of lignin. Dev. Ind. Microbiol. 19, 51–61.

Kirk T K and Farrell R A 1987 Enzymatic combustion: The microbial degradation of lignin. Annu. Rev. Microb. 41, 465–505.

Lyford W H 1963 Importance of ants to brown podzolic soil genesis in New England. Paper Number 7, Harvard Forest, Petersham, Massachusetts, USA.

Lyford W H 1964 Coarse fragments in the Gloucester soils of the Harvard Forest. Paper Number 9, Harvard Forest, Petersham, Massachusetts, USA.

Mariotti A, Pierre D, Vedy J C, Bruckert S and Guillemot J 1980 The abundance of natural nitrogen 15 in the organic matter of soils along an altitudinal gradient (Chablais, Haute Savoie, France). Catena 7, 293–300.

Meentemeyer V 1978 Macroclimate and lignin control of litter decomposition rates. Ecology 59, 465–472.

Melillo J M, Aber J D and Muratore J F 1982 Nitrogen and lignin control of hardwood leaf litter decomposition dynamics. Ecology 63, 621–626.

Melillo J M, Naiman R J, Aber J D and Linkins A E 1984 Factors controlling mass loss and nitrogen dynamics of plant litter decaying in northern streams. Bull Marine Sci. 35, 341–356.

Minagawa M, Winter D A and Kaplan I R 1984 Comparison of Kjeldahl and combustion methods for measurement of nitrogen isotope ratios in organic matter. Anal. Chem. 56, 1859–1861.

Minderman G 1968 Addition, decomposition and accumulation of organic matter in forests. J. Ecol. 56, 355–362.

Nadelhoffer K J and Fry B 1988 Controls of natural ^{15}N and ^{13}C abundances in forest soil organic matter. Soil Sci. Soc. Am. J. 52, 1633–1640.

O'Brien B J and Stout J D 1978 Movement and turnover of soil organic matter as indicated by carbon isotope measurements. Soil Biol. Biochem. 10, 309–317.

Olson J S 1963 Energy storage and the balance of producers and decomposers in ecological systems. Ecology 44, 322–331.

Peterson B J and Fry B 1987 Stable isotopes in ecosystem studies. Annu. Rev. Ecol. Syst.

Reddy C A 1984 Physiology and biochemistry of lignin degradation. pp. 558–571. *In* Microbial Ecology. Proc. 3rd Intl. Symp. Microb. Eds. M J Klug and C A Reddy. Am. Soc. of Microbiol., Washington, D.C.

Riga A, Van Praag H J and Brigode N 1971 Rapport isotopique naturel de l'azote de dans quelques sols forestiers et agricoles de Belgique soumis à divers traitements culturaux. Geoderma 6, 213–222.

Schleser G H and Bertram H G 1981 Investigation of the organic carbon and δ^{13}C profile in a forest soil. *In* Recent Developments in Mass Spectrometry in Biochemistry, Medicine, and Environmental Research, 7. Ed. Afrigerio. pp 201–204. Elsevier, Amsterdam.

Schleser G H and Pohling R 1980 δ^{13}C record in a forest soil using a rapid method for preparing carbon dioxide samples. Int. J. Appl. Rad. Isotopes 31, 769–773.

Shearer G, Kohl D H and Chien S H 1978 The nitrogen-15 abundance in a wide variety of soils. Soil Sci. Sec. Am. J. 42, 899–902.

Smith W H 1969 Release of organic materials from the roots of tree seedlings. For. Sci. 15, 138–142.

Stout J D, Rafter T A and Troughton J H 1975 Possible significance of isotopic ratios in paleoecology. pp 279–286. *In* Quarternary Studies. Eds. R P Suggate and M M Cresswell. The Royal Society of New Zealand.

Swift M J, Heal O W and Anderson J M 1979 Decomposition in Terrestrial Ecosystems. University of California Press, Berkeley and Los Angeles 362 p.

Waksman S A 1926 On the origin and nature of the soil organic matter or soil 'humus' V. The role of microorganisms in the formation of 'humus' in the soil. Soil Sci. 22, 421–436.

Waksman S A and Stevens K R 1930 A critical study of the methods for determining the nature of abundance of soil organic matter. Soil Sci. 30, 97–116.

Waksman S A and Reuszer H W 1932 On the original of the uronic acids in the humus of soil, peat, and composts. Soil Sci. 33, 135–151.

Waksman S A, Tenney F G and Stevens K R 1928 The role of microorganisms in the transformation of organic matter in forest soils. Ecology 9, 126–144.

M. Clarholm and L. Bergström (Eds.), Ecology of arable land, 63–76.

Nitrogen cycling in farming systems derived from savanna: Perspectives and challenges

M. J. SWIFT[1,3], P. G. H. FROST[1], B. M. CAMPBELL[1], J. C. HATTON[1] and K. B. WILSON[2]
[1]*Department of Biological Sciences, University of Zimbabwe, Harare, Zimbabwe and* [2]*formerly Faculty of Agriculture, University of Zimbabwe and Department of Anthropology, University College, London, UK.* [3]*Corresponding author*

Key words: decomposition, farming systems, granite sands, grazing, low-input agriculture, manure, maize, N-balance, savanna, sustainable agriculture

Abstract

A conceptual model, CAFS (Communal Area Farming System), is described for agropastoral peasant agriculture in Zimbabwe. Three interlinked subsystems are depicted; the household, the grazing land, and the arable lands. Estimates are made of some of the major fluxes of N within the system. The grazing subsystem is savanna woodland on inherently infertile, granite derived sandy soils. Quantitatively the most significant transfer is of manure-N from cattle grazing in the savanna which is used as fertiliser in the arable fields. Crop residues are given as supplementary feed in the dry season when both the availability and the quality of fodder in the savanna is low. Additional nutrient transfers from savanna to crops are commonly made in the form of leaf litter and soil from termitaria. On the basis of simple nitrogen budgets it is suggested that under current conditions of cultivation at least 14 t ha of savanna are required to support 2 t ha^{-1} of maize production in each hectare of arable land. Current ratios are below this and the sustainability of the systems therefore depends not only on increased access to fertiliser but also on improvements in nutrient use efficiency. This is typically very low in the arable crop fields. Manipulation of the organic inputs to synchronise nutrient release with plant demand is suggested as one approach to improved management. The sustainability of maize cultivation in the arable subsystem is dependent on a N-subsidy from the grazing subsystem. The maintenance of stable, conservative internal nutrient cycles in the savanna area is thus essential to the viability of the system as a whole. The intensity of grazing may influence these, but the role of trees as nutrient conserving agencies in savanna ecosystems is identified as a crucial area for future investigation. Whilst nutrient cycling research can play an important part in developing self-sustainable systems for this type of agriculture, it is recognized that socio-economic constraints may be the ultimate determining factors.

Introduction

The development of management practices which promote nutrient conservation is a matter of urgency at a time when the fragile ecosystems of the tropics are subject to an apparently accelerating rate of degradation. The African savannas are typically the home of tropical pastoralism, with the exception of those areas where trypanosomiasis is prevalent. In many parts of the continent, including Zimbabwe, extensive livestock management is associated with a limited area of cultivation for subsistence food production. In modern times traditional agropastoral practices have become modified as a result of changed political, social and economic circumstances. In particular farming that was formerly characterised by nomadism or shifting cultivation has tended to become fixed in location. General accounts of these types of practice are given in UNESCO (1979) and Webster and Wilson (1980). Our objectives in this paper are: to present an integrated view of nitrogen cycling within such mixed farming systems, with a particular emphasis on the interactions between the livestock and arable components; to assess the potential self-sustainability of the systems under conditions of low input of inorganic fertiliser; and to identify priorities in nutrient cycling research with the target of developing improved management practices.

Communal area farming systems

Low-input agropastoral farming in Zimbabwe is largely confined to land within the Communal Areas (CA). These are regions where the peasant farming sector is concentrated, derived from the previous politically imposed apportionment of land between European settlers and the indigenous peoples. We shall discuss the role of nutrient cycling studies by reference to a conceptual model of such a farming system, CAFS (Communal Area Farming System model, Fig. 1). The CAFS is centred on one or more households with associated arable fields which are more or less fixed in their location. These croplands are in direct or close proximity to areas of savanna that are used communally, with other household groups, for grazing livestock. The woodland in many area has been opened by tree cutting for timber and fuelwood.

Savannas are defined as tropical or sub-tropical ecosystems characterised by the coexistence of a discontinuous tree or shrub layer and a more or less continuous herbaceous layer dominated by C4 grasses. The most important climatic feature is the strong seasonality of rainfall which results in a soil water deficit during the protracted dry season. The composition and physiognomy of savannas vary globally and locally and various attempts have been made to categorise them (*e.g.* see Johnson and Tothill, 1985; Frost *et al.*, 1986). The equilibrium between the tree and grass layers is commonly recognised as the most important feature in this respect as it both determines and reflects many functional attributes of a particular vegetation type (Walker *et al.*, 1981). The main factor determining this balance appears in most cases to be the annual pattern of soil water availability. It has been proposed that the distinctions between major

savanna types can be accounted for largely on this basis and that of soil nutrient status, with fire and herbivory acting as secondary determinants (Frost *et al.*, 1986).

The annual rainfall in the Communal Areas is generally between 400 and 600 mm with high unpredictability, the coefficient of variation for annual rainfall being between 30 and 40%; intra-seasonal variation is also high. The predominant soils of the region are characterised as 'granite sands' (approximating to Alfisols in the US Soil Taxonomy and Ferralsols or Luvisols in the FAO legend). These are derived from granites rich in quartz but low in ferromagnesian minerals and are generally classified as sands or sandy loams. They are recognised as inherently infertile and requiring fertilisation for any significant level of production (Grant, 1981). Heavier soils are found in some areas associated with intrusions (small dykes) of more basic rocks, but our discussion will be confined to the granite sands. The sandy character of the soil, coupled with moderately high temperatures and low and unpredictable precipitation, means that the effective rainfall is usually below 400 mm (Vincent and Thomas, 1960).

The vegetation of the CAFS area is classified as savanna woodland. The northern part is composed of miombo elements (*Brachystegia* spp., *Julbernardia globiflora* whilst in drier regions, at lower altitudes in the south, *Terminalia*, *Burkea* and *Combretum* species become dominant. *Acacia* spp are generally confined to the more eutrophic soils. The grasses are mostly caespitose, fibrous and of low nutritional quality especially in the dry season. The most important genera are *Hyparrhenia*, *Andropogon*, *Eragrostis*, *Digitaria*, *Aristida* and *Panicum*. The grass layer is generally discontinuous. Removal of trees may produce a denser sward but the grazing intensity is now generally so high that little grass survives the dry season. For this reason fire is of little significance.

Nitrogen cycling in the CAFS

Fig. 1 shows the main flows of nutrients within the communal area farming system model. Three major subsystems are depicted: the arable croplands; the associated savanna grazing lands; and the household. In some cases former arable fields in

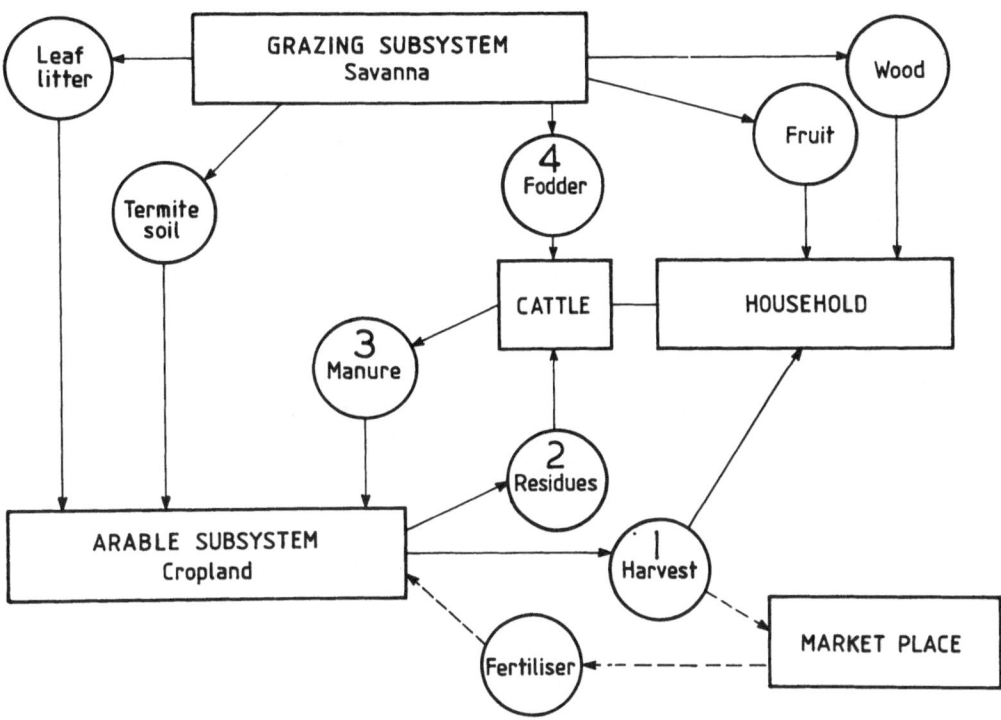

Fig. 1. CAFS, a generalised model of a Communal Area Farming System. The main subsystems are shown as boxes, with cattle as a subset of the household. Transfers of materials, and nutrients, between subsystems are shown in circles. Estimates for the four numbered fluxes are given in the text.

fallow also form a significant area. Detailed integrated nutrient budgets are not currently available but we have made preliminary estimates of four of the major flows of N in Fig. 1.

The interaction between the grazing and arable subsystems is centred round the role of cattle which have a multi-purpose status in the CAFS economy, contributing draught power, fertiliser, milk and meat (Scoones and Wilson, 1988). Livestock are the major capital investment for the farmer and his family by providing an insurance against times of low food availability or high cash demand. Perhaps the most significant single economic benefit is however that of draught power for ploughing the arable lands: possession of cattle permits early ploughing and it has been clearly demonstrated that this has a highly significant effect on the crop yield (Shumba, 1984a, b).

Arable cropping is now dominated by maize cultivation. Crop yields in the first year after clearing are low (2–3 t ha^{-1}, Grant, 1967). There are no precise measurements of the productivity of the natural ecosystems of the area, but on the basis of annual tree leaf fall and changes in herbaceous biomass, the above-ground production may lie between 3 and 5 t ha^{-1} yr^{-1}. This sets some limit to the expectations for agricultural production. Inputs of N and P, and in many cases of S and other micronutrients, are necessary to sustain the initial level of production (Grant, 1967, 1981). Fertiliser recommendations for N are between 80 and 120 kg ha^{-1} (Cooper and Fenner, 1981). Access to chemical fertiliser has increased since Independence but problems still exist in integrating its use with other practices. The availability is also variable from year to year dependent on economic policy. Whilst for many farmers their arable crops are purely a food source, some substantial cash-crop successes have been recorded in the CA sector during years of good rain and a favourable marketing climate. There is thus a growing expectation in many areas of cash profit from arable cultivation. Nonetheless the majority of CA farmers rely on inputs other than inorganic fertiliser for maintenance or improvement of the nutrient status of their soils. Cattle manure is the major source of such fertilisation. Cattle are grazed in the communal savanna area and kept in pens (kraals) overnight. Manure is dug out from the pens and spread on the fields just before ploughing at the end of the dry season. The quality and availability of fodder in the savanna declines during the protracted dry season however and supplementary feeding is necessary to maintain the condition of the cattle. Only the richer farmers have access to supplementary feed of high nutritional quality; the majority utilise the above-ground crop residues from the arable land. These are either consumed directly on the fields, or while the cattle are in the pens. The most significant nutrient flux within the CAFS is thus the transfer of nutrient from the grazing lands to the arable fields in the form of manure (Fluxes 3 and 4, Fig. 1). This can be represented as a subsidy to the arable subsystem at the expense of the grazing land. The recommended rate of application is $10 \, t \, ha^{-1} \, yr^{-1}$ of high quality manure (Grant, 1981), this being a modification of the earlier practice of $37 \, t \, ha^{-1}$ every fourth year in rotation (the so-called 'Alvord' system). This level of application significantly increased maize yields in granite sand soils (Rodel *et al.*, 1980).

Harvest of one tonne of grain removes about 12 kg of N. Thus to sustain a yield level of $2 \, t \, ha^{-1}$ (Flux 1) requires a return of about $24 \, kg \, N \, ha^{-1}$. Cattle feeding on high-quality fodder produce manure with a N content of up to 1.5% but Communal Area manures are frequently of a lower quality and most commonly closer to 1.0% (Mugwira and Mukumbira, 1984; Tanner and Mugwira, 1984). The required replacement manure input would therefore be of the order of $2.4 \, t \, ha^{-1}$. Nutrient use efficiency is however well below replacement level. Grant (1981) quotes a return of about 22% of applied-N in grain yield and in the experiment of Rodel *et al.* (1980) it was only about 13%. Thus the input of manure-N (Flux 3) probably needs to be at least four times the replacement level ie about $96 \, kg \, N \, ha^{-1}$. This is in close agreement with the recommended fertilisation of $10 \, t \, ha^{-1}$.

For each tonne of maize grain yield an approximately equal quantity of above-ground residue will be available to the farmer (Triplett and Mannering, 1978) although the ratio of grain to residue will tend to increase with increased yield. This residue is usually fed to the cattle rather than incorporated into the soil (Flux 2,

$12 \, kg \, N \, ha^{-1} \, yr^{-1}$). The nutrient is eventually returned to the arable subsystem as manure N, although the proportion returned by this means is not easy to estimate. It does imply however, that at least $84 \, kg \, N \, yr^{-1}$ must be transferred from the grazing subsystem in the form of manure to support each hectare of arable land (Flux 4, Fig. 1).

A substantial proportion of farmers lack cattle and thus have no direct access to either draught power or manure. For the former they are forced to borrow or hire animals from owners. Part of the price of this arrangement may be crop residues, representing a significant loss of nutrient to their fields. This further emphasises the role of cattle as a key determinant of socio-economic status in the Communal Areas; a variety of relationships between families, ranging from co-operation to clientship may stem from it (Scoones and Wilson, 1988).

Other sources of nutrient for the arable fields are the transfer of leaf litter from the savanna area and the spreading of soil from termite mounds. The quantities transferred are not known and likely to be highly variable. The benefits may be substantial however. Freshly fallen leaf litter gathered from miombo woodland has a N content of about 1.2% and annual litter fall is about $1.5 \, t \, ha^{-1}$. Thus transfer of 5% of this from each hectare of the grazing area would give a N input of 1.1 kg. Soil from termite mounds located within the grazing area is also sometimes spread on crop fields by farmers. Watson (1977) showed that termitarium soil supported greater plant production than the topsoil from surrounding miombo and maize fields in three granite sand sites in Zimbabwe. Termitarium soil has been shown to have more organic matter and clay, and enhanced nutrient and microbiological status than neighbouring topsoils (Meiklejohn, 1965; Nyamapfene, 1986). With the exception of organic matter these differences appear to derive from clay-rich subsoil being brought to the surface rather than from any direct modifying influence of the termites on soil characteristics (Hesse, 1955). Access to this resource may thus be an important occasional source of input for the arable fields.

Sustainability of N subsidy within the CAFS

Harvest of N from the arable subsystem, and

losses due to leaching and volatilisation from the cattle pen and the croplands, dictates constant replenishment from the grazing subsystem, as calculated above. We now examine some of the implications of this level of internal subsidy in the CAFS.

Rodel *et al.* (1980) suggested that one LSU (livestock unit = 500 kg live body weight) produced about 1.3 t of recoverable manure a year. Recommended stocking levels in the Communal Areas on granite sands vary between about 8 and 11 ha per LSU (Whitsun Foundation, 1978), although current levels are probably three or less (Shumba, 1984b). Thus one hectare of grazing land might be expected to produce about 400 kg of recoverable manure in a year under present stocking levels. Dependent on the quality of the manure this may represent an export from the grazing subsystem of between 2 and 6 kg N ha^{-1} yr^{-1}. This means that between 14 and 42 hectares of grazing are needed in order to supply the computed subsidy of 84 kg of manure-N (Flux 4, Fig. 1) required for production of the target yield of 2 t ha^{-1} of maize. Put in another way this means that the fraction of farming area that could be devoted to arable cultivation would be between 2.3 and 6.7%. This factor is based on a very broad set of assumptions and will vary substantially depending on both the nutrient quality of the grazing and the efficiency of nutrient use in the arable lands, features which we discuss in more detail below. Barnes and Clatworthy (1976) using somewhat different criteria came to the conclusion that about 13% of land could be maintained in productive cultivation in association with improved grazing land. Additional transfers such as leaf compost may amount to a further 1 kg N ha^{-1} yr^{-1} and change the balance slightly.

Each individual farmer (or household) has nonetheless to work within a finite area. Any increase in the area of land cultivated means a corresponding loss in the area of grazing land. As the levels of nutrient offtake from the two are so different, even a small shift can have a dramatic effect on the viability of the system. This is illustrated in Fig. 2 where the effects of varying the ratio of grazing to arable land area are explored for a 'farm' (accessible area) of 40 ha. Four scenarios are considered embracing conditions of low and high transfer from the grazing subsystem to the arable and low and high demand for nutrients within the arable subsystem. The critical point comes when the ratio of grazing area to arable area drops below the ratio of the nutrient offtake from the arable subsystem to the nutrient input from the grazing subsystem. At this juncture there will be net losses of nutrient from the arable subsystem and its nutrient status will rapidly decline. For instance the lowermost curve in Fig. 2 depicts a situation of relatively low transfer (5 kg N ha^{-1}) and high demand (150 kg N ha^{-1}). The critical ratio is thus 30:1 (3.3% cultivation). At this ratio the farmer can cultivate about 1.3 ha and retain 38.7 for grazing. Any drop below this rapidly becomes critical. It is more common however to cultivate two or three times this area (*e.g.* see Truscott, 1986). This requires that the N-output:N-input ratio in a 40 ha farm be in the region of 10–15:1 *i.e.* more closely resembling the middle two curves in Fig. 2. This entails either a marked increase in the efficiency of nutrient use on the arable field (*i.e.* reduction in nutrient losses) or that the input from the grazing subsystem be doubled by increasing the utilisation of the resources of the savanna. The potential for either of these options depends on management of the nutrient cycles within each of the subsystems. In practice the total area accessible to a typical CA household may well be less than half the 40 ha used in our illustration.

Nutrient conservation within the subsystems

The arable subsystem

High nutrient use efficiencies are dependent on the maintenance of efficient and tight nutrient cycles. On the basis of yield-N to input-N ratios the N-use efficiency of arable cropping systems in CAFS is low, irrespective of input quantity. There is little detailed information on nutrient fluxes from which to deduce the origins of the dysfunctions. The potential for improvement by management practices which are within the resources available to the CA farmer must therefore depend on the results of further research.

Initial losses of N from the system may occur even before the manure fertiliser is applied to the field. For instance laboratory experiments have shown that there is a high potential for loss of N from manure by volatilisation under the conditions in which it is stored in the cattle pens (H. Mugwira, personal communication). Losses may also occur

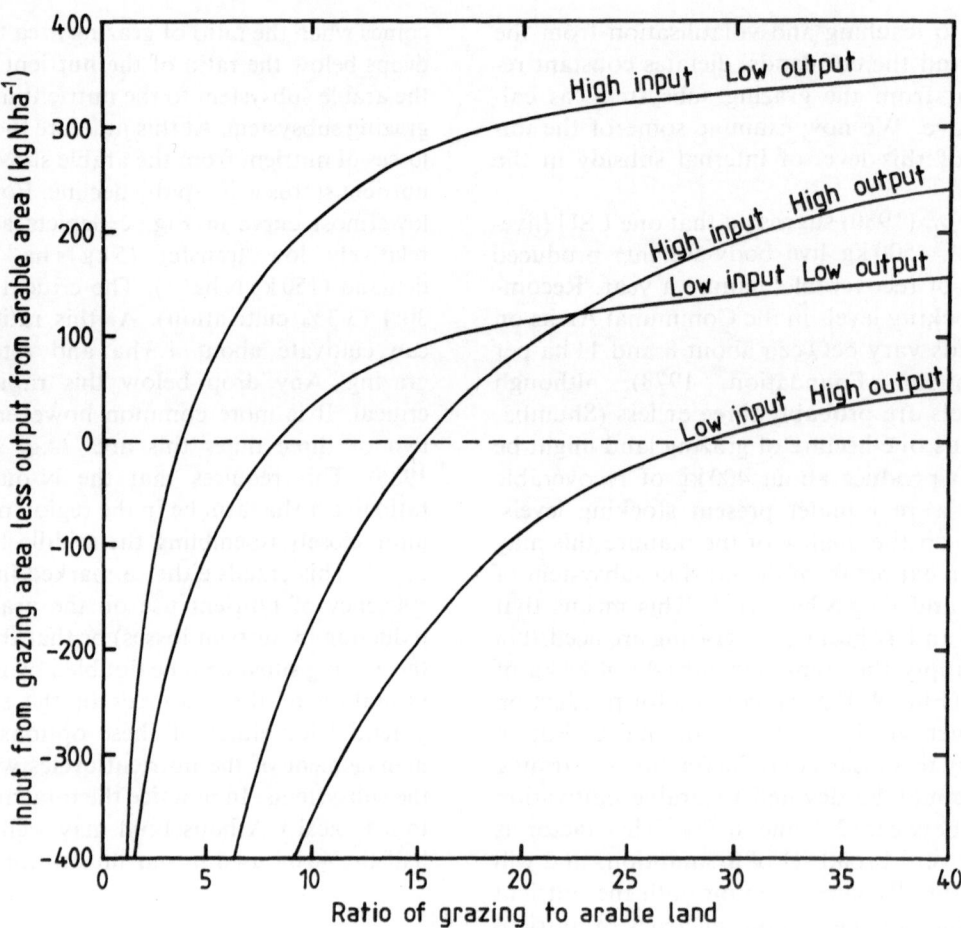

Fig. 2. Nutrient sustainability in the CAFS. The graph shows theoretical relationships between the apportionment of land to arable and grazing areas in a 40 ha farm and the amount of nutrient that could be supplied as subsidy from the grazing subsystem to meet the demands of the arable subsystem. This is shown as four scenarios involving two levels of supply (inputs) and two levels of demand (outputs). Inputs = transfers of nutrient from grazing subsystem to arable subsystem; High = 10 kg N ha^{-1}; Low = 5 kg N ha^{-1}. Outputs = nutrient requirement of arable land to produce minimum yield; High = 150 kg N ha^{-1}; Low = 40 kg N ha^{-1}. The dashed line indicates where input and output are equal. Thus above the line there will be a net gain of nitrogen to the arable subsystem and below the line a net loss.

by both leaching and volatilisation after fertilisation. Grant (1967) showed that most of the available-N was released from manure early in the season. Nitrification rates are high and there is therefore considerable risk of leaching of nitrate out of the rooting zone. This would be minimised in circumstances where the soil organic matter (SOM) is maintained at a high level and where the processes of nutrient release, from SOM, and from organic inputs, are synchronised with plant nutrient demand. Different organic materials (resources) show differing patterns of nutrient release through time (Fig. 3). These differences can be

related to variation in resource quality (Swift *et al.*, 1979). Nutrients released by decomposition processes may be immediately taken up by the plant, immobilised on SOM or retained on the exchange complex of soil to be re-released at a subsequent time or lost through leaching. It has been hypothesised that a highly asynchronous pattern of supply and demand results in greater nutrient loss (Swift, 1984) and experiments to investigate this hypothesis have been proposed (Swift, 1987). In particular it has been advocated that manipulation of the spectrum of resources of differing qualities offer a potential approach to

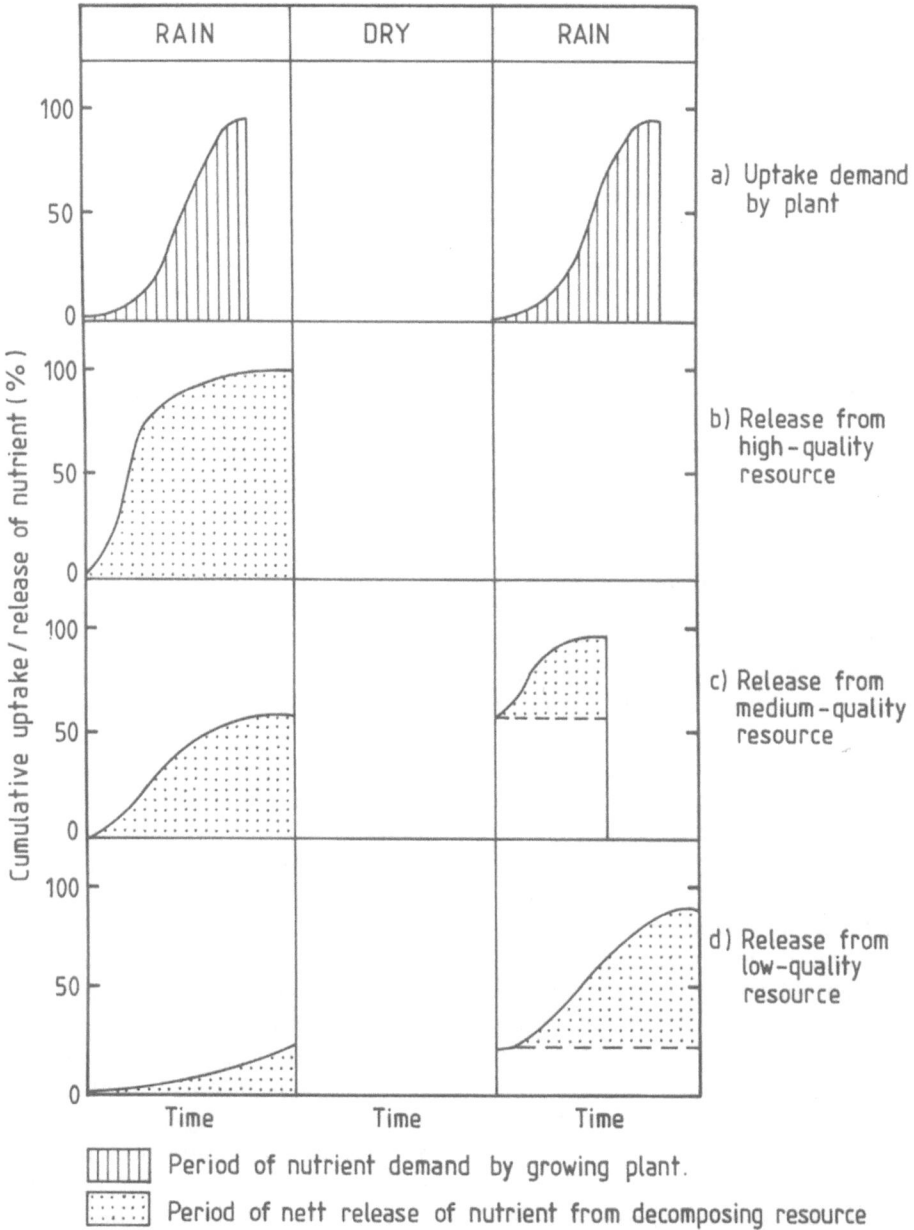

a) Uptake demand by plant

b) Release from high-quality resource

c) Release from medium-quality resource

d) Release from low-quality resource

▨ Period of nutrient demand by growing plant.

▧ Period of nett release of nutrient from decomposing resource

Fig. 3. Synchrony models. The diagram pictures three theoretical examples of relationships between nutrient release by decomposition processes from resources of different quality (b, c and d) and the demand for nutrient by the growing plant (a). The primary control of plant growth, decomposition and nutrient release is moisture, shown here as response to seasonal patterns of wet and dry periods. The secondary control on decomposition and nutrient release is resource quality; high quality resources show a rapid flush of nutrients at the onset of rain (b), whilst in low quality resources the main period of net release may be delayed until the subsequent rainy season (d). Medium quality resources, or mixtures of low and high qualities may show intermediate patterns (c) which may synchronise more closely with plant demands (a).

management of the timing of nutrient release to synchronise more closely with the period of maximum demand by the plant.

Farming systems such as the CAFS, with access to organic resources of varying quality have potential in this respect. For instance maize stover typically shows a pattern of net N-immobilisation during the early stages of decomposition; the nutrient benefit from such a material may lie however in the longer term, in the seasons after that of application (Fig. 3d). Using ^{15}N-labelled residues, Seligman *et al.* (1986) have shown a pattern of this type for another low-quality residue, wheat straw. In contrast high-quality residues (*e.g.* cowpea, Swift *et al.*, 1980) show net nutrient release very soon after incorporation (Fig. 3b). Manure is generally regarded as a high quality resource but Tanner and Mugwira (1984), showed that some Communal Area manures depressed the growth of maize plants in the first four weeks after application; net gain of N to the plants was observed however in the subsequent four-week period; in contrast, responses to manure with lower C:N ratios were seen in the first period. The rate and pattern of flow of N to the crop plant from different types of residue varies considerably and is dependent on a variety of factors. For instance N content alone is not sufficient to predict availability, other resource quality determinants such as the lignin or polyphenol contents, being in some instances regulators of equal or greater importance (Swift *et al.*, 1979).

Rotations of maize with legumes has sometimes been practiced in Zimbabwe but present implementation of this system appears somewhat irregular (Truscott, 1986). The carry over of a residual N effect from season to season after a legume crop has been shown to be significant in a variety of circumstances (L. Mukurumbira, personal communication). The amplification of this type of system by managing mixtures of low and high quality residues is worth investigation. A wider spectrum of resource qualities could be used from the savanna areas. The development of integrated agroforestry systems also offers additional possibilities by introduction of species suitable for crop mulches. Central to any future improvement in nutrient use efficiency is the interaction of inorganic fertilisers with organic resources of differing quality. In circumstances where there are severe economic constraints on synthetic fertiliser availability even a small gain in efficiency would be valuable.

Irrespective of the nature of the above ground inputs, root residues are a constant feature of the system. Maize develops a low shoot:root ratio on sandy soils. In agronomic terms this may be regarded as an undesirable diversion of nutrient

from the grain. It is however also a means of conserving N and other nutrients from season to season against losses by leaching. Root development is much influenced by the type of tillage and practices which encourage deep rooting are beneficial in resisting water stress, conditions which are commonly experienced in CAFS (Willat, 1967). Selection of crop varieties with enhanced rooting characteristics may be beneficial in tightening nutrient cycles. In contrast to the benefits of deep tillage for stimulating root growth however, there is growing evidence of the fertility gains from decreased intensity of tillage (*e.g.* see Elliott *et al.*, 1984). Reduced tillage has become a necessity in many CAs due to limitations on access to draught power (Shumba, 1984b). Investigation of the impact of such changes (*e.g.* only ploughing every second or third season) on soil fertility are thus of considerable significance.

Tillage also influences the SOM content of soil. The low levels of active clay in granite sand soils means that the SOM is responsible for the major part of the CEC. Maintaining a significant SOM level is important also to sustain a reservoir of mineralisable N, P and S in the soil and to increase the moisture holding capacity. Soil organic matter levels are low even in undisturbed savanna (SOM-C is generally less than 1%) and decline rapidly after clearing and cultivation (Greenland, 1980). Management practices that can sustain a higher equilibrium level are therefore highly desirable.

The equilibrium SOM level in soil is determined in the first instance by the relationship between the quantity of the organic input to the soil and the rate of its decomposition (Greenland and Nye, 1959) but it has also been hypothesised that the quality of the organic input influences the quantity, quality and stability of the SOM (Young, 1986; Swift, 1984, 1987). This is a further justification for the use of low quality resource inputs in addition to those that yield a more rapid return of nutrients. Grant (1967) showed a small but significant beneficial effect of manure on the SOM-C and N levels of granite sand soils in this respect. Investigation of the relative effects on residual-N of different inputs is an important research priority, but one which requires a long time scale (Swift, 1987). It should be remembered however that the equilibrium organic matter content of soil commonly shows a direct relationship to that of clay due to physical protection against breakdown (Jones, 1973). The low clay content of granite sand soils may therefore impose a limit to the potential SOM level. The higher clay content of termitarium soil may be beneficial in this respect.

The grazing subsystem

The preceding discussion has demonstrated that the sustainability of arable cultivation in the CAFS, in the absence of nutrient subsidy from external sources, is dependent on the utilisation of the natural resources of the grazing subsystem. The suggestion was also made that a greater quantity and diversity of these resources could be effectively employed to improve the nutrient use efficiency of the arable subsystem. The CA farmer also utilises other savanna products such as wood and fruit (Fig. 1). It is thus clear that a high priority should be given to research directed at understanding the functioning of the nutrient cycles in savanna woodland and their response to differing patterns and intensities of resource utilisation.

There is evidence that neotropical savanna ecosystems on dystrophic soils have tight nutrient cycles and show a strong tendency to immobilise nutrients in the soil, even under conditions of frequent fire (Kellman and Sanmugadas, 1985; Kellman *et al.*, 1985). There is no complete data set for nutrient cycling in CAFS savannas in Zimbabwe, but some general principles can be drawn from the limited data that are available (Campbell *et al.*, 1988) and from studies made elsewhere in Southern Africa (Frost, 1985; Huntley and Walker, 1982).

The plants contain only a small proportion of the total nutrient pool of the savanna ecosystem; in the *Burkea-Terminalia* savanna at Nylsvley in South Africa for instance, only 9% of the total N, and about 1% of the total P, Mg and K (Frost, 1985). The plant-available fraction of the soil pool is small however, though apparently in excess of plant demand. Both mineralisation and uptake are limited by soil moisture for significant periods of the year and in *Brachystegia* woodland we have shown a sharp increase in N-mineralisation at the onset of the rains with minor peaks at other times associated with rewetting of the soil after dry spells. A similar pattern has been demonstrated for *Acacia*

savanna in the Sahel by Bernhard-Reversat (1982).

Savanna soils show great spatial heterogeneity in nutrient distribution in both vertical and horizontal planes. The majority of nutrients are concentrated near the surface, declining sharply with depth. Horizontal variation occurs at a number of different scales. At the level of the landscape, upland and midslope soils are generally less fertile than those of bottom lands or drainage lines; at a more local level there is considerable heterogeneity associated with the presence of termite mounds and the distribution of open and closed phases of woodland development (Campbell *et al.*, 1988). The most profound influence however is that of tree composition and distribution. Higher concentrations of soil organic matter and mineralisable-N are found under tree canopies than outside (Bernard-Reversat and Poupon, 1980; Campbell *et al.*, 1988; du Preez *et al.*, 1983; Kellman, 1979). A variety of factors may contribute to this: In the drier savannas it may be associated with a closed nutrient cycle within the canopy area associated with higher herbaceous growth stimulated by a humid microclimate; litter with high lignin contents may accumulate under trees and contribute to localised zones of high SOM synthesis; many savanna trees also have extensive rooting systems and the extraction of nutrients from surrounding areas and from deep in the soil contributes to the enrichment of the soil beneath the tree canopy; in grazed areas of low tree density livestock seeking shade may contribute to nutrient enrichment under the canopy by deposit of excreta.

This influence of trees on nutrient heterogeneity is one of several key contributions to the stabilisation of nutrient cycles in savannas. Indeed there seems to be good evidence to suggest two distinct but interconnected nutrient cycles in savannas: a relatively rapid cycle through the herbaceous layer and a slower one through the trees (Frost, 1985).

Woody plants may make up more than 80% of the plant biomass in savannas, 30 to 50% of which is below-ground. In *Burkea-Terminalia* savanna in South Africa, however, the productivity of the grass-herb layer was more than 50% that of the trees, indicating the comparatively high rate of turnover of the former as a result of herbivory and decomposition (Frost, 1985; Rutherford, 1982). The trees also contain a greater fraction of the total nutrient pool than the grasses and herbs, as would be expected from the differences in biomass, but when a dynamic view is taken these differences are diminished. In *Burkea-Terminalia* savanna the grass-herb layer accounted for 29% of the total annual flux of N (and 41% of the P), although constituting only 14% of the N in the biomass (21% for the P). Grass litter at the same site is decomposed more than twice as fast as the tree leaves (Frost, 1985). The woody component of the tree input of course constitutes an even slower turnover pool. Preliminary evidence suggests a similar relationship between grass and tree components in Zimbabwean miombo.

The flux of nutrients through the tree layer seems to be more constant and less influenced by disturbance than that of the herbaceous layer. Although there are year-to-year differences in production related to climatic variation, the fluctuation is small relative to the biomass (Rutherford, 1984). Because much of the annual shoot growth occurs before the onset of the annual rains (Campbell *et al.*, 1988; Rutherford and Panagos, 1982), drawing on internal reserves, tree leaf production is much less susceptible to short-term (intra-seasonal) variations in soil water and nutrient availability than grasses. Longer term trends may have significant effects however. Any major decrease in the woody plant biomass (for instance due to tree cutting) could initiate major and long-lasting changes in the pattern of nutrient cycling. Recovery from such changes is likely to be correspondingly long-term.

The influence of grazing. African savannas originally supported a wide diversity of grazing and browsing mammals. These have now been largely replaced by domestic livestock. The stocking rate of the cattle can strongly influence the soil characteristics of the savanna subsystem. Heavy grazing leads to a reduction in plant and litter cover and thence to increased exposure of the soil to sun, wind and rain. Under the impact of raindrops soil aggregates break down, the displaced particles seal the soil surface pores and infiltration is diminished. Compaction of the soil surface by hooves may exacerbate this although this effect is only really significant on clay soils. Reduced infiltration in its turn leads, on slopes, to increased run-off and surface erosion, ultimately creating an environment which is hostile to seedling establishment and growth.

Large herbivores may substantially influence the rate and pattern of nutrient cycling by consuming plant material and returning it as dung and urine. Losses occur because the excreta show high susceptibility to losses of N by volatilisation and leaching (Henzell, 1973). Grazing can also increase the rate of cycling of nutrients through the herbaceous layer by stimulating compensatory growth in grasses and thence enhancing nutrient uptake by the plants. Nonetheless evidence for significant direct effects of grazing on soil chemical status of savannas is equivocal. Indeed it has been proposed that as long as the tree canopy remains intact, chemical aspects of soil fertility may be substantially conserved (Gambiza, 1987). Whether this hypothesis is correct or not, changes in soil physical properties and hydrological processes are certainly detectable earlier than critical changes in nutrient cycling. The two sets of processes are nonetheless inextricably linked and there is an urgent need for research into the relationships between the physical and nutrient components of fertility in savannas (Frost *et al.*, 1986). For instance the surface erosion which may develop as a result of changes in soil surface and hydrological properties may have a significant effect on nutrient status of soil. Most nutrients are located in the top five centimetres of soil in miombo woodland. At current levels of erosion experienced in CAFS grazing areas the total loss of N from soil by this means in Zimbabwe may be as much as 1425 million kg per year; that is an average rate of about $92 \, kg \, N \, ha^{-1} \, yr^{-1}$ (Stocking, 1986). This is in fact nearly double the rate of loss estimated for arable land by the same author, further emphasising the importance of considering nutrient cycles in the grazing systems as components of the nutrient economy of the system as a whole.

Changes in soil characteristics are presumably in some way related to the intensity of use by livestock. Despite a wide range of studies of the environmental effects of grazing (*e.g.* see Barnes, 1979; Gammon, 1978) there is still little consensus in Zimbabwe about the key criteria for assessing maximum stocking levels. These have been generally calculated on the basis of optima for beef production. In the Communal Areas however cattle have a multi-purpose role and this criterion is inadequate. An alternative to strictly economic-return criteria for estimating optimal livestock densities is via some estimate of the ecological carrying capacity. This is usually based on vegetational criteria but an equally valid approach would be to consider the effects on soils. In nutrient cycling terms the limits might be set below the level at which nutrient outputs, including those due to erosion, exceed the capacity of the system to recover them.

One factor in such an equation is the 'harvest' of manure taken for fertilisation of the arable fields. To enable $5\,kg\,N\,ha^{-1}\,yr^{-1}$ to be exported as manure would require consumption of virtually the whole of the grazeable biomass. This has been calculated as between 400 and $1400\,kg\,ha^{-1}$, dependent on live tree density (Ward and Cleghorn, 1964, 1970) probably amounting to 4 to $14\,kg\,N\,ha^{-1}$. Nonetheless removal of $5\,kg\,N\,ha^{-1}$ may not be a very significant amount. At the Nyslsvley savanna it represents only 2% of the total phytomass-N and about the same fraction of the sum of litter-N and available soil-N. This level of removal, even when repeated over a number of years, is readily replaceable by natural inputs from nitrogen fixation (Isichei, 1980). Immediate effects on soil nutrient status are thus likely to be less significant than direct impacts on the plants. The long term effects of disruption are more difficult to predict, although herbaceous plants can recover relatively rapidly from short-term disturbances, and thus have the potential to re-establish this part of the nutrient cycle. One important inference is that nutrient turnover via the root system will become relatively more important.

Concern about the effects of heavy grazing on savannas has concentrated almost exclusively on the herbaceous layer; very little attention has been given to the tree component. Changes in the species composition of grasses and herbs in response to herbivory are readily apparent, although care must be exercised to eliminate the possible confounding effects of climatic-induced changes (O'Connor, 1985). Trees on the other hand are not obviously affected. A common recommendation is to thin or remove the tree layer to promote the development of a grass sward (Barnes, 1979). Although there clearly is a negative relationship between tree density and herbaceous yield (Dye and Spear, 1982; Gammon, 1983) savanna trees have a number of features which favour nutrient conservation and which deserve examination as possible ways of maintaining the stability and sustainability of the system. First, some species provide an important supplementary source of fodder, as browse, par-

ticularly during drought conditions. Second, they have considerable resistance to disturbance due to their ability to sprout if damaged. Third, the large biomass with high root:shoot ratio provides a nutrient reservoir which is largely inaccessible to both herbivores and fire. The stabilising influence of this slow turnover pool on the nutrient dynamics of the subsystem has already been stressed.

These assertions about the value of trees as conservers of soil fertility are however largely speculative (Prinsley and Swift, 1986) and there is a strong case for investigating the benefits of maintaining a significant extent of tree canopy within grazing areas. For instance removal of the above ground tree biomass at Nylsvley would potentially result in the direct loss of 56% of the phytomass-N, a figure which must be taken seriously in terms of the economy of the whole system. In fact the deficit is likely to be higher because of losses by leaching from decaying surface and root litters in the period following cutting.

Cleared pastures on granite sand soils show marked responses to fertilisation with N and other elements (Barnes and Clatworthy, 1976; Smith, 1965). What is not clear is the extent to which the stimulation of grass growth after clearing is due to utilisation of the nutrient reserves released into the soil from residual organic matter from the trees and, if so, how long the effect persists. In any event we suggest that the interaction of nutrient removal by tree clearing with subsequent losses by heavy grazing is likely to be far more significant than grazing losses alone in systems with a tree component. Furthermore, excessive clearing of the tree layer increases the risk of loss of grass through the activities of termites (Dye and Spear, 1982).

Management of the tree component requires the regulation of tree use, for fuelwood and timber, within the regenerative capacity of the woodland. This is dependent on social and economic factors, but clearly, in areas where substantial clearing has already taken place, it may include replanting. In this respect the choice of tree, and the pattern of planting should take into account the effects on the soil and on nutrient cycling, in addition to the suitability of the species of choice as fuelwood (Sanginga et al., 1987). Some species accumulate nutrients more than others; some compete more strongly with grasses; and some are more resilient to damage.

Conclusions

Whilst the CAFS is a valid general model for Communal Area farming practices in much of the granite sand zone of Zimbabwe it is only one of a variety of present practices in the region. The varying structures of these systems, with or without major differences in their ecological context, offer an interesting opportunity for comparison in terms of nutrient use efficiency. Many of the practices have a long history in central-southern Africa (Puzo, 1978), but all of them have been subject to some modification within the era of European settlement. In particular the CAFS is a hybrid rather than a 'traditional' farming practice. The system was developed following the political apportionment of land into European and African Areas (Tribal Trust Lands, TTLS, the predecessors of the Communal Areas). The present practices of farmers are based on prescriptions derived both from traditional insights and from the recommendations of advisers schooled in 'conventional' agricultural research. In some cases the mixture of the heuristic and the scientific is supportive; in other circumstances (*e.g.* in the assessment of appropriate stocking levels) it is in conflict. Nutrient cycling offers a means of appraising both types of wisdom.

Frissel (1978) described agropastoral (mixed) farming systems of similar type to the CAFS as potentially 'self-sustainable'. This analysis of a generalised farming system for Zimbabwe supports this conclusion. The sustainability is achieved by compensating for nutrient removals from the arable fields by subsidy from the nutrient-conservative savanna subsystem. In a study of adjacent grassland and wheat fields in North America, Schimel (1986) showed that whereas the grassland retained N by immobilisation on organic resources, the cropland showed a high nett loss related to their relative lack of low-quality inputs. We have suggested here that the same distinction may exist between the grazing and arable subsystems and furthermore that, in the former, the trees play an important role in nutrient conservation. The nutrient sustainability of the CAFS is evidently dependent on a variety of interacting factors, socioeconomic as well as biological, but for planning purposes these can be reduced to three major categories:

1. the existence of a sufficient area of savanna (the grazing subsystem) to support the necessary nutrient transfer to the arable and household subsystems;
2. the maintenance of a stable and self-renewable nutrient reservoir in the grazing subsystem;
3. a high efficiency of nutrient use in the arable subsystem.

The two latter features are ones to which scientific research can, and should be addressed. In most Communal Areas neither condition is met at present. This can only be rectified by changes in farming system practice, which may in many cases involve a return to methods more closely related to traditional rather than modern agriculture. A number of these have been mentioned in the text; Stocking (1988), among others, has given a useful review of the soil-conservation value of a variety of farming practices. Soil fertility is only one consideration among many in designing farming systems, but the interaction of water and nutrient availability lies at the heart of the productivity of farming systems in the savanna zones and must therefore be given priority in land use planning.

Acknowledgement

We are grateful for the useful comments made by Dr John Clatworthy.

References

Barnes D L 1979 Cattle ranching in the semi-arid savannas of East and Central Africa. *In* Management of Semi-Arid Ecosystems. Ed. B H Walker. pp 9–54. Elsevier, Amsterdam.

Barnes D L and Clatworthy J N 1976 Research in veld and pasture production in relation to the Tribal Trust Lands. Rhod. Sci. News 10, 271–278.

Bernhard-Reversat F 1982 Biogeochemical cycle of nitrogen in a semi-arid savanna. Oikos 38, 321–332.

Bernhard-Reversat F and Poupon H 1980 Nitrogen cycling in a soil-tree system in a Sahelian savanna: Example of *Acacia senegal*. *In* Nitrogen Cycling in West African Ecosystems. Ed. T Rosswall. pp 363–369. SCOPE, Stockholm.

Campbell B M, Swift M J, Hatton J C and Frost P G H 1988 Small-scale vegetation pattern and nutrient cycling in **Miombo woodland**. *In* **Vegetation Structure in Relation to** Carbon and Nutrient Economy. Eds. J T A Verhoeven, G W Heil and M J A Werger. pp 69–85. SPB Academic Publishing, The Hague.

Cooper G R C and Fenner R J 1981 General fertiliser recommendations. Zimbabwe Agric. J. 78, 123–128.

du Preez D R, Gunton C and Bate G C 1983 The distribution of macronutrients in a broad leaf woody savanna. S. Afr. J. Bot. 2, 236–242.

Dye P J and Spear P T 1982 The effects of bush clearing and rainfall variability on grass yield and composition in southwest Zimbabwe. Zimbabwe J. Agric. Res. 20, 103–118.

Elliott E T, Horton K, Moore J C, Coleman D C and Cole C V 1984 Mineralisation dynamics in fallow dryland wheat plots. Colorado. Plant and Soil 76, 149–155.

Frissel M J 1978 Cycling of Mineral Nutrients in Agricultural Ecosystems. Elsevier, Amsterdam.

Frost P G H, Menaut J-C, Walker B H, Medina E, Solbrig O T and Swift M J 1986 Responses of savannas to stress and disturbance: A proposal for a collaborative programme of research. Biol. Int. Special Issue, 10, IUBS, Paris.

Frost P G H 1985 Organic matter and nutrient dynamics in a broadleafed African savanna. *In* Ecology and Management of the World's Savannas. Eds. J C Tothill and J J Mott. pp 200–206. Australian Academy of Science, Canberra.

Gambiza J 1987 Some effects of different stocking intensities on the physical and chemical properties of the soil in a marginal rainfall area of Southern Zimbabwe. MSc Thesis, University of Zimbabwe.

Gammon D M 1978 A review of experiments comparing systems of grazing management on natural pastures. Proc. Grassld. Soc. S. Afr. 13, 75–82

Gammon D M 1983 Effects of bush clearing, stocking rates and grazing systems on vegetation and cattle gains in the south western lowveld of Zimbabwe. Zimbabwe Agric. J. 80, 219–228.

Grant P M 1967 The fertility of a sandveld soil under continuous cultivation. I. The effect of manure and nitrogen on the nitrogen status of soil. Rhod. Zam. Mal. J. Agric. Res. 5, 71–79.

Grant P M 1981 The fertilisation of sandy soils in peasant agriculture. Zimbabwe Agric. J. 78, 169–175.

Greenland D J 1980 The nitrogen cycle in West Africa: Agronomic considerations. *In* Nitrogen Cycling in West African Ecosystems. Ed. T Rosswall. pp 73–81. SCOPE/UNEP, Stockholm.

Greenland D J and Nye P H 1959 Increases in the carbon and nitrogen contents of tropical soils under natural fallows. J. Soil Sci. 10, 254–299.

Henzell E F 1973 The nitrogen cycle of pasture ecosystems. *In* Chemistry and Biochemistry of Herbage. Vol. 2. Eds. G W Butler and R W Bailey. pp 228–247. Academic Press, London and New York.

Hesse P R 1955 A chemical and physical study of the soils of termite mounds in East Africa. J. Ecol. 43, 449–461.

Huntley B J and Walker B H 1982 Ecology of Tropical Savannas. Springer-Verlag, Berlin.

Isichei A O 1980 Nitrogen fixation by blue-green algal soil crusts in Nigerian savanna. *In* Nitrogen Cycling in West African Ecosystems. Ed. T Rosswall. pp 191–198. SCOPE/UNEP, Stockholm.

Johnson R W and Tothill J C 1985 Definition and broad geographic outline of savanna lands. *In* Ecology and Management of the World's Savannas. Eds. J C Tothill and J J Mott. pp 1–13. Australian Academy of Science, Canberra.

Jones M J 1973 The organic matter content of savanna soils of West Africa. J. Soil Sci. 24, 42–53.

Kellman M 1979 Soil enrichment by Neotropical savanna trees. J. Ecol. 67, 565–577.

Kellman M, Miyanishi K and Hiebert P 1985 Nutrient retention by savanna ecosystems. II. Retention after fire. J. Ecol. 73, 953–962.

Kellman M and Sanmugadas K 1985 Nutrient retention by savanna ecosystems. I. Retention in the absence of fire. J. Ecol. 73, 935–951.

Meiklejohn J 1965 Microbiological studies on large termite mounds. Rhod. Zam. Mal. J. Agric. Res. 3, 67–79.

Mugwira L M and Mukurumbira L M 1984 Comparative effectiveness of manures from the communal areas and commercial feedlots as plant nutrient sources. Zimbabwe Agric. J. 81, 241–250.

Nyamapfene K W 1986 The use of termite mounds in Zimbabwe peasant agriculture. Trop. Agric. (Trinidad) 63, 191–192.

O'Connor T G 1985 A synthesis of field experiments concerning the grass layer in the savanna regions of Southern Africa. S. Afr. Nat. Sci. Prog. Rep. 114, Pretoria.

Prinsley R T and Swift M J 1986 Amelioration of Soil by Trees: A Review of Current Concepts and Practices. Commonwealth Science Council, London.

Puzo B 1978 Patterns of man-land relations. *In* Biogeography and Ecology of Southern Africa. Ed. M J A Werger. pp 1049–1112. Junk, The Hague.

Rodel M G W, Hopley J D H and Boultwood J N 1980 Effects of applied nitrogen, kraal compost and maize stover on the yields of maize grown on a poor granite soil. Zimbabwe Agric. J. 77, 229–232.

Rutherford M C 1982 Woody plant biomass distribution in *Burkea africana* savannas. *In* Ecology of Tropical Savannas. Eds. B J Huntley and B H Walker. pp 120–141. Springer-Verlag, Berlin.

Rutherford M C 1984 Relative allocation and seasonal sharing of growth of woody plant components in a South African savanna. Prog. Biomet. 3, 200–221.

Rutherford M C and Panagos M D 1982 Seasonal woody plant shoot growth in *Burkea africana — Ochna pulchra* savanna. S. Afr. J. Bot. 1, 104–116.

Sanginga N, Gwaze D and Swift M J 1987 Soil and fertiliser requirements for *Eucalyptus, Casuarina, Leucaena* and *Acacia* in Zimbabwe. Austr. J. For. Res.

Schimel D S 1986 Carbon and nitrogen turnover in adjacent grassland and cropland ecosystems. Biogeochemistry 2, 345–357.

Scoones I C and Wilson K B 1988 Households, lineage groups and ecological dynamics: Issue for livestock research and development in Zimbabwe's Communal Areas. *In* Socioeconomic determinants of livestock production in Zimbabwe's Communal Areas. Eds. B Cousins, C Jackson and I C Scoones. CASS, UZ and G.T.Z. Zimbabwe.

Seligman N G, Feigenbaum S, Feinerman D and Benjamin R W 1986 Uptake of nitrogen from high C-to-N ratio, ion-labelled organic residues by spring wheat growth under semi-arid conditions. Soil Biol. Biochem. 18, 303–307.

Shumba E M 1984a Yields of maize in the semi-arid regions of Zimbabwe. Zimbabwe Agric. J. 81, 91–94.

Shumba E M 1984b Reduced tillage in the communal areas. Zimbabwe Agric. J. 81, 235–239.

Smith L A 1965 Studies of the *Hyparrhenia* veld. VI. The fertiliser value of cattle excreta. J. Agric. Sci. 64, 403–406.

Stocking M A 1986 The cost of soil erosion in Zimbabwe in terms of the loss of three major nutrients. Consultants' Working Paper No 3, Soil Conservation Programme AGLS, FAO, Rome.

Stocking M A 1988 Tropical red soils: Fertility management and degradation. *In* The Red Soils of East and Southern Africa. Eds. K Nyamapfene, J Hussein and K Asamadu. pp 24–55. IDRC, Nairobi.

Swift M J 1984 Soil biological processes and tropical soil fertility: A proposal for a collaborative programme of research. Biol. Int. Special Issue 5, IUBS, Paris.

Swift M J 1987 Tropical Soil Biology and Fertility (TSBF): Inter-regional research and planning workshop report. Biol. Int. Special Issue 13, IUBS, Paris.

Swift M J, Cook A G and Perfect T J 1980 The effects of changing agricultural practice on the biology of a forest soil in the sub-humid tropics. 2. Decomposition. *In* Tropical Ecology and Development. Ed. J I Furtado. pp 341–348. International Society of Tropical Ecology, Kuala Lumpur.

Swift M J, Heal O W and Anderson J M 1979 Decomposition in Terrestrial Ecosystems. Blackwell Scientific Publications, Oxford.

Tanner P D and Mugwira L 1984 Effectiveness of communal area manures as sources of nutrients for young maize plants. Zimbabwe Agric. J. 81, 31–35.

Triplett G B and Mannering J V 1978 Crop residue management in crop rotation and multiple cropping systems. *In* Crop Residue Management Systems. Ed. W R Oschwald. pp 187–206. American Society of Agronomy, Special Publication, 31, Madison.

Truscott K 1986 Socio-economic factors in food production and consumption: A study of twelve households in Wedza Communal Land, Zimbabwe. Food Nut. 12, 27–37.

UNESCO 1979 Tropical Grazing Land Ecosystems. UNESCO, Paris.

Vincent V and Thomas R G 1960 An Agricultural Survey of S. Rhodesia. Part I: Agro-ecological Survey. Government Printer, Salisbury.

Walker B H, Ludwig D, Holling C S and Peterman R M 1981 Stability of semi-arid savanna grazing systems. J. Ecol. 69, 473–498.

Ward H K and Cleghorn W B 1964 The effect of ringbarking trees in *Brachystegia* woodland on the yield of veld grasses. Rhod. Agric. J. 61, 98–107.

Ward H K and Cleghorn W B 1970 The effects of grazing practices on tree regrowth after clearing indigenous woodland. Rhod. J. Agric. Res. 8, 57–65.

Watson J P 1977 The use of mounds of the termite *Macrotermes falciger* (Gerstäcker) as a soil amendment. J. Soil Sci. 28, 664–672.

Webster C C and Wilson P N 1980 Agriculture in the Tropics. 2nd Edition. Longman, London and New York.

Whitsun Foundation 1978 A Strategy for Rural Development: Data Bank No 2: The Peasant Sector Whitsun Foundation, Salisbury.

Willatt S T 1967 The fertility of sandveld soils under continuous cultivation. Part II: Changes in physical properties. Rhod. Zam. Mal. J. Agric. Res. 5, 129–213.

Young A 1986 The effects of trees on soils. *In* Amelioration of Soils by Trees. Eds. R T Prinsley and M J Swift. pp 10–19. Commonwealth Science Council, London.

M. Clarholm and L. Bergström (Eds.), Ecology of arable land, 77–87.

Influences of elemental interactions and pedogenic processes in organic matter dynamics

J.W.B. STEWART and C.V. COLE
Saskatchewan Institute of Pedology, University of Saskatchewan, Saskatoon, Saskatchewan, Canada S7N 0W0 and Natural Resource Ecology Laboratory, Colorado State University, Fort Collins, CO 80523, USA

Key words: C/N/P/S ratios, weathering index, texture, leaching, humic, fulvic, fractionation

Abstract

This paper reviews progress in understanding the processes which are important in elemental interactions and which influence organic matter composition of soils of the Great Plains in N. America. Comparison of grassland (semiarid) soils along environmental gradients and cultivation chrono- and toposequences with adjacent forest (subhumid) soils and consideration of the C/N/P/S ratios of organic matter of genetic horizons in the solum have emphasized the importance of movement of low molecular weight organic compounds in soil solution in addition to microbial degradation in the formulation of organic matter in soils. Phosphorus forms and transformations help to provide both an index on weathering and insight into textural influences. Use of $\delta^{15}N$ and $\delta^{34}S$ in combination with ^{14}C and other radioisotopes has provided valuable information on processes. Submicroscopy techniques in combination with cytoplasmic staining techniques have focussed attention in a realistic way on the mechanisms of organic matter stability. More attention must be paid to the catalytic role of soil inorganic constituents and selected minerals in the abiotic formation of stable organic matter. Conceptual and mathematical simulation models have an invaluable role in focussing attention on important processes and verifying hypotheses.

Introduction

In all soils the balance between the basic processes by which soils form, *i.e.* additions, removals, transformation and translocations, determine the nature of the soil (Anderson, 1988; Shindo and Huang, 1985). Parent material determines the original supply of those nutrient elements that are released by weathering and influences the balances between nutrient loss and retention. Microbes and plants can influence the rate of weathering by exudation and production of organic acids while utilizing the nutrients so released and storing them in organic matter. Nutrients such as N, not present in large quantities in parent material, are obtained by plants and microorganisms from the atmosphere. Organo-mineral colloids tend to add stability to soil organic matter. Nutrients are lost mainly by leaching of ions and dissolved organic constituents.

During the past decade, we have been studying soils developed on the former grasslands and bordering forest soils of the Great Plains of N. America. The overall objective of these investigations (Stewart et al., 1983a) is to develop concepts and procedures in quantitative pedology and to test these concepts on a small number of soils selected along climo, topo and cultivation chronosequences. This has allowed us to expand on the ideas of Jenny (1980) on driving variables or soil-forming factors and to integrate them with hierarchical perspectives used to describe ecosystems (Anderson et al., 1983; Reiners, 1986; Urban et al., 1987). Much of our efforts have been directed to studying organic matter dynamics. This paper will focus on progress in understanding the interrelationships of carbon (C), nitrogen (N), sulfur (S) and phosphorus (P) in nutrient cycling, and the relationship with organic matter stability and quality as soils develop in the Great Plains area of N. America and change under different cultivation systems.

Background studies on soil organic matter

In humus-rich, base-saturated Great Plains soils, the clay-associated humus fraction is of great significance both in its amount and in its capacity as a medium-term storehouse for organically bound nutrients. Studies of the chemical nature, turnover dynamics and biological significances have followed a number of pathways from classical chemical approaches (extraction and identification of specific chemical compounds) to the use of an alkaline extractant followed by physical-chemical fractionation to produce humic acid, fulvic acid and humin (this concept was expanded to extraction with alkali-pyrophosphate followed by peptization and ultrasonic dispersion (Anderson *et al.*, 1974). Mineral-associated organic compounds have also been physically separated and studied, sometimes in combination with a chemical fractionation of physically separated fractions (Anderson *et al.*, 1981; Tiessen and Stewart, 1983; Tiessen *et al.*, 1983). Biological approaches aided by radio and stable isotopes attempt to isolate fractions that are biological entities in terms of cycling of organically bound C, N, S and P. We have used many of these approaches to understand the nature, stability and composition of soil organic matter. Specifically, the following concepts have been verified: (a) carbon dating has demonstrated the stability and slow turnover of C in these soils (Anderson and Paul, 1984), (b) the ability of texture (clay content) to stabilize soil C through physical and chemical adsorption (Anderson, 1979) has been observed in a range of soils and parent materials (O'Halloran *et al.*, 1985; Schimel *et al.*, 1985) and (c) the amounts and rates of turnover and transformation of soil C, N, S and P fit into a logical framework and explain the properties of an ecosystem in a particular environment. Observation of differences in soil organic matter quality across environmental gradients (Anderson, 1979) in chronology of cultivation sequences (Bettany *et al.*, 1980; Tiessen *et al.*, 1982) have identified key concepts (Stewart 1984; Tiessen *et al.*, 1984c) which relate to:

(1) microbial ability to store P and S in conditions of plentiful supply (Hedley and Stewart, 1982; Saggar *et al.*, 1981a)

(2) nature of bond classes (McGill and Cole, 1981) where the direct bonding of an element to carbon, *i.e.*, C–N and C–S will behave differently to ester bonding C–O–S and C–O–P,

(3) formation of stable organo-mineral complexes (Anderson, 1979) especially by reactive P and S groups (Stewart and Tiessen, 1987)

(4) losses of C and N in greater quantities than P and S when a system at equilibrium is distributed by cultivation processes (Bettany *et al.*, 1980; Schimel *et al.*, 1985; Tiessen *et al.*, 1982)

(5) the possibility of biochemical mineralization supplementing biological oxidation when P and S were in short supply (Maynard *et al.*, 1983) and

(6) the importance of labile P as a control of organic matter accumulation (Cole and Heil, 1981; Tiessen and Stewart, 1985).

The next development arose from our attempt to combine conceptual models of processes (Hunt *et al.*, 1983; Stewart *et al.*, 1983b) and other published information on processes (Cole *et al.*, 1977) into simulation models which accurately depicted soil organic matter turnover on a longer time scale (Parton *et al.*, 1983). Any exercise of this sort immediately focusses attention on processes that are not well described experimentally and spawns a host of investigations to clarify concepts and processes.

Recent developments

Soil organic matter quality

Further insight into changes in soil organic matter quality has been obtained by Bettany and coworkers (Roberts *et al.*, 1985, 1986; Roberts and Bettany, 1985; Schoenau and Bettany, 1987). They expanded consideration of soil quality to include the complete soil profile and examined C, N, S and P relationships in a climotoposequence of soils across narrow environmental gradients. Changes in organic matter composition across a 100 meter catenary sequence of soils (Table 1) were similar to changes associated with midslope positions across several hundred kilometers of an environmental gradient, providing evidence of environmental control of similar processes. The most consistent trends were observed in the lower slope soils, where organic C/N, C/P and C/N/P/S ratios widened from Brown to Gray soils along a gradient of increasing moisture content. Higher amounts of organic S and a lower proportion of total organic S occurring as

Table 1. Organic C/N/C/P and C/N/P/S ratios (Ap horizons of lower slope soils) of the Brown, Dark Brown, Black Chernozemic and Gray Luvisolic soils and of the Chernozemic and Luvisolic soils at upper, mid and lower slope positions*

Soil zone	Slope position	C/N ratio	C/P ratio	C/N/P/S ratio
All soil zones				
Brown Chernozem	lower	9.8	48	68: 6.9:1.4:1
Dark Brown Chernozem	lower	11.2	65	84: 7.5:1.3:1
Black Chernozem	lower	11.8	69	103: 8.7:1.5:1
Gray Luvisol	lower	12.5	78	145:11.6:1.8:1
Grassland soils				
Chernozem	upper	9.5	56	70: 7.4:1.3:1
Chernozem	mid	10.3	60	78: 7.6:1.3:1
Chernozem	lower	10.9	61	85: 7.7:1.4:1
Forest soils				
Luvisol	upper	10.7	66	106: 9.9:1.6:1
Luvisol	mid	13.7	101	157:11.5:1.6:1
Luvisol	lower	12.5	78	145:11.6:1.8:1

* From Roberts *et al*, (1986).

organic sulfate (C–O–S bonds) also were found with increasing availability of moisture. Similarly, trends in the distribution of inorganic and organic P fractions (Roberts *et al*., 1985) across the same catenary sequence were parallel to changes across the wider environmental gradient at the midslope position. Changes in soil forms were closely integrated and were strongly influenced by the moisture regimes across soil zones. Increasing moisture along catenas or across soil zones had a two-fold

Table 2. Organic C/N, C/P, C/S ratios and % of total organic S as organic sulfate in horizons from a Brown Chernozemic and a Gray Luvisolic soil*

Soil zone	Horizon	Slope position	C/N ratio	C/P ratio	C/S ratio	% Organic sulfate
Chernozem	Ah	Upper	10.3	92	61.6	51.4
	Bmk		9.2	66	30.4	70.3
	Cca		9.9	48	21.1	69.5
	Ck		8.7	40	20.9	67.5
	Cksa		9.1	52	—	—
	Ah	Mid	10.4	76	66.4	59.8
	Bm		9.7	58	44.2	67.1
	Bmk		8.6	58	38.1	64.7
	Cca		8.6	35	19.3	68.0
	Cksa		7.9	49	—	—
Luvisol	L-H	Upper	17.5	286	230.8	20.5
	Ae		14.1	87	152.3	24.6
	Bt		11.3	53	100.0	34.8
	Ck		5.8	18	17.0	27.0
	L-H	Mid	16.6	249	240.0	22.0
	Ae		13.2	71	132.9	28.8
	Bt		11.3	36	80.5	37.8
	Ck		7.2	29	19.8	19.8
	L-H	Lower	15.8	256	214.9	25.8
	Ae		12.6	49	126.9	25.4
	Bt		10.3	31	69.4	45.8
	Cca		7.9	38	41.1	43.3

* From Schoenau and Bettany (1987).

effect: it increased biomass production and increased weathering intensity. This work also indicated that (a) organic matter found in the B horizons of a soil profile was different from that in the A horizon suggesting a leaching of part of the organic matter from the A horizon and (b) that the form and availability of P in soil horizons could be used as an index of weathering intensity lending further support to earlier work reviewed by Smeck (1973, 1985) and others (Tiessen and Stewart, 1985; Tiessen *et al.*, 1984c).

Bettany and co-workers (Schoenau and Bettany, 1987) later examined the form and availability of organic matter of soil horizons from profiles selected along an environmental gradient ranging from semiarid to subhumid. Most of the soils examined were developed on calcareous parent material so that leached sulfate ions were precipitated in the C horizon as secondary calcium sulfates. Leaching was also found to occur in organic forms (Table 2), mainly as sulfur-rich low molecular weight compounds (fulvic acids). An increasing proportion of NaOH extractable C, N, P and S was found in the fulvic acid fraction as samples were taken from lower depths in profiles, suggesting that fulvic acids produced in biologically active surface horizons have been translocated to B and C horizons by percolating water. Leaching of nutrient-rich organic matter provides a valid explanation for the observed narrowing of soil organic matter C/N, C/P and C/S ratios with increasing depth. Of the four elements C, N, P and S, P appeared to be the most susceptible to deep leaching in the organic form (Schoenan and Bettany, 1987). In the soils in the semiarid zone little S is lost from the profile but in the more strongly leached soils developed under forest leaching of low molecular weight fulvic acids (containing C, N, S and P) was an important factor in nutrient loss (Roberts *et al.*, 1985; Roberts and Bettany, 1985; Schoenau and Bettany, 1987).

Phosphorus studies

The fact that P is thought to be relatively immobile, slowly weathering and subject to transformations to P_i and P_o compounds within the profile has been reexamined (Smeck, 1973; Smeck, 1985; Stewart and Tiessen, 1987; Tiessen and

Stewart, 1985). Analytical techniques, which chemically fractionate soil P into fractions of biological significance (Hedley *et al.*, 1982) have been used across the wide range of surface soils found in N. America (Tiessen *et al.*, 1984b) to demonstrate that the transformation of P to different forms in the soil was closely related to soil taxonomy criteria and by definition to degree of weathering. Recent studies in our laboratories have used the "pedogenic index" as a means of quantitatively evaluating changes which have occurred in soils (Santos *et al.*, 1986; St. Arnaud *et al.*, 1988) during their development. This index quantifies changes in soil constituents present in relation to original contents in either total horizons or in component soil separates within them. The soils used in deriving the pedogenic index concept have been shown to be developed from uniform deposits as evidence from compositional analyses and verified by the constancy of the ZrO_2/quartz ratios with depth. Using the P in the A horizon as an example the:

Pedogenic Index =

$$\frac{\dfrac{\text{Wt of total P in A horizon}}{\text{Wt of quartz in A horizon}}}{\dfrac{\text{Wt of total P in C horizon}}{\text{Wt of quartz in C horizon}}} \times 100$$

where weights of P and quartz are expressed in grams /cm^2. An index value of 80 indicates that 80% of the original P remains in the horizon; conversely the horizon has lost 20%. Index values for the solum provide an overall evaluation of the net

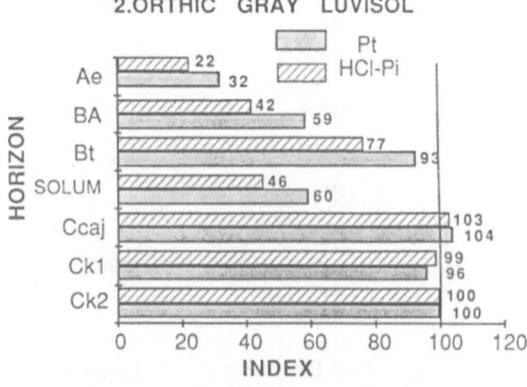

Fig. 1. Pedogenic index values for total phosphorus (P_t) and HCl-soluble inorganic P (HCl-P_i) in horizons of an Orthic Gray Luvisol (from St. Arnaud *et al.*, 1988).

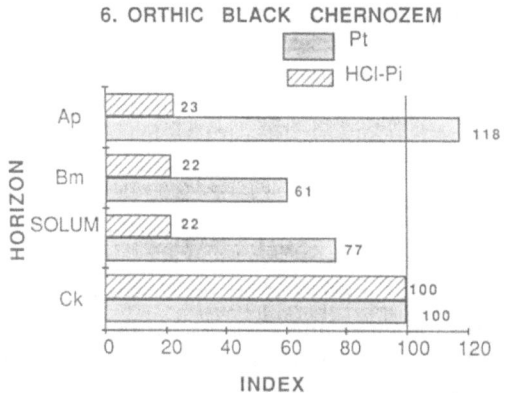

Fig. 2. Pedogenic index values for total phosphorus (P_t) and HCl-soluble inorganic P (HCl-P_i) in horizons of an Orthic Black Chernozem (from St. Arnaud *et al.*, 1988).

changes which have occurred during soil development. Levels of HCl-soluble P_i (dominantly Ca-phosphate P of apatite form) in the original parent material (Ck horizons) may also be used to determine index values which give a measure of the degree of weathering. Calcium-P, which originally accounted for over 90% of the total P in the soil studies, is partly leached and partly converted to other forms, particularly organic P in the grassland

soils where organic matter accumulations are the greatest (St. Arnaud *et al.*, 1988). Soils developed under forest (Luvisols/Boralfs) have lost 40% of their original P (Fig. 1) in comparison to 23% loss from a grassland soil (Fig. 2, Black Chernozem/Cryoboroll). Further investigation of a possible mechanism (Frossard *et al.*, 1988) whereby P is lost from soil profiles developed under forest confirms suggestions (Schoenau and Bettany, 1987) that one loss mechanism is through leaching of low molecular weight organic compounds containing P which are more mobile than inorganic phosphates in subhumid soils. In contrast, leaching processes were not sufficiently strong in semiarid grassland soils to remove significant amounts of P from the solum and losses can mainly be accounted for by crop removal.

Changes in soil P forms as determined by a sequential fractionation procedure also have been used (O'Halloran *et al.*, 1985; O'Halloran *et al.*, 1987a; O'Halloran *et al.*, 1987b) to assess the influence of soil texture and management practices on the forms and distribution of soil P in grassland soils at Swift Current, Saskatchewan (Brown Chernozem/Aridic Haploboroll) and at Sidney, Nebraska (Duroc Loam/Pachic Haplustoll). Much of the

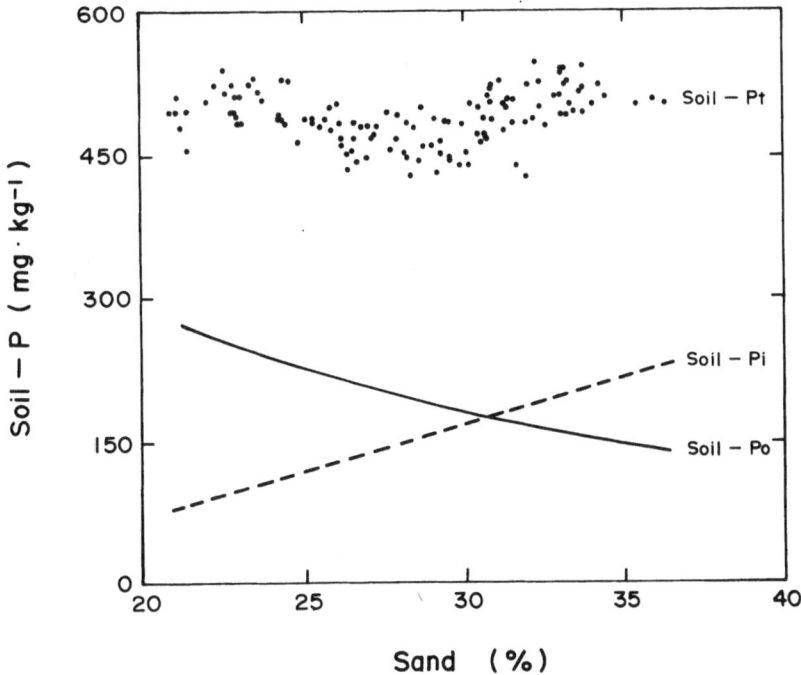

Fig. 3. Values of total soil phosphorus (Soil P_t) with increasing sand content, and predicted organic and inorganic phosphorus levels (Soil P_o and Soil P_i, respectively) (from O'Halloran *et al.*, 1985) derived from regression analysis of actual data.

variability in soil P form could be attributed to changes in texture. At Swift Current where soils had a variable texture due to eolian deposition in the glaciation of the Northern Great Plains, changes in the forms and distribution of soil P with changing texture followed patterns similar to those associated with a weathering sequence. For instance, at the latter site (O'Halloran *et al.*, 1985; O'Halloran *et al.*, 1987a), total P (P_t) content remained constant in samples of different sand content (Fig. 3) whereas total inorganic P (P_i) content increased and total organic P (P_o) decreased with sand content. The effect was two-fold as increasing clay plus silt (decreasing sand) held more moisture, was more weathered and promoted more plant growth, therefore accumulated greater quantities of P_o over the time period since glaciation ($\sim 10,000$ yrs). At Sidney (O'Halloran *et al.*, 1987b), where soils were of more recent origin having developed on fluvial deposits, most of the differences in P forms were associated with erosional processes and mixing in of sandier lower horizons during cultivation.

Elemental interactions

Natural ^{15}N abundance has been used (Tiessen *et al.*, 1984a) as an indicator of soil organic matter

transformations associated with land use. Changes in abundance of different organo-mineral size fractions from soil surface horizons can be ascribed to the following processes or organic matter formation and N transformations. Plant litter of low ^{15}N enrichment enters the soil system as coarse particulates (sand-sized), which are subsequently broken down to smaller particles associated with silt fractions. These secondary fractions show a higher ^{15}N enrichment than the sand, since they undergo biological and chemical transformations that undergo isotope discrimination. A large pulse of low enrichment litter can be followed through the soil system into the silt fractions as well as into microbial products associated with the fine clays. The isotopic composition of fine clays and coarse clays shows little change, even after prolonged cultivation, reflecting their relative stability under all conditions. Similarly, recent work (Table 3), which examined the natural ^{34}S in various soil organic matter and plant S fractions from soil profiles (Schoenau and Bettany, 1988) has proven valuable in integrating concepts regarding soil S flows and transformations. The systematic nature of the variations enabled several hypotheses to be constructed regarding the origin and nature of S found in soil and plant components and its interaction with other terrestrial pools (Schoenau, J.J, Unpublished Ph.D. Thesis, University of Saskatchewan

Table 3. Examples of δ^{34}S values obtained from soil fractions and plant material

| | δ^{34}S wrt* Canon Diablo Troilite (‰) | | | |
| | HI-S Fraction | | Total S Fraction | |
	Mean	Standard deviation ‰	Mean	Standard deviation ‰
Boroll Ap horizon	-3.1	0.2	-4.0	0.4
Boroll Ah horizon			-1.4	0.3
Boroll Bm horizon	-4.2	0.1		
Boroll Bmk horizon	$+0.9$	0.1		
Boroll Cca horizon	-6.0	0.2		
Boroll Ck horizon (saline)	-8.0	0.2	-8.1	0.3
Boroll Ah 0.01 M CaCl$_2$ extract	-3.0	0.3		
Aquoll Ahe horizon			-7.3	0.1
Aquoll Btg horizon	-5.0	0.3		
Boralf L-H leaf mat	$+2.7$	0.1		
Boralf L-H humic acid extract			$+1.8$	0.4
Boralf Ck fulvic acid extract			$+1.6$	0.1
Wheat straw			$+0.3$	0.1
Manure			-6.8	0.1

* wrt = with respect to.

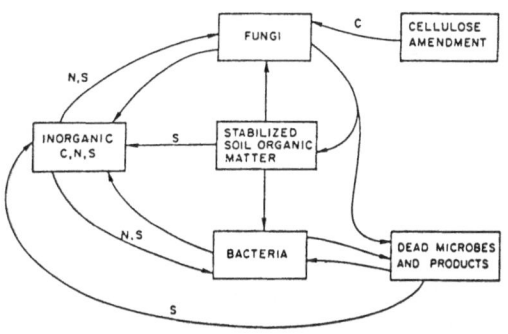

Fig. 4. State variables and processes in the simulation of C, N and S transformations (model of Hunt *et al.*, 1986) Arrows labelled with elements transfer only those elements. Unlabelled arrows transfer C, N and S.

1988). These investigations have supplemented process studies using radioactive tracers (^{14}C, ^{15}N, ^{35}S and ^{32}P) in understanding short and long-term effects of processes.

We have used conceptual and simulation models to integrate concepts and verify hypotheses. Hunt *et al.* (1983) proposed a conceptual model for interactions between C, N, S and P in grassland. The objective was to relate differences in nutrient cycling patterns (Stewart *et al.*, 1983b) among C, N, S and P to their fundamental chemical properties, to information from microbial physiology and trophic ecology, and to current theories on the

formation and decomposition of soil organic matter. The model took account of differences in chemical bonding to account for variation in the composition of C, N and S compounds. Microbial biomass was differentiated into bacteria and fungi and the element ratios in each group were assumed to vary. A simulation model (Hunt *et al.*, 1986) (Fig. 4) evaluated these concepts by applying it to data on microbial biomass, sulfate, nitrate, and CO_2 evolution obtained in 60-day laboratory incubations amended with sulfate and cellulose (Saggar *et al.*, 1981b). Both the data and the simulation model concur in showing that N behaves differently from S (Fig. 5). The model successfully tested an important regulation mechanism: that the stimulation of sulfohydrolase enzyme depended on sulfur stress in microbial biomass. The hypothesis that sulfate is stored as ester sulfate is supported by model calculations. The model has now to be applied to other experimental data where S mineralization/immobilization was measured in the presence of plants (Maynard *et al.*, 1983; Maynard *et al.*, 1985) and the model has now to be adapted to field conditions.

The dynamics of C, N, P and S in cultivated and uncultivated grassland soils also has been modelled (Parton *et al.*, 1983; Parton *et al.*, 1987; Parton *et al.*, 1988) using a monthly time step which allows the dynamics of soil organic matter over long time periods (100 to 10000 yrs) to be simulated (for full

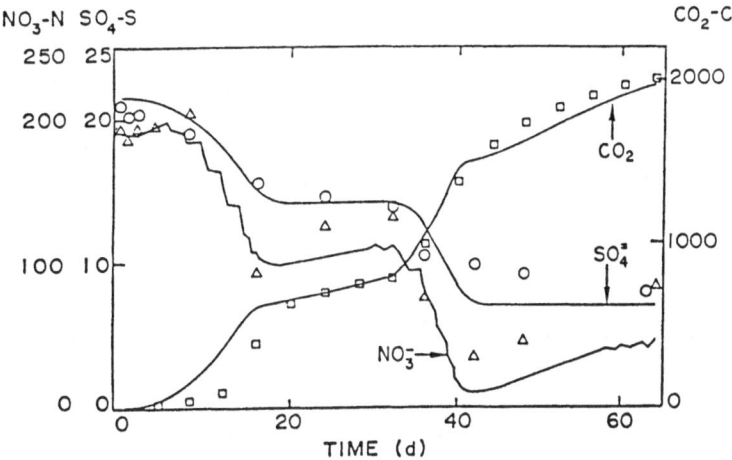

Fig. 5. Patterns of mineralization/immobilization of nitrogen and sulfur in response to cellulose amendments (1500 μg C g^{-1} soil) at the beginning and after 32 days. Data (Saggar *et al.*, 1981b) represented by symbols: $\square$, μg C g^{-1} soil, $\circ$, μg S g^{-1} soil and $\triangle$, μg N g^{-1} soil. Model output lines from Hunt *et al.*, (1986).

description and update on the Century model development see papers by Parton *et al.* and Cole *et al.*, 1989). The Century model was used (Parton *et al.*, 1988) to simulate the impact of cultivation (100 years) on soil organic matter dynamics, nutrient mineralization, and plant production and to simulate soil formation during a 10000 year run. The model correctly predicted that N and P are the primary limiting nutrients for plant production and simulated the response of the system to inorganic N, P and S fertilizer. Simulation results indicate that controlling the C/P and C/S ratios of soil organic matter fractions as functions of the labile P and S levels respectively, allows the model to correctly simulate the observed changes in C/P and C/S ratios in the soil and to simulate the impact of varying the labile P and S levels on soil P and S net mineralization rates.

Organic matter stabilization

Field and laboratory observations (Elliott, 1986) of aggregate structure and C, N and P in native and cultivated soils have corroborated that the hierarchical conceptual model of soil structure (Tisdall and Oades, 1982) extends into the N. American grassland soils. This model can be applied to partially explain accumulation of organic matter under native conditions and its loss upon cultivation. Possible mechanisms of stabilization and immobilization of soil organic matter have been reviewed and refined (Tiessen *et al.*, 1984c). Combined submicroscopy and staining techniques (Foster, 1985), have allowed us to study the organo-mineral and microbial associations in microaggregates of different sizes (Tiessen and Stewart, 1986). Organic matter inputs are (1) relatively coarse plant debris that is deposited within or upon the soil and (2) colloidal and soluble organic components from root exudates, microbial products or litter leachates. In soils, the persistence of organic matter has been classified as transient (mainly polysaccharides), temporary (roots or hyphae) and persistent compounds (strongly humified organic matter). An addition to this broad classification might be recent organic colloids stabilized by interactions with metal precipitates (*cf.* reviews by Emerson *et al.*, 1986; Foster, 1985). Microscopic observations

of transient and temporary cementing agents suggest highly specific processes of formation, deposition and crosslinking with minerals, which also vary in size, shape, crystallinity and electric charge. The stability of organic matter in these porous water-stable aggregates is governed by the accessibility of aggregate or particle surfaces in the pores to microbial activity. Organic matter of moderate stability is the result of interactions of microbial gels and decomposing plant material with soil minerals and metal cations (Tiessen and Stewart, 1986). Both particulate and colloidal organic matter can undergo physical stabilization in the soil environment.

One mechanism that has received scant attention until recently is the abiotic formation of humic substances in the environment. Huang and co-workers (Shindo and Huang, 1982; 1984a; 1984b; 1985; Wang and Huang, 1986; Wang *et al.*, 1986) have clearly demonstrated the importance of the catalytic role of soil inorganic constituents and selected soil minerals in the abiotic formation of humic substances. Manganese oxides, in particular, have been shown to be very effective (more so than iron oxides) in cleavage of aromatic ring compounds. For instance, reacting pyrogallol with manganese oxides in experimental conditions results in the formation of high molecular weight organic compounds similar in IR spectra to soil humic material. The aromaticity or alkalinity of soil humic material needs to be reexamined in conjunction with the mineral composition and the type of inorganic constituents coating soil colloidal material in the soil profile. Ring cleavage of polyphenols as catalyzed by clay minerals or short-range order oxides has to be considered along with other processes such as leaching and microbial decomposition in the formation of soil organic matter.

The reasons for organic matter stabilization are complex and difficult to simplify. Texture has been shown to be important, in part due to the inorganic constituents which are surface coatings of colloidal material in the soil and also due to the moisture holding properties. Polycondensation of lignin type compounds, abiotic and biotic degradation of same, leaching of organic solubles *etc.* all contribute to the formation and properties of soil organic matter. All horizons of the solum including inorganic weathering sequences have to be considered in future studies in this area.

Conclusion

Significant progress has been made in understanding the processes which are important in elemental interactions and which. influence organic matter composition. The use of conceptual and mathematical simulation models has proven to be useful techniques to focus attention on important processes and to verify hypotheses. Further insight has been provided by comparison of grassland (semiarid) soils with adjacent forest (subhumid) soils and by a consideration of the composition of the solum. Use of $\delta^{15}N$ and $\delta^{34}S$ shows promise in combination with ^{14}C and other radioisotopes to understand processes and resulting organic matter stability. Submicroscopy techniques in combination with cytoplasmic staining techniques have focussed attention in a realistic way on the mechanisms of organic matter stability. More attention must be paid to abiotic degradation. Phosphorus transformations can provide information on the degree of weathering in soils, have helped clarify some textural interactions, and can be used to quantify loss mechanisms. A greater understanding of soil processes of organic matter stabilization will depend on the simultaneous use of many of these techniques when examining soils.

References

Anderson D W 1979 Processes of humus formation and transformation in soils of the Canadian Great Plains. J. Soil Sci. 30, 77–84.

Anderson D W 1988 The effect of parent material and soil development on nutrient cycling in temperate ecosystems. Biogeochem. 5, 71–79.

Anderson D W and Paul E A 1984 Organo-mineral complexes and their study by radiocarbon dating. Soil Sci. Soc. Am. J. 48, 298–301.

Anderson D W, Heil R D, Cole C V and Deutsch P C 1983 Identification and characterization of ecosystems at different integrative levels. *In* Nutrient Cycling in Agricultural Ecosystems. Eds. R R Lowrance, R L Todd, L E Asmussen and R A Leonard. pp 517–531 University of Georgia, College of Agriculture Exp. Stn, Special Publication Number 23, Athens, Georgia.

Anderson D W, Paul E A and St. Arnaud R J 1974 Extraction and characterization of humus with reference to clay associated humus. Can. J. Soil Sci. 54, 317–323.

Anderson D W, Saggar S, Bettany J R and Stewart J W B 1981 Particle size fractions and their use in studies of soil organic matter. 1. The nature and distribution of forms of carbon, nitrogen and sulfur. Soil Sci. Soc. Am. J. 45, 767–772.

Bettany J R, Saggar S and Stewart J W B 1980 Comparison of the amounts and forms of sulphur in organic matter after 65 years cultivation. Soil Sci. Soc. Am. J. 44, 70–75.

Cole C V, Innis G S and Stewart J W B 1977 Simulation of phosphorus cycling in semi-arid grasslands. Ecology 58, 1–15.

Cole C V and Heil R D 1981 Phosphorus effects on terrestrial nitrogen cycling. *In* Terrestrial Nitrogen Cycles. Eds. F E Clark and T Rosswall. Bulletin (Stockholm) 33, 363–374.

Cole C V, Stewart J W B, Ojima D, Parton W J and Schimel D S 1989 Modelling land use effects of soil organic matter dynamics in the North American Great Plains. *In* Ecology of Arable Land – Perspectives and Challenges. Eds. M Clarholm and L Bergström. pp. 89–98. Kluwer Academic Publishers, Dordrecht, The Netherlands.

Elliott E T 1986 Aggregate structure and carbon, nitrogen, and phosphorus in native and cultivated soils. Soil Sci. Soc. Am. J. 50, 627–633.

Emerson W W, Foster R C and Oades J M 1986 Organo-mineral complexes in relation to soil aggregation and structure. *In* Interactions of Soil Minerals with Natural Organics and Microbes. Eds. P M Huang and M Schnitzer. pp 521–548. SSSA Special Publication Number 17, Soil Science Society of America, Inc., Madison, Wisconsin.

Foster R C 1985 *In situ* localization or organic matter in soils. Quaestiones Entomologicae 21, 609–633.

Frossard E, Stewart J W B and St. Arnaud R J 1988 Distribution and mobility of phosphorus in grassland and forest soils of Saskatchewan. Can. J. Soil Sci. (in manuscript).

Hedley M J and Stewart J W B 1982 Method to measure microbial phosphate in soils. Soil Biol. Biochem. 14, 377–385.

Hedley M J, Stewart J W B and Chauhan B S 1982 Changes in inorganic and organic soil phosphorus fractions induced by cultivation practices and by laboratory incubations. Soil Sci. Soc. Am. J. 46, 970–976.

Hunt H W, Stewart J W B and Cole C V 1983 A conceptualized simulation model of C, N, S and P interactions. *In* The Major Biogeochemical Cycles and Their Interactions, Chapter 10, SCOPE 21. Eds. B Bolin and R C Cook. pp 303–325. John Wiley & Sons, Sussex.

Hunt H W, Stewart J W B and Cole C V 1986 Concepts of sulfur, carbon and nitrogen transformations in soil: Evaluation by simulation modelling. Biogeochem. 2, 163–177.

Jenny H 1980 The Soil Resource. Springer-Verlag, New York, 377 p.

Maynard D G, Stewart J W B and Bettany J R 1983 Sulfur and nitrogen mineralization compared using two incubation techniques. Soil Biol. Biochem. 15, 251–256.

Maynard D G, Stewart J W B and Bettany J R 1985 The effects of plants on soil sulfur transformations. Soil Biol. Biochem. 17, 127–134.

McGill W B and Cole C V 1981 Comparative aspects of organic C, N, S and P cycling. Geoderma 26, 267–286.

O'Halloran I P, Kachanoski R G and Stewart J W B 1985 Spatial variability of soil phosphorus as influenced by soil texture and management. Can. J. Soil Sci. 65, 475–478.

O'Halloran I P, Stewart J W B and Kachanoski R G 1987a Influence of texture and management practices on the forms and distribution of soil phosphorus. Can. J. Soil Sci. 67, 147–163.

O'Halloran I P, Stewart J W B and de Jong E 1987b Changes

in P forms and availability as influenced by crop growth and environment. Plant and Soil 100, 113–126.

Parton W J, Anderson D W, Cole C V and Stewart J W B 1983 Simulation of soil organic matter formation and mineralization of semiarid agroecosystems. *In* Nutrient Cycling in Agricultural Ecosystems. Eds. R R Lowrance, R L Todd, L E Asmussen and R A Leonard. pp 533–550. University of Georgia, College of Agriculture Exp. Stn, Special Publication Number 23, Athens, Georgia.

Parton W J, Schimel D S, Cole C V and Ojima D S 1987 Analysis of factors controlling soil organic matter levels in Great Plains grasslands. Soil Sci. Soc Am. J. 51, 1173–1179.

Parton W J, Stewart J W B and Cole C V 1988 Dynamics of C, N, P and S in grassland soils: A model. Biogeochem. 5, 109–131.

Parton W J, Cole C V, Stewart J W B, Ojima D S and Schimel D S 1989. Simulating regional patterns of soil C, N, and P dynamics in the U.S. central grasslands region. *In* Ecology of Arable – Perspectives and Challenges. Eds. M Clarholm and L Bergström. pp. 99–108. Kluwer Academic Publishers, Dordrecht, The Netherlands.

Reiners W A 1986 Complementary models for ecosystems. Amer. Naturalist 127, 59–73.

Roberts T L, Bettany J R and Stewart J W B 1986 Level of integration approach to the studies of organic carbon, nitrogen, sulphur and phosphorus in the Canadian prairies. Trans. XIIIth Congr. Interntl. Soc. Soil Sci., Hamburg, FRG., Vol. 3, 932–933.

Roberts T L, Stewart J W B and Bettany J R 1985 The influence of topography on the distribution of organic and inorganic soil phosphorus across a narrow environmental gradient. Can. J. Soil Sci. 65, 651–665.

Roberts T L and Bettany J R 1985 The influence of topography on the nature and distribution of soil sulfur across a narrow environmental gradient. Can. J. Soil Sci. 65, 419–434.

Saggar S, Bettany J R and Stewart J W B 1981a Measurement of microbial sulfur in soil. Soil Biol. Biochem. 13, 493–498.

Saggar S, Bettany J R and Stewart J W B 1981b Sulfur transformations in relation to carbon and nitrogen in incubated soils. Soil Biol. Biochem. 13, 499–511.

Santos M C D, St. Arnaud R J and Anderson D W 1986 Quantitative evaluation of pedogenic changes in Boralfs (Gray Luvisols) of East Central Saskatchewan. Soil Sci. Soc. Am. J. 50, 1013–1019.

Schimel D S, Coleman D C and Horton K A 1985 Soil organic matter dynamics in paired rangeland and cropland toposequences in N. Dakota. Geoderma 36, 201–214.

Schoenau J J and Bettany J R 1987 Organic matter leaching as a component of C, N, P, S cycles in a forest, grassland and gleyed soil. Soil Sci. Soc. Am. J. 51, 646–651.

Schoenau J J and Bettany J R 1988 A method for determining sulfur-34 abundance in soil and plant sulfur fractions. Soil Sci. Soc. Am. J. 52, 297–300.

Shindo H and Huang P M 1982 Role of Mn(IV) oxide in abiotic formation of humic substances in the environment. Nature 298, 363–365.

Shindo H and Huang P M 1984a Catalytic effects of manganese(IV), iron(III), aluminum, and silicon oxides on the formation of phenolic polymers. Soil Sci. Soc. Am. J. 48, 927–934.

Shindo H and Huang P M 1984b Significance of Mn(IV) oxide in abiotic formation of organic nitrogen complexes in natural environments. Nature 308, 57–58.

Shindo H and Huang P M 1985 Catalytic polymerization of hydroquinone by primary minerals. Soil Sci. 139, 505–511.

Simonson R W 1959 Outline of a generalized theory of soil genesis. Soil Sci. Soc. Am. Proc. 23, 152–156.

Smeck N E 1973 Phosphorus — An indicator of pedogenic weathering processes. Soil Sci. 115, 199–206.

Smeck N E 1985 Phosphorus dynamics in soils and landscapes. Geoderma 36, 185–199.

St. Arnaud R J, Stewart J W B and Frossard E 1988 Application of the *pedogenic index* to soil fertility studies. Geoderma (in press).

Stewart J W B and Tiessen H 1987 Dynamics of soil organic phosphorus. Biogeochem. 4, 41–60.

Stewart J W B 1984 Interrelation of carbon, nitrogen, sulphur and phosphorus cycles during decomposition processes in soil. *In* Current Perspectives in Microbial Ecology. Eds. C A Reddy and M J Klug. pp 442–446. Proceedings of the 3rd International Symposium. American Society of Microbiology, Washington, D.C..

Stewart J W B, Cole C V and Heil R D 1983a Agroecosystems of the Great Plains of North America. *In* Nutrient Cycling in Agricultural Ecosystems. Eds. R R Lowrance, R L Todd, L E Asmussen and R A Leonard. pp 97–120. University of Georgia, College of Agriculture Exp. Stn, Special Publication Number 23, Athens, Georgia.

Stewart J W B, Maynard D G and Cole C V 1983 Interaction of biogeochemical cycles in grassland ecosystems. *In* The Major Biogeochemical Cycles and Their Interactions, Chapter 8, SCOPE 21. Eds. B Bolin and R C Cook. pp 247–269. John Wiley & Sons, Sussex.

Tiessen H and Stewart J W B 1985 The biogeochemistry of soil phosphorus. *In* Planetary Ecology (Selected Papers from the Sixth International Symposium on Environmental Biogeochemistry), Santa Fe, New Mexico, 1983, Chapter 39. Eds. D E Caldwell, J A Brierley and C L Brierley. pp 463–472. Van Nostrand Reinhold Company, New York.

Tiessen H and Stewart J W B 1983 Particle size fractions and their use in studies of soil organic matter. II. Cultivation effects on organic matter composition in size fractions. Soil Sci. Soc. Am. J. 47, 507–514.

Tiessen H. and Stewart J W B 1986 Organic matter and soil aggregates — A microscopic study. Trans. XIIIth Congress Interntl. Soc. Soil Sci., Hamburg, FRG. Vol. 2, 644–645.

Tiessen H, Karamanos R E, Stewart J W B and Selles F 1984a Natural nitrogen-15 abundance as an indicator of soil organic matter transformations. Soil Sci. Soc. Am. J. 48, 312–315.

Tiessen H, Stewart J W B and Cole C V 1984b Pathways of phosphorus transformations in soils of differing pedogenesis. Soil Sci. Soc. Am. J. 48, 853–858.

Tiessen H, Stewart J W B and Hunt H W 1984c Concepts of soil organic matter transformations in relation to organo-mineral particle size fractions. Plant and Soil 70, 287–295.

Tiessen H, Stewart J W B and Moir J O 1983 Changes in organic and inorganic phosphorus in particle size fractions of two soils during 60 to 90 years of cultivation. J. Soil Sci. 34, 815–823.

Tiessen H, Stewart J W B and Bettany J R 1982 Cultivation effects on the concentration and amounts of carbon, nitrogen and phosphorus in grassland Soils. Agron. J. 74, 831–835.

Tisdall J M and Oades J M 1982 Organic matter and water stable aggregates in soils. J. Soil Sci. 33, 141–163.

Urban D L, O'Neill R V and Schugart H H 1987 A hierarchical perspective can help scientists understand spatial patterns. BioScience 37, 119–127.

Wang M C and Huang P M 1986 Humic macromolecule inter-layering in nontronite through interaction with phenol mon-omers. Nature 323, 529–531.

Wang T S C, Huang P M, Chou Chang-Hung and Chen Jen-Hshuan 1986 The role of soil minerals in the abiotic poly-merization of phenolic compounds and formation of humic substances. *In* Interactions of Soil Minerals with Natural Organics and Microbes, SSSA Special Publication Number 17. Eds. P M Huang and M Schnitzer. pp. 251–281. Soil Science Society of America, Inc., Madison. Wisconsin.

M. Clarholm and L. Bergström (Eds.), Ecology of arable land, 89–98.
© 1989 Kluwer Academic Publishers.

Modelling land use effects of soil organic matter dynamics in the North American Great Plains

C.V. COLE[1,2], J.W.B. STEWART[3], D.S. OJIMA[1], W.J. PARTON[1] and D.S. SCHIMEL[1]
[1]*Natural Resource Ecology Laboratory, Colorado State University, Fort Collins, CO 80523, USA,*
[2]*USDA-Agricultural Research Service, Fort Collins, CO 80523, USA and* [3]*(Corresponding author)*
Saskatchewan Institute of Pedology, University of Saskatchewan, Saskatoon, Saskatchewan S7N 0W0,
Canada

Key words: carbon, grasslands, nitrogen, phosphorus, simulation modelling

Abstract

The Century soil organic model has been used to simulate land use effects on organic carbon and nitrogen and organic and inorganic phosphorus changes in soils of the Great Plains region. Two methods were used to verify the model: the first was the data from experimental stations where the effects of different rotation and tillage practices on crop products were studied over a 30 to 40 year period until 1947; the second was simulated data on cultivation effects on the soils of different textures in one climatic area over a similar time period. The model results were found to accurately simulate these changes, providing a new direction in the analyses of agroecosystems.

Introduction

Changes in organic matter are the direct result of biological and pedogenic processes in soil. Much of our research in the past decade has been focussed on examining transformations of carbon (C), nitrogen (N), sulfur (S) and phosphorus (P) within a framework of the properties of an ecosystem and its driving variables, defined as external factors that govern the processes that operate in a particular environment. The general hypothesis is that the effects of driving variables are expressed through their effects on ecosystem processes. These processes include above- and below-ground primary production, decomposition and nutrient cycling. Major driving variables include climate (temperature and water), parent material (soil texture), base status, total S and P, topography and management (Stewart et al., 1983a; b). The effects of these controlling factors are expressed over a wide range of resolution from the global down to regional, landscape and field plot levels. Our research has therefore concentrated on understanding the

processes and element interactions in organic matter dynamics (Cole and Heil, 1981; Cole et al., 1986; Coleman et al., 1983; McGill and Cole, 1981; Stewart, 1984; Stewart and Cole, 1989; Stewart and Tiessen, 1987) and the development of conceptual and mathematical simulation models (Hunt et al., 1983; Hunt et al., 1986; Parton et al., 1983a; b) which we have used to focus attention on important processes and to verify hypotheses.

Parton and co-workers have developed and described the Century model which simulates soil organic dynamics in cultivated and uncultivated grassland soils and represents the dynamics of C, N, P and S in the soil-plant system using a monthly time step (Parton et al., 1983a; b; 1987; 1988; 1989). This model has been validated using data that show how C, N, P and S change under different cultivation practices in the northern Great Plains, data that show the effect of adding inorganic P on grain yield and P uptake, and data showing the impact of S mineralization in the soil (Parton et al., 1987; 1988). It further has been used (Parton et al., 1989) to simulate regional patterns for plant production, soil organic C and N, and soil organic and inorganic phosphorus for the U.S. central grassland region. This latter paper demonstrated how climatic variables, soil texture and inputs of nitrogen, control regional patterns of soil organic matter and plant production. We now wish to apply this model to the analysis of land use effects on the dynamics of soil organic matter in the North American Great Plains. Fortunately, a large data base exists (Haas *et al.*, 1957; 1961) for model testing which documents land use effects on organic C, N, and P changes in Great Plains soils at a number of stations representing wide gradients of soil properties and climates.

Documentation of soil carbon and nitrogen losses

The semiarid Great Plains of North America (comprising 250 million hectares including 105 million of cropped land) has a relatively short cropping history in a global sense as it remained as native grassland until being broken from sod in the early 1900's and used in the production of grain crops. Early homesteaders brought with them cultivation practices more suitable for humid regions which could not be extrapolated to harsher environments. Serious management mistakes were made that resulted in great financial and social hardship to such an extent that federal and state experimental stations were established through the regions in the early 1900's to study the effects of different rotations and tillage practices on crop production under dryland conditions and to develop better management practices. The results of this research were documented in USDA publications (Haas *et al.*, 1957; USDA, 1974) and summarized in journal publications (Grunes *et al.*, 1955; Haas *et al.*, 1957; 1961). Soil management practices such as summerfallowing, a cultural practice for controlling weeds and storing one season's moisture for the benefit of the next crop, turned out not only to be inefficient for moisture storage (only 19% on average of the incoming precipitation was stored in the fallow period) but also destructive of soil (Haas *et al.*, 1974). Cultivation over 30–43 years resulted in serious loss of organic matter and nitrogen (average N loss was 39%). The loss of organic matter has been associated with a deterioration of soil physical proper-

ties resulting in less water storage, more runoff, increase in surface crusting and greater susceptibility to water erosion (Haas *et al.*, 1974). This data base from long-term conventional plot research (Haas *et al.*, 1957) documented these changes over the period 1906 to 1947 and showed how they could be overcome or at least compensated for by improved varieties and soil management practices. The semiarid Great Plains is now more productive in terms of area yields of wheat, barley, corn and other adapted crops than it was when the sod was first broken. Our first objective, therefore, was to see if the Century model can be validated by comparison with this long-term conventional plot research data base.

The 17 USDA research stations (Fig. 1) represent a wide area within the Great Plains. The cropping systems were corn, barley, wheat, oats, grass and legumes in various rotations with and

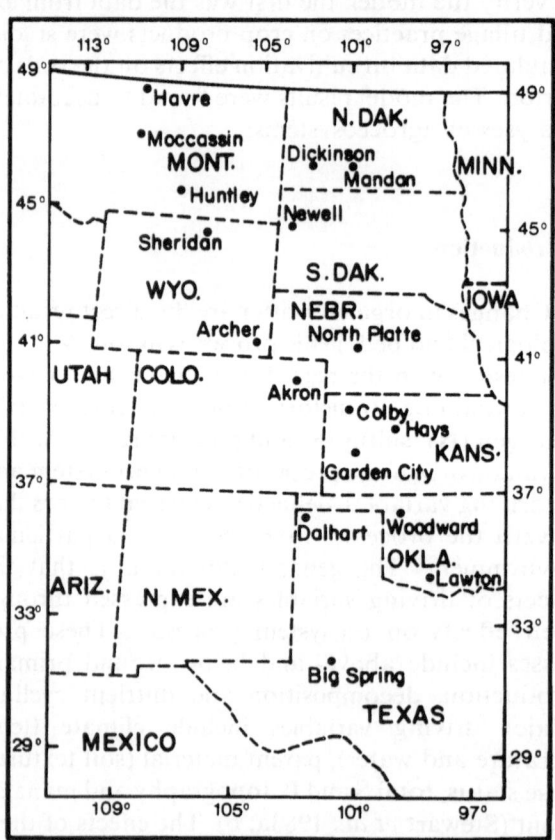

Fig. 1. Map of the Great Plains region showing the location of USDA research stations which documented the effects of crop management systems on losses of soil organic C and N, 1906–1947 (Haas *et al.*, 1957).

without fallow years during the period 1906 through to 1951 (Grunes *et al.*, 1955; Haas *et al.*, 1957; Haas *et al.*, 1961). Annual precipitation at these locations ranged from 290 mm at Havre, Montana to 729 mm at Lawton, Oklahoma with three-quarters of the precipitation falling in the period April through September. Mean annual temperatures ranged from 4.6°C at Dickinson, North Dakota to 17°C at Big Spring, Texas. Potential evaporation ranged from 830 mm at Huntley, Montana to 1306 mm at Garden City, Kansas. Elevations ranged from 339 meters to 1830 meters with an average for all stations of 928 meters.

Our secondary objective was to use the Century model to evaluate textural effects on soil properties and productivity. To achieve this objective we simulated textural effects on soil properties at one site in the Great Plains. We chose Mandan, North Dakota as Bauer, Black and co-workers at the ARS/USDA Northern Great Plains Center have data from experiments where the effect of management has been documented on different textured soils in the same climatic zone (Bauer and Black, 1981). The effects of cropping fine, medium and sandy textured soils to spring wheat under conventional tillage and stubble mulching for a period of 40 years has been compared to similar textured soils under natural grassland. Within each management treatment, soil textures varied from moderately coarse (designated sandy), to medium to moderately fine (designated fine) and sampling sites could be selected on the basis of standard soil classification techniques or uniformity within soil series. The same soil series was sampled within each site for each management. Slope and aspect differed amongst soils but slopes did not exceed 5%.

Results

Regional analysis of the effect of cultivation on soil organic matter in grain yield was conducted using 55 sites distributed in 7 east-west transects in the U.S. Great Plains. The sites ranged from 94.1 to 109.8° longitude and 31.0 to 48.6° north latitude. The climate of this region is quite variable with mean annual temperatures ranging from 3.4 to 18.9°C and the annual rainfall ranged from 24.1 to 128.8 cm. Soil organic matter is controlled by soil texture and to illustrate this over the region we used

the model to simulate organic carbon values in a sandy textured soil (*e.g.*, 75% sand content) and a fine textured soil (*e.g.*, 25% sand content) over the range of mean temperatures and rainfall found in the area (Fig. 2). Each site was established using 10,000 year simulations to establish the initial concentrations for each site (Parton *et al.*, 1989).

Initial soil C values in the simulated grasslands ranged from 2.3 to 7.8 kg/m^2 in the fine textured soil (also shown in Fig. 3) and 1.3 to 4.2 kg/m^2 in the sandy textured soil. Soil C is greatest in the northeastern portion of the region and declines towards the warmer and xeric region in the southwest (Fig. 2). Soil N ratios had a similar regional pattern as soil C and so were not depicted. C/N ratios of the virgin soil was 11.

Simulation of cultivation effects across the Great Plains (Figs. 2 and 3) showed that the greatest loss in soil carbon would be in the wetter and warmer portion of the region. This reflects the greater initial amounts of soil organic matter found in this area and the influence of climatic control on soil organic matter composition. The results shown in Figs. 2 and 3 are for a wheat-fallow rotation. We found that the pattern for soil organic matter under continuous wheat rotation was very similar although the overall losses were significantly less throughout (continuous wheat losses are 82 and 89% of the wheat-fallow for fine and sandy textured soils respectively). The fine textured soils lose greater amounts of soil C compared to the sandy textured soils (Fig. 2) because of the greater initial values of soil C in the fine textured soil (Schimel *et al.*, 1985). On a percentage basis, fine textured soils actually lose less soil organic matter (Fig. 2) than the sandy textured soils. In the north, percent loss in soil C on fine textured soils is 46% whereas on sandy soils the loss is 42–48%. In the south, losses range from 38–54% on the fine textured soils whereas on sandy soil the loss ranges from 44–54%. The pattern for percent of soil C was similar to the pattern shown by the climatic decomposition parameters shown in Parton *et al.* (1989).

When the data base was compared to the simulated model results (Table 1) it can be seen that on average the percent C and N loss due to cultivation were consistent with the observed percentage losses reported by Haas *et al.* (1957). Losses shown in Table 1 were computed on the basis of weight of an element in the top 20 cm of soil. This was necessary

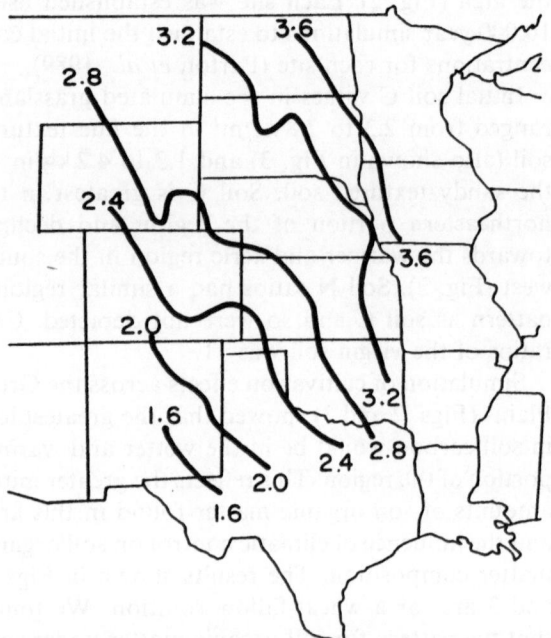

**INITIAL SOIL CARBON
SANDY
(kg m⁻²)**

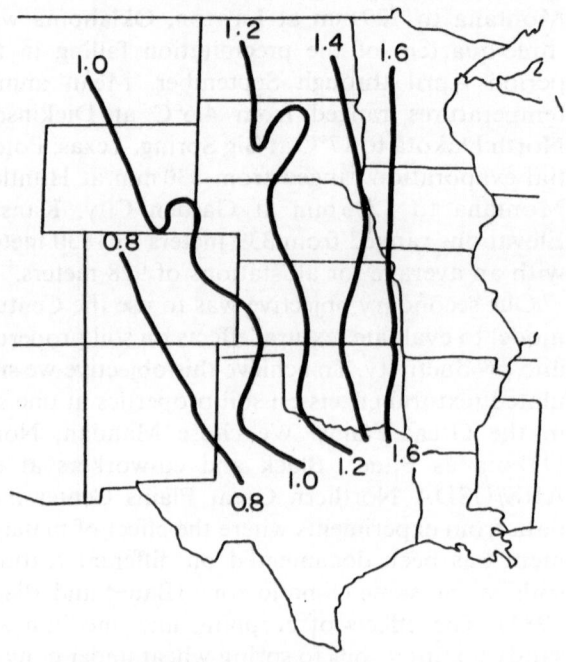

**SOIL CARBON LOSS
SANDY
(kg m⁻²)**

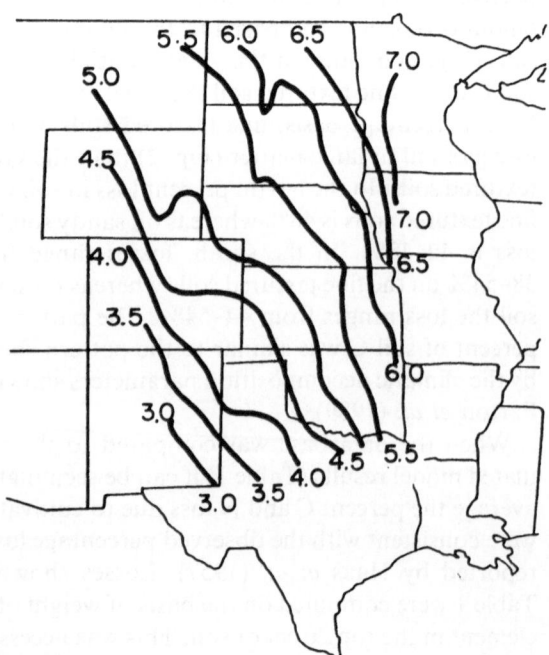

**INITIAL SOIL CARBON
FINE
(kg m⁻²)**

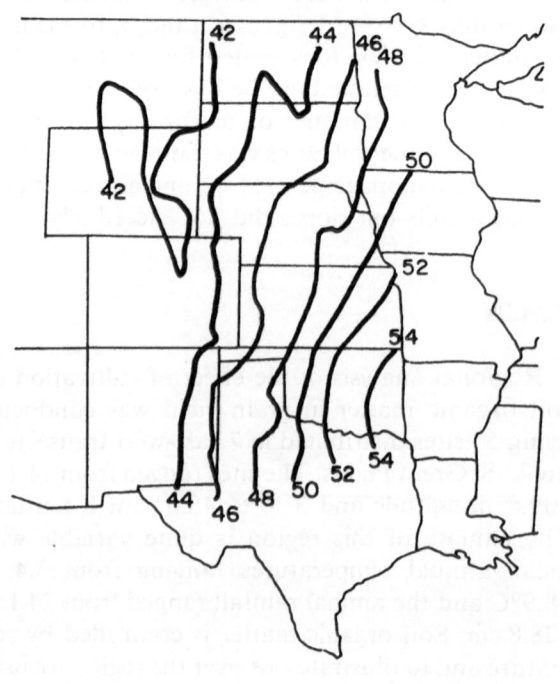

**PERCENT SOIL C LOSS
(sandy)**

Fig. 2.

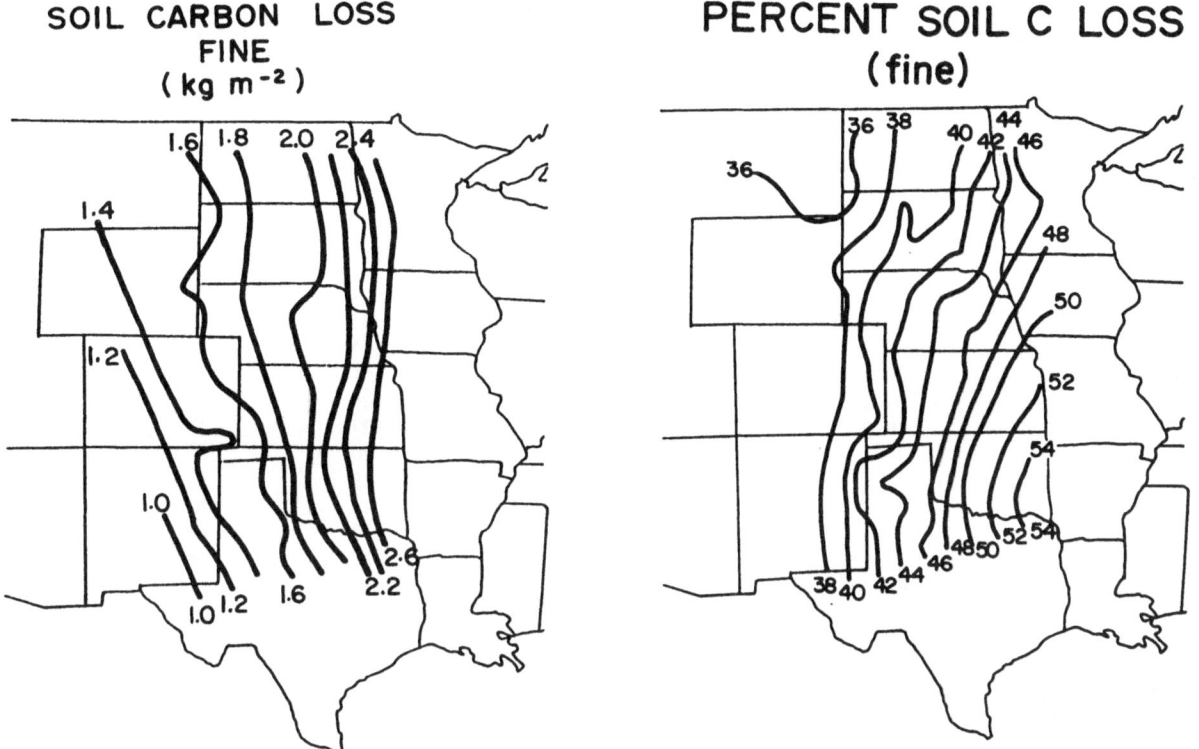

Fig. 2. Regional patterns of initial organic C levels in grasslands, losses after 40 years of cultivation in wheat and percent losses in fine textured and sandy textured soils.

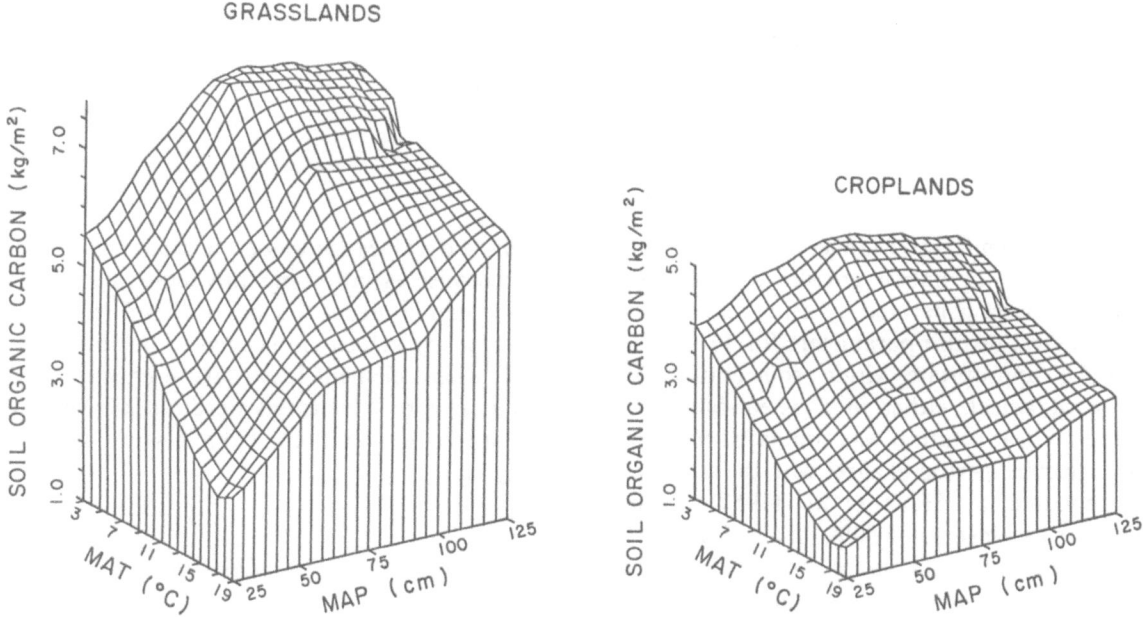

Fig. 3. Simulated levels of soil organic C as a function of mean annual temperature and precipitation in the Great Plains. a) Native grasslands on fine textured soils. b) Croplands on fine textured soils after 40 years of cultivation in wheat-fallow.

Table 1. The effects of cropping in wheat-fallow or continuous wheat on percentage loss of organic C and total N from grassland soils at eleven experimental stations in the U.S. Great Plains region; comparison of data with model results.* Values shown here were computed on the basis of weight of element (g/m^2) in the top 20 cm

	Years of cultivation		Organic C		Total N	
			Data	Model	Data	Model
			- - - - - - - - - - - - % loss - - - - - - - - - - - -			
Havre, MT	31	wf**	49	25	35	20
		wf**	42	17	30	13
Moccasin, MT	39	wf	NR	32	30	26
		ww	NR	24	21	19
Dickinson, ND	40	wf	52	31	46	25
		ww	40	25	40	19
Mandan, ND	30	wf	18	31	15	25
		ww	12	26	9	20
Sheridan, WY	30	wf	18	27	27	22
		ww	12	20	15	16
Archer, WY	34	wf	38	29	28	23
		ww	29	22	20	17
Akron, CO	39	wf	NR	32	29	32
		ww	NR	25	22	25
Colby, KS	41	wf	39	36	34	30
		ww	36	28	28	23
Hays, KS	43	wf	47	41	28	35
		ww	40	35	17	29
Garden City, KS	39	wf	28	39	28	32
		ww	22	34	19	27
Dalhart, TX	39	wf	37	37	35	30
		ww	31	32	31	26
x		wf	33	33	30	27
			(11.7)	(5.0)	(7.5)	(4.8)
sd		ww	29	26	21	21
			(11.7)	(5.7)	(6.7)	(5.1)
		All	31	29	26	24
			(11.5)	(6.2)	(8.5)	(5.7)

*   Straw removal assumed at 50%.
**   wf – wheat-fallow
     ww – wheat-wheat
NR no record

as there were significant changes in bulk density upon cultivation of native grassland soils. Simulated results were generated by using site specific climate and soil texture parameters for 11 sites to achieve the appropriate virgin soil conditions. Cropping of wheat, either in a wheat-fallow or continuous wheat rotation was simulated for each site for a specified period to agree with the cropping period reported by Haas *et al.* (1957). Straw removal was assumed to be 50% on all of the sites. The observed data and simulated results show greater losses of soil C and N from the wheat-fallow rotation (Table 2) than from the continuous wheat rotation. In general, site-specific errors are high because of the difficulties in simulating exact

virgin sod values (discussed in Bauer *et al.*, 1987) and because variation in straw removal at different sites would influence the loss of C and N. Also, soil bulk density values are needed to be able to accurately assess element loss (Parton *et al.*, 1987). In many cases bulk densities were not measured and we had to derive them from texture and other analyses.

Crop yield effects on simulated changes in soil quality

Small grain yields for wheat-fallow rotations reported by Haas *et al.* (1957) indicated that grain

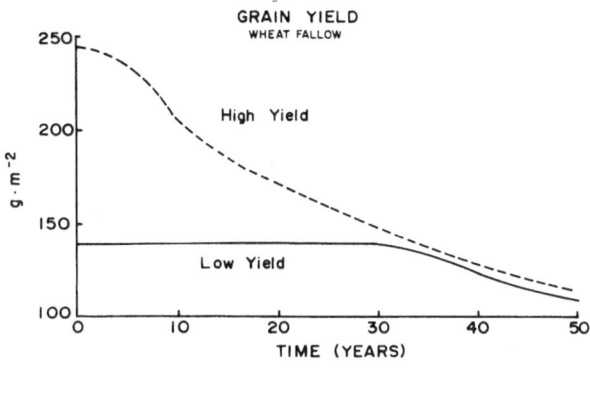

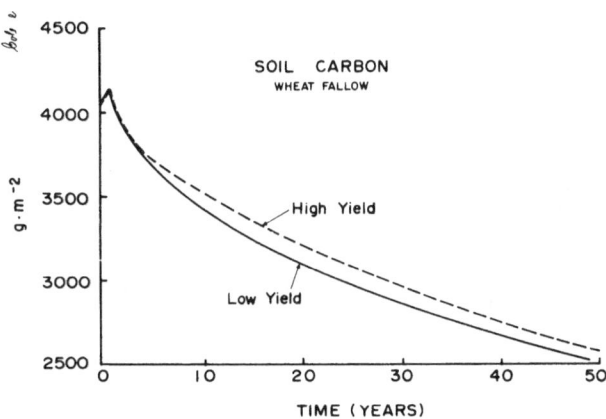

Fig. 4. Simulated comparison of effects of cultivation using low yielding wheat varieties available in the early 1900's with varieties of higher yield potential now in production. a) Changes in yields over the first 50 years after breaking sod without the use of fertilizer. b) Changes in soil organic C.

losses in cultivated soils using present day high yielding varieties, we simulated 50 years of a wheat-fallow rotation using a high yielding wheat variety and compared the simulation to the results of a wheat-fallow rotation using historically low yield variety data from Mandan, North Dakota, USA. Soil carbon was maintained at the higher level in the high yield variety throughout the 50-year simulation compared to the historical variety (Fig. 4b). The reason for this is that the C, N and P inputs back into the system by the high yield variety were much greater than from the historical variety. This greater input accounted for the greater production levels and greater soil C retention during the 50-year simulation.

Textural effects

Simulated effects of cropping fine, medium and sandy textured grassland soils to spring wheat under alternate fallow and continuous cropping at Mandan, North Dakota for a period of 40 years after the sod was broken are shown in Fig. 5. The simulated changes in organic C, N and P during this period under wheat, fallow and the declining grain yields under both alternate fallow and continuous cropping are in agreement with the historical changes which occurred in soils of that locality (Bauer and Black, 1981; Bauer *et al.*, 1987). Grain yields were consistently higher under alternate fallow, representing responses both to stored water carried over after fallow and the additional nitrogen mineralized. The simulations were run under conditions closely matching those recorded by Haas *et al.* (1957) with no fertilizer added and 80% of the straw removed using binder machinery. Grain yield depression occurred first on the sandy soils followed by the medium and fine textured soils.

Values for organic C, N and P in the original grasslands at the end of 40 years are presented in Table 2. Organic matter losses were greatest in the fine textured soils but percentage losses were always greater in the sandy soils. Losses were always greater under alternate wheat-fallow than under continuous cropping in agreement with historical records. Phosphorus changes were simulated assuming the same amounts of total organic and inorganic components for the three soils with

yields obtained from the varieties used in the earlier years of this period (1906–1951) were relatively low compared to present day high yielding varieties (150 g/m² compared to 250 g/m² for modern varieties). Differences in wheat production can potentially affect soil C loss reflecting differential carbon inputs. The simulated grain yield of the historical varieties of Mandan, North Dakota was approximately 140 g/m² and was maintained at this level for 30 years before a noticeable decline in yield was observed (Fig. 4a). The high yield variety initially produced a grain yield of 240 g/m² when the sod was broken and production immediately declined after the breaking of the sod. However, yield levels of the high yield varieties stayed above the historical yield varieties throughout the 50-year simulation.

In order to investigate the alteration of soil C

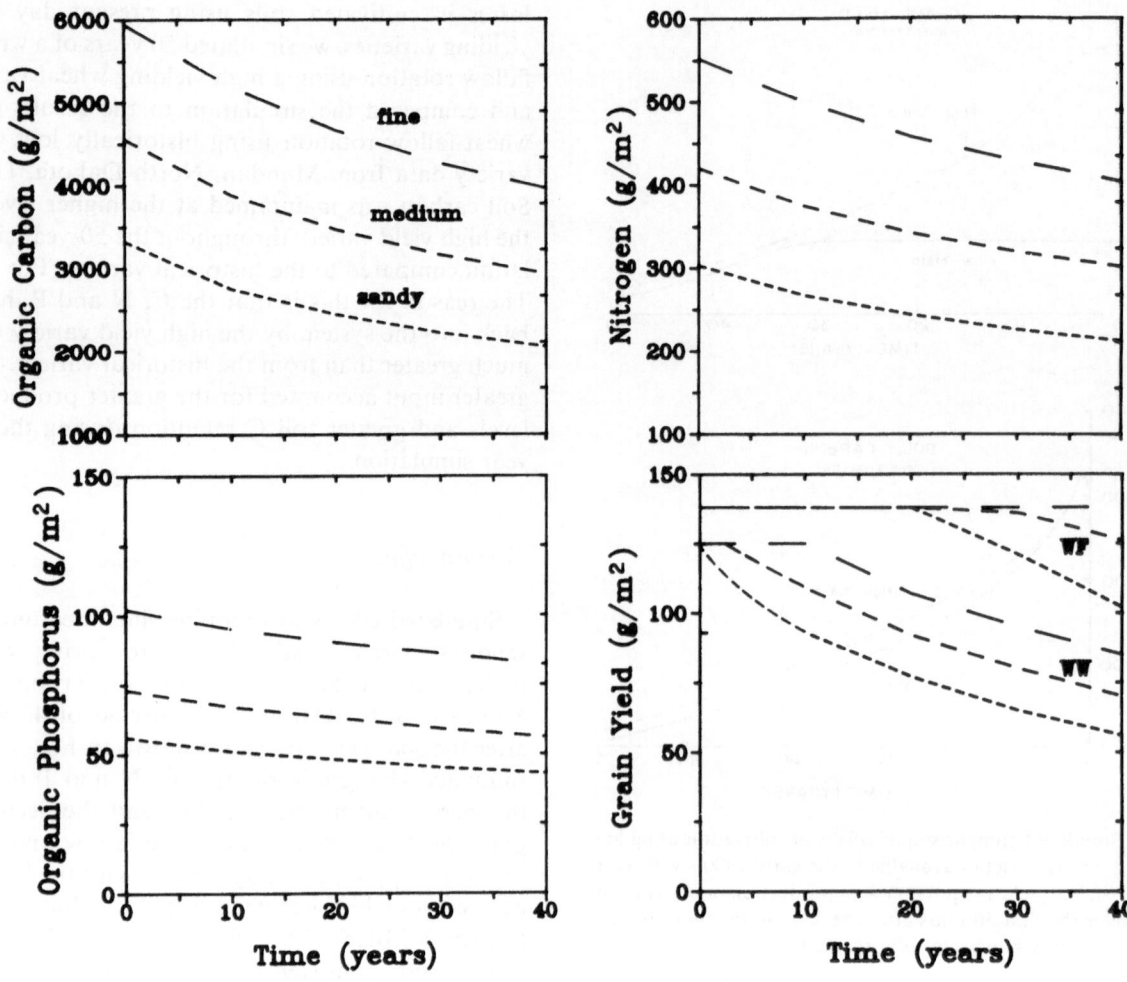

Fig. 5. Simulated effects of soil texture on changes in soil organic C, total N, organic P and wheat yields under wheat-fallow and continuous wheat at Mandan, N.D. during the first 40 years after breaking native sod.

greater transformation to organic forms in the finer textured soils (see Stewart and Cole, 1989; Parton *et al.*, 1989). Percentage losses of organic P were less than those of either C or N and only small amounts of inorganic P were lost during the 40 years simulation. Nitrogen and P budgets based on the simulated results under wheat-fallow are presented in Table 3 to identify the loss mechanisms and compare the behavior of the two elements. Total loss of N was computed from changes in soil N plus the estimated amount of N deposited in the rainfall during the 40-year cropping period. Nitrogen removal in grain and straw amounted to 53, 63 and 73% of the total loss in the fine, medium and sandy soils respectively. Significant amounts of

N were leached below the root zone in agreement with other data in the northern Great Plains (Campbell *et al.*, 1975). In contrast to the N budgets, the P budgets showed that crop removal closely matches declines in organic P in agreement with other studies (Grunes *et al.*, 1955; Haas *et al.*, 1961; Tiessen *et al.*, 1983).

The present effort as summarized in this paper and two others presented at this symposium (Parton *et al.*, 1989; Stewart and Cole, 1989) integrate information on driving variables, processes and properties from individual sites into a regional analysis. These analyses allow direct comparison of management effects on soil properties and crop production in comparable climatic regions across

Table 2. Simulated changes in organic C, total N, organic and inorganic P in topsoils of fine, medium and sandy textured soils at Mandan, ND after 40 years of cropping under wheat fallow (wf) and continuous wheat (ww) management*

			Fine	Medium	Sandy
Organic C					
Virgin sod 0–20 cm		C, g/m^2	5943	4557	3325
Cultivated	wf	C, g/m^2	3959	2930	2053
		% loss	33	36	38
Cultivated	ww	C, g/m^2	4385	3195	2208
		% loss	26	30	34
Total N					
Virgin sod 0–20 cm		N, g/m^2	552	420	303
Cultivated	wf	N, g/m^2	402	300	213
		% loss	27	29	30
Cultivated	ww	N, g/m^2	436	321	225
		% loss	21	24	26
Organic P					
Virgin sod 0–20 cm		P, g/m^2	102	73	56
Cultivated	wf	P, g/m^2	82	56	43
		% loss	20	23	24
Cultivated	ww	P, g/m^2	81	56	43
		% loss	20	23	23
Inorganic P					
Virgin sod 0–20cm		P, g/m^2	62	91	108
Cultivated	wf	P, g/m^2	62	91	103
		% loss	0	0	2
Cultivated	ww	P, g/m^2	60	87	103
		% loss	3	4	5

* Straw removal assumed 80% to reflect harvest with binders instead of combines during this period.

Table 3. Nitrogen and phosphorus budgets for a simulated 40-year cropping period under wheat-fallow management of fine, medium and sandy soils at Mandan, ND

	Fine	Medium	Sandy
Nitrogen			
Nitrogen loss N, g/m^2*	180	150	120
Crop removal			
Grain N, g/m^2	61	60	56
Straw N, g/m^2	34	34	31
Volatile loss	19	16	14
Nitrate leached g/m^2	66	40	19
Phosphorus			
Organic P loss	20	17	13
Inorganic P loss	0	0	3
Crop removal			
Grain P, g/m^2	12	11	10
Straw P, g/m^2	7	6	6

* N losses calculated from change in total N in 0–20 cm layer plus 30 g N/m^2 cumulative atmospheric deposition.

the world. As such this represents a new direction in the analyses of agroecosystems.

Conclusions

The results of this analysis demonstrate the capability of the Century model to simulate management effects on organic C, N, and P in grassland soils of the Great Plains. The original recorded levels under sod as well as historical changes over 40 years of production under wheat were well represented. The effects of alternate management systems were also demonstrated. Model simulations of the effects of a range of soil textures on the original amounts and forms of C, N and P in the soils and changes upon cultivation correspond well with observations of detailed field studies. These comparisons provide a reasonable level of confidence in the model and its application to predict future changes under alternate management systems.

References

Bauer A and Black A L 1981 Soil carbon, nitrogen and bulk density comparisons in two cropland tillage systems after 25 years and in virgin grassland. Soil Sci. Soc. Am. J. 45, 1166–1170.

Bauer A, Cole C V and Black A L 1987 Soil property comparisons in virgin grasslands between grazed and non-grazed management systems. Soil Sci. Soc. Am. J. 51, 176–182.

Campbell C A, Nicholaichuk W and Warder F G 1975 Effects of wheat-summerfallow rotation on subsoil nitrate. Can. J. Soil Sci. 55, 279–286.

Cole C V and Heil R D 1981 Phosphorus effects on terrestrial nitrogen cycling. *In* Terrestrial Nitrogen Cycle, Processes, Ecosystems and Management Impact. Eds. F E Clark and T Rosswall. pp 363–374. Ecological Bulletin 33, Stockholm.

Cole C V, Stewart J W B, Hunt H W and Parton W J 1986 Cycling of carbon, nitrogen, sulfur and phosphorus: Controls and interactions. Trans. XIII Congr. Int. Soc. Soil Sci. (Symposia and Joint Symposia), VI. pp 636–643.

Coleman D C, Reid C P P and Cole C V 1983 Biological strategies of nutrient cycling in soil systems. *In* Advances in Ecological Research, 13. Eds. A MacFayden and E O Ford. pp 1–56. Academic Press, New York.

Grunes D L, Haas H J and Shih S H 1955 Effect of long-term dryland cropping on available phosphorus of Cheyenne fine sandy loam. Soil Sci. 80, 127–138.

Haas H J, Evans C E and Miles E F 1957 Nitrogen and carbon changes in Great Plains soils as influenced by cropping and soil treatments. Technical Bulletin 1164, USDA, 111 p.

Haas H J, Grunes D L and Reichman G A 1961 Phosphorus changes in Great Plains soils as influenced by cropping and manure applications. Soil Sci. Soc. Am. Proc. 25, 214–218.

Haas H J, Willis W O and Bond J L 1974 Summerfallow in the western United States. *In* USDA/ARS Conservation Research Report 17, pp 149–160.

Hunt H W, Stewart J W B and Cole C V 1983 A conceptualized simulation model of C, N, S and P interactions. *In* The major Biogeochemical Cycles and Their Interactions, Chapter 10, SCOPE 21. Eds. B Bolin and R C Cook. pp 303–325. John Wiley & Sons, Sussex.

Hunt H W, Stewart J W B and Cole C V 1986 Concept of sulfur, carbon and nitrogen transformations in soil: Evaluation by simulation modelling. Biogeochem. 2, 163–177.

McGill W B and Cole C V 1981 Comparative aspects of cyling of organic C, N, S and P through soil organic matter. Geoderma 26, 267–286.

Parton W J, Anderson D W, Cole C V and Stewart J W B 1983a Simulation of soil organic matter formation and mineralization of semiarid agroecosystems. *In* Nutrient Cycling in Agricultural Ecosystems. Eds. R R Lowrance, R L Todd, L E Asmussen and R A Leonard. pp 533–550. University of Georgia, College of Agriculture Exp. Stn, Special Publication Number 23, Athens, Georgia.

Parton W J, Persson J and Anderson D W 1983b Simulation of organic matter changes in Swedish soils. *In* Analysis of Ecological Systems: State-of-the-Art in Ecological Modeling, Developments in Environmental Modeling. Eds. W K Lauenroth, G V Skogerboe and M Flug, Chapter 5. pp 511–516. Elsevier Scientific Publisher, Amsterdam.

Parton W J, Schimel D S, Cole C V and Ojima D S 1987 Analysis of factors controlling soil organic matter levels in the Great Plains grasslands. Soil Sci. Soc. Am. J. 51, 1173–1179.

Parton W J, Stewart J W B and Cole C V 1988 Dynamics of C, N, P and S in grassland soils: A model. Biogeochem. 5, 109–131.

Parton W J, Cole C V, Stewart J W B, Ojima D S and Schimel D S 1989 Simulating regional patterns of soil C, N and P dynamics in the U.S. central grasslands region. *In* Ecology of Arable Land — Perspectives and Challenges. Eds. M Clarholm and L Bergstrom. pp 99–108. Kluwer Academic Publishers, Dordrecht, The Netherlands.

Schimel D S, Coleman D C and Horton K A 1985 Soil organic matter dynamics in paired rangeland and cropland toposequences in North Dakota. Geoderma 36, 201–214.

Stewart J W B, Cole C V and Heil R D 1983a Agroecosystems of the Great Plains of North America. *In* Nutrient Cycling in Agricultural Ecosystems. Eds. R R Lowrance, R L Todd, L E Asmussen and R A Leonard. pp 97–120. University of Georgia, College of Agriculture Exp. Stn, Special Publication Number 23, Athens, Georgia.

Stewart J W B, Maynard D G and Cole C V 1983b Interaction of biogeochemical cycle in grassland ecosystems. *In* The Major Biogeochemical Cycles and Their Interactions, Chapter 8, SCOPE 21. Eds. B Bolin and R C Cook. pp 247–269. John Wiley & Sons, Sussex.

Stewart J W B 1984 Interrelation of carbon, nitrogen, sulfur and phosphorus cycles during decomposition processes in soil. *In* Current Perspectives in Microbial Ecology. Eds. C A Reddy and M J Klug. pp 442–446. Proceedings of the Third International Symposium, American Society of Microbiology, Washington, D.C.

Stewart J W B and Tiessen H 1987 Dynamics of soil organic phosphorus. Biogeochem. 4, 41–60.

Stewart J W B and Cole C V 1989 Influences of elemental interactions and pedogenic processes in organic matter dynamics. Plant and Soil 115, 199–209.

Tiessen H, Stewart J W B and Moir J O 1983 Changes in organic and inorganic phosphate in particle size fractions of two soils during 60 to 90 years cultivation. J. Soil Sci. 34, 815–823.

USDA 1974 Summerfallow in the western United States. Conservation Report 17, 160 p.

M. Clarholm and L. Bergström (Eds.), Ecology of arable land, 99–108.

Simulating regional patterns of soil C, N, and P dynamics in the U.S. central grasslands region

W. J. PARTON[1], C. V. COLE[1,2], J. W. B. STEWART[3], D. S. OJIMA[1] and D. S. SCHIMEL[1]
[1]*Natural Resource Ecology Laboratory, Colorado State University, Fort Collins, CO 80523, USA;*
[2]*USDA-Agricultural Research Service, Fort Collins, CO 80523, USA and* [3]*Saskatchewan Institute of Pedology, University of Saskatchewan, Saskatoon, Saskatchewan S7B 0W0 Canada*

Key words: carbon, grasslands, nitrogen, phosphorus, simulation modelling

Abstract

The Century soil organic matter model has been used to simulate regional patterns for plant production, soil organic C and N, and soil organic and inorganic P for the U.S. central grassland region. The results show how climatic variables, soil texture and inputs of N control regional patterns of soil organic matter (SOM) and plant production. Variations of soil texture within any one site generated variations in SOM C, N, and P levels. Effects of soil texture included increased stabilization of soil C and N, weathering of parent P, and formation of organic P as the silt plus clay content increased. Simulated soil formation and parent material P weathering for 10,000-year simulation runs suggested that variations in total soil P have little effect on soil C and N levels after 5000 years of soil formation. However, total P levels less than $100 \, g \, m^{-2}$ caused plant production and soil C and N levels to be reduced because of low P availability during the first 3000 years of soil formation.

Introduction

In recent years a variety of conceptual and simulation models have been developed to simulate soil organic matter dynamics (Anderson, 1979; Jenkinson and Rayner, 1977; Parton et al., 1983; Van Veen and Paul, 1981). The concepts included in these models are quite similar and are based on recent advances in our understanding of the factors that control soil organic matter (SOM) dynamics. The major concepts included in these models include separating SOM into different fractions based on physical and chemical soil fractionations, and identifying the effect of soil texture on microbial turnover and SOM stabilization. Much of the increased understanding has come from long-term incubation studies of ^{14}C-labeled plant material in different soil types (*e.g.*, Ladd *et al.*, 1981; Sorenson, 1981; Stott *et al.*, 1983), soil carbon-dating data (Martel and Paul, 1974), soil particle size fractionation data (Tiessen *et al.*, 1982; Tiessen and Stewart, 1983), and modeling studies at dif-ferent levels of resolution (Cole *et al.*, 1977; Hunt, 1977; Parton *et al.*, 1983; Van Veen *et al.*, 1984).

In this paper we will demonstrate the use of the Century model (Parton *et al.*, 1987) to simulate regional patterns of soil C, N, and P for the U.S. central grasslands region (CGR). The C:N version of the model has already been validated by comparing simulated soil C and N levels and plant production from the central grassland region with observed soil C and N and plant production at 24 sites within the region (Parton *et al.*, 1987). We will describe the P submodel (Parton *et al.*, 1988), including the effect of soil texture.

The improvements in the P submodel are based on a detailed analysis of soil P fractionation data presented by O'Halloran *et al.* (1987) that showed that soil texture has a large effect on the weathering rate of P from soil minerals and that the equilibrium between labile P and secondary P changed as a function of soil texture. The data showed that as the sand content increased, the parent P weathering rate decreased while the ratio of labile P to secondary P increased. The effects of soil texture on the appropriate flows were added to the model. The ability of the model to simulate soil P dynamics was validated by comparing observed and simulated organic P levels for 15 sites in the CGR.

Model description

The model simulates SOM dynamics in natural or agroecosystems and represents the dynamics of C, N, P, and S in the soil-plant system using a monthly time step. We assume that plant residues are decomposed by microbes and that the resulting microbial products are substrate for humus formation. These compounds are capable of various type of chemical and physical bonding with clay minerals and amorphous mineral colloids, and the extent to which they are stabilized in the soil appears to be related to the soil texture.

In the model (Fig. 1a) we divided the soil organic matter into three fractions: (1) an active soil fraction consisting of live microbes and microbial

products (1.5 yr turnover time); (2) a protected fraction that is more resistant to decomposition (25 yr turnover time); and (3) a fraction that is physically protected or chemically resistant and has a long turnover time (1000 yr).

The plant residue is divided into structural (3 yr turnover time) and metabolic (0.5 yr turnover time) pools as a function of the lignin to N ratio of the residue. Decomposition of each of the state variables is calculated by multiplying the decay rate specified for each state variable by the product of soil moisture and soil temperature decomposition terms (climatic decomposition parameter). The soil moisture term is calculated as a function of the ratio of monthly precipitation to monthly potential evapotranspiration, while the soil temperature term is a function of the average monthly soil temperature at the soil surface. The decay rate of the structural material is also a function of the lignin content of the structural pools. The active SOM decay rate changes as a function of the soil silt plus clay content (low values for high silt and clay soils). The respiration loss for each carbon flow is fixed

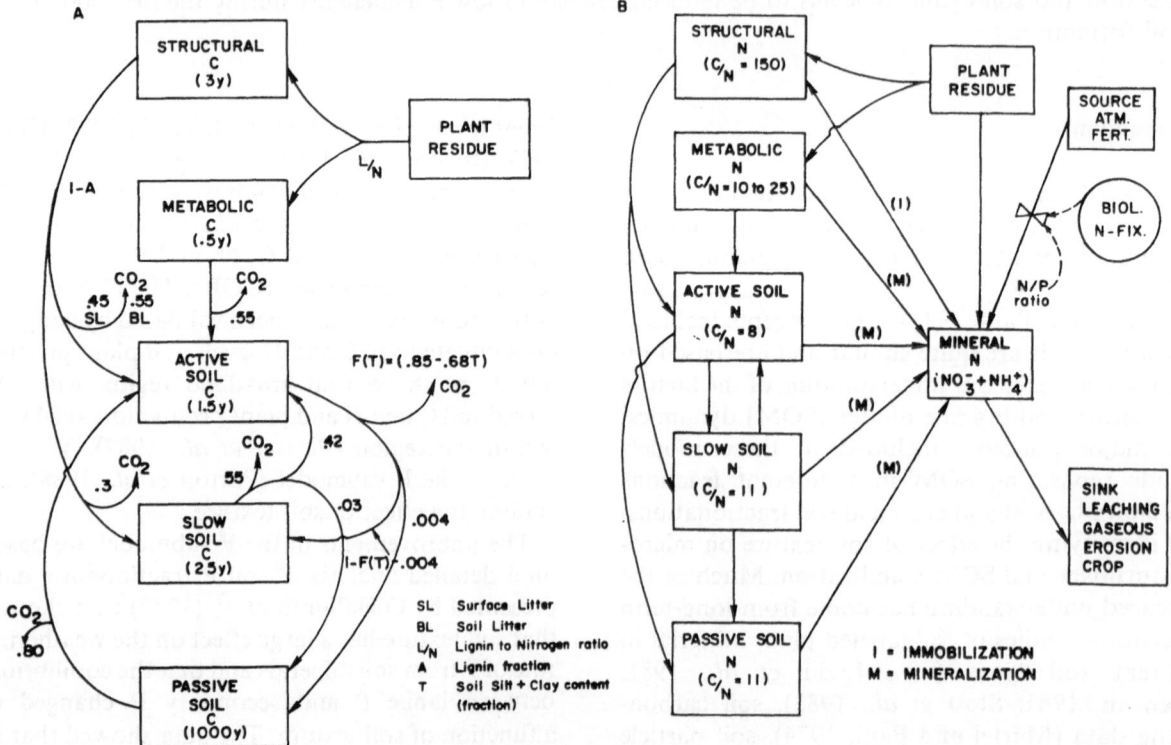

Fig. 1. Flow diagram for the carbon (**A**) and nitrogen (**B**) submodels.

for all of the flows except active soil organic matter, which varied with the soil silt plus clay content (decreasing with high silt plus clay content). The respiration CO_2 losses for each of the carbon flows are shown in Figure 1a and range from 30% for the lignin fraction of the structural residue to 55% for metabolic residue. The detailed description of the structure of the Century model and the C and N submodels is presented by Parton *et al.* (1987).

The model also includes a plant production submodel, which simulates the monthly dynamics of C, N, P, and S in the live and dead aboveground plant material, live roots, and resistant (structural) and labile (metabolic) surface and root detritus pools. Maximum potential plant growth is estimated as a function of the annual precipitation and is reduced if sufficient N, P, or S is not available. Plant production will be reduced by the element that is most limiting and will be reduced if available N, P or S pools are insufficient to produce plant material with C:N, C:P or C:S ratios less than or equal to the maximum values of these ratios (35, 230 and 230, respectively for winter wheat). We assume that 95%, 50%, and 20%, respectively of the available N, P and S pools can be used for plant growth during any given month. The reduction of plant production because of low availability of N, P, or S is an important mechanism of interaction between elements.

Nitrogen submodel

The N submodel (Fig. 1b) has the same basic structure as the carbon model, and we assume that N is bonded mostly to carbon. We assume that the C:N ratio of structural (150), active (8), slow (13), and passive (9) soil fractions remain fixed. The N content of the metabolic pool is allowed to float as a function of the N content of the incoming plant residue. The N flows are assumed to be stochiometrically related to the carbon flows and are equal to the product of the carbon flow rate and fixed C:N ratio of the state variables that receive the carbon. The N attached to carbon lost in respiration (30–80% of the carbon flow) is assumed to be mineralized. Given the C:N ratios of the state variables and the CO_2 losses associated with each flow, decomposition of the metabolic residue and active, slow and passive soil organic matter fractions

results in a net mineralization of nitrogen, while decomposition of structural residue results in immobilization of N. This model also uses simple equations to represent N inputs from atmospheric deposition and soil and plant N fixation. The losses of N from leaching, gaseous losses of N compounds, crop removal and erosion are also represented.

Phosphorus submodel

The P submodel (Fig. 2) includes the organic P flows parallel to those in the C and N submodel, along with flows for the inorganic P variables. The primary mineral source of P in most soils is apatite. Secondary (NaOH-P_i fraction) and occluded P minerals are formed from these phosphorites during the weathering and formation of soils. At the same time as physical and chemical weathering transforms primary P to secondary and occluded P forms, organisms in the soil and plant roots take up P from the solution. As more P becomes fixed through plant and soil microbial uptake, larger amounts become immobilized in organic matter (Cole and Heil, 1981). In the organic-phosphorus

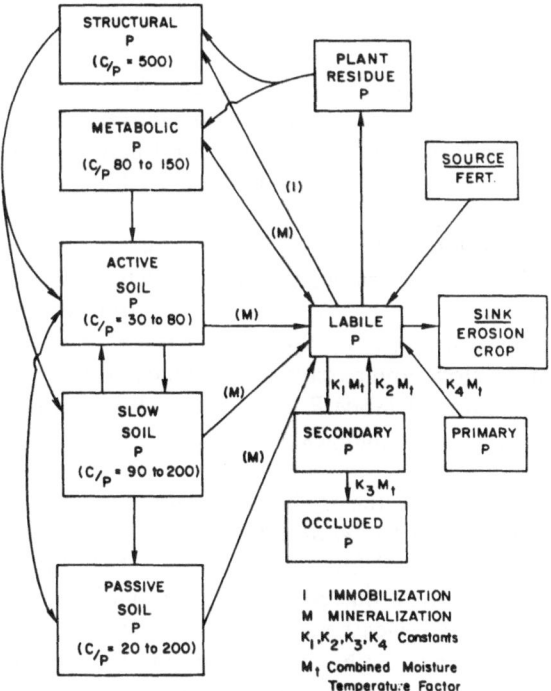

Fig. 2. Flow diagram for P submodel.

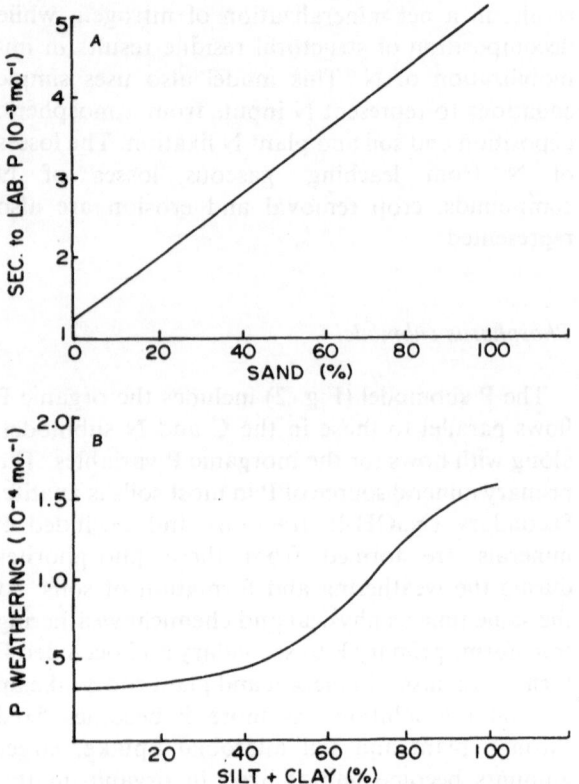

Fig. 3. The effect of soil silt plus clay content on P weathering rates (**B**) and the effect of sand content on the flow of P from secondary P to labile P (**A**).

submodel we followed the C submodel with respect to turnover times of the various active, slow, and passive fractions. The active soil P fraction consisting of microbial biomass and metabolites had C:P ratios ranging from 30 to 80. The most stable organic P forms were placed in the passive fraction, with C:P ratios ranging from 20 to 200, while intermediate forms in the slow soil fraction have C:P ratios ranging from 90 to 200. The C:P ratio of active, slow, and passive soil organic pools varies as a function of labile P level. Labile P is defined as isotopically exchangeable, or resin-extractable, orthophosphate (Sibbesen, 1984). A more complete description of the C, N and P submodels is presented by Parton *et al.* (1987, 1988).

Recent soil P fractionation data (O'Halloran *et al.*, 1987) show that soil texture has an effect on the amount of P weathered from parent P (HCl-P_i soil P fraction), with weathering rates decreasing rapidly as the sand content increases from 20 to 40%. Using their data, we developed a curve to simulate parent P weathering rates as a function of the silt plus clay content of the soil (Fig. 3b). P weathering rates increase slowly as the silt plus clay content increases from 0 to 50% and then increases rapidly as the silt plus clay content increases from 50 to 80%. Most of the data from O'Halloran *et al.* (1987) ranges from 60 to 80% silt plus clay content.

The O'Halloran data also showed that the ratio of labile P (resin and Bicarb P fractions) to secondary P (NaOH-P_i fraction) increased as the sand content increased. These results suggested that the equilibrium between labile and secondary P fractions is controlled by soil texture, with higher relative amounts of labile P in sandy soils. Using these data, we constructed a curve for including the effect of soil sand content on the flow of secondary P to labile P with higher flow rates in sandy soils (Fig. 3a). The actual parameter values used for the effect of soil texture on parent P weathering rates and the flow from secondary P to labile P were determined by fitting the model parameters to O'Halloran's data.

Model results

The Century model was used to simulate regional patterns of soil organic C, N, and P for the CGR. The patterns were simulated for a fine-textured soil (25% sand, 30% clay, and 45% silt) and a sandy soil (75% sand, 10% clay, and 15% silt) by running the model for 56 sites in the region and then using a contouring routine from the S package (Bell Labs, Murray Hill, N.J., USA) to generate the maps. The model was run for 10 000 years at each site for each soil texture, and the plotted variables represent the state of the system after 10 000 years of soil formation. At the beginning of the runs, we assumed that all of the P (100 g m^{-2}) in the soil was in parent P material. Nitrogen inputs into the system were assumed to be proportional to the annual precipitation and were constant throughout the run. The 10 000-yr period was selected because it represents a typical time since glaciation in the northern part of the region (McElroy, 1986). It is also important to note that the climate has been relatively stable during this time period (McElroy, 1986). Thus, we assumed that the present climate represents the average climate during that time period. The average

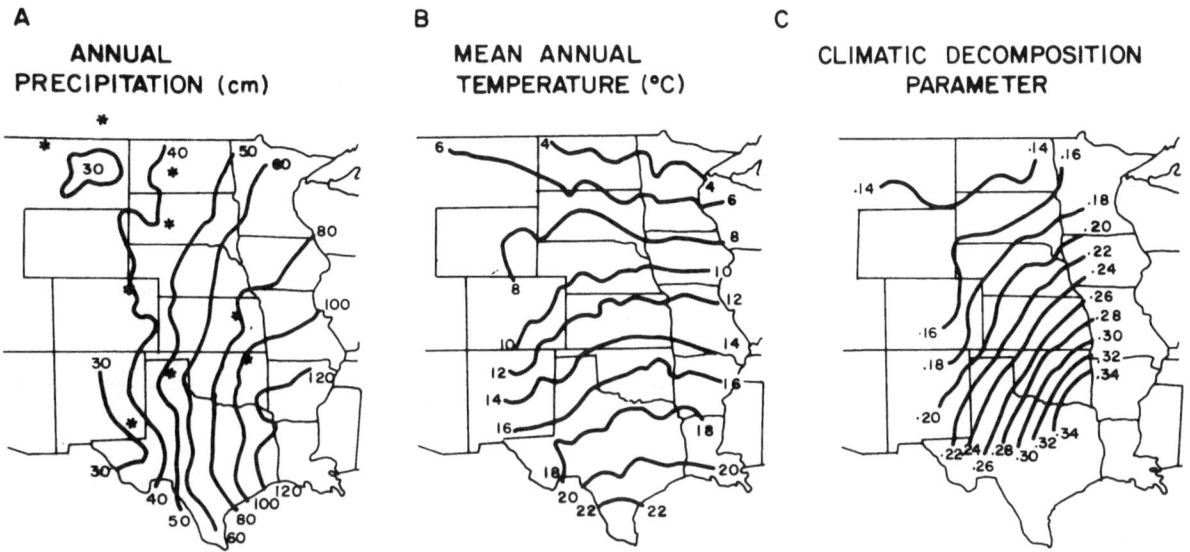

Fig. 4. Regional patterns for annual precipitation (**A**), mean annual air temperature (**B**), and climatic decomposition parameter (**C**) for the central grassland region. The asterisks in Fig. 4a indicate the location of sites used to validate the P model.

monthly maximum and minimum air temperature and average monthly precipitation from 1940 to 1960 were used as driving variables for the model. The climate (Fig. 4) is quite variable spatially in this region, with precipitation ranging from < 30 cm in the western part to > 120 cm in the southeastern part. The mean annual temperature ranges from 22°C in the south to less than 4°C in the northern part of the region. The average annual climatic decomposition parameter (Fig. 4c) was calculated in the model using the input monthly weather data (air temperature and precipitation data) and shows that the climatic decomposition parameter is highest in the southeastern part and decreases to the northwest.

The simulated patterns for soil C and N (0–20 cm depth) were similar for both C and N (Fig. 5), with maximum soil C and N levels in the northeastern part of the region (Minnesota) and minimum values in the southwest (western Texas and New Mexico). The soil C:N ratio for the region is approximately 11. The high soil C and N levels in the northeast are caused by low decomposition rates (low soil temperatures; Fig. 4) and relatively high plant production (Fig. 6). The low soil C and N levels in the southwest are a result of low plant production and relatively high decomposition rates (high soil temperatures). The fine-textured soils have approximately twice the C and N of the sandy soils. The results also show that the absolute difference between the soil C levels in the fine- and sandy-textured soils are directly proportional to SOM levels of the soil, with a $3.5\,\mathrm{kg\,m^{-2}}$ difference in the soil C in the northeast and a $1.1\,\mathrm{kg\,m^{-2}}$ difference in the southwest.

The plant production maps (Fig. 6a, b) show that above- and below-ground production increase from west to east in association with a general increase in annual precipitation along this gradient. The higher plant production in the eastern part is caused by less frequent drought stress and higher N inputs. The ratio of above-ground to below-ground production is lowest in the western part (0.7) and increases to 1.0 in the southeast. Above-ground and below-ground production are similar for the fine and sandy soils (data not shown), with differences of less than 5%.

The net N and P mineralization rates for the region (Fig. 6c, d) increase from west to east, with N mineralization increasing from $3.5\,\mathrm{g\,m^{-2}\,yr^{-1}}$ to $7\,\mathrm{g\,m^{-2}\,yr^{-1}}$, while P mineralization increases from 0.75 to $1.2\,\mathrm{g\,m^{-2}\,yr^{-1}}$. The increase in N mineralization follows the increased N inputs from the atmosphere and soil N fixation. P-mineralization rates follow a pattern that is more similar to the climatic decomposition parameter (Fig. 4c) and increases from the northwest to the southeast. This occurs because weathering of parent P was propor-

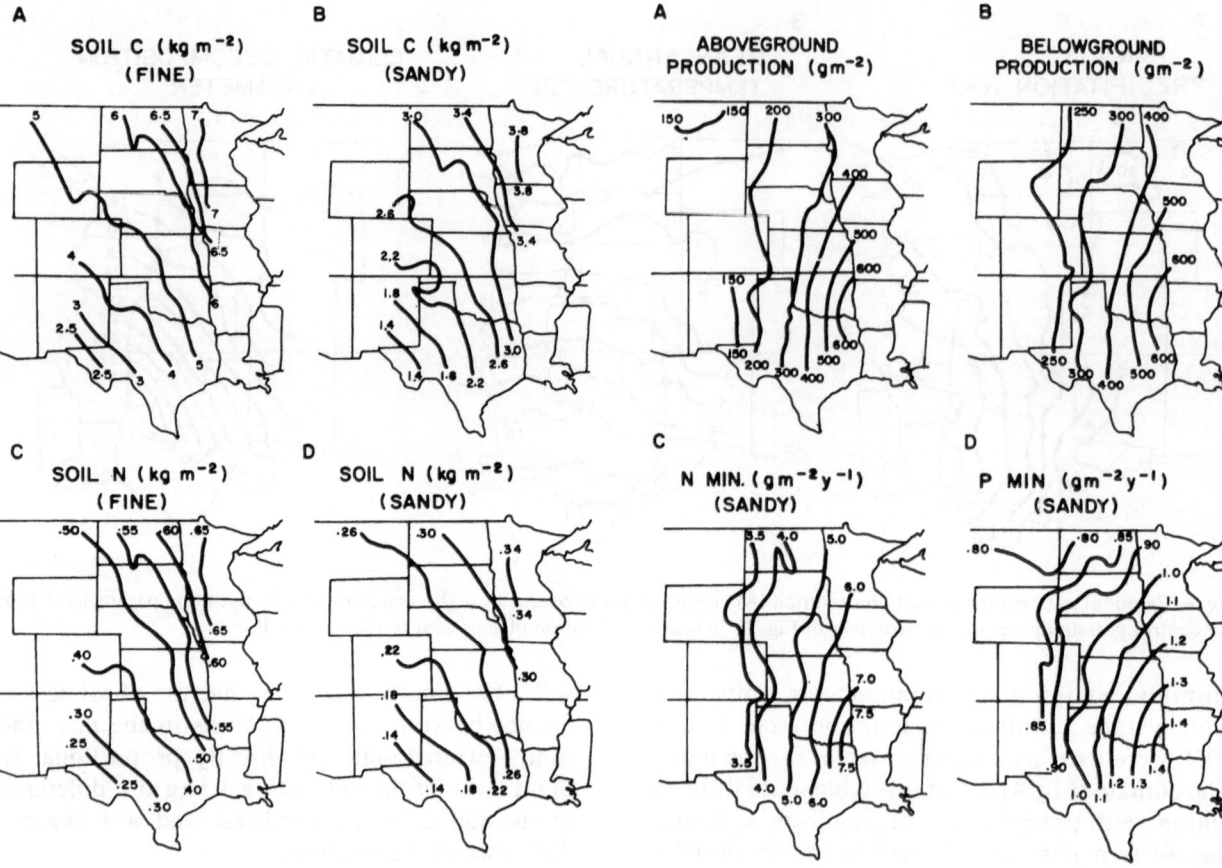

Fig. 5. Simulated soil C to 20-cm depth for (**A**) fine and (**B**) sandy soils, and soil N for a (**C**) fine and (**D**) sandy soil.

Fig. 6. Simulated above-ground (**A**) and below-ground (**B**) plant production, and annual N (**C**) and P (**D**) mineralization rates for the sandy soil.

tional to the climatic decomposition parameter, which increases along the gradient, and thus results in the formation of more organic P (Fig. 7). Soil texture has relatively little effect on N mineralization (data not shown) rates (< 5%), while P mineralization rates in the sandy soils (data not shown) are 15% higher than in the fine-textured soil. The higher P mineralization rates for the sandy soils are caused by higher flows of secondary P to labile P in sandy soils (Fig. 3b).

The simulated patterns of soil organic P (Fig. 7a, b) show that for the sandy soils organic P increases from the northwest to the southeast. The organic P in the fine-textured soil increases only slightly (58 to 68 $g\,m^2\,yr^{-1}$) from west to east. The increase in organic P along the northwest-southeast axis for the sandy soil is caused by a similar gradient in residual parent P, with remaining parent P decreasing from 56 $g\,m^{-2}$ in the northwest to 20 $g\,m^{-2}$ in the southeast (Fig. 7c). Recall that all runs were

initialized with 100 $g\,m^2$ parent P. The higher weathering of parent P in the southeast (lower parent P levels) is caused by the high soil weathering rates (assumed to be proportional to the climatic decomposition parameter; Fig. 4c) and results in the formation of larger amounts of organic P compounds. The higher organic P levels for fine-textured soil are a result of the higher weathering rates for fine-textured soils (Fig. 3). The parent P maps (Fig. 7) show that there is relatively little parent P left (2 to 16 $g\,m^{-2}$) for the fine soils. The increase in the organic P levels to the east is a result of the increases in P weathering and in SOM formation to the east (Fig. 5).

Figure 8 shows the 10 000-yr time series of soil development for a dry sandy soil in eastern Colorado (a) and a wet clay soil in eastern Kansas (b). The parent P weathering rate for the dry sandy soil is low because of the sandy texture of the soil and the low value for the climatic decomposition

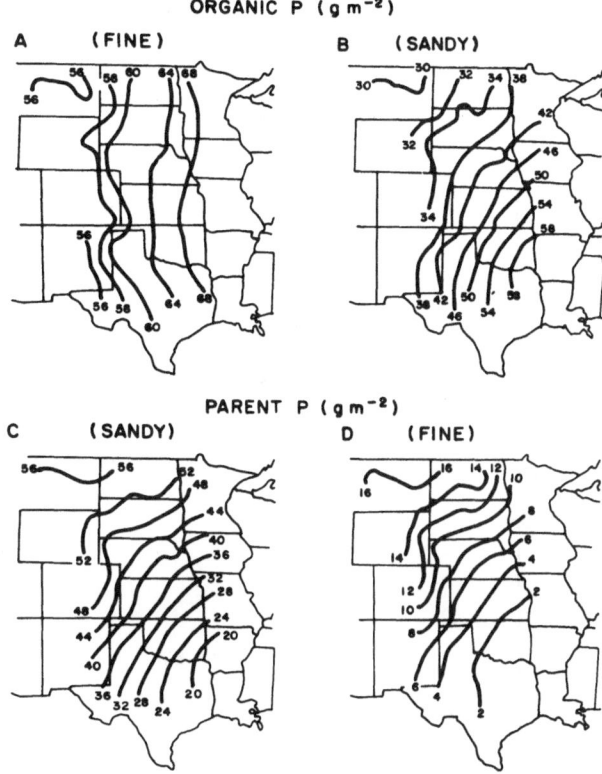

Fig. 7. Simulated patterns for organic P for (**A**) fine and (**B**) sandy soils and simulated pattern for parent P for (**C**) sandy and (**D**) fine soils.

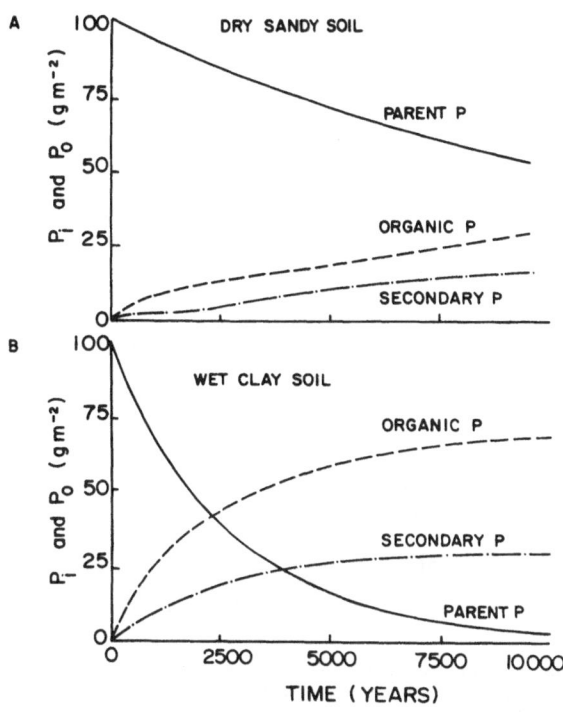

Fig. 8. Simulated soil P development for a dry sandy soil in eastern Colorado (**A**) and a wet clay soil in eastern Kansas (**B**).

parameter (0.16), while the wet clay soil has a high weathering rate because of the fine texture and high values for the climatic decomposition parameter (0.28). Forty percent of parent P has weathered after 10 000 years in the dry sandy soil, with 60% of the weathered parent P going to the organic P fraction and 40% to the secondary P. This contrasts with formation of soil C and N, which reaches near equilibrium soil C and N after 3000 years (data not shown). As expected, parent P in the wet clay soil weathers very rapidly with less than 2% of the original parent P left after 10 000 years. Most of the parent P forms organic P (68%) and secondary P (30%) and the remainder is in labile P (2%). These results suggest that as the parent P weathers, approximately two-thirds of the P released is stabilized into organic P and one-third as secondary P. Note that we assumed that P losses from these soils were minimal during this time period and that the formation of occluded P was small (< 2% of the total after 10 000 years).

We investigated the effect of varying the original parent P levels on plant production (Fig. 9b) and soil C (Fig. 9a) levels during the 10 000-yr model run for a wet clay soil in Kansas (see previous paragraph). The results show that total P levels greater than $100 \, g \, m^{-2}$ had little effect on plant production and soil C formation; however, as total P decreases from 100 to $50 \, g \, m^{-2}$ plant production and soil C formation are reduced during the first 3000 years of soil formation. The model results ($50 \, g \, m^{-2}$ total P run) show that P limits plant production during this time period, while after 5000 years sufficient P has weathered from the soil to support maximum plant production. After 5000 years, soil available N is the primary limiting factor for the system, with N inputs from the atmosphere and soil N fixation setting the limits for plant production. Note that in the model we assume that soil N fixation is a function only of annual precipitation and that soil P levels do not affect N fixation. Cole and Heil (1981) reviewed processes by which soil P levels influence soil N fixation levels. We were unable to successfully incorporate the long-term effect of soil P levels on N fixation in

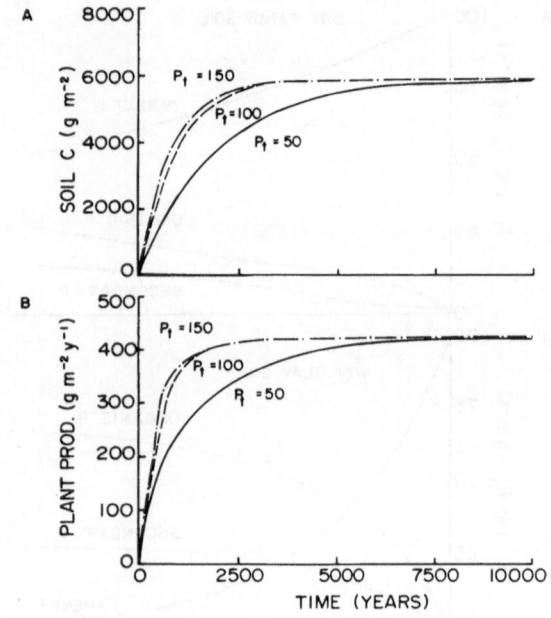

Fig. 9. Simulated soil C (**A**) and plant production (**B**) for soils with different original parent P levels (50, 100, and 150 g m^{-2}) for the wet clay soil in Kansas.

these soils. We are continuing to work on this problem and to investigate whether total P levels have any substantial effect on the steady-state soil C and N levels and plant production for these soils.

Model validation

We validated the model by comparing observed and simulated soil C and organic P for 15 sites in the central grassland region. The location of the sites are designated by an asterisk in Fig. 4a. Several of the sites have more than one data point to include the effect of soil texture. Most of the data came from a paper (Clark *et al.*, 1980) that compared the soil C and organic P levels of grassland sites in the U.S. and Canadian IBP grassland biomes. We also included more recent data collected from sites in eastern Kansas, eastern Colorado, and western North Dakota (Yonker; Schimel, 1985, unpublished data). We ran the model for 10 000 years for each of the sites, using site specific weather data, soil texture data, and observed total-P level assumed to be in parent P at the beginning of the model run. The results (Fig. 10) show that the model did a reasonable job of fitting the

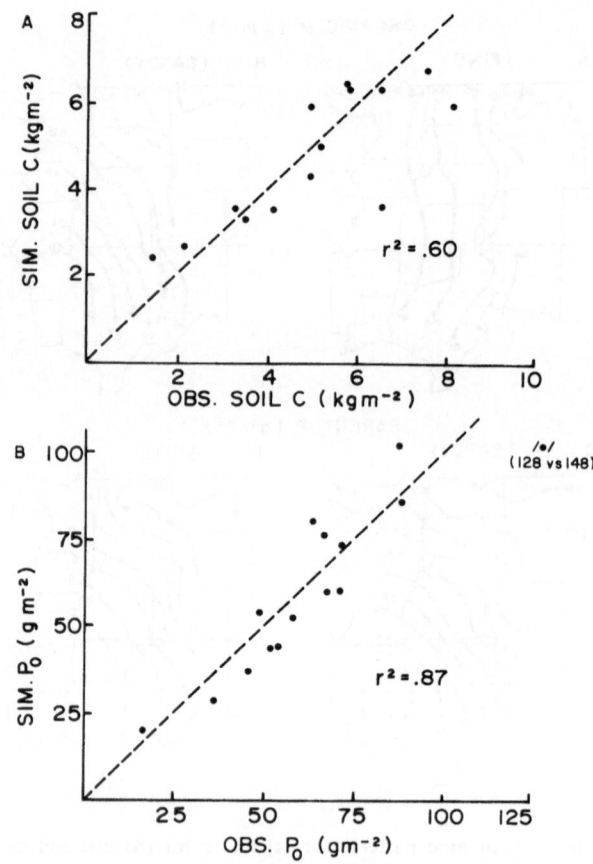

Fig. 10. Comparison of observed and simulated soil C (a) and organic P (b) levels for 15 soils in the central grassland region.

observed data, with observed vs simulated r^2 ranging from 0.60 for soil C levels to 0.87 for the organic P data. We also compared the observed vs simulated ratio of organic P to total P (data not shown), with similar results (r^2 = 0.69).

Conclusions

Our work shows that the Century SOM model can simulate regional patterns of plant production and soil C, N, and P for the U.S. central grassland region. The validity of simulated patterns for soil C and N and plant production have been demonstrated in a previous paper (Parton *et al.*, 1987b), while this paper documents the ability of the model to simulate P dynamics for the region. Soil texture has a large effect on SOM dynamics and the levels of organic C, N, and P in the soil. The weathering rate of parent P increases with increasing silt plus clay content, with the resulting effect that more organic P is formed in soils with high silt plus clay content. Soil organic P levels are also a function of the climatic decomposition parameter, so that higher organic P levels are associated with higher weathering rates. As parent P weathers, approximately 66% of the P is stabilized as organic P, 33% as secondary P, and a small fraction ($< 1\%$) as labile P. The model does a reasonable job of simulating soil formation and weathering of soil parent P for 10 000-yr periods. Total soil P levels less than $100\,g\,m^{-2}$, however, reduce plant production and soil C and N levels during the first 3000 years of soil formation. After 5000 years, total soil P has little effect on plant production and soil C and N levels, because sufficient P has weathered and N becomes the limiting factor for plant growth and soil organic matter formation.

The process of developing the Century model during the last three years has shown that relatively simple ecosystem models can be developed and validated from existing data. The simple models can represent most of ecosystem level dynamics considered in more complex ecosystem models with the advantage of requiring less site specific data to run and test the model. The results from this paper and others (Parton *et al.*, 1987) show that the Century model has been used successfully to simulate regional patterns for nutrient cycling, plant production, and SOM dynamics for the U.S. central grassland region using readily available input data (monthly climate data and soil texture). The results of our efforts and others (Pastor and Post, 1986) suggest that we should continue to develop simple ecosystem models and test the limits of these models for use in ecosystem studies and applied management problems.

Acknowledgements

This work was supported by National Science Foundation grants BSR-8605191 to the Great Plains Agroecosystem project and BSR-8114822 to the Central Plains Experimental Range LTER project. Collaboration with the Saskatchewan Institute of Pedology was supported by the Natural Sciences and Engineering Research Council of Canada.

References

Anderson D W 1979 Processes of humus formation and transformation in soil of the Canadian Great Plains. J. Soil Sci. 30, 77–84.

Clark F E, Cole C V and Bowman R A 1980 Nutrient Cycling. *In* Grassland, Systems Analysis and Man. Eds. A J Breymeyer and G M Van Dyne. pp. 659–712. IBP 19, Cambridge University Press, Great Britain.

Cole C V, Innis G S and Stewart J W B 1977 Simulation of phosphorus cycling in semiarid grasslands. Ecology 58, 1–15.

Cole C V and Heil R D 1981 Phosphorus effects on terrestrial nitrogen cycling. *In* Terrestrial Nitrogen Cycles. Eds. F E Clark and T Rosswall. Ecol. Bull. 33, 363–374.

Hunt H W 1977 A simulation model for decomposition in grasslands. Ecology 58, 469–484.

Jenkinson D S and Rayner J H 1977 The turnover of soil organic matter in some of the Rothamsted classical experiments. Soil Sci. 123, 298–305.

Ladd J H, Oades J M and Amato M 1981 Microbial biomass formed from ^{14}C, ^{15}N-labelled plant material decomposing in soils in the field. Soil Biol. Biochem. 13, 119–126.

Martel Y A and Paul E A 1974 The use of radiocarbon dating of organic matter from a cultivated topsoil in eastern Canada. Soil Sci. Soc. Am. Proc. 38, 501–506.

McElroy M B 1986 Change in the natural environment of the earth: The historical record. *In* Sustainable Development of the Biosphere. Ed. W C Clark. Cambridge Univ. Press, Cambridge.

O'Halloran I P, Stewart J W B and Kachanoski R G 1987 Influence of texture and management practices on the forms and distribution of soil phosphorus. Can. J. Soil Sci. 67, 147–163.

Ojima D S, Parton W J, Schimel D S and Owensby C E 1989 Simulating the long term impact of burning on C, N, and P cycling in a tallgrass prairie. VII Symposium on Environmental Biogeochemistry. (*In press*).

Parton W J, Anderson D W, Cole C V and Stewart J W B 1983 Simulation of soil organic matter formations and mineralization in semiarid agroecosystems. *In* Nutrient Cycling in Agricultural Ecosystems. Eds. R R Lawrence, R L Todd, L E Asmussen and R A Leonard. pp 533–550. Univ. Georgia, College of Agriculture Exp. Sta., Spec. Publ. No. 23, Athens, Ga.

Parton W J, Schimel D S, Cole C V and Ojima D S 1987 Analysis of factors controlling soil organic matter levels in Great Plains grasslands. Soil Sci. Soc. Am. J. 51, 1173–1179.

Parton W J, Stewart J W B and Cole C V 1988 Dynamics of C, N, P and S in grassland soils: A model. Biogeochemistry 5, 109–131.

Pastor J and Post W M. 1986 Influence of climate, soil moisture and succession on forest carbon and nitrogen cycles. Biogeochemistry 2, 3–27.

Schimel D, Stillwell M A and Woodmansee R G 1985 Biogeochemistry of C, N, and P in a soil catena of the shortgrass steppe. Ecology 66, 276–282.

Sibbesen E 1984 Determination of isotopically exchangeable P in soil (L-values) over several crop cuttings. J. Sci. Food Agric. 35, 731–732.

Sørensen L H 1981 Carbon-nitrogen relationships during the

humification of cellulose in soils containing different amounts of clay. Soil Biol. Biochem. 13, 313–321.

Stott E, Kasin G, Jarrell W M, Martin J P and Haider K 1983 Stabilization and incorporation into biomass of specific plant carbons during biodegradation in soil. Plant and Soil 70, 15–26.

Tiessen H, Stewart J W B and Bettany J R 1982 Cultivation effects on the amount and concentration of carbon, nitrogen and phosphorus in grassland soils. Agron. J. 74, 831–835.

Tiessen H and Stewart J W B 1983 Particle size fractions and their use in studies of soil organic matter. II. Cultivation effects on organic matter composition in size fractions. Soil Sci. Soc. Am. J. 47, 509–514.

Van Veen J A and Paul E A 1981 Organic carbon dynamics in grassland soils. I. Background information and computer simulation. Can. J. Soil Sci. 61, 185–201.

Van Veen J A, Ladd J H and Frissel M J 1984 Modeling C and N turnover through the microbial biomass in soil. Plant and Soil 76, 257–274.

M. Clarholm and L. Bergström (Eds.), Ecology of arable land, 109–122.

Management of earthworm populations in agro-ecosystems: A possible way to maintain soil quality?

P. LAVELLE,[1,4] I. BAROIS,[2] A. MARTIN,[1] Z. ZAIDI[1] and R. SCHAEFER[3]
[1] *Laboratoire d'Ecologie, Ecole Normale Supérieure, 46 Rue d'Ulm, F-752 30 Paris Cedex 05, France*
[2] *Instituto de Ecologia, Mexico*
[3] *Université Parix XI*
[4] *Corresponding author*

Key words: earthworms, drilosphere, soil fertility, soil quality

Abstract

Earthworm activities in natural ecosystems are very diverse and vary markedly according to soil and climate conditions. Within their drilosphere, *i.e.* the soil and microflora which they influence, they affect the physical properties of soils through their burrowing activities or by producing above and below-ground casts which are generally resistant macroaggregate structures. They affect the decomposition of organic matter in a number of ways, *e.g.*, by incorporating leaf litter into the soil and activating both mineralization and humification of the soil organic matter. Their overall effect appears to favour short and rapid, rather than long-term turnover of organic matter and nutrients. Finally, their effects on nutrient release and physical properties of soils are generally assumed to be in synchrony with plant demand and regulated so as to promote conservation of the soil structure and organic matter reserves.

In agroecosystems, earthworm communities are deeply —though diversely— affected by cultivation, pesticide and fertilizer application and by the cultural practices. Thus the drilosphere effects are depressed. Facilitation of earthworm activities and/or the introduction of better adapted species to conserve soil quality and fertility is discussed and the needs for further research considered.

Introduction

After the 30 euphoric years of the 'green revolution', during which the productivity of soils had been substantially increased in most temperate regions, modern agriculture is now facing serious environmental problems in a context of economic difficulties. Nitrates leach down to water tables and running waters, and heavy metals accumulate and contaminate the crops. The physical properties of soils are damaged by heavy machinery and decrease of soil organic matter content, leading to erosion, compaction and the formation of thick plough layers. High-input technologies increasingly appear unable to conserve the quality of soils in the long term.

In the humid tropics where high input technologies, although feasible (*e.g.* Sanchez *et al.*, 1982), are still underdeveloped for socio-economic reasons (Fearnside, 1987), traditional cultivation techniques are no longer sufficient for the rapidly increasing demand for food, and soil fertility may be rapidly lost. In such areas, there is an urgent need to develop new low-input techniques for sustained agriculture (Lavelle, 1987a; Swift, 1984).

Problems, though very diverse, result from ecosystem disfunctioning and should be treated as such. Soils are hierarchical systems structured as a cascade of elements with increasing dimensions, distributed along progressively larger time and space scales (*e.g.*, Allen and Starr, 1982; Di Castri, 1985; Lavelle, 1987b; Tisdall and Oades, 1982). An example of such a multilevel organization may be represented by an organization of the soil structure (microaggregates, aggregates, horizon, profile, catena, watershed) or the soil biocenosis (individual microbial population, functional groups of microflora, rhizosphere or 'drilosphere' system, pasture or forest . . .).

In hierarchical systems, each next upper level has emergent properties, i.e. properties of the whole which individual elements do not possess. For example, the existence of regulated cycles of nutrients may be considered as one of these emergent properties. In such systems, observations at one level of resolution only allow comparisons and correlations, while determinants are to be searched for at the next upper level of complexity and the explanation of mechanisms at the next level down. Further, slow and large entities of upper levels constrain the faster and smaller entities at lower levels. As a consequence, climate and substrate factors are the most important determinants of soil processes. At the next level down, biological systems of regulation, which are central to the functioning and conservation of natural systems, must be considered (Lavelle, 1984, 1987a). Their evaluation is necessary to indicate possible reasons for disfunctions; their manipulation would then appear as a possible solution to problems.

Earthworms and their 'drilosphere' (i.e., the soil and microflora submitted to their influence) comprise a major biological system of regulation in soils. They usually constitute the major invertebrate biomass (*ca.* 2/3 of total) in most soils of cultivated regions and they greatly affect the establishment and conservation of the soil physical structure, as well as the cycling of carbon and nutrients (see review in Lee, 1985 and Lavelle, 1988).

Agricultural practices generally affect the populations of earthworms in different ways. It is now necessary first to know precisely what effect such practices have upon populations and communities. Then, the question will arise whether it is worthwhile to have large earthworm populations and change some management practices in order to achieve this. Thus, it is necessary to identify their potentially beneficial effects and be aware of those which could be harmful.

This paper aims to present the available information necessary to answer such questions and identify research needs. The first section recalls the way earthworms adapt to soil constraints through a selected number of different adaptive strategies. The effect on soil properties of the different ecological categories thus defined, is detailed. The second

section is a rapid compilation of the information available on the effect of agricultural practices on earthworm communities. The possible management of earthworm populations is then considered. Examples of present and past experiments are discussed and the research needs to ensure further experiments a better scientific basis are presented.

The drilosphere concept

In order to take into account the different effects of earthworms it is necessary to consider not only the earthworms alone, but also the whole soil and microflora which is affected by their activities, that is to say the drilosphere. As a matter of fact, the effect earthworms have on ecosystems is much more than their direct contribution to carbon and nutrient fluxes by assimilation, production and respiration processes. In the moist savannas of Lamto (Ivory Coast), the overall C-mineralization due to earthworm respiration is equivalent to 5–6% of the C annually incorporated through primary production. However, as much as 1000 to 1250 Mg dry soil.ha^{-1}, may transit through their guts each year. This soil contains *ca* 1/3 of the organic matter (down to − 40 cm depth) of a grass savanna, and as much as 60% of the organic matter of the upper ten centimeters (Lavelle, 1978). Only 9% of this organic matter is assimilated by the worm, but it has been demonstrated that the conditions for further decomposition of this organic matter are modified because microbial activity is changed in the casts (Lavelle *et al.*, 1983; Zaidi, 1985). As a result of the ingestion of such high amounts of soil, the upper 20 cm of the profile in such ecosystems may be considered an accumulation of casts of different ages deposited by the different species.

The drilosphere as a biological system of regulation may be compared to the rhizosphere in which comparable efects are observed *i.e.*, the production of exudates, energetic equivalents of intestinal and cutaneous earthworm mucus, the selective activation of soil microflora and also modifications of the soil structure.

Earthworm adaptive strategies and community structure

Earthworms are currently facing three major constraints in soil: feeding on relatively low quality

Table 1. Ecological categories of earthworms and associated adaptations

	Epigeics	Anecics	Polyhumics	Mesohumics	Oligohumics
Pigmentation	homochr. complete	homochr. ant. dors.	< - - - - - - - - - - none - - - - - - - - - - >		
Feeding regime	litter	litter + soil	rich soil	bulk A₁	deep soil
Size	small	large	small	medium	large
Digestion (hypothetic)	direct- - >	< - - - - ext. rumen - - - - >		< - - - - - mutualistic - - - - - >	
Demographic profile	r	K	r	r–K	K
Resistance to drought	cocoons	diapause	< - - - - - - - - - - quiescence - - - - - - - - - >		
Colonization capacity	high	medium	low	low	extr. low

resources (leaf litter, dead or live roots and soil organic matter); moving in a compact environment; and, adapting occasionally to unfavourable microclimatic conditions. The way they have adapted to these constraints has led to the differentiation of three separate ecological categories of earthworms, whose way of life and effects on soil are different (Bouché, 1977; Lavelle, 1981) (Table 1).

Epigeic earthworms feed on and live in the leaf litter and are red or green-pigmented. Unable to dig the soil, they balance a high mortality due to predation and microclimatic hazards by having a high metabolic rate which they achieve by being small and feeding on leaf litter, qualitatively the best resource. Their digestion seems to be direct, *i.e.*, they extract, with the help of their digestive enzymes, assimilable compounds from the litter.

These earthworms have no effect on the soil structure as they cannot dig; their role, very similar to that of soil macroarthropods, is that of efficient agents of comminution and fragmentation of leaf litter which they finally transform into stabilized organic matter.

Anecic earthworms feed on leaf litter which they mix with soil of the upper horizons, but shelter in semi-permanent vertical burrows dug into the soil. They are typically pigmented on the anterior half of the body and large-sized. They have K-type demographic profiles with relatively slow growth and low fecundity balanced by a high life expectancy. Their digestion would be direct or indirect when they reingest their feces after a microbial incubation described as the 'external rumen' type digestion (Swift *et al.*, 1979).

Anecic earthworms have two main effects on the soil. They build vertical burrow systems which have

demonstrated effects on water infiltration rate (see Assad, 1987 and review by Lee, 1985). They also take surface litter and expose it to the soil decomposition system. This 'anecic effect', comparable to the effect of most termites in the tropics, may be very important in regard to the cycling of carbon and nutrients. Anecic earthworms may also produce surface casts.

Endogeic earthworms are unpigmented, they live in the soil and feed on soil organic matter plus dead or live roots. They may daily ingest large amounts of soil, from 1–2 to 20–30 times their body weight in terms of dry soil (Bolton and Phillipson, 1976; Lavelle, 1978; Lavelle *et al.*, 1987). They cast most of it in the soil, giving it a grumous resistant structure.

They digest the soil organic matter through a mutualistic association with the soil microflora they ingest. In the anterior part of the gut, microorganisms find suitable conditions for their activity, especially high contents of water (100 to 150% of the dry weight of ingested soil) and assimilable organic matter in the form of intestinal mucus (6 to 16% according to the species, Martin *et al.*, 1987) and pH close to neutrality. In the middle part of the gut, microorganisms which had first developed at the expense of the mucus produced by the worm, become able to digest the complex organic matter of the soil while their activity is multiplied by up to 6 to 10 times (Fig. 1). The product of their external digestion is reabsorbed in the posterior part of the gut. Such a digestion system may be very efficient and as much as 9 to 19% of the soil organic matter may be assimilated during the transit (Barois, 1987; Lavelle, 1978; Martin *et al.*, 1987).

Endogeics can be divided into three different

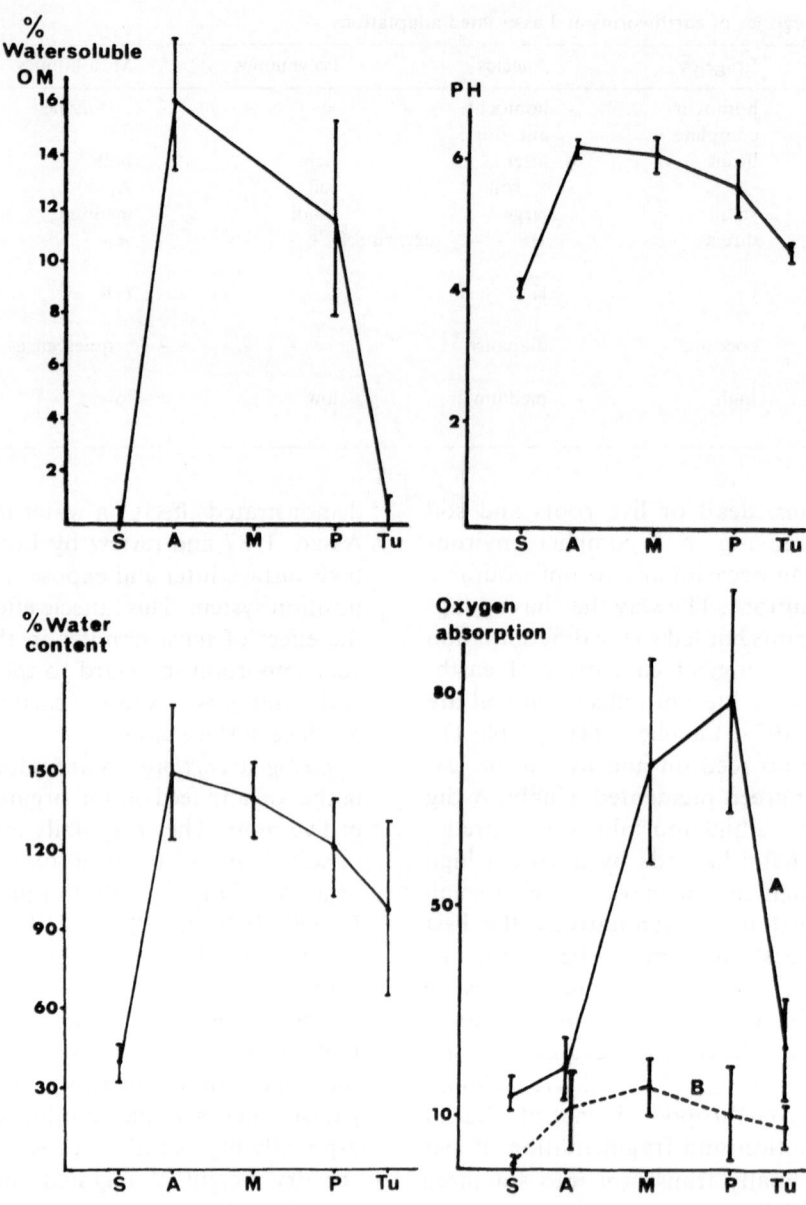

Fig. 1. Changes in some physico-chemical characteristics and oxygen absorption (in nl.mg^{-1} dry wt h^{-1}) in the three successive portions (Anterior, Medium and Posterior) of the gut of *Pontoscolex corethrurus*, the control soil (S) and the casts (Tu). A: overall oxygen absorption; B: oxygen absorption after inhibition by NaN^3. (Barois and Lavelle, 1986).

groups according to the relative organic richness of the soil they feed on. *Polyhumics* ingest a soil with a higher organic content than the bulk A$_1$ horizon. They are top-soil feeders or very small species which only ingest the fine, organic rich, soil particles or live in the rhizosphere. With their limited size, they generally have an r-type demographic profile and balance a high mortality by a rapid growth and a high fecundity. *Mesohumics* ingest the A$_1$ soil as it is, without making any selection. They are medium-sized earthworms, averaging 10 to 20 cm long, with an intermediate r-K demographic profile. Finally, *oligohumic* worms are specialized species of Mediterranean and humid tropical areas which feed on soil from the deep horizons. They are typical K-strategists and balance a low growth and fecundity due to the low energetic value of the soil they ingest by a low mortality.

The role earthworms play in soils depends on the way they have adapted to the soil constraints, i.e. the ecological category they belong to. The overall role of communities may be thus assessed by looking at their abundance and the relative importance of ecological categories (= their functional structure).

The analysis of abundance and functional structure of 53 communities sampled in natural (or semi-natural) ecosystems showed that temperature has a great effect on their composition (Fig. 2) (Lavelle, 1983). With increasing temperatures, earthworms are able to use resources of increasingly lower quality, as epigeics are progressively replaced by anecics and then by mesohumic and oligohumic endogeics. This is interpreted as the result of a better efficiency of the mutualistic digestion system under high temperature conditions, which facilitates the spreading of meso- and oligohumics. At the same time, the decrease of litter availability due to fast decomposition and possible competition with termites would deplete epigeic and anecic populations. The overall earthworm density is increasing towards the tropics whereas their biomass tends to culminate in mild temperate areas and decrease in the tropics.

At a regional scale, the effects of vegetation type is important; forested areas generally have a lower abundance of earthworms and a greater proportion of litter-feeding (epigeics + anecics) individuals. Soil texture, soil depth and nutrient richness are other variables which may affect earthworm communities (see Bouché, 1972; Phillipson *et al.*, 1976; Lavelle, 1978; Fragoso and Lavelle, 1987 and reviews by Edwards and Lofty, 1972 and Lee, 1985).

Drilospheric effects

The way earthworms have adapted to soil constraints through their different ecological categories may greatly affect the physical structure and chemical processes of the soil. Other characteristics of earthworm communities which may also be relevant to soil functioning are their low ability to colonize and the possible synchrony of their activities with plant growth.

Casting and burrowing as mechanisms for elaboration and conservation of the soil structure

Soil physical characteristics are influenced by casting in the soil or on the surface and burrowing activities of earthworms.

Casting. Surface casts are produced by populations

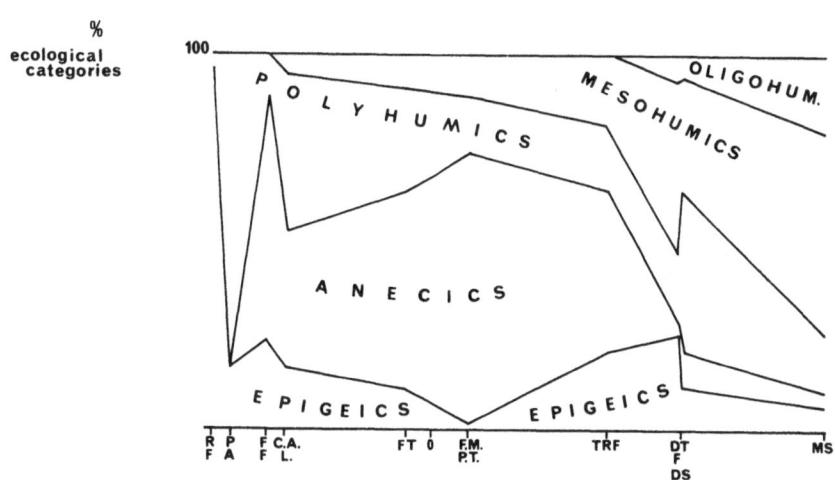

Fig. 2. Changes in the composition of earthworm communities in relation to the thermolatitudinal sequence of humid ecosystems (RF: resinous forest; PA: alpine grasslands; FF: cold deciduous forest; CAL: cold open formations; FT: temperate deciduous forest; FM: mediterranean forest; PT: temperate grassland; TRF: tropical rain-forest; DTF: 'dry' tropical forest; DS: 'dry' savannas; MS: moist savannas) (after Lavelle, 1983).

of the different ecological categories, except the oligohumic endogeics. Their chemical, physical and biological characteristics reflect the feeding regime of the populations which produce them. They have a generally higher content of assimilable nitrogen and phosphorus (Barois *et al.*, 1987; Graff, 1971; Lee, 1983; Lunt and Jacobson, 1944; Parle, 1963; Sharpley and Syers, 1977) then the ingested soil. Greater phosphatase (Satchell and Martin, 1984) and urease (Syers and Springett, 1983) activities have been measured in casts of Lumbricidae. Casts of litter-feeding species are often rich in available nutrients coming from the decomposing litter; ammonium concentrations of up to 222 $\mu g.g^{-1}$ have been recorded in fresh casts of *L. terrestris* in arable plots in Denmark (Andersen, 1983) and still higher concentrations have been measured on a number of occasions (see *e.g.* Parle, 1963, Syers *et al.*, 1979; Barois *et al.*, 1987). Inorganic P concentrations of 32 $\mu g.g^{-1}$ (*i.e.* ca. 4 times the upper 5 cm soil value) have been measured in casts of A. *caliginosa* by Sharpley and Syers (1977). The average production of assimilable nitrogen released in the soil by earthworm populations has been estimated at 18–50 kg.ha^{-1}.yr^{-1}, with maximum values of up to 100 kg.ha^{-1}.yr^{-1}.

On average, casts of litter-feeding species are expected to have a high content of available nutrients coming from the leaf litter. Casts produced by endogeics, on the other hand, are obviously less rich in available nutrients; in some cases, however, their selection of light organic particles or fine organic-rich fractions may lead to the production of casts with high nutrient and organic matter levels.

Microbial activity in the casts may be enhanced or decreased when compared to the non-ingested soil as a result of the different types of interactions between earthworms and microorganisms and the quality of the ingested organic matter (Lavelle *et al.*, 1983 and reviews by Edwards and Lofty, 1972 and Satchell, 1983). In the humid savannas of Lamto (Ivory Coast) microbial activity in the casts of the mesohumic endogeic *Millsonia anomala* was activated when the ingested soil had a low (< 0.1%) concentration in water soluble organic matter. With higher concentrations, microbial activity in the casts was inferior to that in the non-ingested soil. This result led to the hypothesis of a regulation of microbial activity by earthworms depending on the content of soil water soluble organic matter, this latter characteristic being considered as an indicator of the equilibrium between mineralization and humification processes.

Casting activities affect the soil physical properties in three different ways; they create voids in the soil, stabilize the soil structure, and may form a resistant layer protecting the soil against erosion.

When casts are deposited on the surface, an equivalent amount of voids are created in the soil. Though only a low proportion of the total ingested soil (*ca.* 5% in the moist savannas of Lamto, Ivory Coast) is casted on the surface, they may represent up to 20–100 mg dry soil.ha^{-1} in many situations, which is equivalent to 15–75 m^3.ha^{-1}. Casting activity is clearly related to soil compaction as indicated by Bouché (1972) and Lavelle (1988), and therefore may be considered a mechanism of regulation of soil macroporosity.

Casts are often the individual aggregates of solid macroaggregate structures (Fig. 3). When earthworm activity is intense, the soil may be regarded as an accumulation of casts from different species deposited at different times (Lavelle, 1978) and each gut transit of soil particles leads to a rejuvenation of the soil structure (Barois, 1987).

Finally, surface casts, when abundant and resistant, may protect the soil against sheet erosion. This is, however, only true for large spheroid grumous structured casts, and not at all of the small fine granular ones, which, contrary to the former, are easily washed away by running waters and may constitute an additional factor of soil creeping and erosion (see *e.g.*, Nye, 1955). More generally speaking, the selective ingestion of fine particles by earthworms has long been recognized as a mechanism of formation of fine textured surface horizons and progressive burial of the non-ingested large particles and stones (Darwin, 1881).

Burrowing. Burrowing is part of the activity of anecic and endogeic worms. Endogeics dig sub-horizontal burrows which are gradually filled with the digested soil. More or less interconnected voids are left in the soil systems. Anecics build semi-permanent vertical burrow systems. They may represent as much as 890 m and 9 L.m^{-2} (Kretzschmar, 1982 and review by Lee, 1985). Part of these burrows however, may be obstructed. Assad (1987) estimated the respective length and volume of the burrow systems at 6.2 to 67 m.m^{-2} and 0.13

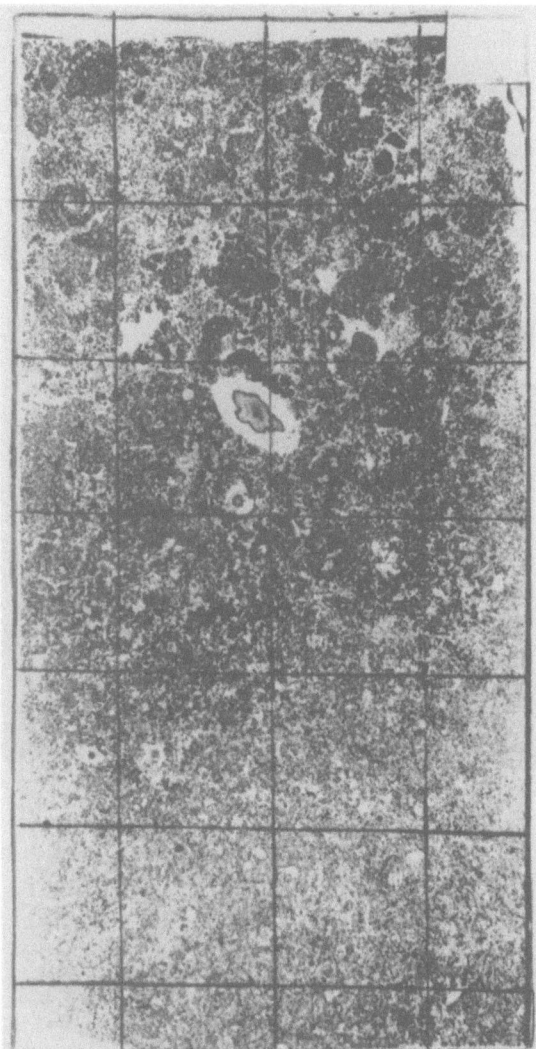

Fig. 3. Thin section of a savanna soil (Lamto, Ivory Coast) showing a macroaggregate structure built by endogeic earthworms (squares are 5 · 5 cm).

to 3 L, the respective length and volume of the burrow systems which allow the infiltration of water in six different soils of France. Though they represented, on the average, less than 1% of the overall porosity, earthworm burrows had a significant effect on water infiltration.

The morphology and average depth of burrow systems depends on the ecological category the populations belong to, and also on the characteristics of the profile. A typology has been proposed according to the depth of the burrow system, its degree of branching and the relative number of obstructed burrows. These three characteristics clearly depend on the proportion of the different ecological cat-

egories represented in the population and the presence of structural discontinuities (such as the presence of a thick clayey B horizon) in the soil profile.

Carbon and nutrient 'strategies' of cycling

The way earthworms affect soil chemical characteristics is, to a large extent, a result of their digestion. They have an evident effect through the assimilation of organic matter which results in the release of CO_2 and increase of assimilable nutrients in the casts as previously mentioned. Another effect is linked with the production of huge amounts of intestinal and cutaneous mucus. The production of intestinal mucus has been estimated at 5 to 16% of the dry weight of the ingested soil for five tropical species. It is a mixture of glycoproteins (ca. 45,000 d. molecular weight) and small glucidic and proteic molecules (Fig. 4) (Martin *et al.*, 1987). The average N content was 3% in *Pontoscolex corethrurus*.

This represents a considerable flux of assimilable organic matter, estimated at up to 50 Mg.ha^{-1}.yr^{-1} and 1.5 Mg Nitrogen in humid savanna soils of the Ivory Coast. The mucus is rapidly disintegrated by the ingested soil microflora which then reaches the critical metabolic level necessary to start digesting the soil organic matter. This mucus is rapidly incorporated to microbial biomass and/or reabsorbed as metabolites in the posterior part of the gut since no mucus is found in the casts (Barois and Lavelle, 1986). This transfer represents, however, a huge flux of quickly metabolized elements whose overall effect in the cycling of C and nutrients is probably important.

Cutaneous mucus is another source of rapidly assimilable organic matter which is produced by earthworms. Unlike intestinal mucus, cutaneous mucus is excreted in the soil to serve as a lubricant and possibly facilitates the cutaneous water absorption, and it is not reabsorbed. Its production is still poorly estimated since no really effective technique is as yet available. Krishnamoorthy (1985) estimated at 600–800 kg.ha^{-1} dry weight the yearly production of cutaneous mucus by a tropical earthworm community in India.

Fluxes of mucus as mentioned above support the hypothesis that earthworms would by their activi-

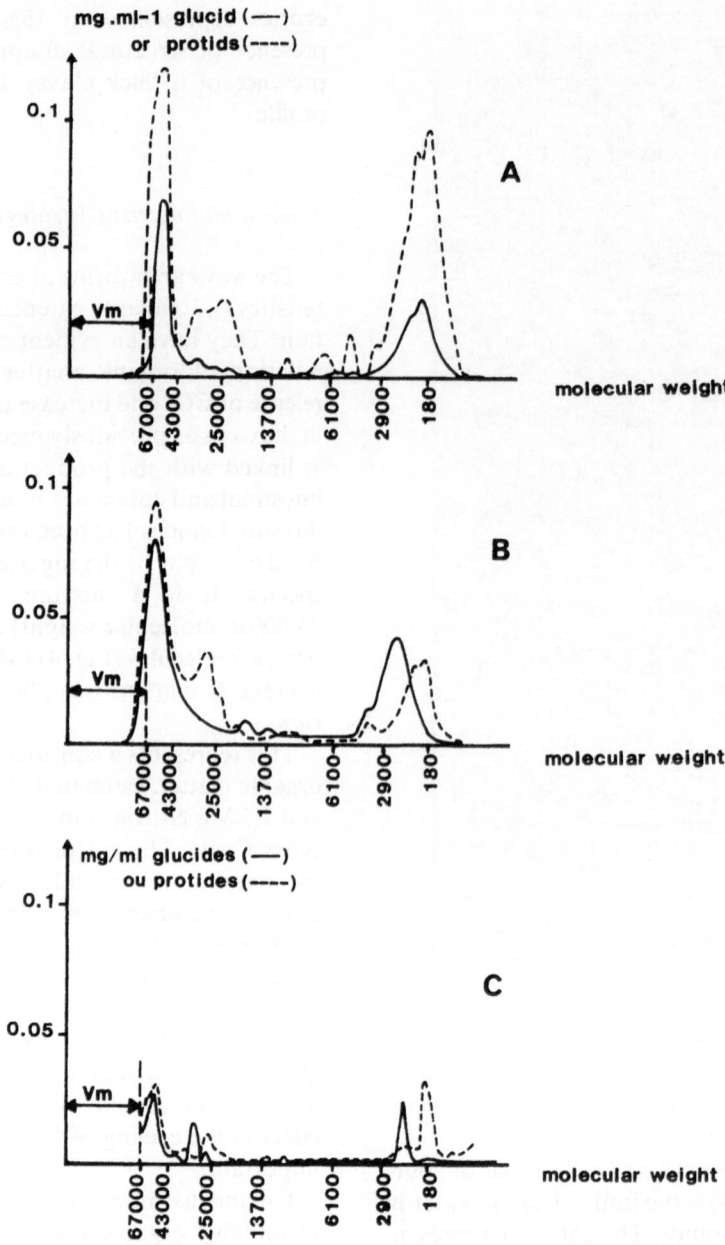

Fig. 4. Column chromatography of mucus taken in the anterior gut of three endogeic tropical earthworms: **A**, *Millsonia anomala* (mesohumic); **B**, *Dichogaster terrae-nigrae* (oligohumic) and **C**, *Millsonia ghanensis* (oligohumic). Vm = void volume of the column.

ties favour fast and short cycles of C and nutrients. Indirect evidence for this hypothesis is given by recent works of Ferrière and Bouché (1985) and Barois *et al.* (1987), which show a very rapid turnover of N in earthworm biomass. In the first experiment, homogeneously ^{15}N labelled lumbricid earthworms had renewed all the N of their bodies in as little time as 40 days. In the second experiment, adult *Pontoscolex corethrurus* fed during one month with ^{15}N labelled microbial biomass had lost 1/3 of their labelling after one month. Such experiments, which need to be repeated and confirmed, indicate that the gross flux of C and nutrients through earthworm populations might well be much higher than the net measured one.

Limitations for earthworms to colonize soil

Earthworms generally have low colonization capacities as a result of both the difficulty to move in such a compact medium as soil and relatively low fecundity. Earthworms are often considered sedentary, and, in natural populations their monthly movements are generally limited to 1 or 2 metres (Mazaud and Bouché, 1980). On some occasions, however, especially after heavy rains, they may find their way to the surface and move much longer distances before entering again into the soil. This relative sedentarity, however, varies with the ecological categories: epigeics may be relatively mobile especially when the litter is wet and some of them are able to climb up the trees, accumulating in palm-tree crowns or in epiphytic Bromeliads (see *e.g.* Vuattoux, 1970, Lavelle and Kohlmann, 1984; Fragoso, 1985).

Epigeics, which are the most mobile, also have high fecundities and some of them are parthenogenetic. These characteristics give them better colonization capacity than species from the other ecological categories, and it is not surprising that they constitute a large proportion of the peregrine species (Lee, 1985). Despite such favourable characteristics, the spread of their population is low: in the subantarctic island La Possession, Frenot (1985) estimated that, on the average, the advancement of a population of the epigeic Lumbricid *Dendrobaena tenuis*, introduced 10 years ago, is ten meters a year.

In temperate regions, anecics generally dig their burrows in autumn and seem to live in them for several months (Kretzschmar, 1982). There are indications that also endogeics are sedentary, especially the mesohumic and oligohumic ones which never appear at the soil surface. Among the few endogeic peregrine species, *Pontoscolex corethrurus* is remarkable as it has invaded all the disturbed (and some undisturbed) lands of the humid tropics. The colonization capacity of this species has been related to a relative tolerance for a wide range of soil conditions (despite a low tolerance of drought), a very efficient digestion system which allows the

allocation of a great part of the assimilated energy to reproduction, and parthenogenetic reproduction which allows one single individual to build up a population (Lavelle *et al.*, 1987).

The high rate of endemism observed in many regions is a result of this sedentarity and of the impossibility for earthworms to cross such geographical barriers as rivers or mountains. As a consequence, perturbations in natural ecosystems which would destroy the local fauna may have long lasting effects if species adapted to the newly created conditions are not found in adjacent areas. In pastures created within tropical forests, the savanna-like conditions are no more suitable for most forest species which disappear. If savanna species are not introduced, these soils are likely to lack one major biological system of regulation which normally operates in comparable natural ecosystems.

On the other hand, due to biogeographical reasons, the species naturally occuring in a determined area are not always the most well-adapted ones. Spectacular increases of grass production have been obtained in New Zealand after the introduction of better adapted anecic Lumbricids taken from European pastures (Stockdill, 1982).

Synchrony of earthworm activities, plant growth and plant nutrient demand

Earthworm activity may lead to the release of significant amounts of assimilable nutrients in their casts and in the linings of burrows. Further mineralization may also occur during up to 30-40 days after casts have been deposited, due to a possible microbial activation. Finally, if gross mineralization due to earthworm activities proves to be as high as preliminary results indicate, then the question of the synchrony between earthworm activities and the resulting release of assimilable nutrients becomes important to consider.

A high plasticity can be observed in earthworm function at the scale of individuals, populations and communities. The activity of individual worms is directly regulated by soil moisture, temperature, and the quality and distribution of feeding resources. The structure of populations respond to such changes within a few weeks, and ultimately, long-term changes of the environment result in subsequent changes in earthworm communities (Lavelle,

1983). Such an aptitude to change qualitatively and quantitatively as environmental conditions change, may lead to a close synchrony between the drilosphere activities (nutrient release and conservation of the soil structure) and plant growth, at least in natural ecosystems. Such an hypothesis, however, has not yet been really tested.

Effects of cultivation on earthworm communities

In a recent review, Lee (1985) distinguishes six main practices which may affect earthworm communities: vegetation changes due to land clearance, ploughing and harrowing, irrigation, mulching and application of fertilizers and biocides. The combination of such effects generally gives contradictory results as far as earthworm communities are concerned, and they are not easily interpreted in the absence of detailed information on the present and past practices.

Clearance

In Europe, deforestation generally affects epigeic populations, whereas anecics and endogeics populations adapt to the new situation (Bouché, 1972). In pastures, the biomass is higher than in forests: Edwards and Lofty (1972) report that earthworm biomasses range from 40 to 68 g fm.m^{-2} in deciduous forests and 51 to 152 g in pastures. Bouché (1978 and pers. comm.) indicates biomasses of up to 300 g in french pastures. Lee (1985) however, observes that, in Europe, most of the species found in the original forest could adapt to the grasslands.

In tropical areas, and perhaps more generally speaking, in areas where glaciations have not had a dramatic effect on earthworm distribution (see Bouché, 1983), this is not the case. Forests generally support low earthworm biomasses which rarely exceed 10–20 g fm.m^{-2} (Fragoso and Lavelle, 1987; Lavelle, 1987a). After clearance, local species mostly disappear except for a few small polyhumics or epigeics, with limited effects on the soil functioning. Recolonization, when made possible by the proximity of adapted species, may lead to the build-up of large earthworm populations, with a generally low species diversity, but biomasses of up to 100–200 g, as measured by Lavelle and Pashanasi (unp. data) in induced pastures of the Amazon forest (Table 2). Similarly, Barois (unp.) measured biomasses of up to 400 g in irrigated pastures of the Martinique island with only one species (*Polypheretima elongata*). In such cases, the dominant species is most often *Pontoscolex corethrurus*, a peregrine endogeic intermediate between polyhumics and mesohumics, originating from South America.

Cultivation

Ploughing generally decreases the abundance of earthworm communities. After 25 years of continuous cultivation, Low (1972 quoted in Lee, 1985) found that earthworm populations were only 11 to 16% of those populations found in equivalent soils used as pastures in England. The populations observed at Kjettslinge by Boström (1988) in different types of land use show equivalent trends.

The mechanical devices used for tillage have often been considered as responsible for such decreases. Cuendet (1979) estimated 5–10% as the percentage of biomass brought to the soil surface by tillage in Swiss regions, and a further 0.2 to 2.5% due to harrowing. 25% of these worms were mortally wounded. Endogeics were proportionally more affected than anecics. It seems thus that the effect of cultivation is mainly a secondary one, due to a decrease of the energetic status of the soil and also, especially in the case of tropical cultures, dramatic microclimatic changes. As a matter of fact, direct drilling, or any reduced ploughing technique leaves a higher earthworm population: Lee (1985) indicates experiments in which the use of direct drilling led to an increase of populations of 1.1 to 6.7 times greater, according to the species.

Mulching and organic fertilizers

By improving the microclimatic conditions in the topsoil and providing nutritive resources, such practices have beneficial effects on earthworm populations (see Lee, 1985 and Table 4).

One indirect effect of mulching is improvement of the soil structure through enhanced earthworm activities (Tisdall, 1978). The nature and abun-

Table 2. Abundance (.m^{-1}) and functional structure (%) of earthworm communities in different types of land-use at Yurimaguas (Amazonia, Peru) (Lavelle and Pashanasi, unp. data)

	Density	Biomass	Epigeics	Anecics	Polyhum.	Mesohum.
Primary forest	122	24.3	5.0	75.8	5.6	13.6
Brachiaria-Desmodium pasture	742	123.1	0.0	0.0	0.1	99.9
Traditional rice	10	0.5	0.0	99.0	1.0	0.0
High input corn	14	1.2	0.0	0.0	0.0	100.0
Pueraria fallow	363	5.9	0.0	0.0	70.0	30.0

Table 3. Earthworm abundances at Kjettslinge (Sweden) in different types of land-use (B0: barley with no N input; B120: barley with 120 kg N.ha^{-1}; GL: fertilized (200 kg.ha^{-1}) grass ley; LL: Luzern ley. (Boström, 1988)

	B0	B120	GL	LL
Density .m^{-2}	30.0	30.6	39.8	141.3
Biomass g fm.m^{-2}	14.7	18.3	27.5	67.4

dance of the mulching material however, greatly influences the development and composition of earthworm communities (Kuhle, 1983).

Inorganic fertilizers

The effect of inorganic fertilizers on earthworm populations seems to be rather variable. On a general basis, the increased production due to fertilizer application is likely to increase earthworm populations as food availability is greater. However, negative effects may be detected when high quantities are applied at the same time. Such fertilizers as ammonium sulfate might eliminate earthworms by increasing the soil acidity (Edwards and Lofty, 1972; Huhta *et al.*, 1967, Lee, 1985; Richardson, 1938). Such effects, however, have not been really proved. Liming of acidic soils generally increases earthworm populations in acidic soils (Huhta, 1979; Toutain *et al.*, 1987).

Biocides

The effect of biocides on earthworm populations is highly variable. Lee (1985) gives a compilation of 50 papers which analyze the effect of 84 different chemicals used for different purposes: insecticides, acaricides, nematicides, fungicides, herbicides and vermicides. The experiments, however, lacked a standardized method for evaluation of effects such as the one proposed by Bouché (1984).

Some pesticides with broad spectrums of activity are toxic to earthworms. Organochloride compounds (Chlordane, Heptachlor, Toxaphene) and carbamates are especially harmful. The most toxic organophosphorous compounds are Ethotrop, Phorate and Fensulphotion. Twenty-three others have been reported to show rather low toxicity.

Most carbamates used as fungicides, insecticides or nematicides, fumigants and lead arsenate or mercuric chloride used as vermicides are very toxic to earthworms. On the other hand, most of the herbicides which have been tested seem to be harmless to earthworms when used at normal doses (Edwards and Lofty, 1977).

Discussion: why, when and how to manipulate earthworm communities?

The earthworm community found in a cultivated soil is the result of many past and present effects, acting with different time scales. The continental drift was the first event which led to the isolation of faunas with different aptitudes to influence the functioning of soils, and glaciations have destroyed the original fauna of temperate areas. These have been recolonized from untouched areas by species with relatively high colonization capacities. Cultivation has greatly affected earthworm communities. Agricultural practices have variable effects, depending on their influence on the microclimate, as well as the energetic and nutrient status of the soil. Finally, mechanical preparation of the soil and application of chemicals, especially biocides, generally have had a dramatic effect, directly or indirectly.

As a result of natural or artificial perturbations, many soils, cultivated or not, have limited or inad-

Table 4. Earthworm populations in three different corn-fields of Bretagne (France) with inorganic fertilization and different inputs of manures. (Binet, unp. data)

	Inorg. fert. alone	+ 100 T.ha^{-1} pork manure	+ 40 Mg.ha^{-1} cow manure
Density m^{-2}	220.0	469.0	862.0
Biomass[a] g fm.m^{-2}	13.1	13.7	59.6
N species	5.0	6.0	11.0
epigeics (%)	0.0	0.1	7.5
anecics (%)	0.5	7.0	3.5
endogeics (%)	99.5	92.9	88.9

[a] weight in 4% formalin.

equate earthworm communities and thus, a poor overall drilospheric effect as compared with the one observed in comparable natural ecosystems. For example, induced pastures in tropical rainforests generally have earthworm communities far less diversified and active as compared to their natural equivalent, i.e., African moist savannas. In some cases accidentally introduced *Pontoscolex corethrurus*, may constitute large populations which develop a strong drilospheric effect.

In temperate regions, where soils often have a greater natural fertility, and physical and chemical processes are developed over much longer time scales, detrimental effects of reduction or disappearance of earthworm communities seem more difficult to demonstrate in the short term. Moreover, the ploughing and mixing of crop residues into the soil simulate part of earthworm activities. In addition, effects of former earthworm activities are most likely conserved in the inner structure of the soil for some time. In the long-term perspective, however, this effect certainly disappears and agricultural practices, though sophisticated, are unlikely to maintain the quality of soils. Moreover, as new techniques using limited tillage develop, instead natural systems of tillage like earthworm activities will increase in importance.

The management of earthworm communities increasingly appears as one possible way to maintain or increase soil fertility. This may be done in new soils (polders or reclaimed land). A number of experiments have been performed, sometimes with spectacular results (see *e.g.* Curry and Cotton, 1983; Hoogerkamp *et al.*, 1983; Trehen and Bouché, 1983; Van Rhee, 1971; Vimmersted, 1983). Introduction of earthworms has been made into soils which had inadapted earthworm communities

for biogeographical reasons. In New Zealand, the introduction of the Lumbricidae species *Aporrectodea caliginosa* has been reported to have dramatic effects on soil physical characteristics and grass production which was increased by 71% (Stockdill, 1959; 1982).

There are three possible means of managing earthworm communities. The minimum option is just to make sure they are not destroyed by the management techniques, especially the application of biocides. Until the real importance of drilosphere effects is clearly established, this seems a reasonable and inexpensive option.

In situations where it seems necessary to enhance earthworm activity, practices which facilitate their development may be envisaged. These are especially practices which improve the microclimatic conditions (irrigation or surface mulching), the energetic status of the soil (organic or inorganic inputs) or the pH (liming).

The introduction of adapted species is an ultimate option which has been applied successfully on a still limited number of occasions. It is increasingly considered, especially in the context of tropical agriculture. Controlled experiments with introduction and a better knowledge of species with promising patterns of activity are needed to take advantage as much as possible of the thousands of species present all over the world. Further research focusing on the drilospheric effects (e.g., microbial regulations, conservation of the soil structure and effects on C and nutrients cycling) are needed. They should take into account the variety of adaptive strategies developed in earthworms and consider the interactions between the drilosphere and other biological systems of regulations, especially the rhizosphere. Finally, the persistence over time of drilosphere effects needs to be investigated.

References

Allen T F H and Starr T B 1982 Hierarchy. Perspectives for Ecological Complexity. The Univ. of Chicago Press, Chicago.

Andersen N C 1983 Nitrogen turnover by earthworms in arable plots treated with farmyard manure and slurry. *In* Earthworm Ecology: From Darwin to Vermiculture. Ed. J E Satchell. pp 139–150. Chapman and Hall, London.

Assad M 1987 Contribution à l'étude de la macroporosité lombricienne de différents types de sols de France. Thèse Université des Sciences et Techniques du Languedoc 240 p.

Barois I 1987 Interactions entre les Vers de terre géophages et la microflore du sol pour l'exploitation de la matière organique du sol. Travaux des chercheurs de Lamto, 7, 151 p.

Barois I, Verdier B, Kaiser P, Lavelle P, Mariotti A and Rangel P 1987. Influence of the tropical earthworm *Pontoscolex corethrurus* (Glossoscolecidae) on the fixation and mineralization of ntirogen. *In* On earthworms. Eds. A M Bonvicini and P Omodeo. pp. 151–159. Mucchi editore, Modena, Italy.

Barois I and Lavelle P (1986) Changes in respiration rate and some physiochemical properties of a tropical soil during transit through *Pontoscolex corethrurus* (Glossoscolecidae, Oligochaeta). Soil Biol. 18, 539–541.

Bolton P J and Phillipson J 1976 Burrowing. feeding, egestion and energy budgets of *A. rosea* Savigny (Lumbricidae). Oecologia (Berl.) 23, 225–245.

Boström U 1988 Ecology of earthworms in arable land: Population dynamics and activity in four cropping systems. Report 34, Swedish University of Agricultural Sciences, Uppsala.

Bouché M B 1972 Lombriciens de France: Ecologie et Systématique. Ann. Zool., Ecol. Anim., No hors série 1972, 671 p.

Bouché M B 1977 Stratégies lombriciennes. Ecol. Bull. 25, 122–132.

Bouché M B 1978 Fonctions des Lombriciens. VII. Recherches françaises et résultats d'un programme forestier coopératif (RCP 40). Bull. Sci. Bourgogne, 20, 143–227.

Bouché M B 1983 The establishment of earthworm communities. *In* Earthworm Ecology: From Darwin to Vermiculture. Ed. J E Satchell. pp. 431–448. Chapman and Hall, London.

Bouché M B 1984 Ecotoxicologie des Lombriciens. I. Ecotoxicologie contrôlée. Acta Oecologia 5, 271–287.

Curry J P, Coton D C F 1983 Earthworms and land reclamation. *In* Earthworm Ecology: From Darwin to Vermiculture. Ed. J E Satchell. pp. 215–228. Chapman and Hall, London.

Cuendet G 1979 Etude du comportement alimentaire de la mouette rieuse (*Larus ridibundus* L.) et de son influence sur les peuplements Lombriciens. Thèse de Doctorat, Lausanne, 111 p.

Darwin C R 1881 The Formation of Vegetable Mould through the Action of Worms, with Observations on Their Habits. Murray, London.

Di Castri F, Ferrar A A, Hansen R A J, Owen-Smith M C, Rutherford M C, Scholes J, Siegfried W R and B Van Wielgen 1985 A hierarchical view of life. Concepts in Southern African terrestrial Ecology: A review workshop. 31/7–3/8/ 1985. Univ. of Witwatersrand (South Africa).

Edwards C A and Lofty J R 1972 Biology of Earthworms. Chapman and Hall, London. 183 p.

Edwards C A and Lofty J R 1977 Biology of Earthworms. Second edition. Chapman and Hall, London. 204 p.

Fearnside P M 1987 Rethinking continuous cultivation in Amazonia. BioScience 37, 209–213.

Ferrière G and Bouché M B 1985 Première mesure écophysiologique d'un débit d'élément dans un compartiment endogé: Le débit d'azote de *Nicodrilus longus* Ude (*Lumbricidae* Oligochaeta) dans la prairie de Citeaux. C.R. Acad. Sci., 301, 111, 17, 789–794.

Fragoso C 1985 Ecologia general de las lombrices terrestres Oligochaeta: Annelida) de la région Boca del Chajul, Selva Lacandona Chiapas, Mexico. Thesis UNAM, Mexico, 133 p.

Fragoso C and Lavelle P 1987 The earthworm communities of a tropical rainforest from Mexico (Chajul, Chiapas). *In* On Earthworms. Eds. A M Bonvicini and P Omodeo. pp 281–297. Mucchi editore, Modena, Italy.

Frenot Y 1985 Etude de l'introduction accidentelle de *Denbrobaena rubida tenuis* Oligochaeta, Lumbricidae) à l'ile de la Possession. Bull. Ecol. 16, 47–54.

Graff O 1971 Stickstoff, Phosphor und Kalium in der Regenwurmlosung auf der Wiesenversuchsfläche des Sollingprojekts. Ann. Zool. Ecol. Anim., Special Publ. 4, 503–512.

Hamilton W E and Vimmersted J P 1980 Earthworms on forested spoil banks. *In* Soil Biology as Related to Land Use Practices. Ed. D L Dindal. pp 409–417. EPA, Washington D.C.

Hoogerkamp M, Rogaar H and Eijsackers H J P 1983 Effect of earthworms on grassland on recently reclaimed polder soils in the Netherlands. *In* Earthworm Ecology: From Darwin to Vermiculture. Ed. J E Satchell. pp. 85–106. Chapman and Hall, London.

Huhta V, Karpinnen E, Nurminen M and Valpas A 1967 Effect of silviculture practices upon arthropod, annelid and nematode populations in coniferous forest soils. Ann. Zool. Fenn. 4, 87–143.

Huhta V 1979 Effects of liming and deciduous litter on earthworm (Lumbriciadae) populations of a spruce forest, with an inoculation experiment on *Allolobophora caliginosa*. Pedobiologia 19, 340–345.

Kretzschmar A 1982 Description des galeries des Vers de terre et variation saisonnière des réseaux, observations en conditions naturelle). Rev. Ecol. Biol. Sol. 19, 579–591.

Krishnamoorthy R V 1985 Nitrogen contribution by earthworm populations from grassland and woodland sites near Bangalore, India. Rev. Ecol. Biol. Sol. 22, 463–472.

Kuhle J C 1983 Adaptation of earthworm populations to different soil treatments in apple orchards. *In* New Trends in Soil Biology. Eds. P Lebrun *et al/*. pp 487–501. Dieu-Brichart, Louvain la Neuve.

Lavelle P 1978 Les Vers de terre de la savane de Lamto (Côte d'Ivoire): Peuplements, populations et fonctions dans l'écosystème. Thèse Doctorat, Paris, VI. Publ. Labo. Zool. E.N.S. 12, 301 p.

Lavelle P 1981 Stratégies de reproduction chez les Vers de terre. Acta Oecol. General. Vol. 2, No. 2, pp 117–133.

Lavelle P 1983 The structure of earthworm communities. *In* Earthworm Ecology: From Darwin to Vermiculture. Ed. J E Satchell. pp 449–465. Chapman and Hall, London.

Lavelle P 1984 The soil system in the humid tropics. Biology International 9, 2–17.

Lavelle P 1987a Biological processes and productivity of soils in the humid tropics. *In* The Geophysiology of Amazonia. Ed. R E Dickinson. pp 175–214. Wiley and Sons, New York.

Lavelle P 1987b Interactions, hiérarchies et règulations dans le sol: À la recherche d'une nouvelle approche conceptuelle. Rev. Ecol. Biol. Sol. 24, 219–229.

Lavelle P 1988 Earthworm activities and the functioning of soils. Biol. Fert. Soils 6, 237–251.

Lavelle P, Zaidi Z and Schaefer R 1983 Interactions between earthworms, soil organic matter and microflora in an African savanna soil. *In* New Trends in Soil Biology. Eds. P Lebrun *et al.* pp 253–261. Dieu Brichart, Louvain-la-Neuve.

Lavelle P and Kohlmann B 1984 Etude quantitative de la ma-

122   *Earthworm populations in agro-ecosystems*

crofaune du sol dans une forêt tropical humide du Mexique (Bonampak, Chiapas). Pedobiologia 27, 377–393.

Lavelle P, Barois I, Cruz I, Fragoso C, Hernandez A, Pineda A and Rangel P 1987 Adaptive strategies of *Pontoscolex corethrurus* (Glossoscolecidae-Oligochaeta), a peregrine geophagous earthworm of the humid tropics. Biol. Fert. Soils 5, 188–194.

Lee K E 1983 The influence of earthworms and termites on soil nitrogen cycling. *In* New Trends in Soil Biology. Eds. P. Lebrun *et al.* p. 35–48. Dieu Brichart, Louvain-la-Neuve.

Lee K E 1985 Earthworms: Their Ecology and Relationships with Soils and Land Use. Acad. Press, London, 411 p.

Lunt H A and Jacobson G M 1944 The chemical composition of earthworm casts. Soil Sci. 58, 367–375.

Martin A, Cortez J, Barois I and Lavelle P 1987 Les mucus de Ver de terre: moteur de leurs interactions avec la microflore. Rev. Ecol. Biol. Sol. 24, 549–558.

Mazaud D and Bouché B 1980 Introduction en surpopulation et migrations de Lombriciens marqués *In* Soil Biology as Related to Land Use Practices. Ed. D Dindal. pp 687–701. EPA, Washington D.C.

Nye P H 1985 Soil forming processes in the humid tropics. IV. The action of the soil fauna. J Soil Sc. 6, 51–83.

Parle J N 1963 Microorganisms in the intestines of earthworms. J Gen. Microbiol. 31, 1–11.

Phillipson J, Abel R, Steel J and Woodell S R J 1976 Earthworms and the factors governing their distribution in an English beechwood. Pedobiologia 16, 258–285.

Rhee J Van 1971 The productivity of orchards in relation to earthworm activities *In* IV Colloquium Pedobiologiae. Ed. J d'Aguilar. pp 99–106. INRA 71-7, Paris.

Richardson H C 1938 The nitrogen cycle in grassland soils, with special reference to the Rothamsted Park grass experiments. J. Agric. Sci. Camb. 28, 73–121.

Sanchez P A, Bandy D E, Villachica J H and J J Nicholaides 1982 Amazon Basin soils: Management for continuous crop production. Science 216, 821–827.

Satchell J E 1983 Earthworm Microbiology: *In* Earthworm Ecology: From Darwin to Vermiculture. Ed. J E Satchell. pp 351–364, Chapman and Hall, London.

Satchell J E and Martin K 1984 Phosphatase activity in earthworm faeces. Soil Biol. Biochem. 16, 191–194.

Sharpley A N and Syers J K 1977 Seasonal variation in casting activity and in the amounts and release to solution of pho-

sphorus forms in earthworm casts. Soil Biol. Biochem. 8, 341–346.

Springett J A 1985 Effect of introducing *Allolobophora longa* Ude on root distribution and some soil properties in New Zealand pastures. *In* Ecological Interactions in Soil. Eds. A Fitter *et al.* pp 399–406. Blackwell, London.

Stockdill S M J 1959 Earthworms improve pasture growth. N.Z. J. Agric., 98, 227–233.

Stockdill S M J 1966 The effects of earthworms on pastures. Proc. N.Z. Ecol. Soc. 13, 68–75.

Stockdill S M J 1982 Effects of introduced earthworms on the productivity of New Zealand pastures. Pedobiologia 24, 29–35.

Swift M J (ed) 1984 Soil Biology and Fertility in the Tropics: A proposal for a collaborative programme of research. Biology International, 5, 1–38.

Swift M J, Heal O W and Anderson J M 1979 Decomposition in Terrestrial Ecosystems: Studies in Ecology, Vol. 5. Blackwell, Oxford, 372 p.

Syers J K, Springett J A and Sharpley A N 1979 The role of earthworms in the cycling of phosphorus in pasture ecosystems. Quoted in Lee, 1985.

Syers J K, Springett J A 1983 Earthworm ecology in grassland soils. *In* Earthworm Ecology: From Darwin to Vermiculture. Ed. J E Satchell. pp. 67–84. Chapman and Hall, London.

Tisdall J M and Oades J M 1982 Organic matter and waterstable aggregates in soils. J. Soil. Sci. 33, 141–163.

Tisdall J M 1978 Ecology of earthworms in irrigated orchards. *In* Modifications of Soil Structure. Ed. W W Emerson, R D Bond and A R Dexter. pp 297–303. Wiley, Chichester.

Toutain F, Diagne A and Le Tacon F 1987 Conséquences d'un apport d'engrais sur le fonctionnement d'un écosystème forestier de l'Est de la France. Rev. Ecol. Biol. Sol. 283–300.

Trehen P and Bouché M B 1983 Place des lombriciens dans les processus de restauration des sols de lande. *In* New Trends in Soil Biology. Eds. P Lebrun *et al.* pp 471–486. Dieu—Brichart, Louvain—la-Neuve.

Vuattoux R 1970 Observations sur l'évolution des strates arborées et arbustives de la savanne de Lamto (Côté d'Ivoire). Ann Univ. Abidjan, E, 3, 285–315.

Zaidi Z 1985 Recherches sur les modalités de l'interdépendance nutritionnelle entre Vers de terre et microflore dans la savane guinéenne de Lamto (Côte d'Ivoire) Esquisse d'un système interactif. Thèse 3 cycle, Paris XI, 111 p.

M. Clarholm and L. Bergström (Eds.), Ecology of arable land, 123–132.

Impact of human activities on nematode communities in terrestrial ecosystems

L. WASILEWSKA
Institute of Ecology, Polish Academy of Sciences, Dziekanów Leśny near Warsaw, PL-05-092 Łomianki, Poland

Key words: bacteriophages, body size, diversity, human impact, manure, nematodes, N-fertilization, omnivores, phytophages, predators, root consumption

Abstract

Soil nematodes use a small proportion of the primary production in ecosystems (0.06–4.3%), but their role in many processes is important. Phytophagous nematodes consume 1.4–21.8% of the primary production of roots, whereas bacteriophages consume, via bacteria and fungi, about 40% of the organic matter decomposed over a year. Examples are given of the effect of addition of mineral nitrogen fertilizers, of nitrogen mineralization after drainage of peatland, and of the effect of different types of landscape on functional groups of nematodes. The proportion of plant parasites and bacteriophages increased after perturbation whereas the proportion of pantophages and predators decreased, the species richness declined, and individual body weight was reduced. All these observations provide evidence of functional changes.

Introduction

Assuming after O'Neill (1976) that each ecosystem consists of three basic components: living plant tissue, dead organic matter, and heterotrophic organisms, it becomes obvious that any impact on the first two components must affect the third. For example, chemicals used against phytophagous nematodes act through the plant and 'per se'. Nematodes are diversified with respect to their diet and may be divided into trophic groups which respond to external factors.

Although nematodes utilize only a small part of the primary production, they play a significant role in many ecosystem processes. Phytophages, especially those which are true parasites of plants, may reduce plant production. They are also largely responsible for the ways of energy flow: by consuming and digesting living tissues, they increase the input of root litter to the detritus pathway and stimulate decomposition by contributing to the soil nitrogen pool by their excretion. The role of nematodes feeding on bacteria and fungi, which was investigated in microcosm experiments, has already been reviewed by several authors (Anderson et al., 1981; Coleman et al., 1983; Ingham et al., 1985; Yeates and Coleman, 1982). They showed that this group of nematodes were responsible for the release of nutrients and an increase in the rate of their mineralization and cycling. As a result, these nutrients can be used by producers and decomposers, hence these nematodes indirectly stimulate the growth of plants. The ecology of both omnivores and predators (which can be considered as K-strategists when compared with other nematodes) is such that they extend the length of trophic chains, thereby reducing the rate of nutrient cycling in ecosystems. We may suspect that they enhance the stability on community level (Wasilewska, 1985).

Nematodes versus net primary production of ecosystems

Soil nematodes account for 6–15% of the soil invertebrate biomass (Wasilewska, 1986). Soil nematode communities use only a small part of the

Table 1. Root consumption by phytophagous nematodes in different ecosystems

Ecosystem	Consumption ($g\,d\,wt\,m^{-2}\,year^{-1}$)	Percent consumed primary production
Mixed-grass prairie South Dakota USA (Smolik, 1974)	39–57	17.2–21.8
Short-grass prairie Colorado USA (Scott, 1979)	34.8	7–26
Mixed-grass prairie North America USA (Ingham and Detling, 1984)	14.8–17.9	5.8–12.6
Mountain pasture with sheep, Jaworki, Poland (Wasilewska, 1979)	11.4	10
Meadows on drained peatland, Wizna, Poland (Wasilewska, 1989a)	16.0–50.3	1.6–19.6
Rye crop, Poland (Wasilewska, 1979)	6.2	4.2
Potato crop, Poland (Wasilewska, 1979)	5.3	1.4

net primary production (NPP) of ecosystems to cover the maintenance costs. The data available so far indicate that the consumption (C) of nematode communities in the temperate zone ranges from 3.7 to 61.5 g d. wt. m^{-2} year^{-1} in grasslands, from 0.9 to 29 g d. wt. m^{-2} year^{-1} in forests, and from 7.1 to 32.2 g d. wt. m^{-2} year^{-1} in cropland (Coulson and Whitaker, 1978; Petersen and Luxton, 1982; Philipson *et al.*, 1977; Sohlenius, 1979; Wasilewska, 1974a; 1976; 1979; 1989a; Witkowski, 1985). This consumption accounts for 0.26–4.3% of NPP in grassland ecosystems, 0.06–2.0% of NPP in forest ecosystems, and 0.6–2.7% of NPP in cropland ecosystems if the NPP of the ecosystems is taken after Ryszkowski (1985a).

A more detailed analysis indicates that the importance of nematodes in terrestrial ecosystems is greater than suggested by the above figures. Recent papers provide evidence that phytophagous nematodes alone consume from 5.3 to 57 g d. wt. of root biomass m^{-2} year^{-1}, which accounts for 1.4–21.8% of the root primary production (Table 1). Root consumption by phytophagous nematodes is high in grasslands (11.4–57 g d. wt. of roots m^{-2} year^{-1}), and much lower in cropland (5.3–6.2 g d. wt. of roots m^{-2} year^{-1}). The role of another functional group of nematodes, bacterial and fungal grazers, can be estimated from experiments on decomposition. It has been shown that bacteriophagous and mycophagous nematodes consume via bacteria and fungi from 34 to 44% of the organic matter decomposed during one year (Wasilewska and Bieńkowski, 1985).

Structure of nematode communities in agroecosystems

As compared with other ecosystems, crop fields are characterized by an increased proportion of phytophages in the standing crop, the total edaphon, reduction of body size of soil animals, increase in respiration rate per unit biomass, and more rapid decomposition (Ryszkowski, 1985b). Mean results for 30 different ecosystems in Poland show that agroecosystems are characterized by high numbers, biomass, and metabolic activity of bacteriophagous and phytophagous nematodes whereas pantophages and predators are scarce (Wasilewska, 1979). Similar observations were made by other authors (Anderson, 1979; Boström, 1982; Ferris and Ferris, 1974; Freckman and Caswell, 1985; Winslow, 1964). In cropfield ecosystems, nematodes have the smallest mean individual body weight (Wasilewska, 1979). Species diversity of nematodes seems to be less in croplands than in other ecosystems (Ferris and Ferris, 1974; Wasilewska, 1979; Yeates, 1979). Nematode communities in different agroecosystems seem to be less variable than those in different forest ecosystems or in different grassland ecosystems (Wasilewska, 1979).

Nematode communities in agroecosystems and in natural and semi-natural ecosystems are influenced by increasing human impact, which they reflect. 1. Nematodes are useful for evaluating human impact because they have high diversity in species and trophic levels, high abundance and occur in every type of soil. 2. Nematodes have high colonization ability, are easy to sample and can be sampled in every season. 3. Their permeable cuticle permits a range of reactions to pollutants.

The effect of human activity on nematode communities

In the temperate zone, two basic types of stresses affecting terrestrial ecosystems manipulated by man seem to be most important: 1. increasing chemicalization of soil due to the application of mineral fertilizers, some organic fertilizers and accumulation of industrial dust and 2. progressive drying of soil. The responses of nematode trophic groups seem to be symptomatic in such cases.

Reaction of phytophagous nematodes

Phytophagous nematodes are more abundant in cultivated than in uncultivated soils (Ferris, 1982). The application of mineral fertilizers, especially nitrogen, is followed by a clear response of this group. Numbers of phytophagous nematodes, as measured by oxygen consumption of this group, increased by a factor of three in a wheat field fertilized with $240\,kg\,N\,ha^{-1}$ as compared with an unfertilized field (Curve 1 in Fig. 1). Phytophagous nematodes increased from 19% to 43% and total numbers increased from 3.8 to 5.6×10^6 individuals m^{-2} (Witkowski, 1985). Similar results were obtained for barley. The dose of $120\,kg\,N\,ha^{-1}$ increased the abundance of obligate plant parasites (Curve 2 in Fig. 1) by a factor of over 2.5, the proportion of this group in the community increased from 16.3% to 30.2%, and the abundance of the total nematode community increased from 5.1 to 7.2×10^6 individuals m^{-2}, respectively

(Sohlenius and Boström, 1986). Although one of the study plots was located in Poland and the other in Sweden, the increase in the proportion of these groups after fertilization with $80-120\,kg\,N\,ha^{-1}$ was strikingly similar, accounting for 16–18% in the unfertilized fields to about 30% in the fertilized fields (Fig. 1). The most spectacular increase in obligate plant parasites was observed in drained fens managed as meadows, in which mineralization of soil nitrogen occurred. At a low level of nitrogen mineralization in soil ($65\,kg\,N\,ha^{-1}$ per growing season), the abundance of these nematodes was 0.45×10^6 individuals m^{-2}, whereas at a very high mineralization rate ($350\,kg\,N\,ha^{-1}$ per growing season) it averaged 8.1×10^6 individuals m^{-2}, thus it increased by a factor of about 18 (Curve 3 in Fig. 1). The dominance of this group in the total nematode community increased from 11.8% to 48.8%, respectively, though the increase in the abundance of the total community was less impressive, from 3.8×10^6 to 16.6×10^6 individuals m^{-2} (Wasilewska, 1989a).

The response of obligate plant parasitic nematodes to increased intensity of farming and to heavy industrialization will be illustrated for potato fields. The potato fields under study were located in landscapes of different types. Recreational landscape in northern Poland (Mazuria in the region of Mikołajki) consisted of lakes covering more than 20% of the area, forests and meadows ca. 20%, whereas arable land accounted for less than 40%. The agricultural landscape of western Poland (Wielkopolska region), characterized by the driest climate in Poland, is predominated by cropland (64%), and had an old tradition of farming. The degree of mechanization in this region was high. A simplified crop rotation was applied in the potato field under study (potato, rye, and rye in the aftercrop in between). In central Poland (Mazovia, closer and more distant regions of Warsaw) farming was less intense, crop fields were smaller, and the crop rotation longer. In the intensely industrialized region of Silesia (coal mining and heavy industry) the concentration of SO_2 in the air was $0.24\,mg\,m^{-3}$, on the average, and the concentration of heavy metals ranged from 8.9 to $15.2\,t\,km^{-2}$ $month^{-1}$ (data from the Rybnik Coal District). Organic and mineral fertilization of the potato fields under study was similar. Numbers of obligate plant parasites were lowest in the fields located in

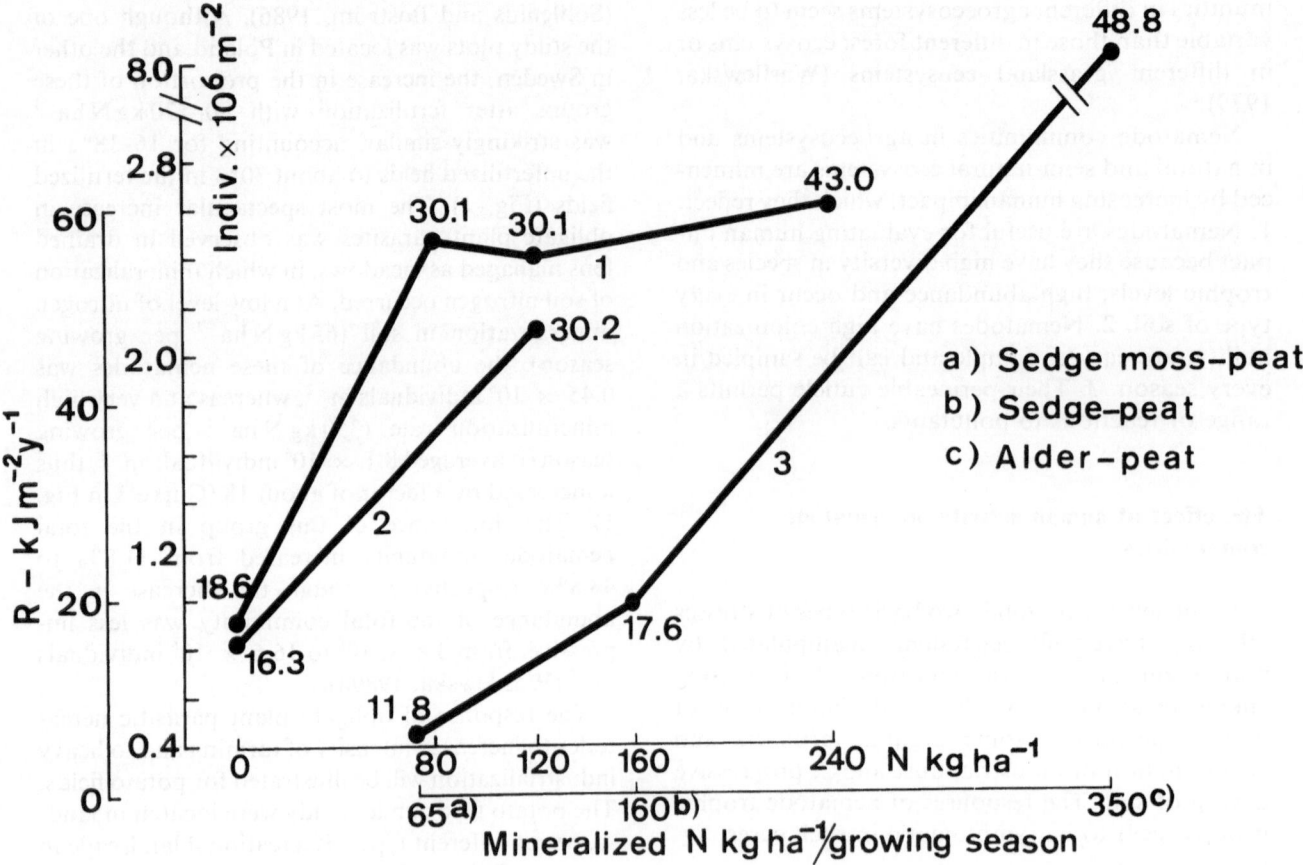

Fig. 1. The effect of fertilizing with nitrogen on nematodes: Curve 1, respiratory metabolism of phytophagous nematodes in a wheat field after Witkowski (1985); curve 2, number of obligate plant parasites in a barley field after Sohlenius and Boström (1986), and curve 3, number of obligate plant parasites in soil of managed meadows on drained peatland (Wasilewska, 1989a). Figures at points indicate percentage of the group in the total nematode community.

the recreational landscape and the highest in the region with intense farming. The dominance of this group in the total nematode community increased, also when the abundance of other trophic groups of nematodes increased markedly (Fig. 2).

Drainage of fens and their management as meadows also affected nematodes. The extent of the change mostly depended on the origin of peat (sedge moss-peat, sedge-peat, and alder-peat). The mineralization of dried peat, including nitrogen mineralization, depended on soil moisture, which, in turn, was related to the water-holding capacity of peat. Nitrogen mineralization was most rapid in drained alder-peat. This process, together with symptoms of aridity led to a serious soil degradation. Even when an alder peatland is carefully managed as a meadow, drainage should be con-

sidered as the most stressing factor. It is less stressing in sedge peatland and the least in sedge moss-peatland. One of the most important reactions observed in the soil nematode community in these meadows was an increase in numbers and proportions of phytophagous nematodes after drainage, which was particularly high on alder peatland (Curve 3 in Fig. 1). Figure 1 shows the situation 15–20 years after drainage with respect to peat origin. The response of phytophages only on alder peatland was even more pronounced 1–5 years after drainage, and it was still maintained even 50 years after drainage. The numbers of phytophagous nematodes on alder peat-soil were still higher 100 years after drainage than on natural, not drained peatlands (Fig. 3). The nematode community after drainage was totaly dominated by one .

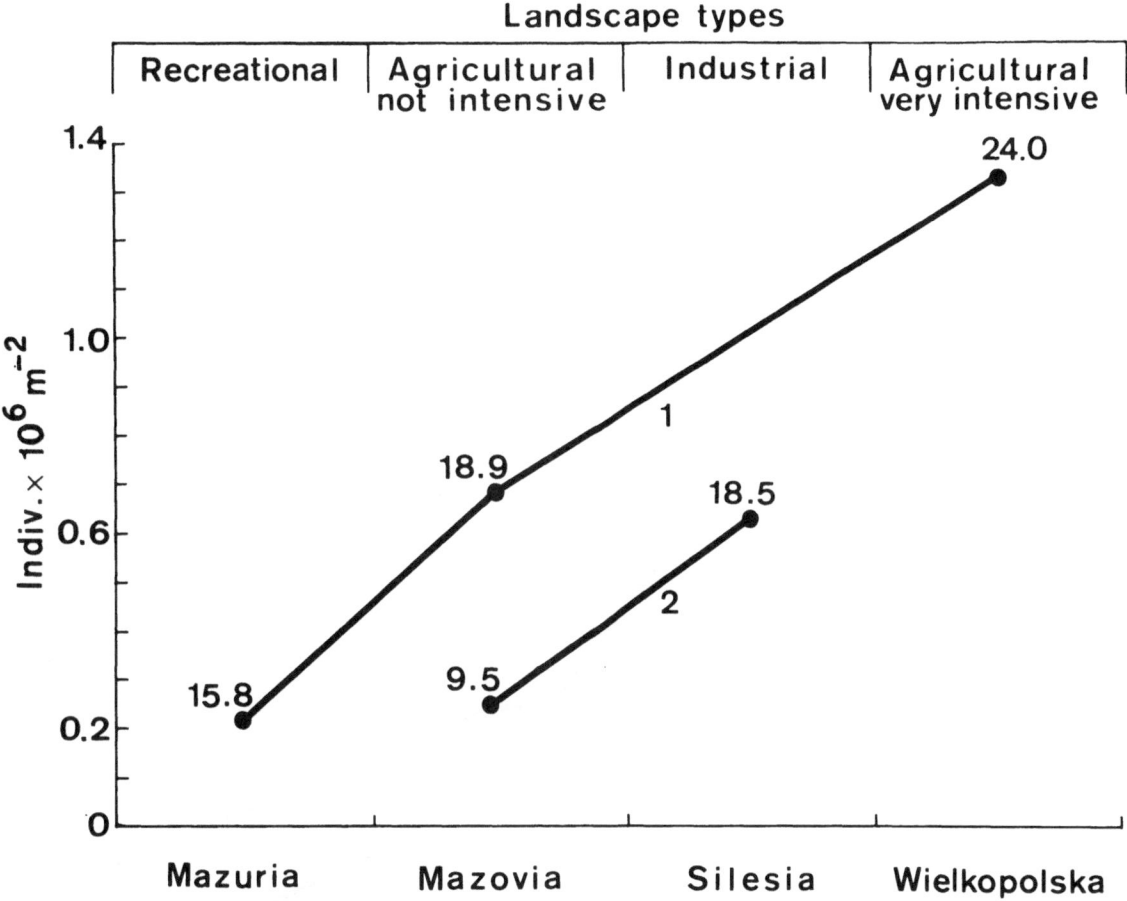

Fig. 2. Number of obligatory plant parasite nematodes in potato fields located in different landscape types. Curve 1, after Wasilewska (1987), 8–20 samples; Curve 2, modified after Kozłowska and Domurat (1980), 4–8 samples. Figures at points indicate percentage of the group in the total nematode community.

group of ectoparasitic nematodes, namely *Paratylenchus* sp. It has been proposed to use the dominance of *Paratylenchus* sp. as an indicator of degradation of drained alder-peat soils (Wasilewska, 1989a). The degradation is a result of changing physical and chemical soil properties in meadows. Grasses are gradually replaced by nitrophilous herbs, more weeds appear in the plant cover, the sward is deteriorating, and excessive concentrations of nitrates in ground waters are observed (Okruszko, 1977). The increased dominance of phytophages in the nematode community as an effect of drainage is also expressed by a higher proportion of respiration from this group as compared to the respiration from the total nematode community (Wasilewska, 1989a). A similar trend was recorded in the consumption of

NPP of roots by these nematodes (Wasilewska, 1989a). In alder peat, one-fifth of the grass root biomass produced per year was consumed by phytophagous nematodes.

Response of bacteriophagous nematodes

Bacteriophagous nematodes can be considered as indicators of bacterial activity (Bååth *et al.*, 1981; Freckman and Caswell, 1985; Wasilewska and Bieńkowski, 1985). Thus, introduction into the soil of mineral fertilizers, organic fertilizers (liquid manure from cattle, sheep faeces) (Fig. 4 and Dmowska and Kozłowska, 1986), and drainage of peatland (Fig. 4, Curve 2) have a positive effect on the growth of microorganisms and thereby increas-

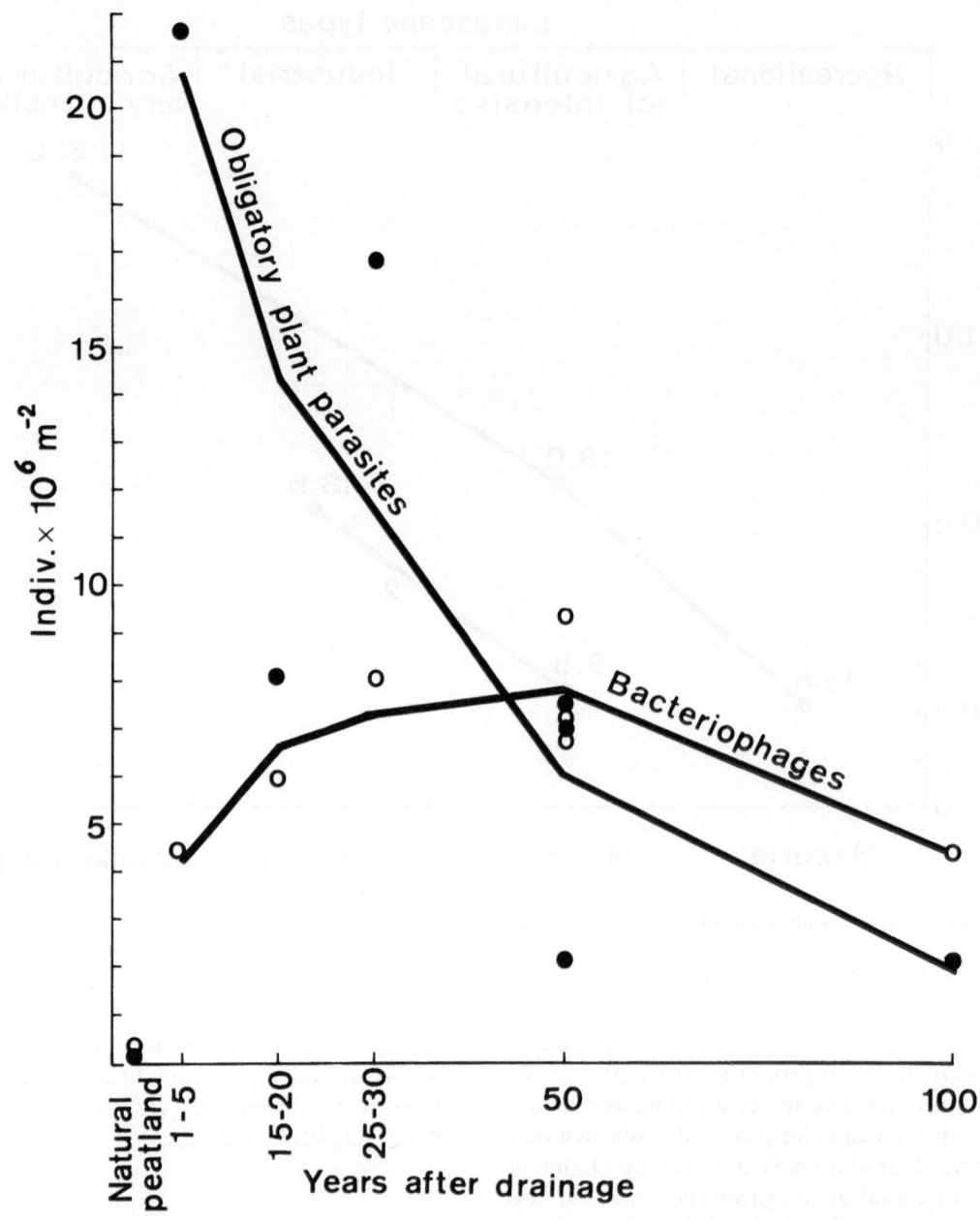

Fig. 3. Number of two trophic groups of nematodes in relation to time after the drainage of an alder peatland. Managed meadows in the ice marginal valley of the river Biebrza, Poland. Mean values for 6 years, sampled 2–21 times on every of 6 drained stations and of one not drained, natural peatland; curves were fitted by polinomials.

ing the importance of this group of nematodes. This increase in numbers of bacteriophages ends after the exhaustion of resources. For example,

even if high doses of cattle liquid manure are applied, increase of bacteriophages ceases after 5 months (Dmowska and Kozłowska, 1986). This

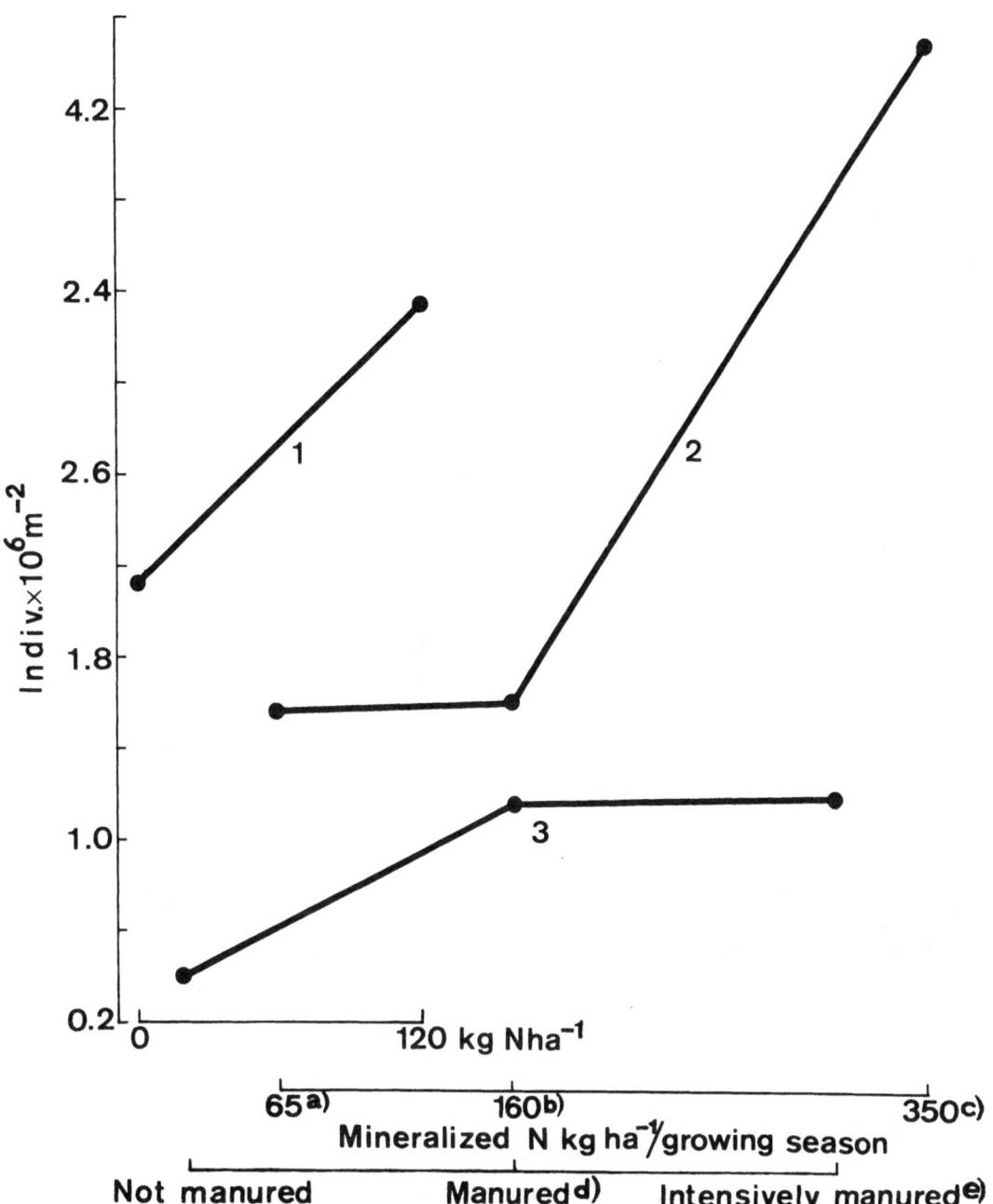

Fig. 4. The effect of different fertilizers on bacteriophagous nematodes. Curve 1, N-fertilization, barley after Sohlenius and Boström (1986), curve 2, mineralization of organic matter in managed meadows on drained peatland after Wasilewska (1989a) and curve 3, manuring by sheep faeces, pasture after Wasilewska (1974a).

Explanation: **a**, sedge moss-peat, **b**, sedge-peat, **c**, alder-peat, **d**, with $360\,g\,d$ wt of dung$\cdot m^{-2}$ per season in small daily doses and **e**, for 3 days with $500\,g\,d$ wt of dung$\cdot m^{-2}$ during penning-up sheeps and subsequently as in d.

should be compared with alder peat mineralization, where bacteriophages continue to increase in numbers for more than 50 years after drainage (Fig. 3). Estimates of the limits to which a habitat can be transformed are needed.

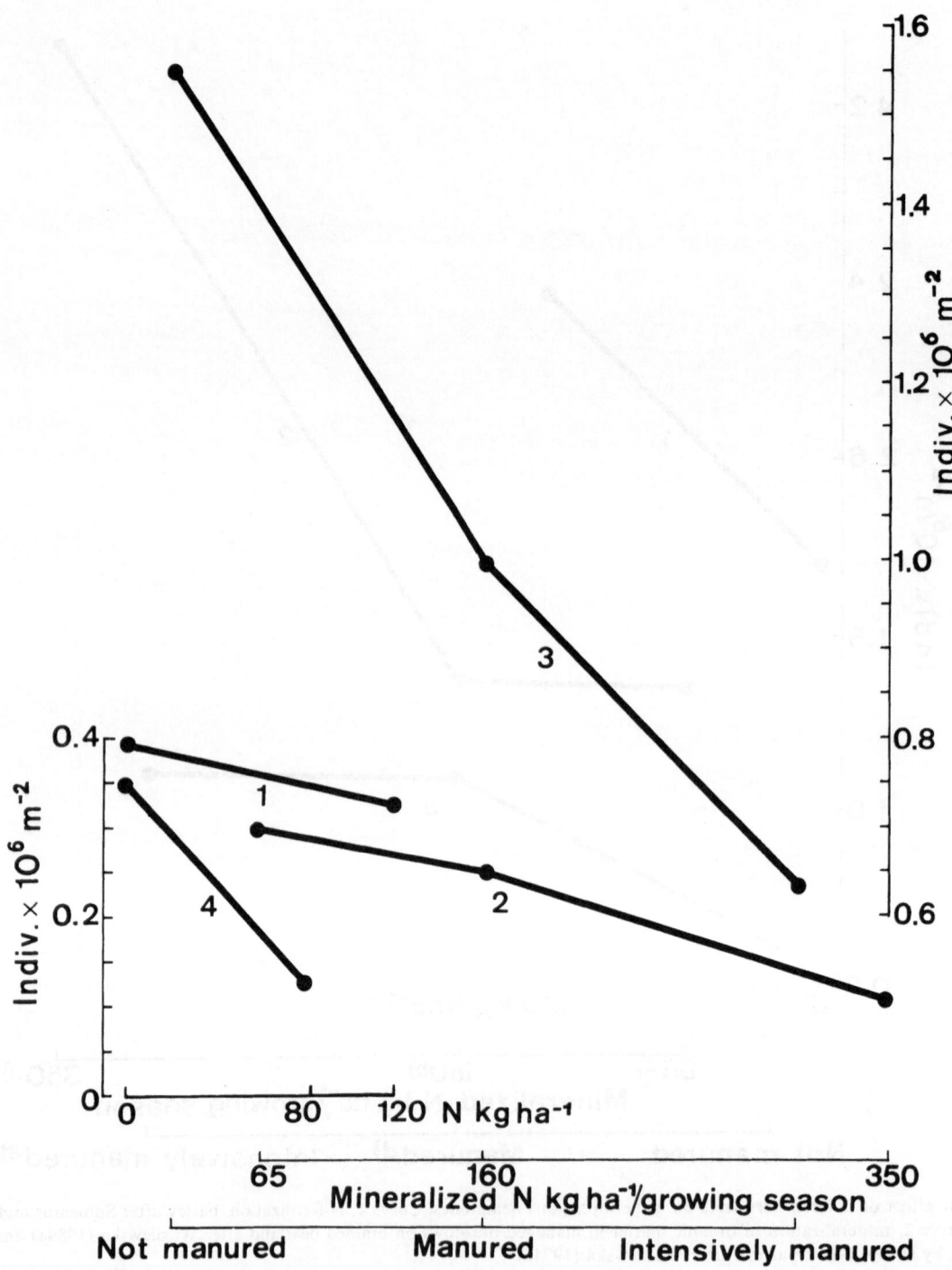

Fig. 5. The effect of different fertilizers on omnivorous and predacious nematodes. Curve 1, N-fertilization, barley after Sohlenius and Boström (1986), curve 2, mineralization of organic matter in managed meadows on drained peatland after Wasilewska (1989a), curve 3, manuring by sheep faeces, pasture after Wasilewska (1974a) and curve 4, N-fertilization, dry pine forest after Sohlenius and Wasilewska (1984). Additional explanation as in Fig. 4.

Response of pantophages and predators

Reduction of omnivorous and predatory nematodes as an effect of human pressure has already been reported (Ferris and Ferris, 1974; Nusbaum and Ferris, 1973; Wasilewska, 1974b; 1989b). The application of mineral fertilizers, organic fertilizers such as sheep manure, and nitrogen mineralization in meadow soils after drainage of fens are treatments reducing numbers of this group (Fig. 5). Thus, these are treatments eliminating K-strategists (dorylaimids and mononchids) from the community. But opposite effects have also been recorded (Kozłowska and Domurat, 1980; Witkowski, 1985).

Changes in individual size

Individual body size is considered as a distinct feature of the community. Communities dominated by smaller organisms have to transform energy, nutrients, and pollutants at a much higher rate than communities dominated by large organisms (Peters, 1983). So far we do not have many examples concerning nematodes. Additions of sheep manure, and stress caused by peatland drainage followed by nitrogen mineralization cause a reduction of the mean body weight in the total nematode community (Wasilewska, 1974a; 1989a).

Changes in diversity

Both fertilization and microbiological activation of the organic matter (due to peatland drainage) are forms of increasing the content of readily available nutrients in soil. In such a situation Tilman (1982) predicted that species diversity of plant communities and small sessile invertebrates living in the soil water phase will decline. Increased nitrogen mineralization caused by peatland drainage and nitrogen fertilization of a barley crop, respectively, reduced the index of diversity (Sohlenius and Boström, 1986; Wasilewska, 1989a), while manuring of a pasture with sheep dung caused an increase in diversity (Wasilewska, 1974a). This aspect requires further, more thorough study.

Final remarks

Different rates of application of fertilizers, especially high doses, drainage of natural peat grasslands, and intense farming all caused stress to ecosystems. Nematode communities, responded to these stresses. The human impacts investigated did not limit the occurrence of nematodes, on the contrary, it increased their two most abundant components. Their response did differ from the response of humifying organisms (*e.g.* earthworms), the abundance of which is reduced (Kajak *et al.*, 1985). The response of nematodes caused changes in the trophic structure of communities; increases in plant parasites and bacteriophages, and decreases in pantophages and predators, a reduction of species richness and of mean body weight of individuals. Functional consequences of these structural changes were increased plant infestation and increased rate of mineralization caused by abundant bacteriophages and phytophages which both speeded up nutrient cycling. A faster turnover of organic matter, and a reduced stability, due to the elimination of K-strategists; was also found.

References

Anderson H J 1979 Migratory nematodes in Danish barley fields. II. Population dynamics in relation to continuous barley cropping. Tidskrift Planteavl. 83, 9–27.
Anderson R V, Coleman D C and Cole C V 1981 Effects of saprotrophic grazing on net mineralization. *In* Terrestrial Nitrogen Cycles. Eds. F E Clark and T Rosswall. Ecol. Bull., Stockholm. 33, 201–216.
Bååth E, Lohm U, Lundgren B, Rosswall T, Söderström B and Sohlenius B 1981 Impact of microbial-feeding animals on total soil activity and nitrogen dynamics: A soil microcosm experiment. Oikos 37, 257–264.
Boström S 1982 A study of the nematode fauna under barley and grass ley in an experimental field in Central Sweden. MSc-thesis, Dept. Zoology and Appl. Entomology, Imperial College of Science and Technology, Univ. of London. 51 pp.
Coleman D C, Reid C P P and Cole C V 1983 Biological strategies of nutrient cycling in soil systems. *In* Advances in Ecological Research. Eds. A Macfadyen and E D Ford. Academic Press 1–55.
Coulson J C and Whittaker J B 1978 The ecology of moorland animals. *In* Production Ecology of British Moors and Montane Grasslands. Eds. O W Heal and D F Perkins. Ecol. Studies 27, 52–593. Springer Verlag, Berlin.
Dmowska E and Kozłowska J 1986 Experiments on the effect of

cattle liquid manure on changes in communities of soil nematodes. Pol. ecol. Stud. 12, 155–161.

Ferris H 1982 The role of nematodes as primary consumers. *In* Nematodes in Soil Ecosystems. Ed. D W Freckman. Univ. Texas Press, Austin, pp 3–13.

Ferris V R and Ferris J M 1974 Interrelationships between nematode and plant communities in agricultural ecosystems. Agro-Ecosystems 1, 275–299.

Freckman D W and Caswell E P 1985 The ecology of nematodes in agroecosystems. Annu. Rev. Phytopathol. 23, 275–296.

Ingham R E and Detling J K 1984 Plant-herbivore interactions in a North American mixed-grass prairie. III. Soil nematode populations and root biomass on *Cynomys ludovicianus* colonies and adjacent uncolonized areas. Oecologia, Berlin. 63, 307–313.

Ingham R E, Trofymow J A, Ingham E R and Coleman D C 1985 Interactions of bacteria, fungi and their nematode grazers: Effects on nutrient cycling and plant growth. Ecol. Monogr. 55, 119–140.

Kajak A, Andrzejewska L, Chmielewski K, Ciesielska Z, Kaczmarek M, Makulec G, Pętal J and Wasilewska L 1985 Long-term changes in grassland communities of heterotrophic organisms on drained fens. Pol. ecol. Stud. 11, 21–52.

Kozłowska J and Domurat K 1980 Communities of soil nematodes in some crops of Silesia and Masovia regions. Pol. ecol. Stud. 6, 645–654.

Nusbaum C J and Ferris H 1973 The role of cropping systems in nematode population management. Annu. Rev. Phytopath. 11, 423–440.

Okruszko H 1977 Kinds of changes in site conditions on the Wizna fen as influenced by reclamation. Pol. ecol. Stud. 3, 85–95.

O'Neill 1976 Ecosystem persistence and heterotrophic regulation. Ecology 57, 1244–1253.

Peters R H 1983 The Ecological Implication of Body Size. Cambridge Univ. Press, 329 pp.

Petersen H and Luxton M 1982 A comparative analysis of soil fauna populations and their role in decomposition processes. Oikos 39, 286–388.

Philipson J, Abel R, Steel J and Woodell S R J 1977 Nematode numbers, biomass and respiratory metabolisms in a beech woodland — Wytham Woods, Oxford. Oecologia, Berlin. 27, 141–155.

Ryszkowski L 1985a Primary production in agroecosystems. Intecol. Bull. 11, 25–34.

Ryszkowski L 1985b Impoverishment of soil fauna due to agriculture. Intecol. Bull. 12, 7–17.

Scott J A 1979 An ecosystem-level trophic-group arthropod and nematode bioenergetics model. *In* Perspectives in Grassland Ecology. Ed. N R French. pp 107–116. Springer Verlag, New York.

Smolik J D 1974 Nematode studies at the Cottonwood site-grassland biome. Colorado State Univ. Fort Collins, Colo. Techn. Rep. 251, 80 pp.

Sohlenius B 1979 A carbon budget for nematodes, rotifers and tardigrades in a Swedish coniferous forest soil. Holarctic Ecology 2, 30–40.

Sohlenius B and Boström S 1986 Short-term dynamics of nematode communities in arable soil — Influence of nitrogen fertilization in barley crops. Pedobiologia 29, 183–191.

Sohlenius B and Wasilewska L 1984 Influence of irrigation and fertilization on the nematode community in a Swedish pine forest soil. J. Appl. Ecol. 21, 327–342.

Tilman D 1982 Resource Competition and Community Structure. Princeton Univ. Press, 297 pp.

Wasilewska L 1974a Analysis of a sheep pasture ecosystem in the Pieniny mountain (the Carpathians). XIII. Quantitative distribution, respiratory metabolisms and some suggestion on production of nematodes. Ekol. pol. 22, 651–668.

Wasilewska L 1974b Rola wskaźnikowa wszystkożernej grupy nicieni glebowych (The role of the omnivorous group of soil nematodes as ecological indicators). *Summary in English*, Wiad. ekol. 20, 385–390.

Wasilewska L 1976 The role of nematodes in the ecosystem of meadow in Warsaw environs. Pol. ecol. Stud. 2, 4, 137–156.

Wasilewska L 1979 The structure and function of soil nematode communities in natural ecosystems and agrocenoses. Pol. ecol. Stud. 5, 2, 97–145.

Wasilewska L 1985 Nicienie pantofagi i drapieżce a stabilność biocenozy (Pantophagous and predatory nematodes versus the biocenose stability). *Summary in English*, Prace Kom. nauk. Pol. Tow. Glebozn. Warszawa. 90/III/30/, 125–131.

Wasilewska L 1986 Wpływ antropopresji na strukturę i funkcjonowanie zespołów nicieni glebowych (Effect of anthropogenic pressure on the structure and functioning of soil nematode communities). *Summary in English*, Zesz. probl. Post. Nauk. roln. 323, 11–31.

Wasilewska L 1987 Nicienie glebowe w agroekosystemach mazurskiego krajobrazu rolniczego na tle rejonów o większej intensywności uprawy (Soil nematodes in agroecosystems of the Masurian agricultural landscape against the background of regions with higher cultivation intensity of crops). Zesz. probl. Post. Nauk roln. 322, 275–310.

Wasilewska L 1989a Soil nematode communities of peat meadows. I. Soil nematode communities of drained fen 'Wizna' differentiated by origin of peat. Ekol. pol. *In press.*

Wasilewska L 1989b The role of nematodes in agroecosystems. Zesz. probl. Post. Nauk roln. *In press.*

Wasilewska L and Bieńkowski P 1985 Experimental study on the occurrence and activity of soil nematodes in decomposition of plant material. Pedobiologia 28, 41–57.

Winslow R D 1964 Soil nematode population studies. I. The migratory root Tylenchida and other nematodes of the Rothhamsted and Woburn six-course rotation. Pedobiologia 4, 65–76.

Witkowski T 1985 Zgrupowania nicieni glebowych w warunkach chemizacji agrocenoz (Soil nematodes groupings under conditions of chemicalization in agrocenoses). *Summary in English*, UMK Rozprawy Toruń. 100 pp.

Yeates G W 1973 Nematoda of a Danish beech forest. II. Production estimates. Oikos 24, 179–185.

Yeates G W 1977 Nematoda of a Danish beech forest — a correction. Oikos 28, 309.

Yeates G W 1979 Soil nematodes in terrestrial ecosystems. J. Nematol. 11, 213–229.

Yeates G W and Coleman D C 1982 Role of nematodes in decomposition. *In* Nematodes in Soil Ecosystems. Ed. D W Freckman. Univ. Texas Press, Austin, pp 55–80.

M. Clarholm and L. Bergström (Eds.), Ecology of arable land, 133–137.

Cycling of nutrients from dying roots to living plants, including the role of mycorrhizas

E. I. NEWMAN and W. R. EASON
Department of Botany, University of Bristol, Bristol BS8 1UG, UK and Welsh Plant Breeding Station, Plas Gogerddan, Aberystwyth SY23 3EB, UK

Key words: *Lolium perenne*, mycorrhiza, nitrogen, nutrient cycling, phosphorus, roots

Abstract

This paper presents information about the release of nitrogen and phosphorus from dying grass roots and the capture of phosphorus by other, living plants. We have paid particular attention to the part played by mycorrhizas in this phosphorus capture, and the possible importance of mycorrhizal links between dying and living roots.

When *Lolium perenne* plants were grown with ample nutrients and their roots then detached and buried in soil, about half the nitrogen and two-thirds of the phosphorus was lost in three weeks, but only one-fifth of the dry weight. The C:N and C:P ratios suggest that microbial growth in the roots would at first be C-limited but would become N- and P-limited within three weeks.

Rapid transfer of ^{32}P can occur from dying roots to those of a living plant if the two root systems are intermingled. The amount transferred was substantially increased in two species-combinations that are known to form mycorrhizal links between their root systems. In contrast, in a species-combination where only the living ('receiver') plant could become mycorrhizal no significant increase of ^{32}P transfer occurred. This evidence, although far from conclusive, suggests that mycorrhizal links between dying and living roots can contribute to nutrient cycling. This research indicates a major difference in nutrient cycling processes between perennial and annual crops.

Introduction: The role of mycorrhizas in ecosystems

Most higher plants can form mycorrhizas. Fungi closely resembling present-day vesicular-arbuscular (VA) mycorrhizas have been found in underground organs of fossils of early land plants (Wagner and Taylor, 1981), suggesting that the mycorrhizal association has been in existence almost as long as plants have been on land, and that roots and mycorrhizal fungi have evolved together. A plant root system without mycorrhizal infection is the exception rather than the norm.

Most plants of economic importance can form mycorrhizas. This includes nearly all forest trees and pasture plants. Many arable crop plants can also form mycorrhizas, though Cruciferae such as cabbage and rape are important exceptions. There has been much interest in the potential use of mycorrhizal fungi to promote crop production, particularly emphasising their role in increasing uptake of elements which are relatively immobile in soil, phosphorus, copper, zinc and perhaps others (*e.g.* Lambert *et al.*, 1979). Much of the research has studied individual plants sown into bare soil in pots. This simulates the arable crop, but not the mixed-species, mixed-age plant stands that make up most non-crop vegetation. Since mycorrhizas evolved long before there were any arable crops, it is relevant to ask what role mycorrhizas play in natural and semi-natural vegetation. One clue can be found by considering where non-mycorrhizal plants occur; these must be habitats in which mycorrhizas are not essential. In disturbed habitats non-mycorrhizal species are common, though not universal. Mobile areas of sand dunes are an example. Ernst *et al.* (1984) found in a Dutch dune system that many annual species, some of them grasses, had little or no mycorrhizal infection, though some of the perennial species were infected. Arable crop land is also a disturbed habitat, and this raises two questions. (1) Are mycorrhizas likely to be beneficial in arable crops, if so many species persist in natural disturbed areas without them? (2) In less disturbed vegetation, do mycorrhizas have functions other than uptake of immobile elements from soil? A mycorrhizal fungus can infect more than one plant, and thus link them together by interconnecting hyphae (Newman, 1988). This could influence relationships between the linked plants, including competition and nutrient cycling.

This paper considers nutrient cycling, especially the release of nitrogen and phosphorus from dying roots and the part played by mycorrhizal fungi in the capture of these nutrients by living plants. In the past, study of nutrient cycling has paid far more attention to above-ground parts than to roots as sources of nutrients for recycling; yet roots contain a substantial proportion of the nutrients in dying plants. It has been assumed that release of nutrients from dying plants can be studied quite separately from the subsequent capture of these nutrients by living plants. This paper will suggest that such a division may miss important features of the cycling process in perennial vegetation, including the possible role of mycorrhizal links between plants.

Loss of nutrients from ryegrass roots

This paper considers principally nutrient cycling in perennial grassland, using perennial ryegrass, *Lolium perenne* cv S23, as the main experimental species. Several experiments were performed to investigate the rate of loss of phosphorus, and in some experiments nitrogen, from dying roots.

Lolium perenne was grown in solution culture at 20°C in a growth room for six weeks. For 'high-P plants' the solution was half-strength Hoagland's solution throughout; for 'low-P plants' it was the same except that phosphate was omitted for the last three weeks. The orthophosphate in the solutions was labelled with ^{32}P, so the plants became uniformly labelled. At age 6 weeks the shoots were cut

off and removed; the roots were left hanging in aerated solution which was the same as before except now not labelled with ^{32}P, so loss of ^{32}P from roots to solution could be determined by measurements on samples of the solution. Fig. 1 shows the amount of ^{32}P lost during the first three weeks after the roots were detached. The rate of loss was fastest at the start and gradually declined. High-P roots lost about two-thirds of their phosphorus in 22 days, low-P roots about a quarter.

Since this experiment had shown the rate of phosphorus loss becoming quite slow by three weeks, further experiments were carried out to measure the loss of nitrogen and phosphorus during three weeks under more natural conditions. *L. perenne* was grown in solution culture as before, except that the solution was not labelled with ^{32}P. Roots were then cut off the plant, some parts analysed for concentration of nitrogen and phosphorus, other parts placed in fine-mesh nylon bags which were buried at 10 cm depth in soil in pastureland. After three weeks the buried roots were recovered and analysed for nitrogen and phosphorus. Table 1 shows results from one of the experiments, conducted in May when the soil was moist and its mean temperature 11.6°C. A large proportion of not only the phosphorus but also the nitrogen in the rye-

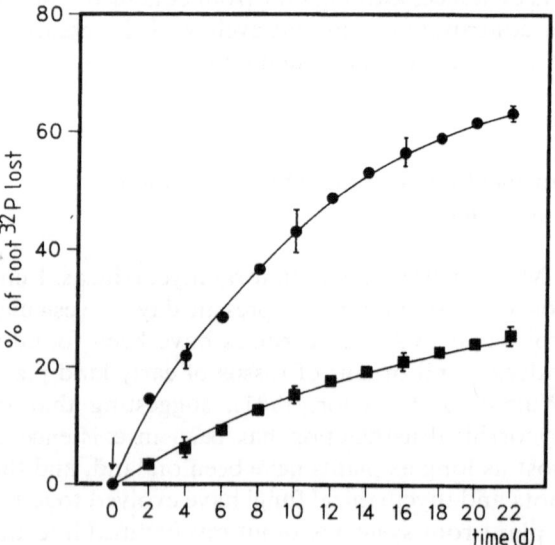

Fig. 1. Cumulative loss of ^{32}P from detached *Lolium perenne* roots suspended in aerated nutrient solution, expressed as a percentage of the amount of ^{32}P in the roots at the start. ● high-P roots, ■ low-P roots. Means of 6 replicate plants. Vertical bars are standard errors. For more details see text.

grass roots was lost within three weeks. The high-P roots lost a far greater percentage of their nitrogen and phosphorus then of their dry weight. The low-P roots, which had similar nitrogen concentration to the high-P roots at the start, lost dry weight and nitrogen at a similar rate to them, but (in agreement with Fig. 1) lost phosphorus more slowly.

Thus the phosphorus status of roots can influence not only the amount of phosphorus available for release but the rate at which it is released. In this experiment the high-P roots contained 4.2 mg P per g dry weight at the start and released 65% of it in three weeks, *i.e.* 2.7 mg per g, whereas the low-P roots contained only 1 mg P per g and released only 19% of it, *i.e.* 0.19 mg per g. We have investigated what classes of phosphorus-containing compounds in the roots contribute to the rapid loss. In high-P roots much of the phosphorus was in water-soluble forms, presumably mainly inorganic phosphate plus phosphate esters, and most of this was lost within three weeks. Most of the phospholipid was also lost within three weeks, but

other water-insoluble forms, including DNA and RNA, decreased little during this period.

Nutrient transfer from dying roots to living plants

In the nutrient cycling process, plants can be viewed as competing against other 'sinks' for the nutrients being released. Other main sinks for nitrogen are loss by leaching and as gases and immobilisation by microorganisms; sinks for phosphorus are chemical fixation, surface adsorption and microorganisms. Saprophytic microorganisms can ramify on the surface of or within the dying roots and are therefore likely to be better positioned to capture nutrients than are living roots. Because labile compounds are often in short supply in soil these microorganisms are likely to depend mainly on the dying root for carbon sources as well as mineral nutrients. Table 1(b) shows that the C:N ratio of *Lolium* roots was initially in the range below 20 conventionally taken to indicate carbon-limitation to microbial growth (Swift *et al.*, 1979). During three weeks of decomposition, however, the loss of nitrogen was so much more rapid than of carbon that the C:N ratio by then indicates N as limiting, rather than C. If 200 is taken as the critical value for C:P ratio (Hayman, 1975), the high-P roots also became P-limited during the three weeks. In the low-P roots the high C:P ratio would normally be taken to predict P immobilization, yet loss of P did occur. Our results suggest, therefore, that there is a short period during the initial decomposition of roots when release of nitrogen and phosphorus compounds can be so fast that saprophytic microorganisms are unable to capture a substantial proportion of them, which therefore go to other sinks. Timing in capture of the released mineral nutrients could therefore be crucial in determining what proportion goes into living plants, and this may be where mycorrhizal fungi play an important part.

To study phosphorus transfer from dying roots of *Lolium perenne* to living *Plantago lanceolata* (a common herb of British grasslands), the two plants were grown together in a pasture soil in pots, either with or without mycorrhizas. Carrier-free ^{32}P was fed to the shoots of the *Lolium*. After allowing four days for the roots of the *Lolium* to become labelled its shoot was removed. The time-course of ap-

Table 1. Results from an experiment in which roots of *Lolium perenne*, grown with either ample or deficient phosphorus supply, were detached from the plants and buried in soil for three weeks. Values are means of 11 replicates, with standard errors in brackets. (Eason, 1987)

(a) Percentage of initial dry weight, nitrogen and phosphorus in roots that was lost during three weeks

	Percentage loss		
	Dry weight	Nitrogen	Phosphorus
High-P roots	21.9 ($\pm$3.8)	51.5 ($\pm$5.3)	65.2 ($\pm$3.9)
Low-P roots	17.6 ($\pm$5.4)	53.1 ($\pm$3.7)	19.0 ($\pm$5.8)

(b) C:N and C:P ratios of roots at start and end of three-week period. Calculations assume weight of carbon = 0.4 × dry weight

	C:N		C:P	
	At start	After 3 weeks	At start	After 3 weeks
High-P roots	16.1 ($\pm$1.5)	26.2 ($\pm$1.3)	105 ($\pm$11)	242 ($\pm$13)
Low-P roots	16.6 ($\pm$1.3)	29.4 ($\pm$1.8)	410 ($\pm$25)	422 ($\pm$29)

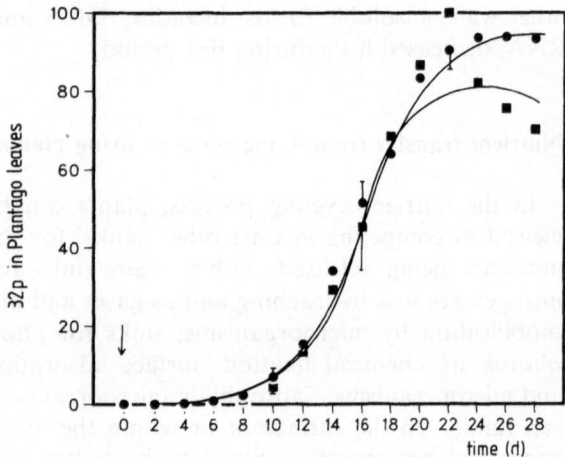

Fig. 2. Time-course of transfer of ^{32}P from roots of *Lolium perenne*, whose shoots were detached on Day 0, to leaves of intact *Plantago lanceolata* plants. ● VA-mycorrhizal, ■ non-mycorrhizal. ^{32}P is expressed in arbitrary units, standardized so as to make the curves meet at 50. All values corrected for isotope decay. Vertical bars are standard errors. Data of Eason (1987).

pearance of ^{32}P in the living *Plantago* was measured by placing leaves against a scintillation probe. Fig. 2 shows that detaching the *Lolium* shoot resulted, after a few days' lag, in a rapid transfer of ^{32}P into the *Plantago*, but this declined to a slow rate after about three weeks. The shape of the time-course was similar for mycorrhizal and non-mycorrhizal plants. The total amount of ^{32}P transferred from *Lolium* to *Plantago* was substantially greater if the plants were mycorrhizal than if they were non-

mycorrhizal. Mycorrhizas cause a similar increase of ^{32}P transfer if both the 'donor' and the 'receiver' plant are *Plantago* (Table 2).

These two species, *Lolium perenne* and *Plantago lanceolata*, can when growing in this soil form mycorrhizal links between their root systems (Heap and Newman, 1980a), and links have also been reported between several other pairs of species (Newman, 1988). This raises the possibility that mineral nutrients could be transported between plants via these hyphal links. This would require transfer of the nutrient from host to fungus, the reverse of the normal direction. There is conflicting evidence (Newman, 1988) on whether substantial amounts of nutrients are transported in this way between living plants. When roots of one plant are dying, movement of nutrients from root to mycorrhizal fungus seems more possible. Evidence on whether direct transfer of phosphorus between plants via mycorrhizal links can then occur is so far limited, and we summarize here only one piece of evidence. We have carried out two experiments similar to the *Lolium*-to-*Plantago* and *Plantago*-to-*Plantago* ^{32}P transfer experiments, except that the donor plant was cabbage (*Brassica oleracea*). Cabbage and *Plantago* were grown together in pots of soil. Since cabbage does not form mycorrhizas (confirmed in these experiments) no hyphal links between donor and receiver plant were possible. In some pots *Plantago* was mycorrhizal, so its associated hyphae could ramify around the dying roots of the cabbage but not penetrate them; in other

Table 2. Influence of VA-mycorrhizal infection on transfer of ^{32}P from dying roots of 'donor plants' to living *Plantago lanceolata*, expressed as ratio of amount transferred when plants mycorrhizal: amount transferred when non-mycorrhizal. In experiments involving *L. perenne* or *P. lanceolata* as the donor plant, both species became mycorrhizal in the 'mycorrhizal' treatment, and hyphal links between the plants would form. In contrast, when cabbage was the donor only the *Plantago* could become mycorrhizal. Each figure is from a different experiment, but all used the same soil and similar growth conditions

Donor plant	^{32}P transfer, mycorrhizal / ^{32}P transfer, non-mycorrhizal	Statistical significance[a]	Source of data[b]
Lolium perenne	2.2	$P < 0.05$	1
	3.2	$P < 0.02$	1
	4.2	$P < 0.001$	1
	3.1	ns	2
Plantago lanceolata	2.8	$P < 0.05$	1
	2.5	$P < 0.05$	3
Brassica oleracea	0.8	ns	3
(cabbage)	0.8	ns	3

[a] Indicates whether ratio significantly different from 1. ns = not significant.
[b] 1, Heap and Newman (1980b); 2, Ritz (1984); 3, Eason (1987).

pots both species were non-mycorrhizal. ^{32}P was fed to cabbage through some of its leaves. Later its shoot was removed but the root left in the soil. The amount of ^{32}P in the *Plantago* was determined four weeks later. Table 2 shows that mycorrhizal infection caused no increase in the amount of ^{32}P transferred. This contrasts with the *Lolium-Plantago* and *Plantago-Plantago* experiments where hyphal links did occur, and mycorrhizas consistently increased transfer.

Clearly these experiments do not provide conclusive evidence. The question merits further study, since if mycorrhizal hyphae do allow direct transfer of nutrients from dying to living roots this could be important in allowing the nutrients to bypass other potential sinks in the soil.

Conclusions

This paper has concentrated on perennial grassland, where dying and living roots are likely to intermingle. We have shown that ryegrass roots detached from their shoot lose nitrogen and phosphorus rapidly, and that when dying ryegrass roots intermingle with living roots of another plant rapid transfer of phosphorus to the living plant can occur. Mycorrhizal fungi probably play an important role during this period, especially in cases where saprophytic microorganisms are limited by supply of labile carbon compounds. We have presented evidence that direct mycorrhizal links between roots are involved in phosphorus transfer from dying to living roots. It is likely, however, that mycorrhizal hyphae in soil, attached to living plants, also increase nutrient cycling by being well positioned to capture water-soluble nutrients as they leak out of dying roots.

Our results show that there is a crucial period of a few weeks during the initial decomposition of dying roots when success of living roots in capturing the released nutrients is crucial in determining how much of the nutrients remain in the plant biomass rather than being lost to other 'sinks'. This could have a fundamental influence on the productivity of the whole ecosystem. We suggest that it is important to find out whether similar rapid nutrient transfer from dying to living roots occurs in other types of vegetation. Our research indicates a fundamental difference in nutrient cycling between

perennial crops, where dying and living roots are likely to intermingle, and annual crops many of whose roots die at a time when no other plants are present to capture the released nutrients. The tendency of annual crop-land to lose more nitrogen by leaching than perennial grassland has been reported but, it has not been suggested how much of the lost nitrogen may come from dying roots. We suggest that arable cropping may also result in more phosphorus being lost from plant-available pools to less available pools in the soil that turn over only very slowly. Our results emphasize the potential value of undercropping practices in retaining nutrients in available form in the system, and suggest that it may be advantageous for the crop and the undercrop species to share the same mycorrhizal fungus.

Acknowledgement

This research was supported by grants from the Natural Environment Research Council.

References

Eason W R 1987 The Cycling of Phosphorus from Dying Roots Including the Role of Mycorrhizas. Ph.D. Thesis, University of Bristol.

Ernst W H O, van Duin W E and Oolbekking C T 1984 Vesicular-arbuscular mycorrhiza in dune vegetation. Acta Bot. Neerl. 33, 151–160.

Hayman D S 1975 Phosphorus cycling by soil microorganisms and plant roots. *In* Soil Microbiology. Ed. N Walker. pp 67–91. Butterworths, London.

Heap A J and Newman E I 1980a Links between roots by hyphae of vesicular-arbuscular mycorrhizas. New Phytol. 85, 169–171.

Heap A J and Newman E I 1980b The influence of vesicular-arbuscular mycorrhizas on phosphorus transfer between plants. New Phytol. 85, 173–179.

Lambert D H, Baker D E and Cole H 1979 The role of mycorrhizae in the interactions of phosphorus with zinc, copper, and other elements. Soil Sci. Soc. Am. J. 43, 976–980.

Newman E I 1988 Mycorrhizal links between plants: Their functioning and ecological significance. Adv. Ecol. Res. 18, 243–270.

Ritz K 1984 Phosphorus Transfer between Grassland Plants. Ph.D. Thesis, University of Bristol.

Swift M J, Heal O W and Anderson J M 1979 Decomposition in Terrestrial Ecosystems. Blackwell, Oxford.

Wagner C A and Taylor T N 1981 Evidence for endomycorrhizas in Pennsylvanian Age Plant fossils. Science 212, 562–563.

M. Clarholm and L. Bergström (Eds.), Ecology of arable land, 139–148.

Mycorrhizal mycelia and their role in soil and plant communities

R.D. FINLAY and B. SÖDERSTRÖM
Department of Microbial Ecology, University of Lund, Helgonavägen 5, S-223 62 Lund, Sweden

Key words: biomass, carbon cycling, mycorrhiza, mycelia, nitrogen, nutrient cycling, plant communities

Abstract

Studies of mycorrhizal plant roots during the last hundred years have yielded much information concerning the structure and physiology of mycorrhizal associations. Great emphasis has been placed upon the nutritional effects of infection on individual plants or roots, but study of the mycelial phase of the symbiosis in the soil is less well developed, and our understanding of the wider significance of mycorrhizal associations at the level of the plant community or ecosystem has remained relatively restricted. Adequate methods to quantify and study the mycorrhizal mycelium must be developed and further research is required to investigate the processes of mycelial uptake and translocation to plants and to identify the full spectrum of nutrients which may be important. Non-nutritional effects of mycorrhizal infection may also be of considerable importance and the extent to which mycorrhizas influence nutrient cycling and plant interactions within communities merits further attention. We discuss some recent experimental results and attempt to outline some areas of research which will broaden our understanding of the role of mycorrhizal symbiosis in this wider context.

Introduction

Symbioses between plant roots and mycorrhizal fungi are almost universal in terrestrial ecosystems and much research effort has been concentrated on the generally beneficial effects of these associations on plant growth and nutrient uptake. It is now well established that, in the case of vesicular-arbuscular (VA) and ecto-mycorrhizas at least, these effects arise in part from the provision of an increased and more efficiently dispersed nutrient absorbing surface area, in the form of an extramatrical mycelium, and numerous pot experiments have demonstrated increased plant yield and improved nutrient status in mycorrhizal plants growing in soils of low fertility.

Mycorrhizal plants dominate a wide range of ecosystems and, although the most commonly documented effect of infection is an enhancement of tissue phosphorus concentration, Read (1986) has pointed out that, throughout the wide range of edaphic conditions in which mycorrhizal symbiosis prevails, it is unlikely that its major effect arises solely from the increased inflow of P to roots. The full range of roles which mycorrhiza may play in soil microbial and plant communities has yet to be revealed and further research is required to refine our understanding of their significance in this wider context.

In this paper we do not attempt to provide an exhaustive review of the effects of mycorrhizal infection, but rather to present a personal view of specific areas of ecological interest where information is lacking and in which further research is likely to broaden our understanding of the role which mycorrhizal symbiosis plays. Firstly we consider interactions between mycorrhizas and soil microbial and faunal populations. These interactions include those between different mycorrhizal populations themselves and require the study of microbial biomass and dynamics, competition between, and energy flow through, microbial populations. Many of these processes will exert an indirect effect on plant communities whose roots are influenced by the outcome of interactions between microbial populations and their effects on the abiotic environment. Secondly we discuss the nutritional and non-nutritional interactions between mycorrhizal fungi and plant roots and their influence on the structure, development and functioning of plant communities. The discussion relates principally to vesicular-arbuscular and ectomycorrhizal fungi, although ericoid mycorrhizal associations are briefly considered also. Our emphasis is on natural and semi-natural (plantation) vegetation systems rather than on plant monocultures in the strict agricultural sense.

Interactions between mycorrhizal fungi and soil microbial and faunal populations

Carbon flow and biomass

Despite the large number of studies of the structure and function of mycorrhizal roots, the mycelial phase of the symbiosis in the soil has received much less attention. The delicate nature and inaccessibility of the mycelium make it difficult to study in a non destructive manner and reliable methods with which to quantify it have still not yet been developed.

It is well established that the fungal symbiont derives supplies of energy rich carbon compounds from its host plant and the significance of this carbon input to the soil ecosystem has been discussed by a number of authors (Harley and Smith, 1983; Read *et al.*, 1985). Loss of energy rich compounds from plant roots to soil microbial populations constitutes an important supply process and mycorrhiza can be regarded as a highly specialized form of rhizosphere association which has become partly invasive. Mycorrhizal fungi are unique in that they have direct access to host assimilates and their widespread distribution is testament to the fact that they possess a competitive advantage over other components of the soil microbial community in this respect. Study of these fungal mycelia is important not only in that they play an important absorptive role in plant nutrition but that they must also constitute an important nutrient resource for the whole soil biota.

One question of fundamental importance is the amount of carbon that host plants invest in

monstrated that plants-infected with VA mycorrhizal fungi allocated 8% more carbon to their roots hizal fungi allocated 8% more carbon to their roots than did uninfected ones. In a study of Soybean-VAM-Rhizobium systems, Harris *et al.* (1985) concluded that 10–20% of net assimilate may be invested in the external mycelium alone. Early estimates of annual carbon allocation to ectomycorrhizal fungi by Romell (1939) suggested that it was equivalent to 10% of potential timber production. This figure was regarded as an upper limit by Harley and Smith (1983) despite the fact that the calculations were based solely on fruitbody biomass and did not take account of the vegetative mycelial phase. More recent field studies by Fogel and Hunt (1979) and Vogt *et al.* (1982) have indicated that higher amounts (15–25% of net primary production) of carbon may be consumed by the fungi, and Reid *et al.* (1983) estimated that 22–65% of the carbon assimilated by *Pinus radiata* was translocated to ectomycorrhizal roots and it is possible that half of this may have been used for fungal growth. In a recent study by Söderström and Read (1987) it was suggested that, under specific circumstances, as much as 30% of the total root respiration may be accounted for by respiration of the extramatrical mycelium alone.

Much of the carbon allocated to mycorrhizal roots and subsequently to the fungi themselves will be used for fungal biomass production and estimation of this biomass has provided one way to estimate the carbon flow mentioned above. The biomass of VA fungi within infected roots has been estimated to be between 4 and 17% of total root weight by Hepper (1977) and 10% of total root weight by Tinker (1978). Estimates of the extraradical mycelium, although few, indicate that it is a significant component of the total fungal biomass. Bevege *et al.* (1975) estimated that it constituted 0.9% of total root weight and estimates of mycelial length have varied between 80 cm per cm root length (Sanders and Tinker, 1973) and 2.5–14 m per cm root length (Abbot and Robson, 1985), while Tisdall and Oades (1979) measured up to 50 m of hyphae per gram of grassland soil. Estimates of ectomycorrhizal mycelial length are even fewer. Read and Boyd (1986) measured between 10 and 80 m of hyphae per cm root length in laboratory systems containing ectomycorrhizal mycelia. If data on mycelial respiration rates in similar systems

(Söderström and Read 1987) are applied to relationships between fungal biomass and respiration observed by Bååth and Söderström (1988) figures in excess of 200 m per gram dry weight of soil are obtained and these are similar to those of 100–700 m/g presented earlier by Söderström (1979) for total fluorescein diacetate stained hyphal length in the F/H horizon of Swedish pine forests.

Clearly VA and ecto- mycorrhizal mycelia constitute important components of the soil microbial community and more reliable methods of quantifying biomass, hyphal length and activity are required to assess their role in the carbon economy of soil ecosystems. Refinement of existing microscopical, physiological and immunological methods is an important priority for the development of field studies. Further progress in understanding the biology of mycorrhizal mycelia and their interactions with other components of the soil microbial community can also be made using intact plant/mycelial systems grown under semi-natural conditions in the laboratory. Whilst these systems will not fully reflect natural conditions, they have the advantage that controlled growth conditions can be provided for comparative studies using a range of natural, non-sterile substrates and that manipulation of the mycelium is possible.

Mycelial dynamics and interactions with microbial populations

Although the dynamics of the infection process and the spread of infection within VA mycorrhizal plant roots have received some study (Buwalda *et al.*, 1982; Sanders, 1986; Smith and Walker, 1981), little is currently known about the dynamics of the mycelial phase within the soil. Further information about its temporal and spatial dynamics are still required. In many soils young, uninfected plant roots may grow into a fungal matrix composed of the mycelia of different mycorrhizal fungi. The outcome of interactions between these fungi will initially be influenced by competition for infection of these roots and subsequent supply of host assimilates. It is generally accepted that low levels of host specificity are important in this process and indirect methods of assessing the outcome of these interactions have led to the study of colonization of uninfected roots and the notion of mycorrhizal

succession (Fleming, 1983). We still know very little, however, about the subsequent mechanisms of interaction between different mycorrhizal fungi and between mycorrhizal mycelia and other microbial populations.

Interactions between ectomycorrhizal fungi and bacteria (Bowen and Theodorou, 1973; 1979) show no clear pattern and the effect of bacteria on fungal root colonization can be stimulatory, inhibitory or neutral. Competition for energy sources and antibiosis may exert negative effects and positive effects may be direct or indirect via stimulation of increased leakage of nutrients from the roots. Interactions between mycorrhizal mycelia and saprophytic organisms will depend on competition for energy rich substrates and ectomycorrhizal fungi possess a competitive advantage in this respect in that they have direct access to host assimilates. The depression of rates of litter decomposition by ectomycorrhizal fungi shown by Gadgil and Gadgil (1971, 1975) lends evidence to this theory, although antibiotic effects cannot be ruled out as a possible explanation of the results. It has recently been demonstrated that some species of ectomycorrhizal fungi possess the ability to utilize complex organic nitrogen sources that are not directly available to plants (Abuzinadah and Read, 1986a; b; Abuzinadah *et al.*, 1986). The full extent of this saprophytic capability in different species is not yet known but it will clearly influence competitive interactions between ectomycorrhizal fungi and other populations and may be of great significance for nutrient cycling as a whole since the mycorrhizal fungi may return assimilated protein to the host tree, thereby reducing reimmobilization of nutrients in decomposer populations.

Pure culture studies may reveal important physiological differences between different species and isolates but it is clearly important to study mycelial interactions in the presence of the host plants where there is a natural carbon balance. The use of mycelial growth chambers provides one possible method of studying mycelial growth patterns and interactions in different substrates in semi-natural conditions using different combinations of known mycorrhizal fungi and host plants. Mycelial growth, differentiation and substrate colonization by *Suillus bovinus* has been observed in this way by Finlay and Read (1986a). In differentiated mycelial systems colonization of peat is not uniform and it is tempting to speculate that ecto-mycorrhizal mycelial systems are capable of differential exploitation of soil heterogeneity in a manner similar to that described by St John *et al.* (1983a, b). An ability to colonize areas of soil rich in organic material would be of particular importance on those species which have been shown to possess proteolytic capability.

Interactions with soil animals

In spite of the possibility that mycorrhizal hyphae may often constitute a significant component of the soil fungal biomass and that there is an extensive literature relating to fungal feeding by soil animals, most laboratory experiments on the effects of mycorrhizal infection have excluded the soil fauna. The effects of invertebrate populations have been considered largely in terms of the overall influence of grazing on decomposition processes and nutrient cycling (Anderson and Ineson, 1984; Coleman *et al.*, 1984), but direct interactions with mycorrhiza are less frequently considered (Finlay, 1985; Shaw, 1985). Mycorrhizal mycelia may constitute a nutrient resource of great significance for the soil biota and the direct impact of grazing on plant growth, as well as its wider implications for nutrient cycling, deserve further attention.

Nutritional and non-nutritional interactions between mycorrhizal fungi and plant roots: Influences on structure, development and functioning of plant communities

Nutritional interactions

Whilst it is generally assumed that mycorrhizal mycelia play a major role in the process of nutrient absorption and transport to plants there have been relatively few studies involving the direct observation and quantification of these processes. Translocation of phosphorus has been demonstrated in VA- (Rhodes and Gerdemann, 1975) and in ecto- (Melin and Nilsson, 1950) mycorrhizal hyphae. Quantitative studies of hyphal uptake and translocation of nutrients by VA mycorrhizal mycelium (Cooper and Tinker, 1978; 1981) have enabled the direct measurement of flux rates through individual hyphae but less quantitative information is available for ectomycorrhizal systems. Finlay and Read (1986b) demonstrated movement of ^{32}P over distances in excess of 40 cm in ectomycorrhizal mycelia of *Suillus bovines* and measured flux rates similar to those obtained in VA systems, suggesting that transport is primarily by symplastic flow. Further, more detailed studies are necessary to determine the transport pathway at the cellular level and to investigate the influence of different environmental variables on translocation rates.

Information concerning the role mycorrhiza play in nitrogen nutrition is less readily available but Ames *et al.* (1983) have shown that *Glomus mosseae* hyphae can transport ammonium ions to plants and Melin and Nilsson (1952, 1953) demonstrated the transfer of labelled nitrogen through ectomycorrhizal mycelia to pine seedlings. Almost 100 years ago Frank (1894) proposed that ectomycorrhizal infection might provide access to organic N compounds and pointed out that these constituted the major source of nitrogen in organic forest soils. There have been few studies of nitrogen uptake and translocation since then and most of these have involved mineral nitrogen sources. Recent experiments by Abuzinadah and Read (1986a, 1986b) and Abuzinadah *et al.* (1986) have demonstrated proteolytic capability in a number of ectomycorrhizal fungi and increased supply of N from protein sources to infected plants. Utilization of N from protein sources has also been demonstrated in mycorrhizal ericaceous plants (Bajwa *et al.* 1985) and suggests that greater attention should be paid to the dynamics of organic nitrogen sources in field studies of mycorrhizal plants. Detailed studies of the utilization of inorganic and organic ^{15}N-labelled nitrogen sources in a range of ectomycorrhizal associations are now being carried out in our laboratory (*e.g.* Finlay *et al.*, 1988).

Non-nutritional interactions

Mycorrhizal infection occurs in many different terrestrial ecosystems characterized by a wide range of soil conditions. Distinct types of association are found under different edaphic conditions along altitudinal or latitudinal gradients which have been recognized by Read (1984). The dominant vegetation types along these gradient are determined by a range of factors among which non-nutritional ones such as soil pH and water availability may be of prime importance.

Non-nutritional effects of mycorrhizas have received much less attention than nutritional effects but there is a growing body of evidence to suggest that these may be of major importance in certain environments. Recent experiments (Allen *et al.*, 1981; Allen, 1982; Hardie, 1986) suggest that water flow through VA mycorrhizal mycelium may constitute a significant component of the transpirational flux. In ectomycorrhizal associations the vessel hyphae of the mycorrhizal mycelium are well suited to the transport of water and Duddridge *et al.* (1980) have shown that water can be transferred through the mycelial strands. Experiments by Brownlee *et al.* (1983) and Read and Boyd (1986) have shown that reductions of transpiration in excess of 40% and of photosynthesis in excess of 25% are obtained when mycelial systems connecting pine and birch seedlings to moist peat are severed, suggesting that mycelial uptake of water may be of great importance. Further work is required in this area to assess the importance of mycelial water uptake under different field conditions and to determine the transport pathway at the cellular level.

Resistance to heavy metal toxicity has been demonstrated in mycorrhizal ericaceous plants (Bradley *et al.*, 1982) and in ectomycorrhizal plants (Jones and Hutchinson, 1986; Denny and Wilkins, 1987) and may be important in acid environments where the availability of these metals is likely to be high. Such environments are exemplified by acid heathland communities where concentrations of phyto-toxic organic acids are also high and recent experiments by Leake (1987) suggest that the ability of the ericoid mycorrhizal fungus *Hymenoscyphus ericae* to metabolize phenolic acids may provide ericaceous plants with a critical defence against the high levels of these compounds usually found in mor humus.

Specificity and compatibility

In systems where the roots of different plants are linked by the mycelial networks of different fungal species, the degree of host specificity shown by the fungi may have important consequences for the plant community as a whole. In general the degree

of specificity shown by mycorrhizal fungi is apparently low and where it does exist it generally occurs at the genus level or above (Duddridge, 1986a). The low host specificity of VA mycorrhizal fungi has been shown in a range of experiments where inoculation has been carried out without the use of organic amendments, but in many ectomycorrhizal experiments the high levels of carbohydrate frequently added to the synthesis medium can cause abnormal development of the host-fungus interface and Duddridge (1986b, c) has stated the importance of omitting carbohydrate in studies of compatibility, Molina and Trappe (1982) studied the patterns of host specificity in 27 fungal species with seven host species and distinguished three groups of fungi, those with a broad host range, those with an intermediate host range which were nevertheless specific or limited in their sporocarp-host associations, and those which had only a narrow host potential. In studies of incompatible associations such as those between *Suillus grevillei* and both *Pseudotsuga menziesii* and *Pinus sylvestris* (Duddridge, 1986b, c) phenolic substances appear to be formed at the host-fungus interface, host wall appositions, lignification or thickening may occur, and loose, unstructured sheaths and irregular Hartig net development have also been demonstrated. Investigation of such incompatible interactions will provide more information about compatibility and recognition in mycorrhizas and has already provided much ultrastructural information but is is not known whether the host-fungus interface is functional in such associations and further studies of the efficiency of assimilate translocation and nutrient uptake and transport are required in such situations.

Influence of infection on plant-plant interactions and community structure and development

Many studies of the effects of mycorrhizal infection have considered the responses of individual roots or of individual plants, but in many vegetation systems the roots of different plants grow in close proximity to each other. One consequence of the extensive colonization of soil by the extramatrical mycelium of both VA and ectomycorrhizal fungi is that young, uninfected roots germinating into soil rapidly become infected by mycelium growing from adjacent plants. This infection process leads to the formation of mycelial connections both within individual plant root systems and between the root systems of adjacent plants. The low levels of host specificity shown by many mycorrhizal fungi permit both intra-specific plant connections (Hirrel and Gerdemann, 1979) as well as inter-specific connections (Heap and Newman, 1980a) and there has been much interest in the functional significance of these as pathways for the transfer of nutrients between plants. Hirrel and Gerdemann (1979) showed that mycorrhizal infection enhanced transfer of carbon between onion plants and Heap and Newman (1980b) have shown that ^{32}P transfer from the roots of dying plants to those of neighbouring plants was enhanced if they were mycorrhizal, but they were not able to determine the extent to which this transfer was by means of direct hyphal connections or by facilitated hyphal uptake of leaked ^{32}P from the soil solution since the roots were beginning to decompose. Transfer of ^{32}P between living mycorrhizal plants has been demonstrated in the field (Chiariello, *et al.*, 1982), and in pot experiments by Whittingham and Read (1982), who demonstrated that the flux of nutrients from nutrient-enriched source plants was sufficient to produce significant yield responses in younger mycorrhizal 'sink' plants. Only indirect evidence of a mycorrhizal pathway was available from the above studies but autoradiographic evidence of direct hyphal transfer of ^{14}C has been demonstrated by Francis and Read (1984), who also showed transfer of labelled assimilate could be enhanced by inducing carbon sinks in recipient plants by different levels of shading. The relative importance of direct hyphal transfer of nutrients and of facilitated uptake will depend upon a number of conditions but Read *et al.* (1985) point out that direct transfer will be more efficient than facilitated uptake because the nutrients are absorbed from intracellular pools to which the fungus alone has access, and are transported to distant sinks without being released to the saprophytic flora. Intra-specific transfer of N and P between mycorrhizal *Plantago lanceolata* and *Festuca ovina* plants, resulting in significant yield increases and significantly elevated contents of these nutrients have been shown by Francis *et al.* (1986). Recent experiments by Occampo (1986) involving 'host' and 'non-host' plants also suggest that mycelial linkage may be an important mechanism of nutrient transfer in plant communities.

Mycelial connections between ectomycorrhizal plants have also been shown to occur and direct transfer of nutrients in these systems has been postulated (Woods, 1970; Woods and Brock, 1964). Natural connections between ectomycorrhizal pine plants have been studied by Brownlee *et al.* (1983), who demonstrated autoradiographically the capacity of mycelial strands to act as functional pathways for the transfer of labelled plant assimilate. This phenomenon was studied further by Finlay and Read (1986a) who presented limited evidence that transfer of assimilate could be increased in some cases by shading plants to induce metabolic sinks. Such experiments, in common with those of Francis and Read (1984), demonstrate that the mycelial system provides a pathway for the transport of substances along chemical concentration gradients from areas of high substrate availability to areas of low substrate availability, but do not provide unequivocal evidence of net transfer. Similar patterns of transfer have also been demonstrated in ectomycorrhizal systems in the field (Read *et al.*, 1985) but these authors have pointed out (Francis *et al.*, 1986) that the full significance of such results cannot be revealed by relatively short term experiments using isotopes. The existence of mycelial connections between plants may have profound consequence in terms of its influence on plant-plant interactions both in natural situations and in agricultural systems where two species are intercropped and should be the subject of further intensive investigation. There have been few studies of the influence of mycorrhizal infection on plant competition. Fitter (1977) showed that mycorrhizal infection radically altered the outcome of competition between two grasses *Holcus lanatus* and *Lolium perenne* and Hall (1978) demonstrated that *Trifolium repens* was better able to compete with *Lolium perenne* in the presence of mycorrhizal infection. These competitive interactions were assumed to have a nutritional basis, but other factors may also be important and in recent experiments by Leake (1987) involving competition between Festuca and Calluna plants, the superior competitive ability of mycorrhizal Calluna plants was partly attributed to the ability of the mycorrhizal endophyte to metabolize potentially toxic phenolic acids. In situations where disturbance of climax communities leads to successional process mycorrhizal infection may play an important role in influencing rates of the succession since primary colonists of disturbed sites are frequently 'non-mycotrophic' and usually replaced by obligately mycorrhizal species. The influence of mycorrhiza on succession has been discussed by Janos (1980) and Allen and Allen (1980, 1984) who conclude that inoculum density may influence rates of succession. Further study of these important processes is necessary and of practical as well as academic importance, especially in relation to revegetation practices of arid land (Reeves *et al.*, 1979).

The importance of nutrient transfer between connected hosts will clearly vary according to the species composition of the community, the age and nutrient status of the plants, and of the environment. Ritz and Newman (1984) observed exchange of P between intact grassland plants of the same age, but concluded that there was no net transfer. In situations involving plants of different ages and nutrient status net transfer may be of greater importance since many grassland herbs and grasses germinates in a restricted volume of rooting space and deplete their seed reserves rapidly. Integration of young seedlings into both VA and ectomycorrhizal mycelial networks can occur within 48 hours and may be of great importance in situations where plant development takes place in closed stands of established vegetation, seed reserves are limited and access to light is restricted by established plants. In addition to receiving supplies of assimilate from the overstorey vegetation the developing root system gains access to soil nutrients supplied by an extensive mycelial network. In a recent experiment using turf microcosms (Grime *et al.*, 1987) the presence of VA mycorrhizal infection has been shown to influence seedling survivorship and promote species diversity. The implications of the above processes in terms of survivorship and community structure may be great and further experiments of this type are needed to improve our understanding of mycorrhizal effects at the community level.

Concluding remarks

Direct studies of the mycorrhizal mycelium have been neglected because it is difficult to observe and manipulate in a non-destructive manner. The mycelial phase of the symbiosis is of fundamental importance however, since it is in intimate contact with the soil and represents a major pathway for the flow of nutrients to, and between, plants and of energy-rich carbon compounds to the soil ecosystem. In many soils mycorrhizal fungi constitute a significant component of the microbial community but we still know little about their interactions with other microbial populations or with soil animals. Further research is required to refine methods of quantifying mycelial biomass and dynamics in the field. Laboratory studies of intact mycorrhizal systems are also necessary to provide more detailed information about the biology of the mycelium, the processes involved in nutrient absorption and translocation and the way these are influenced by different environmental variables.

In studies of the nutrition of mycorrhizal plants there has been great emphasis on mineral nutrients in general and on phosphorus in particular but less information is available concerning other growth limiting nutrients such as nitrogen. Recent research suggests that ericaceous and ectomycorrhizal plants are able to utilize complex organic nitrogen compounds. This may be of great significance in environments where mineralization rates are restricted by low temperatures and further research is required to determine the extent of this proteolytic capability in different mycorrhizal fungi.

The past hundred years of mycorrhizal research have provided much information concerning the structure and physiology of the symbiosis and its nutritional effects on individual plants. It is now necessary to refine our understanding of the effects of mycorrhizal infection in a broader sense. Detailed studies of the biology of the mycelium, together with studies of specific habitats will reveal the importance of non-nutritional effects of infection in different environments. The consequences of mycorrhizal infection at the community and ecosystem level may have profound implications for nutrient cycling and plant interactions in different environments and further study of these phenomena should provide important developments in our understanding of the dynamics of terrestrial ecosystems.

References

Abbott L K and Roson A D 1981 Infectivity and effectiveness of vesicular-arbuscular mycorrhizal fungi: Effect of inoculum

146   *Finlay and Söderström*

type. Aust. J. Agric. Res. 32, 631–639.

Abuzinadah R A and Read D J 1986a The role of proteins in the nitrogen nutrition of ectomycorrhizal plants. I. Utilization of peptides and proteins by ectomycorrhizal fungi. New Phytol. 103, 481–493.

Abuzinadah R A and Read D J 1986b The role of proteins in the nitrogen nutrition of ectomycorrhizal plants. III. Protein utilization by Betula, Picea and Pinus in mycorrhizal association with *Heboloma crustiliniforme*. New Phytol. 103, 507–514.

Abuzinadah R A, Finlay R D and Read D J 1986 The role of proteins in the nitrogen nutrition of ectomycorrhizal plants. II. Utilization of protein by mycorrhizal plants of *Pinus contorta*. New Phytol. 103, 495–506.

Allen M F 1982 Influence of vesicular-arbuscular mycorrhizae on water movement through *Bouteloua gracilis* (H.B.K.) Lag ex Steud. New Phytol. 91, 191–196.

Allen E B and Allen M F 1980 Natural re-establishment of vesicular-arbuscular mycorrhizae following strip-mining reclaimation in Wyoming. J. Appl. Ecol. 17, 139–149.

Allen E E and Allen M F 1984 Competition between plants of different successional stages: Mycorrhizae as regulators. Can. J. Bot. 62, 2625–2629.

Allen M F, Smith W K, Moore T S and Christensen M 1981 Comparative water relations and photosynthesis of mycorrhizal and non-mycorrhizal *Bouteloua gracilis* H.B.K. lag ex steud. New Phytol. 88, 683–693.

Ames R N, Reid C P P, Porter L K and Cambardella C 1983 Hyphal uptake and transport of nitrogen from two ^{15}N labelled sources by *Glomus mosseae*, vesicular-arbuscular mycorrhizal fungus. New Phytol. 95, 381–396.

Anderson J M and Ineson P 1984 Interactions between microorganisms and soil invertebrates in nutrient flux pathways of forest ecostystems. *In* Invertebrate – Microbial Interactions. Eds. J M Anderson *et al.* pp 413–432. Cambridge University Press, Cambridge.

Bååth E and Söderström B 1988 FDA-stained fungal mycelium and respiration rate in reinoculated sterilized soil. Soil Biol. Biochem. 20, 403–404.

Bajwa R, Abuarghub S and Read D J 1985 The biology of mycorrhiza in the Ericaceae. IX. Peptides as nitrogen sources for the ericoid endophyte and for mycorrhizal and non-mycorrhizal plants. New Phytol. 101, 469–486.

Bevege D I, Bowen G D and Skinner M F 1975 Comparative carbohydrate physiology of ecto- and endomycorrhizas. *In* Endomycorrhizas. Eds. F E Sanders, B Mosse and P B Tinker. pp 149–174. Academic Press, London.

Bowen G O and Theodorou C 1973 Growth of ectomycorrhizal fungi around seeds and roots. *In* Ectomycorrhizae. Eds. G C Marks and T T Kozlowski. pp 107–150. Academic Press, London.

Bowen G D and Theodorou C 1979 Interactions between bacteria and ectomycorrhizal fungi. Soil Biol. Biochem. 11, 119–126.

Buwalda J G, Ross G J S, Stribley D P and Tinker P B 1982 The development of endomycorrhizal root systems. III. The mathematical representation of the spread of vesicular-arbuscular mycorrhizal infection in root systems. New Phytol. 91, 669–682.

Bradley R, Burt A J and Read D J 1982 The biology of mycor-

rhiza in the Ericaceae. VIII. The role of mycorrhiza in heavy metal resistance. New Phytol. 91, 197–209.

Brownlee C, Duddridge J A, Malibari A and Read D J 1983 The structure and function of mycelial systems of ectomycorrhizal roots with special reference to their role in forming inter-plant connections and providing pathways for assimilate and water transport. Plant and Soil 71, 433–443.

Chiariello N, Hickman J C and Mooney H A 1982 Endomycorrhizal role for interspecific transfer of phosphorus in a community of annual plants. Science 217, 941–943.

Coleman D C, Ingham R E, McClellan J F and Trofymow J A 1984 Soil nutrient transformations in the rhizosphere via animal-microbial interactions. *In* Invertebrate – Microbial Interactions. Eds. J M Anderson *et al.*. pp 35–58. Cambridge University Press, Cambridge.

Cooper K M and Tinker P B 1978 Translocation and transfer of nutrients in vesicular-arbuscular mycorrhizas. II. Uptake and translocation of phosphorus, zinc and sulphur. New Phytol. 81, 43–52.

Cooper K M and Tinker P B 1981 Translocation and transfer of nutrients in vesicular-arbuscular mycorrhizas. IV. Effect of environmental variables on movement of phosphorus. New Phytol. 88, 327–339.

Denny H J and Wilkins D A 1987 Zinc tolerance in *Betula* spp. IV. The mechanism of ectomycorrhizal amelioration of zinc toxicity. New Phytol. 106, 545–554.

Duddridge J A 1986a Specificity and recognition in mycorrhizal associations. *In* Physiological and Genetical Aspects of Mycorrhizae. Eds. V Gianinazzi-Pearson and S Gianinazzi. pp 45–58. Proceedings of the 1st European Symposium on Mycorrhizae, INRA, Paris, 1986.

Duddridge J A 1986b The development and ultrastructure of ectomycorrhizas. III. Compatible and incompatible interactions between *Suillus grevillei* (Klotzsch) Sing. and 11 species of ectomycorrhizal hosts in vitro in the absence of exogenous carbohydrate. New Phytol. 103, 457–464.

Duddridge J A 1986c The development and ultrastructure of ectomycorrhizas. IV. Compatible and incompatible interactions between *Suillus grevillei* (Klotzsch) Sing. and a number of ectomycorrhizal hosts in vitro in the presence of exogenous carbohydrate. New Phytol. 103, 465–471.

Duddridge J A, Malibari A and Read D J 1980 Structure and function of mycorrhizal rhizomorphs with special reference to their role in water transport. Nature 287, 834–836.

Finlay R D 1985 Interactions between soil micro-arthropods and endomycorrhizal associations of higher plants. *In* Ecological Interactions in Soil: Plants, Microbes and Animals. Eds. A H Fitter *et al*. British Ecological Society Special Publication No. 4, pp. 319–331.

Finlay R D, Ek H, Odham G and Söderström B 1988 Mycelial uptake, translocation and assimilation of nitrogen from ^{15}N-labelled ammonium by *Pinus sylvestris* plants infected with four different ectomycorrhizal fungi. New Phytol. 110, 59–66.

Finlay R D and Read D J 1986a The structure and function of the vegetative mycelium of ectomycorrhizal plants. I. Translocation of ^{14}C-labelled carbon between plants interconnected by a common mycelium. New Phytol. 103, 143–156.

Finlay R D and Read D J 1986b The structure and function of the vegetative mycelium of ectomycorrhizal plants. II. The uptake and distribution of phosphorus by mycelial strands interconnecting host plants. New Phytol. 103, 157–165.

Fitter A H 1977 Influence of mycorrhizal infection on competition for phosphorus and potassium between two grasses. New Phytol. 79, 119–125.

Fleming V 1983 Succession of mycorrhizal fungi on birch: Infection of seedling plants around mature trees. Plant and Soil 71, 263–267.

Fogel R and Hunt G 1979 Fungal and arboreal biomass in a western Oregon Douglas fir ecosystem: Distribution patterns and turnover. Can. J. For. Res. 9, 245–256.

Francis R and Read D J 1984 Direct transfer of carbon between plants connected by vesicular-arbuscular mycorrhizal mycelium. Nature 307, 53–56.

Francis R, Finlay R D and Read D J 1986 Vesicular-arbuscular mycorrhiza in natural vegetation systems. IV. Transfer of nutrients in inter- and intra-specific combinations of host plants. New Phytol. 102, 103–111.

Frank A B 1984 Die Bedeutng der Mykorrhiza-Pilze für die gemeine Kiefer. Fortwiss. Zbl. 16, 1852–1890.

Gadgil R L and Gadgil P D 1971 Mycorrhiza and litter decomposition. Nature 233, 133.

Gadgil R L and Gadgil P D 1975 Suppression of litter decomposition by mycorrhizal roots of *Pinus radiata*. NZ. J. For. Sci. 5, 33–41.

Grime J P, Mackey J M L, Hillier S H and Read D J 1987 Mechanisms of floristic diversity: evidence from microcosms. Nature 328, 420–422.

Hall I R 1978 Effects of endomycorrhizas on the competitive ability of white clover. NZ. J. Agric. Res. 21, 509–515.

Hardie K 1986 The role of extraradical hyphae in water uptake by vesicular-arbuscular mycorrhizal plants. *In* Physiological and Genetical Aspects of Mycorrhizae. Eds. V Gianinazzi-Pearson and S Gianinazzi. pp 651–655. Proceedings of the 1st European Symnposium on Mycorrhizae, INRA, Paris, 1986.

Harley J L and Smith S E 1983 Mycorrhizal Symbiosis. Academic Press, London.

Harris D, Packovsky R S and Paul E A 1985 Carbon economy of Soybean-Rhizobium-Glomus associations. New Phytol. 101, 427–440.

Heap A J and Newman E I 1980a The influence of vesicular-arbuscular mycorrhizas on phosphorus transfer between plants. New Phytol. 85, 173–179.

Heap A J and Newman E I 1980b Links between roots by hyphae of vesicular-arbuscular mycorrhizas. New Phytol. 85, 169–171.

Hepper C M 1977 A colorimetric method for estimating vesicular-arbuscular mycorrhizal infection in roots. Soil Biol. Biochem. 9, 15–18.

Hirrel M C and Gerdemann J W 1979 Enhanced carbon transfer between onions infected with a vesicular-arbuscular mycorrhizal fungus. New Phytol. 83, 731–738.

Janos D P 1980 Mycorrhizae influence tropical succession. Biotropica 12, 56–64.

Jones M D and Hutchinson T C 1986 The effect of mycorrhizal infection on the response of *Betula papyrifera* to nickel and copper. New Phytol. 102, 429–442.

Leake J E 1987 Metabolism of phyto- and fungitoxic phenolic acids by the ericoid mycorrhizal fungus. 7th North American Conference on Mycorrhizae, Gainsville, Florida.

Melin E and Nilsson H 1950 Transfer of radioactive phosphorus to pine seedlings by means of mycorrhizal hyphae. Physiol. Plant. 3, 88–92.

Melin E and Nilsson H 1952. Transport of labelled nitrogen from an ammonium source to pine seedlings through mycorrhizal mycelium. Svensk Bot. Tidskr. 46, 281–285.

Melin E and Nilsson H 1953 Transfer of labelled nitrogen from glutamic acid to pine seedlings through the mycelium of *Boletus variegatus* (Sw.) Fr. Nature 171, 134.

Molina R and Trappe J M 1982 Patterns of ectomycorrhizal host specificity and potential among Pacific northwest conifers and fungi. For. Sci. 28, 423–458.

Ocampo J A 1986 Vesicular-arbuscular mycorrhizal infection of 'host' and 'non-host' plants: Effect on the growth responses of the plants and competition between them. Soil Biol. Biochem. 18, 607–610.

Read D J 1984 The structure and function of the vegetative mycelium of mycorrhizal roots. *In* The Ecology and Physiology of the Fungal Mycelium. Eds. D H Jennings and A D M Rayner. pp 215–241. Symposium of the British Mycological Society, Cambridge University Press.

Read D J 1986 Nonnutritional effects of mycorrhizal infection. *In* Physiological and Genetical Aspects of Mycorrhizae. Eds. V Gianinazzi-Pearson and S Gianinazzi. pp 169–176. Proceedings of the 1st European Symposium on Mycorrhizae, INRA, Paris, 1986.

Read D J and Boyd R 1986 Water relations of mycorrhizal fungi and their host plants. *In* Water, Fungi and Plants. Eds. P G Ayres and L Boddy. pp 287–303. British Mycological Society Symposium No. 11., Cambridge University Press.

Read D J, Francis R and Finlay R D 1985 Mycorrhizal mycelia and nutrient cycling in plant communities. *In* Ecological Interactions in Soil: Plants, Microbes and Animals. Eds. A H Fitter *et al.* pp. 193–217. British Ecological Society Special Publication No. 4.

Reeves F B, Wagner D, Moorman T and Kiel J 1979 The role of endomycorrhizae in revegetation practices in the semi-arid west. I. A comparison of incidence of mycorrhizae in severely disturbed vs. natural environments. Am. J. Bot. 66, 6–13.

Reid C P P, Kidd F A and Ekwebelam S A 1983 Nitrogen nutrition, photosynthesis and carbon allocation in ectomycorrhizal pine. Plant and Soil 71, 415–421.

Ritz and Newman E I 1984 Movement of ^{32}P between intact grassland plants of the same age. Oikos 43, 138–143.

Rhodes L H and Gerdemann J W 1975 Phosphate uptake zones of mycorrhizal and non-mycorrhizal onions. New Phytol. 75, 555–561.

Romell L G 1939 The ecological problem of mycotrophy. Ecology 20, 163–167.

Sanders F E 1986 Quantitative approaches to the analysis of the development of mycorrhizal root systems. *In* Physiological and Genetical Aspects of Mycorrhizae. Eds. V Gianinazzi-Pearson and S Gianinazzi. pp 209–216. Proceedings of the 1st European Symposium on Mycorrhizae, INRA, Paris, 1986.

Sanders F E and Tinker P B 1973 Phosphate flow into mycorrhizal roots. Pest. Sci. 4: 385–395.

Shaw P J A 1985 Grazing preferences of *Onychiurus armatus* (Insecta: Collembola) for mycorrhizal and saprophytic fungi of pine plantations. *in* Ecological Interactions in Soil: Plants, Microbes and Animals. Eds. A H Fitter *et al.* pp. 333–337. British Ecological Society Special Publication No. 4.

Smith S E and Walker N A 1981 A quantitative study of mycorrhizal infection in Trifolium: separate determination of the rates of infection and mycelial growth. New Phytol. 89, 225–240.

Snellgrove R C, Splitstoesser W E, Stribley D P and Tinker P B 1982 The carbon distribution and the demand of the fungal symbiont in leek plants with vesicular-arbuscular mycorrhizas. New Phytol. 92, 75–81.

Söderström B 1979 Seasonal fluctuations of active fungal biomass in horizons of a podsolized pine-forest soil in central Sweden. Soil Biol. Biochem. 11, 149–154.

Söderström B and Read D J 1987 Respiratory activity of intact and excised ectomycorrhizal mycelial systems growing in unsterilized soil. Soil Biol. Biochem. 19, 231–236.

St. John T V, Coleman D C and Reid C P P 1983a Growth and spatial distribution of nutrient-absorbing organs: Selective exploitation of soil heterogeneity. Plant and Soil 71, 487–493.

St. John T V, Coleman D C and Reid C P P 1983b Association of vesicular-arbuscular mycorrhizal hyphae with soil organic particles. Ecology 64, 957–959.

Tinker P B 1978 Effects of vesicular-arbuscular mycorrhizas on plant nutrition and plant growth. Physiol. Veg. 16, 743–751.

Tisdall J M & Oades J M 1979 Stabilization of soil aggregates by the root systems of ryegrass. Aust. J. Soil Res. 17, 429–441.

Vogt K A, Grier C C, Meier C E & Edmunds R L 1982 Mycorrhizal role in net primary production and nutrient cycling in *Abies amabilis* stands. Oecologia (Bot.) 50, 170–175.

Whittingham J and Read D J 1982. Vesicular-arbuscular mycorrhiza in natural vegetation systems. III. Nutrient transfer between plants with mycorrhizal interconnections. New Phytol. 90, 277–284.

Woods F W 1970 Interspecific transfer of inorganic materials by root systems of woody plants. J. Appl. Ecol. 7, 481–486.

Woods F W and Brock K 1964 Interspecific transfer of ^{45}Ca and ^{32}P by root systems. Ecology 45, 886–889.

M. Clarholm and L. Bergström (Eds.), Ecology of arable land, 149–160.
© 1989 Kluwer Academic Publishers.

Activity of nitrifiers in relation to nitrogen nutrition of plants in natural ecosystems

J.W. WOLDENDORP and H.J. LAANBROEK

Institute for Ecological Research, Boterhoeksestraat 22, 6666 GA Heteren, The Netherlands

Key words: chemolithotrophs, natural ecosystems, nitrate, nitrate reductase, nitrification, population, production, rhizosphere, uptake

Abstract

Three aspects of the nitrate production in natural ecosystems are discussed, *i.e.* the population biology of nitrifying bacteria, the nitrate-producing activity of these organisms and the uptake of nitrate by higher plants. It is concluded that the three methods used in enumerating the nitrifying bacteria, *i.e.* the Most Probable Number method, the Fluorescent Antibody technique and the Potential Nitrification Rate, all have serious drawbacks and count different segments of the nitrifying populations.

From the number of nitrifying bacteria no reliable estimate of the rate of nitrate production can be obtained and also estimates that are made using field-incubation and $^{15}N-NH_4^+$ techniques do not yield reliable data. Possibly the best results can be obtained using Schimel's method to estimate the actual nitrification rate using $^{15}N-NO_3^-$, but this method has still not been tested under different sets of soil conditions.

From the nitrate reductase activity and the chemical composition of the plant a picture can be obtained of the quantities of nitrate and ammonium that have been taken up. However, it is shown that nitrate and ammonium are taken up in different proportions that they are produced. It is concluded that the various parameters have to be studied simultaneously, preferably in defined systems with plants, in which the participating organisms are known.

Introduction

The productivity of high-input agro-ecosystems during long periods of the growing season is not limited by the availability of nutrients but by the growth rate of the crop. As soon as the soil surface is completely covered by plants the productivity is determined by the light intensity. This situation rarely occurs in many natural ecosystems where plant growth is usually not limited by photosynthesis but by the availability of water or nutrients. In nutrient-limited ecosystems the productivity of higher plants is often determined by phosphorous or potassium. The level of nitrogen is meanwhile regulated by biological processes (Pomeroy, 1970). As long as there is an excess of energy, the energy-requiring steps of the nitrogen cycle dominate, such as fixation and incorporation into organic com-

pounds. A surplus of nitrogen is lost from the system by nitrification and subsequent denitrification or leaching of nitrate (Woldendorp, 1981). This means that in balanced ecosystems the productivity in practice is not only determined by phosphorous or potassium but also by nitrogen. In such balanced ecosystems a gradual accumulation of stable organic nitrogen compounds occurs for long periods of time and the process of nitrification seems to be of less importance than in arable soils although high nitrification rates may also be detected in mature ecosystems such as old-growth forest soils (Robertson and Vitousek, 1981). From this point of view the nitrifiers have the function of eliminating a surplus of reduced nitrogen and to store it as nitrate as was much earlier discussed by Jansson (1958). In the competition for ammonium with immobilizing micro-organisms and the roots

149

of higher plants the nitrifying bacteria are supposed to be the weaker ones. However, when the heterotrophic bacteria and the plants are not N-limited, nitrification may occur (Riha *et al.*, 1986).

In the above discussion the ecosystem is considered to be a cybernetic, *i.e.* regulated system. However, in the neo-Darwinian sense ecosystems consist of collections of populations. Processes in these populations are determined at the level of the individual and it is at this level that selection takes place. Thus, "regulation" is not a property of populations and ecosystems and cannot be compared with homeostatis in an individual organism. The fact that processes in populations and ecosystems sometimes have regulating properties does – of course – not mean that these are their function (Bakker, 1980). Such processes need to be described using stochastic models, that have an uncertain outcome, in contrast to processes in organisms that can be described with deterministic models that have a definite outcome (Rao, 1989). Consequently, to give a causal explanation of processes in the vegetation, they have to be studied at the level of the individual plant and the supply of nutrients has also to be studied at this level, *i.e.* the rhizosphere. As will be discussed below, within the vegetation and even within the rhizosphere of the individual plant the supply of nutrients is very inhomogeneous and, although generally spoken nitrification is not a predominant process, locally there may be a considerable production of nitrate. Consequently, nearly all species of higher plants, with the exception of most ericaceous species, are able to cope with this nitrate supply by means of the inducible enzyme nitrate reductase (Runge, 1983).

Several aspects of the nitrate production and the nitrification process in the rhizosphere of grassland plants will be discussed in this paper. This contribution does not aim at reviewing the literature on the subject (for reviews see Belser, 1979; Berg, 1986; Focht and Verstraete, 1977; MacDonald, 1986a; Schmidt, 1982) but to draw attention to some bottlenecks in the research on nitrification. The opinions we express are based on studies of the nitrate production in some semi-natural grassland soils in the Netherlands. Three aspects of the nitrification process will be discussed, *i.e.* the population biology of nitrifying bacteria, the nitrate producing activity of these organisms and the uptake of nitrate by higher plants.

Methods used in studying nitrate production

In studying the three aspects of the nitrification process in the field situation cited above, various techniques have been used (Table 1). All of them deal only with certain aspects of nitrate formation in the rhizosphere and to make full use of the information they provide, they should be studied simultaneously. Up till now these comprehensive sets of data have not been collected and this makes the data on numbers and composition of nitrifying populations difficult to combine with estimates of their activity; also, data on nitrate production by these organisms have rarely been compared with those on nitrate uptake and assimilation by higher plants. A second problem is that questions on the population biology and activity of nitrifying bacteria have been approached largely by extrapolation of pure culture studies (Schmidt, 1982). The combination of these factors means that our insight into the nitrification process is still far from complete.

Population biology of nitrifying organisms

Organisms

Apart from the chemolithotrophic ammonium-oxidizing organisms, methanotrophs (Whittenbury and Kelly, 1977) and a number of denitrifiers (*e.g.* Kuenen and Robertson, 1987) are also able to oxidize ammonium, be it at a low rate and high levels of substrate. That these organisms contribute significantly to the nitrification process under natural conditions may, however, be questioned (Schmidt, 1982; Verstraete, 1981). In addition to these organisms, in acid soils heterotrophs have

Table 1. Parameters used to characterize the nitrate production in the soil

Soil parameters
1. Most Probable Number (MPN)-count
2. Fluorescent Antibody (FA)-counts
3. Potential NH_4^+- and NO_2^--oxidation rates
4. Mineral nitrogen
5. Mineralization rate (*in situ*, ^{15}N)
6. Actual NH_4^+-oxidation rate

Plant parameters
7. Nitrate accumulation
8. Nitrate reductase activity (NRA)
9. Chemical composition, Cations–Anions (C–A)-contents, C–A/N_{org}-ratio, H^+, HCO_3^--excretion

Table 2. The growth of three *Nitrobacter* species on 0.1 m*M* nitrite and variable amounts of nitrate (Both, unpubl. results).

m*M* NO$_3^-$	10	20	30	40	50
N. agilis	+	−	−	−	−
N. winogradskyi	+	+	+	+	+
N. hamburgensis	+	+	+	+	+

been claimed to participate in nitrate production (for a recent review see Kilham, 1987) but the arguments are not always convincing. In a number of situations in the Netherlands there seems to be an association between ammonifying heterotrophs and chemolithotrophic nitrifiers (De Boer, in press). Due to these uncertainties, only the chemolithotrophs will be discussed below.

During the last decade a much higher diversity has been shown to exist among the latter organisms than was originally thought. Among 96 strains of ammonium oxidizers analysed for G + C-content and DNA-homology, Koops and Harms (1985) differentiated between at least 8 *Nitrosomonas*, 3 or 4 *Nitrosococcus*, 5 *Nitrosospira*, 2 *Nitrosolobus* and 2 *Nitrosovibrio* species. Many of them, all chemolithotrophs that could also to some extent incorporate organic compounds (mixotrophic growth), were isolated from soil.

From the four nitrite-oxidizing genera *Nitrobacter, Nitrococcus, Nitrospira* and *Nitrospina*, only the first one was encountered in soil. So far only two species in this genus have been described, *i.e.* *N. winogradskyi* (*N. agilis*) and *N. hamburgensis* but a further differentiation seems to be likely in the near future (Bock *et al.*, 1987). In the 8th edition of Bergey's Manual *N. winogradskyi* and *N. agilis* were combined by Watson (1974) into one species but at least some strains exhibit considerable physiological and serological differences. In contrast to the ammonium oxidizers, *Nitrobacter* species are able to grow heterotrophically (Bock, 1976), but mixotrophy has only been observed in *N. hamburgensis*.

All nitrifiers are slow growing organisms with generation times varying from 8 h for *Nitrosomonas* and 10 h for *Nitrobacter* to 60 h for *Nitrosospira*. Nevertheless, despite their lower growth rate, the genera *Nitrosospira* and *Nitrosolobus* seem to be more common in soil than *Nitrosomonas*. Apparently, multiple niches exist in the soil which lead to many coexisting species. However, little is known about the specific requirement of these spe-

cies and it is not known for example whether a differentiation between specialists and generalists exists as has been observed in other chemolithotrophic bacteria (Kuenen and Harder 1982). Apparently, the species differ in sensitivity to concentrations of their substrates and products as shown in Table 2 for the *Nitrobacter* species. In this respect a strong interaction with the pH may be expected to occur because uptake as well as the toxic effects of NH$_3$ and HNO$_2$ take place as undissociated molecules. Differences in sensitivity towards inhibitors also exist (Keeney, 1987; Powell, 1987), *e.g. Nitrosomonas* being less sensitive to N-Serve than *Nitrosolobus* and *Nitrosospira*.

All nitrifying bacteria are long-living organisms that can persist under a variety of adverse environmental conditions such as drought and anaerobiosis. Such behaviour agrees with the hypothesis that in slowly growing organisms there will be a strong selection for longevity (K-selection). Thus, populations of nitrifying bacteria are expected to vary comparatively less in time than those of most other bacteria, in numbers that can cope with the maximum supply of ammonium in a particular soil and that are not obliged to lag, due to their slow growth, far behind this ammonium supply.

Enumeration methods

MPN-enumeration. Among the various enumeration techniques the MPN-method has been most frequently applied. Several modifications of the method have been used, that differ in medium composition and incubation time (for discussions see Belser, 1979; Berg, 1986; Schmidt, 1982). In enumerating nitrite oxidizers we used the medium of Schmidt and Belser (1982) with incubation times of 3 months. Different nitrite concentrations (0.05 and 5 m*M*, respectively) were used to obtain maximum counts in the soils that were studied. In Junner Koeland, the highest counts were obtained with 5 m*M* concentration (Fig. 1) but in Merrevliet counts at 0.05 m*M* were considerably higher (Table 3). Apparently, the dominating species in these grassland soils differed in affinity and sensitivity to their substrate (Both, in prep.).

The efficiency of the MPN-method is low when compared to other methods, (Belser, 1979; Berg, 1986). This low efficiency has possibly to do with

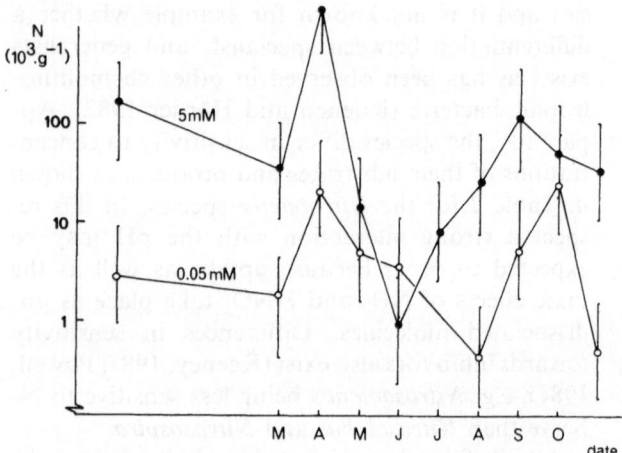

Fig. 1. Most probable numbers of nitrite-oxidizing bacteria at two nitrite concentrations (solid circles: 5 mM; open circles: 0.05 mM) in a natural grassland soil (Junner Koeland)(Both, in prep.).

the selectivity of the growth medium, and the presence of cell aggregates in the dilutions. According to Bock (pers. comm.) resting cells are sometimes difficult to re-activate and often more than one cell is required to start growth. In comparing various enumeration techniques, Berg (1986) advanced the hypothesis that mainly actively growing cells are counted, but there have been no model studies to investigate this suggestion. However, in our opinion, the interpretation of many results obtained with the MPN-techniques is still puzzling. Robertson and Vitousek (1981) for example, studying the relation of soil nitrification and the succession in the vegetation, observed a bad correlation between the results of the MPN-techniques and the potential nitrification activity.

Table 3. The Most Probable Number (MPN)-counts of nitrite oxidizers at two nitrite concentrations (5 mM and 0.05 mM) in a natural grassland (Merrevliet) (Both, in prep.)

	Date	mM NO_2^-	MPN
Site 1	May 1986	5	4.0×10^3
		0.05	1.3×10^5
	Nov. 1986	5	1.1×10^4
		0.05	3.3×10^6
Site 2	May 1986	5	10
		0.05	1.2×10^3
	Nov. 1986	5	180
		0.05	1.7×10^3

Fluorescent-Antibody (FA) Enumeration. The immunofluorescence (FA) technique developed by Schmidt (1973) makes it possible to enumerate specific micro-organisms in the soil (for a review see Schmidt 1982). The technique has been applied in studying ammonium oxidizers among which a high diversity at the serotype level was found. This is not surprising in the light of the high number of different species that have been described by Koops and Harms, (1985), see above. Also among the nitrite oxidizers the differentiation at the serotype level is considerable. Stanley and Schmidt (unpublished results in Schmidt, 1982) found at least 5 serotypes in *Nitrobacter* other than *N. winogradskyi* and *N. agilis*. In studying the *Nitrobacter* strains present in a number of grassland soils in the Netherlands *N. agilis*, *N. winogradskyi* and *N. hamburgensis* were found to be simultaneously present in nearly all soils. However, in dilution series of these soils 38 out of 46 tubes contained strains, that did not react with the antisera of the above 3 strains (Both, unpublish.). The ubiquitous presences of *N. hamburgensis* in the grassland soils is surprising since until now this species has been encountered only in one garden soil from Hamburg and in an arable soil from Mexico (Bock, pers. comm.).

The cell recovery with the FA technique is much higher than with the MPN-method. At least 50 per cent of added cells were recovered from soil when, depending on the soil type, flocculation techniques or dispersion methods are used (MacDonald, 1986b). Lower efficiencies of soil-born cells, however, are possible, for example by failure of the antiserum to react with encapsulated cells, *e.g.* zoogloea of *Nitrosomonas*.

To use the FA-technique to obtain an estimate of the number of nitrifying bacteria in soil, antisera of the full complement of ammonium- and nitrite oxidizers — or at least of the dominating strains — have to be available. In this way systematic comparisons with the MPN-method have as yet not been made but a number of laboratories are working on them. However, with the antisera of only a few strains substantially higher counts were obtained than with the MPN-method (Rennie, 1975; Rennie and Schmidt, 1977).

A special problem of applying the FA-technique in soil studies is that the specificity of the antisera has not yet been wholly established without any doubt. According to Bock (pers. comm.) no cross-reactions occur but Josserand and Navarro (pers. comm.) found differences in DNA-homology between strains of the same serotype.

Potential nitrification activity. Short-term measurements of the nitrification activity aim at estimating the nitrifying capacity of the indigenous population at the time of sampling (Schmidt, 1982). In these measurements soil samples are incubated over a period of a few hours under optimal conditions with regard to pH, substrate and oxygen (*e.g.* in slurry experiments). The total nitrifying capacity is measured by periodically measuring the $NO_2^- + NO_3^-$ yield, and it is assumed that no denitrification takes place.

The ammonium-oxidizing potential can be estimated separately by inhibition of the nitrite oxidation with appropriate concentrations of chlorate and measuring the NO_2^--production (Belser and Mays, 1980). In this method chlorate is reduced to the toxic chlorite by nitrite oxidizers. The potential nitrite oxidation rate can be measured by adding nitrite instead of ammonium.

By relating these potential activities to conversion rates of individual cells, derived from pure culture studies, an estimate of the nitrifying population can be made (Schmidt and Belser, 1982). Such estimates yield numbers that generally are several orders of magnitude higher than MPN-numbers (see *e.g.* Berg and Rosswall, 1985). However, in our opinion cell numbers estimated by measurement of potential nitrification rates are difficult to interpret, because a variety of organisms are involved whose activities are largely unknown. The ammonium oxidation rates of *Nitrosomonas europaea* and *Nitrosospira briensis*, for instance, differ considerably (0.32 and 0.056 pg N cell$^{-1} \cdot$h^{-1}, respectively). Moreover, because chlorite is not only toxic to *Nitrobacter* but also to other bacteria such as *Nitrosomonas*, numbers of nitrifiers calculated from the potential ammonium oxidation capacity should be considered with caution.

Another inherent problem is the reactivation of resting cells of the nitrifiers. In heterotrophically growing cells of *N. hamburgensis* the nitrite oxidizing enzyme is repressed and reactivation of the ability to grow chemolithotrophically takes 3–4 weeks (Bock, 1976). Consequently, the linearity in the oxidation rate that is found in measurements of

the potential nitrification activity suggests that only adapted cells are included in the estimation.

Population dynamics

Changes in numbers of nitrifyers during a year have been followed using the MPN method (Berg, 1986; Blacquière, 1986; Both, in prep.) or measurements of the potential ammonium and nitrite oxidation rates (Berg, 1986). In agricultural soils as well as in semi-natural grassland soils, the numbers were highest in spring and autumn (Fig. 1 and 2) and correlated well with a high ammonification rate. The numbers obtained from potential activity measurements varied less (Berg, 1986). According to this author the latter values should represent the number of microcolonies in potential microsites in

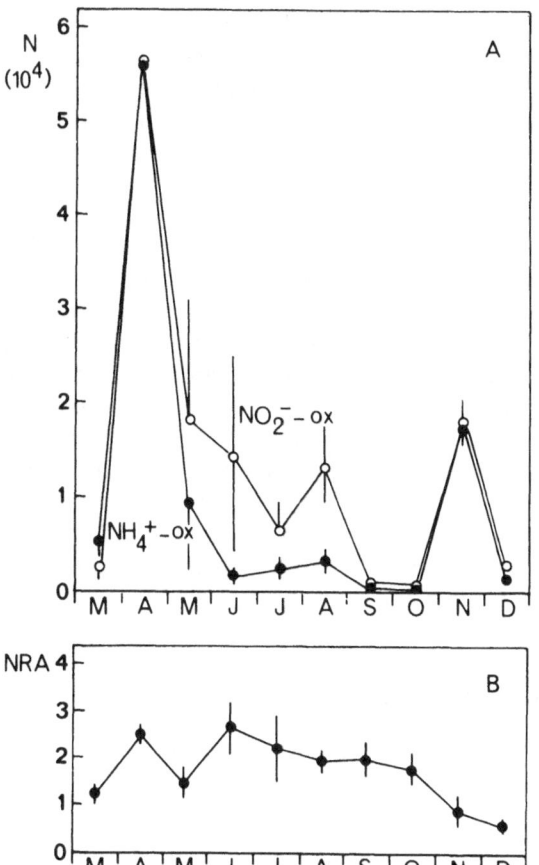

Fig. 2. The numbers of nitrifying bacteria in the rhizosphere (**A**) and the nitrate reductase activity (m*M* NO_2^- (g dry weight)$^{-1}$ h^{-1}) in the leaves (**B**) of *Plantago lanceolata* (Blacquière, 1986).

a particular soil, whereas in MPN-counts, in addition to these microcolonies, also recently produced active cells are estimated. In our opinion the results are difficult to harmonize with each other, as with the two methods results are obtained, that differ a hundred-fold. Measurements of the potential nitrification rates are of too short duration to allow reactivation of resting cells and, therefore, will present nitrification rates of already adapted cells. On the other hand, MPN counts agree quite well with other parameters of the N-cycle such as the mineralization rate but they probably represent only a small fraction of the nitrifying population. The fluctuations in MPN numbers that occur during the annual cycle are not in accordance with the idea of nitrifyers being organisms that compensate a low growth rate by high longevity (*i.e.* a so-called K-strategy), as has been outlined above. Therefore, more evidence is needed before Bergs' hypothesis can be accepted. Such evidence should come from model experiments in which MPN- and FA-enumerations are compared with simultaneous measurements of the potential activities.

In general a high correlation was found between MPN-numbers of ammonium-oxidizing and nitrite-oxidizing bacteria (Berg, 1986; Smit and Woldendorp, 1981), see Table 4. This suggests that both processes are coupled. Because both nitrification steps yield different amounts of energy (272 kJ per mole in ammonium oxidizers and 74 kJ per mole in nitrite oxidizers), the more or less equal numbers of cells that are generally observed, are surprising. Sometimes even higher numbers of nitrite oxidizers are found (*e.g.* Belser, 1977; Berg, 1986), see also Fig. 2. Apart from the unreliability of the MPN-method, there are several possible mechanisms that can explain this result:

1. Ammonium oxidizers may possess a higher death-rate than nitrite oxidizers. This may occur particularly in soil, where ammonium oxidizers may be killed by their own acid production (De Boer, unpubl. results).
2. Heterotrophic growth of *Nitrobacter* strains (see above).
3. Anaerobic growth of nitrite oxidizers on nitrate and organic substrates (Bock *et al.*, 1987).
4. NO_2^-/NO_3^--turnover between nitrite oxidizers and denitrifiers (Belser, 1979; Berg, 1986).

The spatial distribution of nitrifiers outside the rhizosphere has hardly been studied, Smit and Woldendorp (1981) found different numbers of nitrifiers in the rhizosphere of 3 *Plantago* species that occured together in a dune grassland (Table 5). The differences apparently were related to the pH and the general fertility of the spots where the respective *Plantago* species were growing and are supposed not to be due to specific rhizosphere effects. As will be described below, the differences in nitrate supply of the species will result in differences in excretion of hydroxyl (bicarbonate) ions. This mechanism will further develop differences in soil parameters.

Because of the differences in humidity, oxygen supply, and grazing activity of protozoa and nematodes between the interior and the surface of soil aggregates, a considerable differentiation is expected to occur at this level. However, there are no data available on *e.g.* species differentiation, just as there are none on the spatial distribution of cells of ammonium oxidizers that provide the substrate of nitrite-oxidizing cells.

Nitrifiers in the rhizosphere

The nitrification process in the rhizosphere has been discussed by Woldendorp (1975, 1981, 1983) and only a few aspects need to be mentioned. It has been concluded that the nitrification rate in the rhizosphere is often lower than in the bulk of the soil, due to a higher availability of organic carbon. Under such conditions the nitrogen cycle is directed more towards immobilization and uptake by roots of available ammonium than towards nitrification. The evidence of suppression of nitrification in the rhizosphere of plant species from climax vegetation by means of allelopathic compounds, that has been

Table 4. The correlation between ammonium content, MPN-numbers of ammonium- and nitrite oxidizers in the rhizosphere of 13 *Plantago lanceolata* plants and the nitrate reductase activity (NAR) in the leaves (Smit and Woldendorp, 1981).

		r_s
Soil–NH_4^+	: NH_4^+-oxidizers	0.289
Soil–NH_4^+	: NO_2^--oxidizers	0.180
NH_4^+-oxidizers	: NO_2^--oxidizers	0.872
NH_4^+-oxidizers	: NaR-leaves	0.807
NO_2^--oxidizers	: NaR-leaves	0.794

Table 5. Rhizosphere characteristics and nitrate reductase activity of three *Plantago* species growing together in a dune grassland (Smit and Woldendorp, 1981)

	pH–H$_2$O	NH$_4^+$–N (ppm)	NO$_3^-$–N (ppm)	NH$_4^+$- oxidizers*	NO$_2^-$- oxidizers*	NRA**
P. lanceolata	4.4	6.2	0.5	5 500	4 600	0.38
P. coronopus	6.6	4.2	0.5	11 900	6 600	0.88
P. major	7.6	17.5	1.2	64 700	44 000	1.02

* Numbers per g dry soil.
** Nitrate reductase activity (NRA'): μM NO$_2^-$ ·(g dry weight)$^{-1}$ h^{-1}
Note: NO$_3^-$ content shoots always < 20 mmol (kg dry weight)$^{-1}$

advocated by Rice (1974), was considered to be unconvincing. Rhizosphere counts of nitrifiers have often yielded conflicting results, even with the same plant species. Plants of different ages had different effects on nitrifiers. Generally, in the rhizosphere of mature plants and in permanent grassland soils, low numbers of nitrifying organisms are found. These results were recently confirmed by Berg (1986) who found a stimulation of MPN-numbers in the rhizosphere of young barley plants but not in that of older ones. Nitrite oxidizers were stimulated particularly. This stimulation was ascribed by Berg to NO$_2^-$/NO$_3^-$ turnover between nitrite oxidizers and denitrifiers but in our opinion heterotrophic growth of *Nitrobacter* species in the rhizosphere can by no means be excluded.

It was found by Smit and Woldendorp (1981) and by Blacquière (1986) that in the rhizosphere of many plant species, even in that of aerenchyma-containing species from waterlogged soils, considerable numbers of nitrifyers are present. The importance of differences in local rhizosphere conditions for the occurrence of nitrification was demonstrated by correlating MPN-numbers of nitrifiers in the rhizosphere of individual *Plantago lanceolata* plants with the nitrate reductase activity of these plants. A good correlation was found to exist between both parameters (Table 4). These results demonstrated that the nitrogen nutrition of individual plants is directly related to the nitrification process in their rhizosphere. A further analysis of the data revealed that a direct relationship is obtained only when plants are sampled during periods of uniform weather conditions (Blacquière, in prep.). Over longer periods of time the correlation is much weaker because the rate of response to periods of rainfall is much higher in the induction of nitrate reductase than in growth of the nitrifiers.

Nitrification rates

Mineral nitrogen

Several approaches have been advanced to estimate nitrification rates. As has been discussed by Berg (1986), determinations of mineral nitrogen (ammonium and nitrate) provide little information on the rates of production, as they represent steady state concentrations.

Incubation techniques

In incubations of soil samples in the laboratory or under field conditions estimates of the net ammonium and nitrite mineralization can be obtained. However, the values that are found represent the resultant of various processes, such as mineralization, immobilization, nitrification and denitrification. In our opinion, field incubations in combination with data on nitrogen uptake by the vegetation can be used to estimate flushes in mineral nitrogen over the year. The method has been applied by Troelstra and Wagenaar (1980) who followed the mineralization in a dune grassland by incubation of undisturbed soil cores for 4–5 weeks *in situ* (Fig. 3). The total amount of nitrogen that was found to be mineralized in this way over a two-year period agreed fairly well with those taken up by the vegetation in both the dry and wet sites during this period. In the wet sites a decrease of mineral nitrogen was found during the winter months; apparently nitrogen is lost from these sites by leaching or denitrification. The highest rates occurred in early summer, but during a dry spell mineralization came to a standstill. When calculated over the total period, in the wet sites the

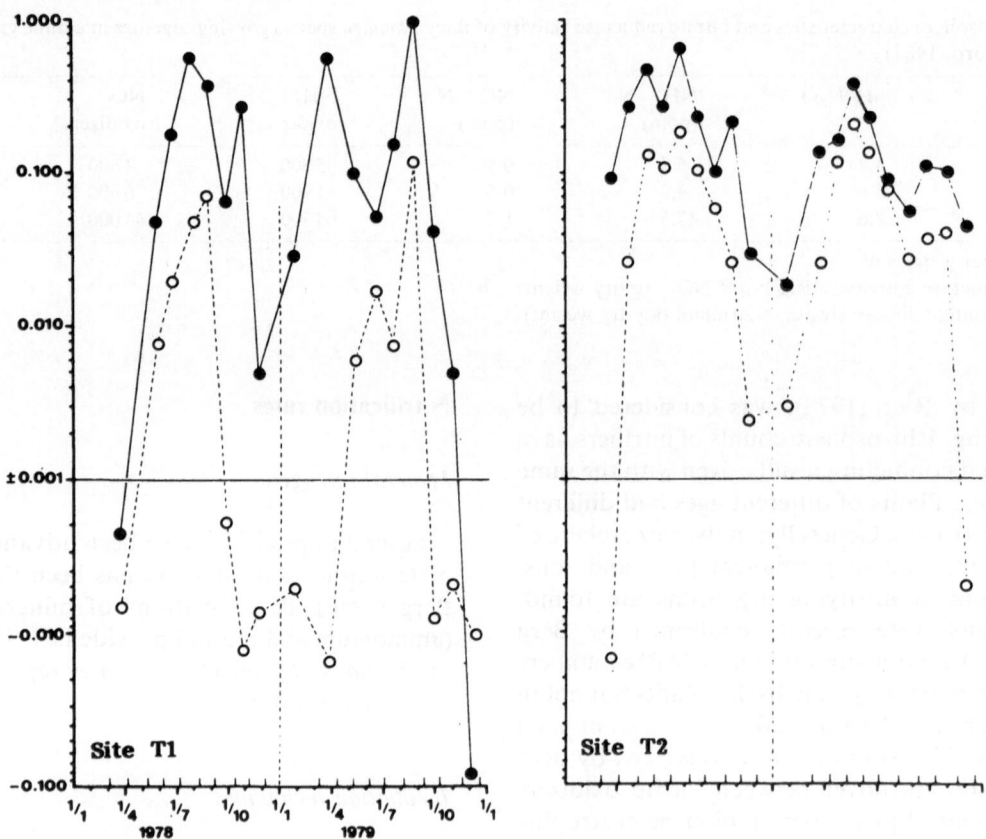

Fig. 3. The rates of net mineralization (g N m^{-2} week^{-1}) during field incubation at a wet site (T1) and a dry site (T2) in a dune grassland. Ammonium-N: solid circles; nitrate-N: open circles (Troelstra and Wagenaar, 1980).

ammonium/nitrate-ratio was much higher than in the dry sites (Table 6). This does not mean that both ions are taken up in the same proportions by the plant roots. On the contrary, data on the chemical composition of the plant suggest that nitrogen uptake in the wet sites was proportionally more in the nitrate form than in the dry sites (see below).

Table 6. The *in situ* nitrogen mineralization in a dune grassland soil during a 2-year period (Troelstra and Wagenaar, 1980)

Site	g N m^{-2} week^{-1}		Ratio NH$_4^+$/NO$_3^-$
	NO$_3^-$–N	NH$_4^+$–N	
1 (wet)	− 0.01–1.11	− 0.09–0.99	16.0
2 (dry)	− 0.02–0.21	0.02–0.64	2.0
3 (wet)	− 0.02–0.08	− 0.03–0.40	14.5
4 (dry)	− 0.01–0.20	− 0.01–0.51	2.8

^{15}N-techniques

The drawbacks of the incubation methods can be largely overcome in experiments with ^{15}N- and ^{13}N-labelled nitrogen. However, the half-lifetime value of ^{13}N is too short to allow use of this isotope in the field. The high number of replicates needed in a natural ecosystem and the time-consuming analyses often makes ^{15}N-experiments too laborious. Moreover, in natural ecosystems the quantities of mineral nitrogen are low, and nitrogen is not present in excess (see section 1). Therefore, enlarging the mineral nitrogen pool with ^{15}N-enriched ammonia, may shift the various processes of the nitrogen cycle in the direction of nitrification. The latter problem can be overcome possibly by a method of short duration that has been described by Myrold and Tiedje (1986). This method simultaneously estimates several N-cycle rates. However, it may be questioned whether in using this method the actual or the potential nitrification rate is determined.

Recently, Schimel (1987) suggested to determine the actual nitrite oxidation rate by calculating the rate by which added $^{15}N-NO_3^-$ is diluted by $^{14}N-NO_3^-$ produced by the process of nitrification. This approach, that has not the drawbacks of other methods, looks very promising.

Actual ammonium oxidation rate

A modification of the determination of the potential ammonium-oxidation rate by means of chlorate inhibition has been published by Berg and Rosswall (1985). In the modified method no ammonium was added to undisturbed soil cores that were incubated in the laboratory. The accumulation of nitrite in the soil cores was believed to represent the actual nitrification rate at the moment of sampling. Generally, the actual rates were lower and fluctuated more than the potential ammonium oxidation rates. Since a negative correlation was found between nitrite concentration and the rate of nitrite accumulation, it was suggested that nitrite would locally reach concentrations that might inhibit nitrification. At present no further data on the suitability of the method are available.

Calculations based on cell numbers

Finally, nitrification rates can be calculated from the numbers of nitrifying cells and data on their oxidation rate obtained in pure culture studies. However, the value of such calculations is still highly uncertain because all estimates of the numbers and species composition are incomplete (see above) and it is questionable whether data on growth kinetics that have been collected in the laboratory also apply to soil conditions.

Nitrate uptake by plants

Nitrate in plant material

In addition to soil parameters, data on N-metabolism in plants also can yield information on the extent of nitrification in soil and on the relative proportions of ammonium and nitrate that are taken up. In our opinion, data on N-metabolism in plants have too rarely been combined with other information, *e.g.* to substantiate claims of a suppressed nitrification in the rhizosphere of plant species growing in climax vegetation.

The presence of nitrate in plant material from natural habitats would constitute direct proof of nitrate uptake. However, as contrasted to plants grown in waterculture, such plants rarely contain nitrate (Table 7). In natural grassland soils, nitrate only accumulates in plants growing on spots rich in nutrients such as around cow-droppings. Because

Table 7. Ionic balances of *Plantago* species grown in water cultures with ammonium or nitrate as the nitrogen source, compared with those of plants from dune grassland and other natural habitats (Troelstra and Smant, 1980)

Species	Growth conditions	C-A* (m-equiv. kg dry wt.$^{-1}$)	C-A/ organic N	NO$_3^-$ (mmol kg dry wt.$^{-1}$)
P. major	Water culture NO$_3^-$	980–2300	0.40–1.84	200–1100
	Water culture NH$_4^+$	680–1050	0.26–0.35	< 20
	Dune grassland	1200–1300	0.75–0.87	0–40
	Other natural habitats	1140–1970	0.60–2.60	0–120
P. lanceolata	Water culture NO$_3^-$	890–2240	0.38–1.31	200–1850
	Water culture NH$_4^+$	320–590	0.13–0.25	< 20
	Dune grassland	710–910	0.49–0.61	0–12
	Other natural habitats	650–1590	0.45–1.58	00–260

* The difference between total amounts of cations and anions; for details, see text.

the nitrate level in plants can be considered more or less as a steady state concentration, its absence does not necessarily mean that no nitrate has been taken up.

Nitrate reductase activity

The activity of the inducible enzyme nitrate reductase (NRA) is a much better measure of nitrate uptake. In Table 4 it is shown that in a period of uniform weather conditions the NRA is correlated with the MPN-numbers (see above). Over the year NRA-levels varied less than the MPN-numbers (Fig. 2), but the high level in May corresponded with a high NRA-level. From NRA-levels in May it could be calculated, using data from a number of pot- and water culture experiments, that the quantities of nitrate that are reduced, allowed a relative growth rate of $0.015 \, g \cdot g^{-1} \cdot day^{-1}$ (Blacquière, in prep.). In the field a growth rate of $0.02 \, g \cdot g^{-1} \cdot day^{-1}$ was measured. The corresponding MPN-numbers in the rhizosphere of these plants produced 35 μmol nitrate per plant per week (assuming a nitrate-production of $0.32 \, pg \, N \, cell^{-1} \, h^{-1}$). With this nitrate-supply a growth rate of $0.023 \, g \cdot g^{-1} \, day^{-1}$ could have been realized. This is higher than the rate, actually measured in the field. Because the chemical composition of the plant material indicates that not all organic nitrogen originates from nitrate (see below), it can be concluded that either not all nitrate produced in the rhizosphere is taken up by the plant or that the nitrate production is overestimated. Because MPN-values certainly constitute only a fraction of the real numbers, it can be concluded that available data on the NO_3^--producing activity of the cells are in any case too high and need re-investigation under more natural conditions.

Chemical composition

An impression of the nature of nitrogen nutrition of plants can also be obtained from an analysis of the chemical composition of the plant material. Generally, the difference between the amounts of cations (Na^+, K^+, Mg^{2+}, Ca^{2+}, NH_4^+) and anions (Cl^-, SO^{-2}, $H_2PO_4^-$, NO_3^-), so called C–A value, has a positive value and is a measure of the amounts of organic anions (malate, oxalate, ci-

trate) that are accumulated in the plant tissue to maintain electroneutrality. The size of this carboxylate pool is influenced by the proportions of cations and anions that are taken up, the reduction of nitrate and sulphate, and by the incorporation of ammonium into amino acids (Troelstra, 1983). In the absence of nitrate, roots always absorb more cations than anions, and excrete more H^+ than HCO_3^-, the cation excess inside the plant being compensated by the synthesis of organic anions. Ammonium uptake is therefore attended by a decrease of the rhizosphere pH, and a low ratio between C–A and organic nitrogen. Upon reduction, nitrate is replaced by organic anions, which leads to higher C–A values than with ammonium as the nitrogen source. In principle the level of C–A should be about equal to the organic nitrogen content (C–A/N_{org} = 1) but ratios < 1 (decarboxylation) or > 1 (carboxylate formation in excess of nitrate and sulphate reduction) are commonly found, with resulting rhizosphere pH effects of an increase or a decrease, respectively.

The levels of C–A values and the C–A/N_{org} ratio are dependent on factors such as plant species, the part of the plant and the age of the plant. Typical data on 2 *Plantago* species, which are derived from a large number of water culture experiments and plants from a wide range of natural habitats are given in Table 7. The C–A values and the C–A/N_{org} ratios of field material are higher than those of ammonium plants. This suggests that, under field conditions, at least part of the nitrogen has been taken up as nitrate.

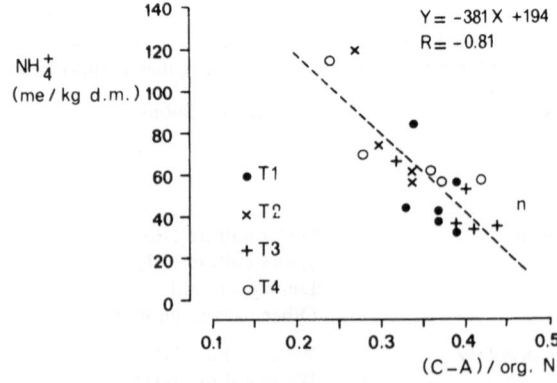

Fig. 4. The relationship between the free ammonium content and the (C–A)/Org N in the above-ground plant material at two wet sites (T1 and T3) and two dry sites (T2 and T4) in a dune grassland.

In Table 6 it was recorded that the highest ratios of ammonium and nitrate were found in wet sites. Nevertheless, the C–A/N$_{org}$ ratios on these sites were higher than those on dry sites (Fig. 4) This suggests that on the wet sites proportionally more nitrate was taken up. These results demonstrate that the proportions in which nitrate and ammonium are taken up from the soil can be different from those in which they are produced in the soil.

Concluding comments

Three different aspects of nitrate production in the rhizosphere have been discussed, *i.e.*, cell numbers of nitrifiers, the rate of nitrate production, and effects of N-sources on plant metabolism. The three enumeration methods of nitrifiers, *i.e.* the Most Probable Number Method, the Fluorescent Antibody technique and measurement of the potential nitrification rates are difficult to compare with each other. They possibly estimate different sections of the nitrifying populations but it is not clear which these are and all methods probably underestimate the real numbers. As far as we know, the three methods have never been applied simultaneously to soil samples in which the dominating organisms are known or in model experiments using defined systems.

The techniques that are used to measure the nitrification rate in natural ecosystems are unsatisfactory because either several simultaneously-occurring processes are measured (*in situ* net mineralization rate) or they establish potential nitrification rates rather than actual rates (*e.g.* short-term activity measurements using $^{15}NH_4^+$). There still is insufficient experience estimating the actual nitrification rate using the method of Schimel (1987). This promising method needs to be tested under a different range of soil conditions and compared with other data on soil nitrification.

Information on the nitrate reductase activity, and the ionic composition of plants gives in combination with the organic nitrogen content a quite accurate picture of the quantities of nitrate and ammonium that have been taken up by the plant to produce the organic nitrogen. However, in such data, the internal nitrogen turnover within the plant is not taken into account. Therefore, the real uptake of nitrate and ammonium is under-estimated. Moreover, it has been shown that nitrate and ammonium are taken up in different proportions than they are formed in the soil. This is due to the different diffusion rates of nitrate and ammonium and is, possibly, also the result of the distribution of ammonifying and nitrifying bacteria in soil of which nothing is known.

In conclusion, we repeat that there is a great need for experiments in which most of the various aspects of nitrification in the soil are studied simultaneously.

Model systems with plants and micro-organisms, in which the participating organisms are known, will also add to our knowledge of a process into which we still have insufficient insight.

References

Bakker K 1980 A place on the planet: Some reflections on population ecology. Neth. J. Zool. 30, 151–160.

Belser L W 1977 Nitrate reduction to nitrite, a possible source of nitrite for growth of nitrite-oxidizing bacteria. Appl. Environm. Microbiol. 34, 403–410.

Belser L W 1979 Population ecology of nitrifying bacteria. Annu. Rev. Microbiol. 33, 309–333.

Belser L W and Mays E L 1980 Specific inhibition of nitrite oxidation by chlorate and its use in assessing nitrification in soils and sediments. Appl. Environm. Microbiol. 39, 505–510.

Berg P 1986 Nitrifier populations and nitrification rates in agricultural soil. Ph.D. Thesis. Swedish University of Agricultural Sciences, Uppsala, Sweden.

Berg P and Rosswall T 1985 Ammonium oxidizers numbers, potential and actual oxidation rates in two Swedish arable soils. Biol. Fert. Soils 1, 131–140.

Blacquière T J 1986 Nitrate reduction in the leaves and numbers of nitrifiers in the rhizosphere of *Plantago lanceolata* growing in two contrasting sites. *In* Fundamental, Ecological and Agricultural Aspects of Nitrogen Metabolism in Higher Plants. Eds. H Lambers, J J Neeteson and I Stuler. pp 347–350. Martinus Nijhoff Publishers, Dordrecht, The Netherlands.

Bock E 1976 Growth of *Nitrobacter* in the presence of organic matter. II. Chemo-organotrophic growth of *Nitrobacter agilis*. Arch. Microbiol. 108, 305–312.

Bock E, Koops H-P and Harms H 1987 Cell biology of nitrifying bacteria. *In* Nitrification. Ed. J I Prosser. pp 17–38. IRL Press, Oxford, UK.

De Boer W, Duyts H and Laanbroek H J 1988 Autotrophic nitrification in a fertilized acid heath soil. Soil Biol. Biochem. 20, 845–851.

Focht D D and Verstraete W 1977 Biochemical ecology of nitrification and denitrification. *In* Adv. Microb. Ecol. Ed. M Alexander 1, 135–214.

Jansson S L 1958 Tracer studies on nitrogen transformations in soil with special attention to mineralization–immobilization relationships. Kungl. Lantbr. Ann. 24, 101–361.

Keeney D R 1987 Inhibition of nitrification in soils. *In* Nitrification. Ed. J I Prosser. pp 99–115. IRL Press, Oxford, UK.

Killham K 1987 Heterotrophic nitrification. *In* Nitrification. Ed. J I Prosser. pp 117–126. IRL Press, Oxford, UK.

Koops H-P and Harms H 1985 Deoxyribonucleic acid homologies among 96 strains of ammonia-oxidizing bacteria. Arch. Microbiol. 141, 214–218.

Kuenen JG and Harder W 1982 Microbial competition in continuous culture. *In* Experimental Microbial Ecology. Eds. R G Burns and J H Slater. pp 342–367. Blackwell Scientific Publishers, Oxford, London, Edinburgh, Boston, Melbourne.

Kuenen J G and Robertson L A 1987 Ecology of nitrification and denitrification. *In* The Nitrogen and Sulphur Cycles. Eds. J A Cole and S Ferguson. pp 161–218. Cambridge University Press.

MacDonald R M 1986a Nitrification in soil: an introductory history. *In* Nitrification. Ed. J I Prosser. pp 1–16. IRL Press, Oxford, UK.

MacDonald R M 1986b Sampling soil microfloras: dispersion of soil by ion exchange and extraction of specific micro-organisms from suspension by elutriation. Soil Biol. Biochem. 18, 399–406.

Myrold and Tiedje 1986 Simultaneous estimation of several nitrogen cycle rates using ^{15}N: Theory and application. Soil Biol. Biochem. 18, 559–568.

Pomeroy L R 1970 The strategy of mineral cycling. Annu. Rev. Ecol. Syst. 1, 171–190.

Powell S J 1987 Laboratory studies of inhibition. *In* Nitrification. Ed. J I Prosser. pp 79–97, IRL Press, Oxford, UK.

Rao P S C, Jones J W and Kidder G 1989 Development, validation and applications of simulation models for agroecosystems: Problems and perspectives. *In* Ecology of Arable Land – Perspectives and Challenges. Eds. M Clarholm and L Bergström. pp 253–259. Kluwer Academic Publishers, Dordrecht, The Netherlands.

Rennie R J 1975 The autecology and enumeration of *Nitrobacter* in soils. Ph.D. Thesis, University of Minnesota.

Rennie R J and Schmidt E L 1977 Immunofluorescence studies of *Nitrobacter* populations in soil. Can. J. Microbiol. 23, 1011–1077.

Rice E L 1982 Allelopathy. Academic Press, London and New York, 353 p.

Riha S J, Campbell G S and Wolfe J 1986 A model of competition for ammonium among heterotrophs, nitrifiers and roots. Soil Sci. Soc. Am. J. 50, 1463–1466.

Robertson G Ph and Vitousek P M 1981 Nitrification potentials in primary and secondary succession. Ecology 62, 376–386.

Runge M 1983 Physiology and ecology of nitrogen nutrition. *In* Physiological Plant Ecology. III. Responses to the Chemical and Biological Environment. Eds. O L Lange, P S Nobel, C B Osmond and H Ziegler. pp 163–200. Springer Verlag, Berlin, Heidelberg, New York.

Schimel J P 1987 Plant-microbial competition for nitrogen in a California forest and grassland. Ph.D. Thesis, University of California, Berkeley.

Schmidt E L 1973 Fluorescent antibody techniques for the study of microbial ecology. Bull. Ecol. Res. Comm. (Stockholm) 17, 67–76.

Schmidt E L 1982 Nitrification in soil. *In* Nitrogen in Agricultural Soils. Ed. F J Stevenson. pp 251–288. Agronomy 22, Madison, Wisconsin.

Schmidt E L and Belser L W 1982 Nitrifying bacteria, pp 1027–1041. Methods of Soil Analyses II. Chemical and Microbiological Properties. Agronomy Monograph, 2nd ed., 9.

Smit A J and Woldendorp J W 1981 Nitrate production in the rhizosphere of *Plantago* species. Pland and Soil 61, 43–52.

Troelstra S R 1983 Growth of *Plantago lanceolata* and *Plantago major* on a NO_3/NH_4 medium and the estimation of the utilization of nitrate and ammonium from ionic balance aspects. Plant and Soil 70, 183–197.

Troelstra S R and Smant W 1980 The ionic balance of some plant species from natural vegetations: Comparison of plants grown in the greenhouse and in the field. Verh. Kon. Ned. Akad. Wetensch., Afd. Natuurk., 2^e Reeks 75, 46–51.

Troelstra S R and Wagenaar R 1980 Seasonal patterns in the availability of mineral nitrogen in a relatively old dune area (Westdiunen) on the island of Goeree. Verh. Kon. Ned. Akad. Wetensch., Afd. Natuurk., 2^e Reeks 75, 37–41.

Verstraete W 1981 Nitrification. *In* Terrestrial Nitrogen Cycles Eds. F E Clark and T Rosswall. Ecol Bull. (Stockholm) 33, 303–314.

Watson S W 1974 *Nitrobacteriaceae* Buchanan. *In* Bergey's Manual of Determinative Bacteriology, 8th Edn. Eds. R E Buchanan and N E Gibbons. pp 450–456. Williams and Wilkins Co., Baltimore.

Whittenbury R and Kelly D P 1977 Autotrophy: A conceptual phoenix. Symp. Soc. Gen. Microbiol. 27, 121–149.

Woldendorp J W 1975 Nitrification and denitrification in the rhizosphere. Societé Botanique de France, Colloque 'La rhizosphère' 122, 89–107.

Woldendorp J W 1981 Nutrients in the rhizosphere. pp 99–125. Proceeding of the 16th Colloquium of the Potash Institute, Bern.

Woldendorp J W 1983 The relation between the nitrogen metabolism of *Plantago* species and the characteristics of the environment. *In* Nitrogen as an Ecological Factor. Eds. J A Lee, I McNeil and I H Rorison. pp 137–166. 22nd Symp. Brit. Ecol. Soc. Blackwell, Oxford, UK.

M. Clarholm and L. Bergström (Eds.), Ecology of arable land, 161–171.
© 1989 Kluwer Academic Publishers.

Nitrogen availability and nitrification during succession: Primary, secondary, and old-field seres

PETER M. VITOUSEK, PAMELA A. MATSON and KEITH VAN CLEVE
Department of Biological Sciences, Stanford University, Stanford, CA 94305, USA; Ecosystem Science and Technology Branch, NASA-Ames Research Center, Moffett Field, CA 94035, USA and University of Alaska, Fairbanks, AL 99701, USA

Key words: nitrogen availability, nitrification, soil development, succession

Abstract

Suggestions that nutrient cycles become more strongly regulated and that nitrification is progressively inhibited in the course of ecological succession have stimulated numerous field measurements. Results of these are inconsistent; in some cases nitrogen turnover and nitrification decrease during succession, while in others both increase substantially.

Consideration of the nature of disturbance which initiates each succession explains much of the difference in nitrogen dynamics. Primary succession (the development of ecosystems on wholly new substrates) invariably involves a low nitrogen availability and nitrification early in succession. In contrast, destructive disturbance followed by immediate regrowth (the 'pure case' of secondary succession) invariably increases nitrogen availability (and generally nitrate production) in recently disturbed sites; it is followed by a decline during later stages of succession. Succession following a period of chronic disturbance (*i.e.* prolonged agriculture) does not follow such clear patterns; the duration and intensity of disturbance may control whether nitrogen availability and potential nitrification increase or decrease early in such seres.

Introduction

Theoretical publications generally agree that the availability of soil nutrients changes during ecological succession, but there is rather less agreement on the direction and significance of such changes. One line of argument suggests that the availability of all of the major resources (light, water, nutrients) is elevated at the soil surface shortly after destructive disturbance (*sensu* Grime, 1979), and that succession thereafter involves a diminution in resource availability (at least initially) (Bormann and Likens, 1979; Odum, 1969; Vitousek and Reiners, 1975). Alternatively, changes in plant species composition during ecological succession have been interpreted as reflecting gradients from high light/ low nutrient conditions early in succession to low light/high nutrient conditions later (Tilman, 1982; 1986).

Successional changes in the availability of nit-

rogen have received particular emphasis, in part because nitrogen often limits primary production in terrestrial ecosystems. Moreover, successional changes in nitrate production can have substantial effects on system-level nitrogen losses to both the hydrosphere and the atmosphere (Gorham *et al.*, 1979; Robertson and Tiedje, 1984). Rice and Pancholy (1972) proposed that nitrification is inhibited systematically by plant-produced allellochemical substances in climax ecosystems, but such that inhibition is relaxed early in succession. Subsequent publications have disagreed with the generality of this pattern (*cf* Lamb, 1980; Robertson and Vitousek, 1981).

Generalizations concerning successional patterns of nutrient availability or nitrification can be tested empirically, and many measurements of nitrogen availability and transformations during ecological succession have been reported. Our initial purpose here is to review results of these studies in relation to theoretical predictions. We show that patterns of change in nitrogen transformations differ widely among seres, and that therefore no single generalization adequately explains the range of variability encountered in the field. We then use the results of this review to explore why seres differ in their patterns of successional change in nitrogen availability.

Earlier discussions suggested that nitrogen transformations differ systematically between primary and secondary succession (Robertson and Vitousek, 1981; Vitousek and Walker, 1987), with nitrogen availability increasing during primary succession but maximum nitrogen availability occurring as a pulse early in secondary succession. However, primary and secondary succession themselves represent extremes on a gradient of disturbance intensity. In this paper, we examine patterns of nitrogen availability for clear examples of primary and secondary succession as well as for examples in which the disturbance itself is intermediate in intensity.

Approach and limitations

We collected published information and unpublished data on changes in nitrogen availability during succession. The task was not straightforward – to start, there is no consensus on how either 'succession' or 'nitrogen availability' is defined, much less measured.

Succession

We made no attempt to examine stages of succession or mechanisms of species replacement during succession (*cf* Connell and Slatyer, 1977), but rather considered successional age to be equivalent to time since disturbance. We classified seres as representing primary, secondary, or 'intermediate' succession. Primary succession occurs following disturbance which leaves no trace of the pre-existing community (if any) (Clements, 1916); the most common forms follow volcanic eruptions, glacial recession, and movement of sand dunes. Early stages of primary succession are generally harsh environments, and soils and biological communities develop slowly.

In a strict sense, secondary succession encompasses all other cases of succession; we will use it in the narrower sense of representing situations where disturbance kills or removes vegetation but leaves the soil intact, and in which vegetation regrowth can begin immediately following disturbance. Clearcut logging and natural windthrows are clear examples of secondary succession in this sense. 'Intermediate' successions occur where some influence of a previous community (most often, some soil organic matter) is present following disturbance, but substantially less than in a comparable intact ecosystem. The most common intermediate form is succession following abandonment of agricultural fields; soil organic matter and nitrogen decline under this chronic disturbance, but there is always a residual influence of a pre-existing biological community. Other intermediate seres include those initiated by silt deposition on river bars or by particularly intense fires. We will consider only post-agricultural succession here, and term this category 'old-field succession'.

The time-scale of succession is long relative to any individual's research career. Consequently, information on nutrient availability during succession is rarely derived from longitudinal studies of a particular ecosystem as it changes with time. Rather, most often an array of sites which represent different successional ages is selected. The sites must then be assumed to be identical except for their age; any variation resulting from other factors such as differences in parent material or topography, intensity of disturbance, or even interannual variations in weather or biological colonization is very difficult to separate from age-related effects. Therefore, even when extreme care is taken in selecting sites for successional studies, it is difficult to be certain that any differences observed between plots are due solely to successional processes.

Nitrogen availability

The term 'nitrogen availability' itself can be defined in many different ways; there is no primary standard except perhaps for N uptake by an N-limited plant, which cannot be measured directly under field conditions in communities dominated by perennial plants. We will consider N availability as equivalent to the total amount of nitrogen which annually enters forms accessible to plant uptake. We assume that this is approximately equivalent to net nitrogen mineralization, although there are certainly circumstances in which N availability could be either greater or less than N mineralization. (For example, some low-molecular-weight organic nitrogen can be taken up by plants – Bowen and Smith, 1981). We distinguish this aspect of 'nitrogen availability' from measurement of the instantaneously-available pool of ammonium and nitrate in soil, from plant (or plant plus microbial) demand for N, and from the extent to which primary production is limited by nitrogen. All of these represent important aspects of nitrogen cycling during succession, but we will focus on net nitrogen mineralization here.

Even given this restricted definition of nitrogen availability, it is difficult to obtain information that can be compared among studies. In part, this is true because we believe that net nitrogen mineralization itself cannot be measured directly within intact ecosystems. It is defined as the difference between the release of nitrogen from organic forms and the uptake of inorganic nitrogen by microorganisms. Net nitrogen mineralization is measured by incubating soils in the absence of plant uptake of nitrogen and measuring the consequent accumulation of inorganic nitrogen, yet any method of preventing plant uptake inevitably affects processes of mineralization and/or immobilization in some way (Clarholm, 1985). Removing roots from soil eliminates the potential priming or inhibitory effects of root exudates and disrupts soil structure; cutting them and leaving them in place adds a pulse of organic matter to the system which can release or immobilize nitrogen. The accumulation of ammonium and/or nitrate within incubated samples can feed back to affect nitrogen mineralization itself. Moreover, the uptake of water by roots is prevented within incubated soils, so moisture status and distribution cannot precisely match that in the field. The importance of these concerns may be minor in a particular soil (Raison *et al.*, 1987), but unfortunately there is no primary standard against which that can be tested. Accordingly, we are uncomfortable with conclusions that net nitrogen mineralization can be measured accurately enough to calculate plant nitrogen uptake directly, and that this information can then be used

to calculate production of fine roots in forest ecosystems (Nadelhoffer *et al.*, 1985).

In the absence of direct methods for measuring net nitrogen mineralization, a wide variety of methods that provide a comparative index of mineralization have been developed (*cf* Ellenberg, 1977; Eno, 1960; Keeney, 1982; Raison *et al.*, 1987; Stanford and Smith, 1972). These include incubations carried out in the field versus under controlled conditions in the laboratory, under aerobic or anaerobic conditions, in mixed or intact soils, and at varying temperatures and moisture contents. While measurements which require minimal disturbance are undoubtedly best (Raison *et al.*, 1987), results obtained with all of these methods are usually well correlated with other nitrogen cycling parameters (within a study) (Birk and Vitousek, 1986; Pastor *et al.*, 1984; Powers, 1980). They can therefore be used to examine patterns of nitrogen availability within a successional sequence where that sequence is studied using a single method.

Results

We located 23 seres for which information on N mineralization, nitrification, or both (measured in the field or the laboratory) was available for at least 3 ages. Of these, 19 reported results in such a way that mineralization and/or nitrification could be summarized on a concentration basis, as $\mu g N$ produced per gram dry weight per month. These included 5 primary (Table 1), 6 secondary (Table 2), and 8 old-field (Table 3) seres.

While results of incubation studies are most often reported on a concentration basis, systematic successional differences in forest floor mass or soil bulk density could cause volumetric or areal estimates of nitrogen availability to differ from concentration-based estimates. Accordingly, wherever information on bulk density or horizon mass was available, we calculated N mineralization on an areal basis (grams N per square meter per month). Results of these calculations are summarized in Figs. 1, 2 and 3 for primary, secondary, and old-field succession respectively. Finally, the total quantity of forest floor and soil nitrogen usually varies during succession. Consequently, expressing nitrogen mineralization in terms of nitrogen released per gram of soil organic matter (Thorne and

Table 1. Summary of potential net nitrogen mineralization (Min.) and nitrate production (Nit.) in primary seres. Results of either field or laboratory incubations are reported. B.D. is bulk density

Site	Type	Results							Reference
Michigan, USA	Sand Dune, soil	Age (yrs)	20	1000?	10,000?				Robertson and
		Min. (µg/g)	− 1.6	28.7	23.8				Tiedje, 1984
		Nit. (µg/g)	− 1.2	19.8	18.7				
		Total N (%)	0.003	0.173	0.102				
		B.D. (g/ml)	1.6	1.01	1.14				
Indiana, USA	Sand Dune forest floor + soil	Age	1	10	50	100	1000	10,000	Robertson and Vitousek, 1981
		Min.	2.7	2.6	3.4	13.7	17.6	13.1	
		Nit.	1.3	2.2	3.2	10.7	16.7	1.2	
		Total N	0.002	0.008	0.005	0.041	0.112	0.110	
		B.D.	1.66	1.69	1.71	1.15	1.04	1.07	
Hawaii, USA	Pahoehoe lava, organic soil over lava	Age	126	1500	2500	3500			Vitousek *et al.*, 1983
		Min.	1.3	3.6	1.6	27			
		Nit.	0	1.5	3.7	27			
		Total N	1.03	0.73	0.56	1.64			
		Horizon (T/ha) Mass	20	326	178				
Hawaii, USA	volcanic cinder forest floor	Age	24	200	1000	4000			Vitousek *et al.*, 1983;
		Min.	−*	0.3	3.8	73			Matson and
		Nit.	−	0.2	4.2	12			Vitousek
		Total N	−	0.76	0.83	1.71			unpublished
		Horizon Mass	0	296	426	105			
	Volcanic cinder mineral soil	Age	24	200	1000	4000			Vitousek *et al.*, 1983;
		Min.	1.4	0.1	3.9	81			Matson and
		Nit.	1.4	0	4.1	73			Vitousek
		Total N	0.008	0.14	0.14	0.89			unpublished
		B.D.	0.8	1.03	0.73	0.30			
Alaska, USA	Floodplain, forest floor	Age	5	25	50	130	250		Van Cleve
		Min.	1	180	28	175	4.8		unpublished
		Nit.	−	17	0.5	0.6	0.1		
		Total N	−						
		Horizon Mass	0	37	50	65	30		
	Floodplain, mineral soil	Age	5	25	50	130	250		Van Cleve
		Min.	0.6	6.2	2.5	6.1	0.03		unpublished
		Nit.	0.01	5.8	0.3	4.1	0.05		
		Total	0.04	0.11	0.07	0.05	0.18		
		B.D.	1.13	0.60	0.95	1.11	0.78		

Others: Roberts *et al.* (1980) and Reeder and Berg (1977) measured changes in N mineralization following 'reclamation' of mining spoils – N mineralization increased with time in the reclaimed sites.
*No forest floor at this stage of succession.

Hamburg, 1985), or better, per gram of soil organic nitrogen (Skeffington and Bradshaw, 1981; Sollins *et al.*, 1984; Vitousek *et al.*, 1983) yields insight into the relative recalcitrance of soil nitrogen from different stages of succession. Where they are available, we have reported concentrations of total nitrogen in forest floor and mineral soil in Tables 1–3.

Comparing all of the seres together (Tables 1–3), there are clearly very wide differences in patterns of nitrogen availability during succession. Five of the 19 have their maximum nitrogen mineralization in the oldest stage sampled. Nine have maximum mineralization either in the first stage sampled, or within the first 10 years of succession, while the remainder have more complex patterns. However, this variation can be accounted for reasonably well within our *a priori* categories of seres. All of the primary seres start with very low N availability (Table 1, Fig. 1), which then increases with age; two have subsequent declines later in succession. All of the secondary seres resulting from logging or stand dieback have their highest nitrogen availability

Table 2. Potential nitrogen mineralization and nitrate production in secondary seres.

Site	Type	Results							Reference
Turrialba, Costa Rica	Slash/burn/release	Age yrs)	0.01	0.6	5	70			Matson *et al.*, 1987;
		Min. (μg/g)	124	45	62	82			Matson, unpublished
		Nit. (μg/g)	124	45	62	82			data
		T.N. (%)	0.62	0.62	0.65	0.64			
		B.D. (g/ml)	~ 0.74	~ 0.74	~ 0.74	~ 0.74			
Indiana	clearcut logging	Age	1	4	9	> 60			Matson and
		Min.	12.4	20.1	9	15.2			Vitousek, 1981
		Nit.	11.3	16.8	7.5	5.1			
		T.N.	0.17	0.14	0.17	0.16			
		B.D.	0.57	0.66	0.81	0.69			
Oregon	wave form dieback forest floor (O_2) Final values only (end of incubation)	Age	< 10	18–50	65–90	> 200			Matson and
		Min.	91	54	35	32			Boone, 1984
		Nit.	trace	trace	0	0			
		T.N.	0.56	0.43	0.45	0.53			
		Horizon (T/ha) Mass	47	55	50	66			
	mineral soil	Age	< 10	18–50	65–90	> 200			Matson and
		Min.	1.2	1.5	0.9	0.4			Boone, 1984
		Nit.	trace	trace	0	0			
		T.N.	0.09	0.09	0.10	0.08			
		B.D.	0.64	0.66	0.64	0.66			
North Carolina	Logging	Age	1	2	22	40–60	80–100	> 150	Vitousek and
		Min.	5.9	3.9	1.5	1.7	0.6	3.7	Matson, 1985
		Nit.	5.4	1.6	0.2	0.1	0	0.15	Saterson
		B.D.	1.1	1.1	1.1	1.07	1.06	0.81	1985
Michigan	Logging	Age	0.2	60	120				Robertson
		Min.	36	5.6	13.0				and Tiedje,
		Nit.	10.2	1.4	4.3				1984
		T.N.	0.19	0.15	0.13				
		B.D.	0.88	0.95	0.98				
Alaska	Logging forest floor only	Age	1	2	3	110			Gordon and
		Min.	29	16	32	19			Van Cleve,
		Nit.	15	11	16	trace			1983
		T.N.	NR	NR	NR	NR			
		B.D.	NR	NR	NR	NR			

Others: Montagnini *et al.* (1986) measured mineralization and nitrification in a clearcut and a control watershed in North Carolina; nitrification was higher 2–6 years after clearcutting than in the control forest, especially in successional sites dominated by the symbiotic nitrogen-fixer *Robinia pseudoacacia*. Wheeler and Donaldson (1983) measured N mineralization in a range of sites in Arkansas; nitrification was generally greater in younger sites, but the sites sampled were not all related successionally.

within 10 years after disturbance (Table 2, Fig. 2). All then decline as succession proceeds, although there is a smaller increase later in succession in several seres. Finally, the old-field seres (Table 3, Fig. 3) are variable; several have an early maximum, others a late one.

Successional changes in nitrification are generally similar to changes in mineralization; both mineralization and nitrification usually increase during primary succession and decrease during secondary succession. However, the decline in nitrification later in secondary succession is sometimes more substantial and more sustained than that in mineralization (*cf* Gordon and Van Cleve 1983; Matson and Vitousek 1981; Saterson 1985). Put another way, the fraction of mineralized nitrogen that is oxidized to nitrate declines in the older stages in 3 or the 6 secondary seres (and at least one old-field sere); it does not vary systematically in the other 3.

Table 3. Potential net nitrogen mineralization and nitrate production during succession following agricultural use

Site	Type	Results						Reference
Michigan	old-field	Age (yrs)	12	18	> 100			Robertson
		Min. (μg/g)	− 11	12	1.5			and Tiedje, 1984
		Nit. (μg/g)	0.8	1.1	3.5			
		T.N. (%)	0.13	0.10	0.16			
		B.D. (g/ml)	1.29	1.32	1.06			
Costa Rica	old-field and	Age	0	3	8	16	31	60 +	Robertson, 1984;
	repeatedly-disturbed	Min.	0	30	66	99	34	84	
	experimental plots	Nit.	25	33	66	96	84	84	Werner, 1984
		T.N.	0.42	NR	0.50	0.37	0.38	0.37	
		B.D.	0.62	NR	0.54	0.64	0.57	0.58	
Australia	old-field	Age	10	15	43	53	> 100		Lamb, 1980
		Min.	18	31	51	54	90		
		Nit.	15	26	51	51	81		
		T.N.	0.52	0.74	0.78	0.83	0.98		
		B.D.	NR	NR	NR	NR	NR		
India Ramakrishnan	degraded grassland	Age	young	intermediate older		forest		
		Min.	−	−	−	−		and Saxena,
		Nit.	0.5	0.5	1.5	9		1984
		T.N.	0.12	0.20	0.27	0.39		
		B.D.	NR	NR	NR	NR		
India	shifting cultivation	Age	< 0.5	1	3	20		Saxena
	cycle-fallows	Min.	−	−	−	−		and
		Nit.	17.5	11.4	9.5	5.4		Ramakrishnan,
		T.N.	NR	NR	NR	NR		1986
		B.D.	NR	NR	NR	NR		
North Carolina	old-field	Age	6–8	40–50	> 150			Montes
		Min.	3.1	0.4	1.1			and Christensen,
		Nit.	3.6	0.1	2.1			1979
		T.N.	0.15	0.14	0.20			
		B.D.	NR	NR	NR			
North Carolina	old-field	Age	4–7	40–50	> 150			Christensen
		Min.	75	54	60			and McAller,
		Nit.	39	15	9			1985
		T.N.	NR	NR	NR			
		B.D.	NR	NR	NR			
New Jersey	old-field	Age	1	4	19	> 200		Robertson and
		Min.	10	12	11	25		Vitousek, 1981
		Nit.	10	12	11	28		
		T.N.	0.12	0.13	0.14	0.16		
		B.D.	1.32	1.33	1.22	1.01		

Others: Pastor *et al.* (in press) measured N mineralization in the field in 4 Minnesota old-fields. Only the sum of May–September mineralization was reported; it increased monotonically from 44 to 65 kg/ha from a 16 yr old to a never-ploughed site. Thorne and Hamburg (1985) measured nitrate production in forest floor from an old-field chronosequence in New Hampshire; nitrification was delayed > 30 days in all, but the delay was shorter and the rate more rapid in the young sites. Chandler and Goosem (1982) reported additional results from the sere studied by Lamb (1980).

Discussion

Overall, the results summarized here suggest that generalizations concerning nitrogen availability during succession can be useful only where they specify a particular type of succession. Even then, the existing data will support only a limited number of conclusions. We can conclude that nitrogen availability (in the sense used here) is low early in primary succession and high early in pure secondary succession; as is discussed below, we can also suggest plausible mechanisms for these patterns.

The overall results also support the generalization that nitrification declines during pure secon-

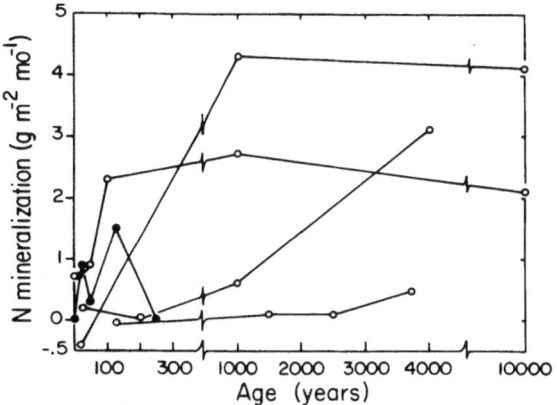

Fig. 1. Potential net nitrogen mineralization in primary seres. Results are reported in $gm^{-2}mo^{-1}$; where data are available, the contributions of forest floor and mineral soil are summed. Both field- and laboratory-based measurements are included. Data from Table 1.

The Alaskan flood-plain sere, which we consider to be intermediate between primary and secondary succession, is represented with closed symbols.

dary succession, but make it clear that the validity of this generalization is restricted to certain types of seres. They do not support the generality of an inhibition of nitrification in climax ecosystems (Rice and Pancholy, 1972; Rice 1984). Many seres increase in nitrate production during succession, and even in secondary seres the proportion of mineralized nitrogen oxidized to nitrate is constant in half of our seres. On the other hand, the results also fail to support Robertson and Vitousek's (1981) suggestion that nitrification is generally regulated by ammonium availability; nitrate production does

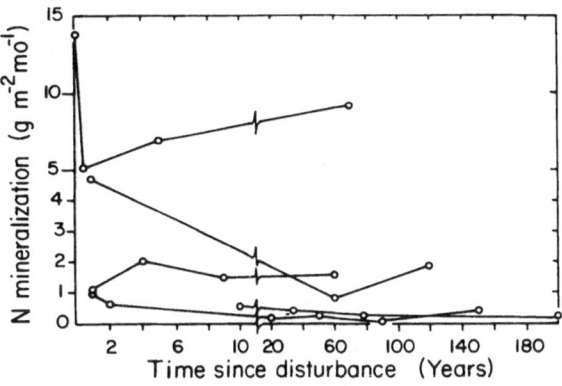

Fig. 2. Net nitrogen mineralization in secondary seres following clearcut logging or stand dieback (in $gm^{-2}mo^{-1}$). Data from Table 2.

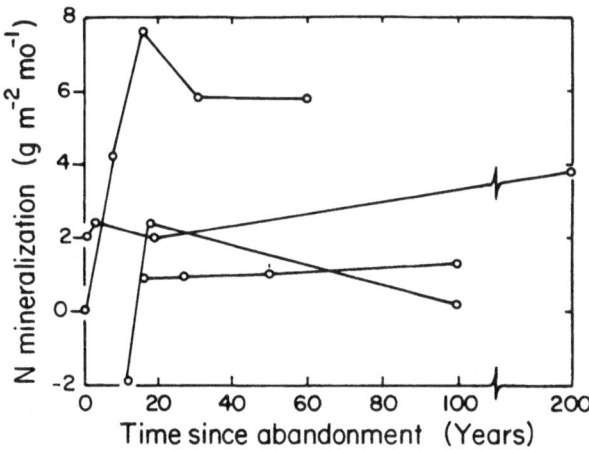

Fig. 3. Net nitrogen mineralization in old-field seres (in $gm^{-2}mo^{-1}$). Data from Table 3.

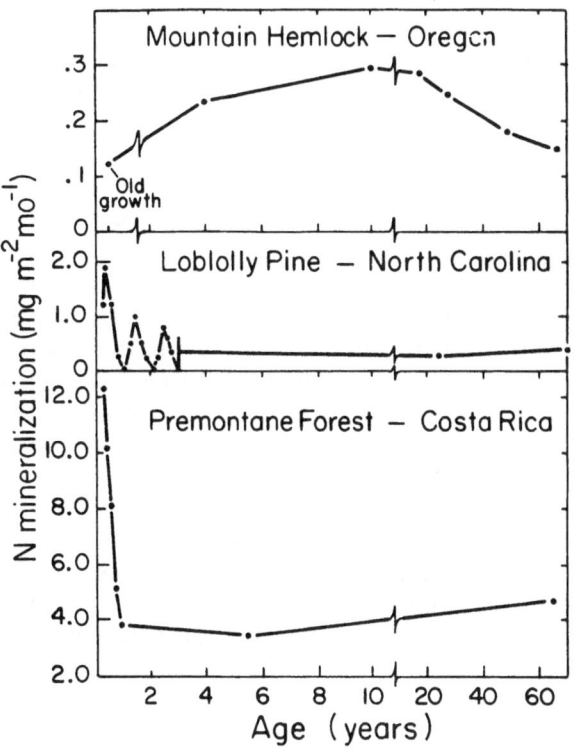

Fig. 4. Timing and magnitude of the increase in potential net nitrogen mineralization measured in the field immediately following disturbance and during secondary succession. Oregon data from Matson and Boone (1984); North Carolina from Vitousek and Matson (1985) and Saterson (1985); and Costa Rica from Matson *et al.* (1987).

decline as a fraction of mineralization in half the pure secondary seres.

Changes during primary succession

The pattern of nitrogen availability in primary succession is consistent with earlier suggestions based on mechanisms of soil development (Robertson and Vitousek, 1981; Vitousek and Walker, 1987; Walker and Syers, 1976). Wholly new substrates usually contain phosphorus and other essential plant nutrients in abundance, but no organic matter or nitrogen. Both the total amount and the availability of nitrogen thus literally have nowhere to go but up. The pattern of resource availability that Tilman (1982, 1986) suggests for succession in general – a change from high light, low nitrogen conditions at the soil surface early in succession to low light, high nitrogen conditions late – applies well to primary succession.

A number of additional suggestions about nitrogen dynamics in primary succession can be drawn from a review of the publications summarized in Table 1 and from our understanding of processes of soil development. First, succession on a floodplain might be expected to proceed differently from other types of primary succession because the 'new' substrate is new only to that location – it contains organic carbon and nitrogen transported from upstream, and hence should have more total and available nitrogen than most other primary substrates. In fact, the development of both a closed forest canopy and of a 'climax' forest occurs more rapidly in the Alaskan floodplain site (Table 1) than in the other primary seres, despite the climatic severity of Central Alaska. Moreover, soil nitrogen availability also increases more rapidly (but see below).

Second, it has long been argued that nitrogen-fixers have a strong competitive advantage early in primary succession because of the lack of fixed nitrogen and adequate levels of phosphorus in such sites (Gorham *et al.*, 1979; Stevens and Walker, 1970; Vitousek and Walker, 1987; Walker and Syers, 1976), and that most primary seres are in fact dominated by symbiotic N fixers at some point in early or mid-succession. The majority of the natural seres summarized here do not have such a stage. However, where N fixers are dominant (alder

in the Alaska floodplain, as plantings in mining spoils), they affect N availability substantially. Montagnini *et al.*, (1986) showed that the presence of nitrogen fixers early in secondary succession could also greatly increase local nitrogen availability.

Third, Grubb (1986) pointed out that the earliest pioneers in primary succession often have non-sclerophyllous leaves and appear not to be nutrient-limited. He suggested that N in precipitation might satisfy the requirement of such colonists, particularly because nitrate is mobile in soils and because the lack of organic matter precludes substantial microbial immobilization of N. Information on N mineralization is not useful for testing this suggestion directly, but it seems to us a reasonable one where nitrate is an important constituent of precipitation. Nitrogen availability may become limiting during primary succession only when the early plant colonization is sufficiently advanced that potential plant uptake exceeds the supply of nitrogen in precipitation, which may take several years where plant colonization is slow.

A final (and related) suggestion is that the proportion of soil nitrogen mineralized might be expected to decline with soil age, even though the total amount mineralized clearly increases (Hirose and Tateno, 1984; Skeffington and Bradshaw 1981). Such a decline could occur as a consequence of increasing microbial immobilization of nitrogen as soil organic matter accumulates (or as the C/N ratio of plant litter increases and its carbon quality decreases when plants become nitrogen-limited), or because organic nitrogen is physically protected from mineralization by combination with soil colloids and/or aggregates (Sollins *et al.*, 1984). We calculated nitrogen mineralized/unit of soil N for each of the primary seres; it clearly declined with successional age in two cases, but increased in two others.

Changes during secondary succession

A number of earlier studies suggested that disturbances which initiate secondary succession cause increased nitrogen availability in the disturbed site (Bormann and Likens, 1979; Stone, 1973; Vitousek and Melillo, 1979). Removal of the dominant vegetation increases soil temperature and moisture, thereby increasing decomposition and nitrogen release. Microbial immobilization of nitrogen and lags in nitrification can delay net mineralization and nitrification in such sites (Vitousek *et al.*, 1982; Vitousek and Matson, 1984), but sooner or later they increase above predisturbance levels. Moreover, the initial colonists of secondary seres are often rapidly-growing species with high nutrient contents and relatively high rates of decomposition and nutrient release (Covington, 1981; Melillo *et al.*, 1982), so the substrate quality for decomposers could also be elevated shortly after disturbance (Matson and Vitousek, 1981).

These predictions are supported by the information summarized in Table 2 and Fig. 2. Mineralization and nitrification are invariably at a maximum shortly after disturbance; they invariably decline thereafter.

A prediction that the timing and magnitude of increased nitrogen availability differs among sites in predictable ways (Vitousek *et al.*, 1982) can also be supported from these data. Measurements of 3 of the seres in Table 2 were undertaken by the same investigator in much the same way (field incubations carried out for a complete growing season or an annual cycle), and the detailed results for these seres are summarized in Fig. 4. The Oregon mountain hemlock sere is in a severe subalpine environment in which nitrogen is strongly limiting; it has a small and long-delayed response to disturbance (and very slow revegetation). In contrast, the premontane forest in Costa Rica had extremely high nitrogen mineralization prior to disturbance – and a rapid, large, and short-lived increase following clearing and burning. The North Carolina site was intermediate in both respects.

Changes in nitrogen availability during secondary succession could continue after the initial pulse of increased availability disappeared (Bormann and Likens, 1979; Miller, 1981; Vitousek and Walker, 1987). Several publications have suggested that nitrogen availability could increase late in succession after an intermediate minimum caused by intense competition for nitrogen (Coats *et al.*, 1976; Vitousek and Reiners, 1975); others have suggested progressive decreases in nitrogen availability caused by the immobilization of nitrogen in forest floor and humus (Weetman, 1962). Overall, the evidence in Table 2 is insufficient to evaluate the generality of these or other patterns of change. At least the North Carolina sere clearly increases in N availability after an intermediate minimum; the Oregon sere may well decrease progressively, at least unless stand-level dieback intervenes (Matson and Waring, 1984). It seems reasonable to speculate that the degree of climatic severity (through its effect on rates of decomposition) may control whether nitrogen availability increases or decreases late in secondary succession.

The total amount of nitrogen in forest floor and soil changes relatively little during 'pure' secondary succession, while nitrogen mineralization generally increases substantially shortly after disturbance. Consequently, the *proportion* of soil organic nitrogen mineralized increases in recently-disturbed sites.

Changes during old-field succession

There has been less discussion of expected patterns of change in nitrogen availability in old-field succession than for primary or pure secondary succession, perhaps in part because old-field succession is in fact a type of secondary succession. However, we anticipated that it might be different from secondary succession following logging because sites which are cropped annually lose substantial amounts of soil organic carbon and nitrogen in relatively few years (Coleman *et al.*, 1984; Haas *et al.*, 1957); they maintain a lower level of soil nitrogen as long as they are cultivated. Moreover, the nitrogen lost is drawn disproportionately from the more labile fractions of soil nitrogen, so the availability of native soil nitrogen should decline even more than total nitrogen (but see Schimel, 1986). Consequently, we expected that nitrogen availability would start low and then gradually increase during old-field succession.

The pattern of changes that we found (Table 3) was not that simple, but at least half of the seres do start low and increase in N availability during succession. Others have rapid mineralization early in succession, a difference which could reflect the duration or intensity of agricultural land use, the effect of residual fertilizer or manuring, or a change in the ratio of gross to net mineralization in agricultural systems (Schimel, 1986).

Nitrogen dynamics in old-field succession are clearly more variable than those in either primary or pure secondary succession, but there is a substantial number of seres for which nitrogen availability increases with successional age in accordance with Tilman's (1982) suggestion. This increase is much more rapid than that in most primary seres (excluding perhaps river bars, which are themselves intermediate). It also differs from primary succession in that the fraction of soil nitrogen mineralized often increases during old-field succession as labile forms of nitrogen are reintroduced into the soil.

Conclusions

The information summarized here demonstrates that patterns of nitrogen availability during succession vary among seres, and that much of this variability can be explained by considering the types of disturbance which initiate succession. Primary succession generally involves an increase in N availability over long periods of time, secondary succession a rapid increase in N availability shortly after disturbance and then a decline, and old-field succession is more variable. The lack of an accurate technique for measuring N availability directly means that these comparisons are primarily qualitative.

Acknowledgements

P.M.V. was supported by NSF grant BSR 8415821 during manuscript preparation; Cheryl Nakashima cheerfully typed several drafts of the manuscript under deadline pressure.

References

Birk E M and Vitousek P M 1986 Nitrogen availability and nitrogen use of efficiency in loblolly pine stands. Ecology 67, 69–79.

Bormann F H and Likens G E 1979 Pattern and Process in a Forested Ecosystem. Springer-Verlag New York.

Bowen G D and Smith S E 1981 The effects of mycorrhizae on nitrogen uptake by plants. *In* Nitrogen Cycling in Terrestrial Ecosystems: Processes, Ecosystem Strategies, and Management Implications. Eds. F E Clark and T H Rosswall. Ecol. Bull. (Stockholm) 33, 237–247.

Chandler G and Goosem S 1982 Aspects of rainforest regeneration. III. The interaction of phenols, light, and nutrients. New Phytol. 92, 369–380.

Christensen N L and MacAller T 1985 Soil mineral nitrogen transformations during succession in the Piedmont of North Carolina. Soil Biol. Biochem. 17, 675–681.

Clarholm M 1985 Interactions of bacteria, protozoa, and plants leading to mineralization of soil nitrogen. Soil Biol. Biochem. 17, 181–187.

Clements F E 1916 Plant Succession: An Analysis of the Development of Vegetation. Carnegie Inst. of Washington 242, 1–512.

Coats R N, Leonard R L and Goldman C R 1976 Nitrogen uptake and release in a forested watershed, Lake Tahoe Basin, California. Ecology 57, 995–1004.

Coleman D C, Cole C V and Elliot E T 1984 Decomposition, organic matter turnover, and nutrient dynamics in agroecosystems. *In* Agricultural Ecosystems: Unifying Concepts. Eds. R Lowrance, B R Stinner and G J House. John Wiley and Sons, New York.

Connell J H and Slatyer R D 1977 Mechanisms of succession in natural communities and their role in community stability and organization. Am. Nat. 111, 1119–1144.

Covington W W 1981 Changes in forest floor organic matter and nutrient content following clearcutting in northern hardwoods. Ecology 62, 41–48.

Ellenberg H 1977 Stickstoff als Standortsfaktor, inbesondere für mitteleuropäische Pflanzengesellschaften. Oecol. Plant. 12, 1–22.

Eno C F 1960 Nitrate production in the field by incubating the soil in polyethylene bags. Soil Sci. Soc. Am. Proc. 24, 277–279.

Gordon A M and van Cleve K 1983 Seasonal patterns of nitrogen mineralization following harvesting in the white spruce forests of interior Alaska. *In* Resources and Dynamics of the Boreal Zone. Eds. R W Wein, R R Pierce and I R Methuene. pp 119–130. Association of Canadian Universities for Northern Studies, Sault Ste Marie, Canada.

Gorham E, Vitousek P M and Reiners W A 1979 The regulation of chemical budgets over the course of terrestrial ecosystem succession. Annu. Rev. Ecol. Syst. 10, 53–88.

Grime J P 1979 Plant Strategies and Vegetation Processes. John Wiley and Sons, Chichester, England.

Grubb P J 1986 The ecology of establishment. *In* Ecology and Landscape Design. Eds. A D Bradshaw, D A Goode and E Thorpe. Blackwell Scientific, Oxford, England.

Haas H J, Evans C E and Miles E F 1957 Nitrogen and carbon changes in Great Plains soils as influenced by cropping and soil treatments. Tech. Bull. 1164, US Dept. of Agriculture, Washington, D.C. USA.

Hirose T and Tateno M 1984 Soil nitrogen patterns induced by colonization of *Polygonum cuspidatum* on Mt Fuji. Oecol. (Berlin) 61, 218–223.

Keeney D R 1982 Nitrogen-availability indices. *In* Methods of Soil Analysis, 2nd Edition. Eds. A L Page, R H Miller and D R Keeney. pp 711–734. Am. Soc. Agronomy, Madison, Wisconsin, USA.

Lamb D 1980 Soil nitrogen mineralisation in a secondary rainforest succession. Oecol. (Berlin) 47, 257–263.

Matson P A and Boone R D 1984 Natural disturbance and nitrogen mineralization: Wave-form dieback of moutain hemlock in the Oregon Cascades. Ecology 65, 1511–1516.

Matson P A and Vitousek P M 1981 Nitrification potentials following clearcutting in the Hoosier National Forest, Indiana. Forest Science 27, 781–791.

Matson P A and Waring R H 1984 Effect of nutrient and light limitation on mountain hemlock: Susceptibility to laminated root rot. Ecology 65, 1517–1524.

Matson P A, Vitousek P M, Ewel JJ, Mazzarino M J and Robertson G P 1987 Nitrogen transformations following tropical forest felling and burning on a volcanic soil. Ecology 68, 491–502.

Melillo J M, Aber J D and Muratore J F 1982 Nitrogen and lignin control of hardwood leaf litter decomposition dynamics. Ecology 63, 621–626.

Miller H G 1981 Forest fertilization: Some guiding concepts. Forestry 54, 157–167.

Montagnini F, Haines B, Boring L and Swank W 1986 Nitrification potentials in early successional black locust and in mixed hardwood forest stands in the southern Appalachians, USA. Biogeochemistry 2, 197–210.

Montes R A and Christensen N L 1979 Nitrification and succession in the Piedmont of North Carolina. Forest Science 25, 287–297.

Nadelhoffer K J, Aber J D and Melillo J M 1985 Fine roots, net primary production and soil nitrogen availability: A new hypothesis. Ecology 66, 1377–1390.

Odum E P 1969 The strategy of ecosystem development. Science 164, 262–270.

Pastor J J, Aber J D, McClaugherty C A and Melillo J M 1984 Aboveground production and N and P cycling along a nitrogen mineralization gradient on Blackhawk Island, Wisconsin. Ecology 65, 256–268.

Pastor J J, Stillwell M A and Tilman D 1987 Nitrogen mineralization and nitrification in four Minnesota old fields. Oecol. (Berlin). 71, 481–485.

Powers R F 1980 Mineralizable soil nitrogen as an index of nitrogen availability to forest trees. Soil Sci. Soc. Am. J. 44, 1314–1320.

Raison R J, Connell M J and Khanna P K 1987 Methodology for studying fluxes of soil mineral-N *in situ*. Soil Biol. Biochem. 19, 521–530.

Ramakrishnan P S and Saxena K G 1984 Nitrification potential in successional communities and desertification of Cherrapunji. Current Science 53, 107–109.

Reeder J D and Berg W A 1977 Nitrogen mineralization and nitrification in a Cretaceous shale and coal mine spoils. Soil Sci. Soc. Am. J. 41, 922–927.

Rice E L 1984 Allelopathy. Academic Press, Orlando, Florida, USA.

Rice E L and Pancholy S K 1972 Inhibition of nitrification by climax ecosystems. Am. J. Bot. 59, 1033–1040.

Roberts R D, Marrs R H and Bradshaw A D 1980 Ecosystem development on reclaimed china clay wastes. II. Nutrient compartmentation and nitrogen mineralization. J. Appl. Ecol. 17, 719–725.

Robertson G P 1984 Nitrification and nitrogen mineralization in a lowland rainforest succession in Costa Rica, Central America. Oecol. (Berlin) 61, 99–104.

Robertson G P and Tiedje J M 1984 Denitrification and nitrous oxide production in successional and old-growth Michigan forests. Soil Sci. Soc. Am. J. 48, 383–389.

Robertson G P and Vitousek P M 1981 Nitrification potentials in primary and secondary succession. Ecology 62, 376–386.

Saterson K A 1985 Nitrogen availability, primary production, and nutrient cycling during secondary succession in North Carolina Piedmont forests. Ph.D. Dissertation, University of North Carolina, Chapel Hill, USA.

Saxena K G and Ramakrishnan P S 1986 Nitrification during slash and burn agriculture (Jhum) in north-eastern India. Acta Oecol-Oecol Plant. 7, 307–319.

Schimel D S 1986 Carbon and nitrogen turnover in adjacent grassland and cropland ecosystems. Biogeochemistry 2, 345–357.

Skeffington R A and Bradshaw A D 1981 Nitrogen accumulation in kaolin wastes in Cornwall. IV. Sward quality and the development of a nitrogen cycle. Plant and Soil 62, 439–451.

Sollins P, Spycher G and Glassman C A 1984 Net nitrogen mineralization from light- and heavy-fraction forest soil organic matter. Soil Biol. Biochem. 16, 31–37.

Stanford G and Smith S J 1972 Nitrogen mineralization potential of soils. Soil Sci. Soc. Am. Proc. 36, 465–472.

Stevens P R and Walker T W 1970 The chronosequence concept and soil formation. Q. Rev. Biol. 45, 333–350.

Stone E L 1973 The impact of timber harvest on soil and water. *In* Report of the President's Advisory Panel on Timber and the Environment. pp 427–467. Government Printing Office, Washington, D.C. USA.

Thorne J F and Hamburg S P 1985 Nitrification potentials of an old-field chronosequence in Campton, New Hampshire. Ecology 66, 1333–1338.

Tilman D 1982 Resource Competition and Community Structure. Princeton University Press, Princeton, N.J., USA.

Tilman D 1986 Nitrogen-limited growth in plants from different successional stages. Ecology 67, 555–563.

Vitousek P M and Matson P A 1984 Mechanisms of nitrogen retention in forest ecosystems: a field experiment. Science 225, 51–52.

Vitousek P M and Matson P A 1985 Disturbance, nitrogen availability, and nitrogen losses in an intensively-managed loblolly pine plantation. Ecology 66, 1360–1367.

Vitousek P M and Melillo J M 1979 Nitrate losses from disturbed ecosystems: Patterns and mechanisms. Forest Science 25, 605–619.

Vitousek P M and Reiners W A 1975 Ecosystem succession and nutrient retention: A hypothesis. BioScience 25, 376–381.

Vitousek P M and Walker L R 1987 Colonization, succession, and resource availability: Ecosystem-level interactions. *In* Colonization, Succession and Stability. Eds. A M Gray, M Crawley and P J Edwards. pp 207–223. Blackwell Scientific, Oxford, England.

Vitousek P M, Gosz J R, Grier C C, Melillo J M and Reiners W A 1982 A comparative analysis of potential nitrification and nitrate mobility in forest ecosystems. Ecol. Monogr. 52, 155–177.

Vitousek P M, Van Cleve K, Balakrishnan N and Mueller-Dombois D 1983 Soil development and nitrogen turnover on recent volcanic substrates in Hawaii. Biotropica 15, 268–724.

Walker T W and Syers J K 1976 The fate of phosphorus during pedogenesis. Geoderma 15, 1–19.

Weetman G F 1962 Establishment report on a humus decomposition experiment. Woodland Research Index 134, Pulp and Paper Research Institute of Canada, Montreal.

Werner P 1984 Changes in soil properties during tropical wet forest succession in Costa Rica. Biotropica 16, 43–50.

Wheeler G L and Donaldson J M 1983 Nitrification in an upland forest sere. Soil Biol. Biochem. 15, 119–121.

M. Clarholm and L. Bergström (Eds.), Ecology of arable land, 173–184.

The influence of invertebrates on soil fertility and plant growth in temperate grasslands

J. P. CURRY

Department of Agricultural Zoology and Genetics, University College, Dublin, Ireland

Key words: decomposition, earthworms, grass growth, grassland invertebrates, nutrient cycling, reclamation, soil fertility

Abstract

Invertebrate biomass typically exceeds 100 g fresh mass m^{-2} in base rich fertile grasslands where earthworms are abundant; in acid soils the biomass is usually in the 10–20 g m^{-2} range, consisting mainly of enchytraeid worms and tipulid larvae. Invertebrates often account for less than 10% and never more than 20% of total heterotrophic respiration. Their contribution to primary litter decomposition is small but they promote litter decomposition and mineralization by ingesting and fragmenting litter, by incorporating it into the soil and by stimulating microbial activity. These effects are not well understood at present and require further study. Earthworms influence soil physical and chemical properties through burrowing, soil ingestion and mixing. Their activity is a major factor in the development of mull soils and stimulates plant growth. Earthworms can play an important role in land reclamation but better methods of introduction are required.

Invertebrate herbivores at non outbreak densities normally consume less than 10% of net primary production in grassland. Their effects on sward productivity may be positive or negative and are considerably greater than their levels of consumption would suggest. Chemical pest control in established grassland is generally undesirable and more environmentally acceptable approaches are required.

Introduction

Assessment of the role of invertebrates in ecosystems has been mainly based on their contribution to energy metabolism, but it is increasingly apparent that an adequate evaluation must take into account the broad range of activities in which these organisms engage.

In this paper the composition of the invertebrate fauna of temperate grassland and its contribution to energy metabolism is briefly reviewed, and some ways in which invertebrates may influence soil fertility and productivity through their feeding, through habitat modification and through interactions with other organisms are considered. In the concluding section some potentially important areas for future research in grassland invertebrate ecology are discussed.

The invertebrate fauna

The composition of the grassland fauna, factors influencing the distribution and abundance of the major groups and their effects on sward productivity have recently been reviewed (Curry, 1987a, b, c) while Petersen and Luxton (1982) give a comprehensive account of the population densities, biomass and energy metabolism of soil invertebrates in a range of terrestrial habitats studied during the International Biological Programme. Table 1 gives population densities and biomass of the major groups in selected sites. There is considerable variation related to site characteristics, time of year and management regime; however, temperate oceanic grasslands on base-rich mineral soils may support invertebrate biomass approaching or exceeding 100 g fresh mass m^{-2}. Earthworms are

Table 1. Mean population densities and biomass of some invertebrates in selected temperate grassland soils (− = not estimated)

Habitat	Numbers						Biomass (g m^{-2}) F = fresh mass D = dry mass
	Nematoda (× 10^6)	Enchytraeidae (× 10^3)	Lumbricidae	Acari (× 10^3)	Collembola (× 10^3)	Other Arthropoda (× 10^3)	
Old grassland, Co. Kildare, Ireland (Curry, 1969)	—	—	364	106.1	105.4	10.6	—
Permanent pasture, Co. Meath, Ireland (Curry, 1976; Bolger and Curry, 1980)	—	—	311	21.6	28.3	0.4	c. 100 F
Three grass leys on reclaimed fen peat, Ireland (Curry and Momen, unpublished)	—	—	268	17.2	39.8	4.3	64.6 F
	—	—	257	14.5	38.0	3.7	70.2 F
	—	—	<10	66.3	73.8	7.7	29.7 F
Permanent pasture, England (Salt et al., 1948)	—	—	—	164.4	61.3	38	—
Moorland limestone grassland, England (Coulson and Whittaker, 1978)	3.3	80	390	33	46	0.26	32 D
Moorland alluvial grassland, England (Coulson and Whittaker, 1978)	—	120	390	36	40	0.51	35.5 D
Juncus moorland (peat), England (Coulson and Whittaker, 1978)	3.9	200	4	45	23	3.05	7.2 D
Old grassland on fen peat, Sweden (Persson and Lohm, 1977)	—	23.8	133	106.9	108.7	7.4	10.6 D
Mown meadow, Poland (Breymeyer, 1978)	—	—	154	46.9	73.1	11.8	10 D
Unmanaged meadow, Poland (Breymeyer, 1978)	—	—	97	105.9	52.5	36.1	6.5 D
Lightly grazed sheep pasture, NSW Australia (King and Hutchinson, 1976; Hutchinson and King, 1980)	0.3	6	74	24.5	20.6	0.5	37.5 F
Pasture on reclaimed peat, New Zealand (Luxton 1983)	—	1.3	—	113.9	81.4	4.7	

dominant in such sites, but are scarce or absent from wet, acid tundra or moorland where enchytraeid worms, tipulid larvae and Collembola are often abundant. Under these conditions, and in semiarid prairie and steppe grasslands, the invertebrate biomass rarely exceeds 10–20 g fresh mass m^{-2}. Temperate grasslands may support up to 30×10^6 nematodes (c. 18 g fresh mass) m^{-2}, with average values of about 9×10^6 ind. and 3.8 g fresh mass m^{-2} (Sohlenius, 1980). Protozoa are rarely included in faunal studies: a biomass of about 5 g m^{-2} may be typical of European grasslands (Clarholm, 1984; Stout and Heal, 1967). Table 2 lists the major groups usually found in temperate grasslands and the activities which determine their ecosystem roles. There is considerable variation in feeding habits even within closely related taxa, and many are not restricted to one trophic level. However, detritivores feeding on

decaying organic matter and associated microflora usually comprise 60—90% of the invertebrate biomass; herbivores normally constitute less than 30% and predators/parasites less than 20% (Breymeyer, 1978; 1980). Only 1–2% of the grassland invertebrate biomass occurs above ground but this can include significant numbers of sap-feeding aphids and other Hemiptera, shoot-mining Diptera, Thysanoptera and Orthoptera; microbivorous Collembola and Acari, detritivorous Diplopoda and Isopoda in rough grassland, predatory Carabidae and Staphylinidae (Coleoptera), Phytoseiidae and Bdellidae (Acari); parasitic Hymenoptera and polyphagous Dermaptera.

Classification on the basis of size (Table 2) is somewhat arbitrary but does appear to have some significance in terms of trophic relationships. Three trophic systems may be recognized within the soil decomposer community separated on size (Heal

Table 2. The main invertebrate groups in temperate grassland and their ecosystem roles

Group	Organic matter decomposition/ mineralization	Soil mixing/ affecting soil properties	Predation/ parasitism	Herbivory	Disease transmission
Microfauna					
Protozoa	+		+		
Nematoda	+		+	+	+
Mesofauna					
Enchytraeidae	+	+			
Acari	+		+	+	+
Collembola	+			+	
Protura	+				
Diplura	+		+		
Pauropoda	+				
Symphyla	+			+	
Macrofauna					
Lumbricidae	+	+			
Mollusca	+			+	
Isopoda	+				
Diplopoda	+			+	
Chilopoda			+		
Araneae			+		
Coleoptera	+		+	+	
Lepidoptera				+	
Diptera	+		+	+	
Thysanoptera				+	
Hymenoptera	+	+	+		
Hemiptera			+	+	+
Dermaptera	+		+		
Orthoptera	+			+	
Neuroptera			+		

and Dighton, 1985). The microtrophic system is found in the water film on organic residues, roots and soil particles and comprises bacteria and yeasts utilizing readily available carbon sources and consumed by protozoa and nematodes which in turn may be preyed on by other protozoa and nematodes. The mesotrophic system is confined to soil pores and small air spaces and encompasses fungi utilizing a broad range of organic substrates, fungal grazing Collembola and mites, predatory gamasid mites and small surface predators such as linyphiid spiders. The macrotrophic system comprises large invertebrates such as earthworms and millepedes which are not constrained by the physical structure of the soil habitat and which ingest both the basic organic resources and their associated microflora and fauna. These three systems overlap, and their relative importance depends on the quality of the organic matter input and the physical-chemical nature of the environment.

Role in energy metabolism

Energy flow through the invertebrate community may be calculated from field population and biomass data coupled with energy utilization coefficients derived from studies on representative species (Andrzejewska and Gyllenberg, 1980; Heal and MacLean, 1975; Humphreys, 1979; McNeill and Lawton, 1970; Petrusewicz, 1967). Alternatively, carbon budgets can be calculated and estimates of nutrient utilization can be made based on C:nutrient ratios of food, body tissues and faeces. In practice these budget calculations are subject to considerable errors associated with variability in food ingestion rates, assimilation efficiencies and production efficiencies (Andrzejewska, 1979; Andrzejewska and Gyllenberg, 1980; Petersen and Luxton, 1982; Reichle, 1968).

Invertebrate respiration normally accounts for a relatively modest proportion of annual heterotrophic metabolism ranging from 6–7% in Swedish fen grassland (Persson and Lohm, 1977) to almost 20% in grassland with large invertebrate biomass (Macfadyen, 1963). Energy consumption by invertebrate herbivores represents a fairly modest proportion of the energy fixed in primary production. This may be as low as 1% in managed grass

and lucerne leys (Paustian *et al.*, 1987), while 6–10% may be more typical for more mature unmanaged grasslands (Persson and Lohm, 1977; Van Hook, 1971). Detritivores have been estimated to ingest 20–30% of the total organic matter input in temperate soils (Petersen and Luxton, 1982), but this is probably an underestimate for earthworm rich soils.

Effects on decomposition, nutrient cycling and soil properties

Invertebrate exclusion studies

Manipulative experiments in which animal activity is suppressed or excluded by physical or chemical means have given variable results but generally point to a greater influence of soil invertebrates on decomposition processes than their energy metabolism would suggest. Seastedt (1984) attempted to quantify the effects of microarthropods on decomposition by partitioning the decomposition constants calculated from literature data into abiotic, microbial and microarthropod components based on selective exclusion experiments. On average, the decomposition rate was 23% greater (Range 0–70%) when microarthropods were present than when they were excluded. The faunal effect was attributed to enhanced microbial respiration and loss of experimental litter in the form of faeces. Dickinson (1983) reported 65% mass loss in 85 days from grass litter accessible to earthworms compared with 30–40% from litter confined in finer mesh bags which excluded earthworms.

Exclusion experiments have a number of limitations. The exclusion techniques currently available are not very selective, and confining experimental litter in mesh bags or boxes limits the scope for measuring the contribution of macroinvertebrates to litter dissipation and incorporation into the soil.

Litter consumption and decomposition

Earthworms are capable of consuming most of the detritus input in managed mesic grasslands and when they are absent or suppressed dead grass and animal dung accumulate on the surface and soil fertility declines (Clements, 1982; Hoogerkamp *et al.*, 1983; Keogh, 1979; Stockdill and Cossens, 1966). Isopods and millepedes can ingest significant quantities of litter in rough grassland (Hassall and Sutton, 1978; Macfadyen, 1963), while enchytraeid worms are often the main detritus consumers in acid moorland (Standen, 1973). Many invertebrate groups can digest plant structural polysaccharides to a limited degree (Hartenstein, 1982; Nielsen, 1962), but their role in primary litter decomposition is probably small. Their main effect may be one of facilitating decomposition through litter fragmentation and incorporation into the soil and through stimulating microbial activity in various ways (van der Drift and Witkamp, 1960; Harding and Stuttard, 1974; Hassal and Sutton, 1978).

Coprophagous invertebrates including dipterous larvae, dung beetles and earthworms enhance dung decomposition partly through their own metabolism, but mainly by promoting aeration and microbial activity and by incorporating dung into the soil (Holter, 1979; Olechowicz, 1976; Valiela, 1974). When an adequate coprophagous fauna is absent pasture fowling, nutrient immobilization, sward deterioration and problems with nuisance dung breeding flies occur (Ferrar, 1973; Hughes *et al.*, 1978).

Role in nutrient cycling

Carbon and nutrient transformations in ecosystems are linked but there are no simple correlations between these processes because of differences in the efficiencies with which different nutrients are utilized by different organisms and developmental stages.

Earthworms have an important role in the transfer of N from decaying organic matter into the soil in plant available form. Quantitative estimates include 100 kg N ha^{-1} yr^{-1} mineralized by *Lumbricus terrestris* L. in deciduous woodland in Britain (Satchell, 1963) and 109–147 kg N ha^{-1} yr^{-1} mineralized by *Aporrectodea caliginosa* (Sav.) in New Zealand grassland, representing about 20% of the total quantity mineralized from organic matter annually (Keogh, 1979). Lee (1983) concluded that the total earthworm contribution could exceed 300 kg ha^{-1} yr^{-1} in earthworm rich soils. Earthworms have been shown to increase plant availabil-

ity of P in litter (Mansell *et al.*, 1981) and in mineral rock phosphate (Mackay *et al.*, 1982), while large increases in assimilable P, K, Ca and Mg have been demonstrated in grass litter after passing through the earthworm gut (Czerwinski *et al.*, 1974). Syers and Springett (1983) place more emphasis on the potential role of earthworms in producing frequent small changes in nutrient availability rather than on their contribution to the total turnover of the nutrient pool in grassland.

Laboratory and field microcosm studies have demonstrated the significance of microbial-faunal interactions in stimulating microbial activity and in releasing nutrients immobilized in microbial biomass into the soil solution (Anderson *et al.*, 1981; Anderson *et al.*, 1983; Anderson *et al.*, 1985; Clarholm, 1985). While the macrotrophic system would appear to be the major pathway for nutrient release in earthworm-dominated grasslands, the microfauna and the microtrophic system appear to play a major role in arable land (Clarholm, 1985; Rosswall and Paustian, 1984) and in semiarid grasslands where macrofauna are scarce. (Hunt *et al.*, 1987).

Anderson *et al.* (1985) developed regression models from microcosm data to quantify the effects of the macrofauna on N mineralization from litter of varying quality and from different sites as a function of temperature and animal biomass but the model had poor predictive value when tested against field data. Ingham *et al.* (1986) have also attempted to bridge the gap between simplified laboratory and more complex field conditions and were able to utilize the results of microcosm experiments in interpreting data from field soil cores on root and shoot production and on populations of microbes and animals in prairie grassland in Colorado.

Ausmus *et al.* (1976) suggested that invertebrates may regulate rates of nutrient immobilization and release from microbial biomass and enhance nutrient retention in mature forest ecosystems and Anderson *et al.* 1985 obtained evidence from field lysimeter studies in deciduous woodland that N losses were lower when animals were present. However, it is doubtful whether soil animals could have much influence in preventing nutrient losses from heavily fertilized, intensively managed agricultural crops and grasslands.

Influence on soil properties

Earthworms play a major role in determining the structure of temperate soils by mixing in dead organic matter from the surface and by continuously working the top 10–20 cm soil layer. The time taken to completely work this layer probably varies from 40–60 years in mesic soils to 100 years in moisture limited mollisols (Barley, 1959; Burl *et al.*, 1973; Curry and Bolger, 1985). Large channels 2–11 mm in diameter created by earthworm burrowing influence water infiltration, aeration and root penetration while medium sized pores (0.003–0.06 mm diameter) created by casting influence water holding capacity (Syers and Springett, 1983). Burrows opening to the soil surface improve water infiltration and drainage (Hoogerkamp *et al.*, 1983; Sharpley *et al.*, 1979; Stockdill, 1966). Earthworm activity is a major factor in the development of the aggregated structure of mull soils (Kubiena, 1953; Rogäar and Boswinkel, 1978), while incorporation of lime, fertilizers and pesticides by earthworms are considered to make an important contribution to the productivity of New Zealand grassland (Springett, 1983; Stockdill and Cossens, 1966).

Other invertebrates which have the capacity to influence soil properties to some degree include nest building ants in unmanaged grasslands (Petal, 1978; 1980) and enchytraeids in acid moorlands (O'Connor, 1967).

Invertebrates and plant growth

Earthworms

Marked effects of earthworms on plant growth have been demonstrated in pot and small scale field enclosure experiments, with dry matter increases ranging from 20–1000% (Curry and Boyle, 1987; Hopp and Slater, 1948; Van Rhee, 1965; Waters, 1951). An initial increase in grass yield of 70% was recorded in New Zealand pasture following earthworm establishment and the breakdown of an accumulated surface mat (Stockdill and Cossens, 1966); subsequently an average yield increase of 25–30% was sustained (Lacy, 1977). Enhanced plant growth has generally been attributed to improved soil structure and aeration, to channels for root growth created by earthworms and to readily available plant nutrients present in the linings of the burrows. Indole compounds with plant hormonal activity are known to be produced by earthworms (Nielsen, 1965; Springett and Syers, 1979), but their effects on plant growth in the field are unknown. Earthworms consume significant amounts of plant root material (Baylis *et al.*, 1986) and this root pruning effect may be implicated in stimulating growth (Davidson, 1979).

Herbivores

Invertebrate herbivores at non outbreak population levels normally consume less than 10% of net primary production in natural ecosystems (Gibbs, 1976; Sinclair, 1975; Wiegert and Evans, 1967), but plant biomass consumption is a poor indicator of overall effect on plant productivity. For example, grasshoppers may destroy several times more plant tissue than they actually consume (Andrzejewska and Wojcik, 1970; Mitchell, 1973), while removal of large quantities of water by xylem feeding Hemiptera can cause severe wilting and growth check (Andrzejewska, 1967). Elimination of herbivores by insecticides can result in yield increases of up to 30% in European grasslands even when there is no overt sign of pest damage (Andrzejewska and Wojcik, 1971; Blackshaw, 1984; Henderson and Clements, 1977). However, such increases in yield could be partly due to temporary increases in available nutrients following death and mineralization of decomposer invertebrates (Davidson *et al.*, 1979).

Regrowth potential and tolerance for invertebrate damage depend on factors such as the size and health of the plants and the degree to which the sward is already under stress. Moderate levels of invertebrate feeding can actually increase grass growth by removing moribund tissue and by stimulating the production of new roots or shoots (Andrzejewska and Wojcik, 1970; Davidson, 1979). Ridsdill Smith (1977) found that root feeding by scarabaeid larvae only reduced green foliage yield when the plants were also regularly defoliated. The time of feeding is important: Lutman (1977) suggested that winter feeding by slugs in upland grasslands in Britain could delay spring growth, while Rodell (1978) suggested that grasshopper feeding during periods of poor plant productivity in Spring and Autumn could have a potentially important effect on overall production in shortgrass prairie.

Selective feeding on preferred plant species can alter botanical composition leading to sward deterioration. This occurs in New Zealand alpine tussock grassland when grasshopper feeding suppresses important ground cover plant species (White, 1974), while control of frit fly and other pests can improve establishment, growth and persistence of Italian and hybrid ryegrass leys (Henderson and Clements, 1979; 1981).

The role of invertebrates as vectors of disease in grassland is not well understood, but significant losses in ryegrass yeilds have been attributed to aphid-borne barley yellow dwarf virus (BYDV), while *Javesella pellucida* (F) (Delphacidae) and *Abacarus hystrix* Nalepa (Acari, Eriophyidae) can also be of significance as vectors of virus disease (Plumb, 1978; Raatikainen, 1967).

The contribution of invertebrate herbivores to organic matter turnover and nutrient cycling in grassland is probably small, but Rodell (1977) considered that the destructive feeding habits of grasshoppers might have a significant influence on rates of nutrient cycling in the longterm in prairie grassland, while the excreta of phytophagous invertebrates are readily mineralized and contain significant amounts of P and other plant nutrients (Andrzejewska, 1979; Hutchinson and King, 1982). It has been suggested that invertebrate herbivores can influence succession and regulate ecosystem processes (Mattson and Addy, 1975; Schowalter, 1981), but there is little evidence at present to suggest that this is the case in grassland.

Future prospects

Estimates of energy utilization by grassland invertebrates can be considerably improved, but metabolic activity is a poor indicator of overall ecosystem role and further broadly based studies of community energetics are likely to be of limited value in relation to the effort and resources required. The autecological approach is likely to prove more rewarding, focussing on selected species and their habitat and community interactions. Energy budgets can give useful insights into aspects of life history strategy in such studies, but

cannot themselves determine the "importance" of organisms in the community.

The development of more effective strategies for managing grassland fertility requires a greater understanding of nutrient transformations and the role of the fauna in these processes. Interactions between soil fauna and microflora and the implications of these interactions for nutrient cycling are important topics for further research. Perturbation studies in which components of the decomposer community are suppressed can be useful in evaluating the role of different organisms, but more selective techniques than these currently available are required. Microcosm studies have given important insights into the role of microbial grazing in decomposition and mineralization processes, but the relevance for field conditions of mechanisms demonstrated in highly simplified systems remains open to question. Indeed the complexity of the soil decomposer food web, involving hundreds of species undergoing continuous population change and functioning in a constantly changing physicochemical environment limits the degree to which understanding can be gained through any single experimental approach. Conceptual models and the use of data from all available sources including field population studies, radiotracer studies, and litter bag and microcosm studies to simulate the various processes appear to offer considerable scope for progress. An example of this approach is afforded by Hung et al., (1987) in their study of N transfer rates through the detrital food web in a shortgrass prairie.

It has been argued that ecosystem process studies might be conducted more fruitfully in arable cropping systems because of their relative simplicity compared with grassland. However, it is important to bear in mind that the structure and functioning of the biological communities are quite different in those two situations. Grassland which is not intensively managed supports a complex invertebrate community adapted to conditions of slow organic matter decomposition and efficient nutrient conservation. By contrast the decomposer fauna of arable land consists mainly of relatively small short lived species with high multiplication rates capable of tolerating repeated disturbance and adapted to exploit periodic flushes of crop residues and associated microflora. These animals

Table 3. Effects of earthworms on grassland in reclaimed cutaway peat in central Ireland (Curry and Bolger, 1985; Curry and Boyle, 1987; and unpublished data)

Carrying capacity	100 g fresh mass m^{-2}
Litter consumption	365 g dry mass m^{-2} yr^{-1}
Soil consumption	1.3 kg m^{-2} yr^{-1} or 20 cm layer worked in 45 y
Soil properties affected	Bulk density Water infiltration Degree of humification Micromorphology
Increased grass/clover shoot production	25–50% in field microplots receiving cattle slurry 30% over 1 y in glass house

exhibit the opportunistic strategy of 'r' selected species and their activities are likely to accentuate rather than to dampen patterns of nutrient immobilization and release.

The rapidly increasing literature on earthworm ecology reflects the increasing awareness of the importance of these animals in maintaining soil fertility and their potential contribution to the management of the earths' labile carbon resources (Hartenstein, 1986). Investigations on earthworms introduced into reclaimed soils afford unique opportunities to study the role of these animals in soil development as well as their potential contribution to practical soil restoration and amelioration (Curry and Cotton, 1983). The main effects of introduced earthworms on reclaimed cutaway peat in central Ireland are summarized in Table 3. Similarly marked effects have been reported for improved soils in New Zealand and for Dutch polders (Hoogerkamp *et al.*, 1983; Stockdill, 1982). The scale of earthworm introduction projects is limited at present, and techniques for mass rearing soil-dwelling species need to be developed if this aspect of earthworm technology is to make progress.

There is increasing interest in the pest management approach for the control of recurring pasture pests in countries like New Zealand where cheap and effective but environmentally unacceptable pesticides such as DDT are no longer allowed (Kain, 1979). This approach requires detailed knowledge of the biology and ecology of the species concerned, and economic thresholds need to be established in terms of pest density — pasture damage — animal production relationships. Life table data have been collected on the more important species such as *Costelytra zealandica* White (the grass grub) and *Heteronychus arator* F. (the black beetle), and models constructed from these data are considered to be potentially useful in devising pest management strategies (East *et al.*, 1981). More complex simulation models have been developed for pasture scarabs (Davidson *et al.*, 1970) and grasshoppers (Rodell, 1977; White, 1979); these are primarily research tools at present, and for the immediate future more empirical methods are likely to be favoured. Classical biological control, altering botanical decomposition, adjusting stocking density and cutting regimes, and more effective timing of insecticidal application are approaches which appear promising in different situations (Crosby and Pottinger, 1979; Lee 1982).

Insecticides are rarely applied to established grassland in Europe but Blackshaw (1985) concluded that routine autumn application of chlorpyrifos to control leatherjackets (Tipulidae) would yield significant economic benefits over much of Northern Ireland and similar climatic areas. More extensive use of pesticides in grassland appears ill-advised in the absence of information on long term effects on non-target organisms, implications for soil fertility and environmental consequences generally. While modern insecticides are less persistent than the organochlorines, materials such as chlorpyrifos are non specific and highly toxic when applied at concentrations needed to kill soil pests. Recent data indicating progressive depletion of polyphagous predators in arable land receiving prophylactic treatment with pesticides (Burn, 1987;

Vickerman, 1987) point to the need for a general reduction in pesticide use and to the importance of grassland as a reservoir for natural enemies in intensive arable areas.

References

Anderson J M, Huish S A, Ineson P, Leonard M A and Splatt P R 1985 Interactions of invertebrates, micro-organisms and tree roots in nitrogen and mineral element fluxes in deciduous woodland soils. *In* Ecological Interactions in Soils. Plants, Microbes and Animals. Eds. A H Fitter, D Atkinson, D J Read and M B Usher. Blackwell, Oxford, pp 377–392.

Anderson J M, Ineson P and Huish S A 1983 Nitrogen and cation mobilization by soil fauna feeding on leaf litter and soil organic matter from deciduous woodlands. Soil Biol. Biochem. 15, 463–467.

Anderson R V, Coleman D C and Cole C V 1981 Effects of saprotrophic grazing on net mineralization. *In* Terrestrial Nitrogen Cycles. Eds. F E Clark and T Rosswall. Ecol. Bull. (Stockholm) 33, 201–216.

Andrzejewska L 1967 Estimation of the effects of feeding of the sucking insect *Cicadella viridis* L. (Homoptera-Auchenorrhyncha) on plants. *In* Secondary Productivity of Terrestrial Ecosystems, Vol. 2. Ed. K. Petrusewica. Polish Academy of Sciences, Warsaw, pp 791–805.

Andrzejewska L 1979 Herbivorous fauna and its role in the eonomy of grassland ecosystems. II. The role of herbivores in trophic relationships. Pol. Ecol. Stud. 5, 45–76.

Andrzejewska L and Gyllenberg G 1980 Small herbivore subsystem. *In* Grasslands, Systems Analysis and Man. Eds. A I Breymeyer and G M Van Dyne. Cambridge University Press, Cambridge, pp 201–267.

Andrzejewska L and Wojcik Z 1970 The influence of Acridoidea on the primary production of a meadow (field experiment). Ekol. Pol. 18, 89–109.

Andrzejewska L and Wojcik Z 1971 Productivity investigation of two types of meadows in the Vistula Valley. VII. Estimation of the effect of phytophagous insects on the vascular plant biomass of the meadow. Ekol. Pol. 19, 173–182.

Ausmus B S, Edwards N T and Witkamp M 1976 Microbial immobilization of carbon, nitrogen, phosphorus and potassium: implications for forest ecosystem processes. *In* The Role of Terrestrial and Aquatic Organisms in Decomposition Processes. Eds. J M Anderson and A Macfadyen. Blackwell, Oxford pp 397–416.

Barley K P 1959 The influence of earthworms on soil fertility. II. Consumption of soil and organic matter by the earthworm *Allolobophora caliginosa* (Savigny). Aust. J. Agric. Res. 10, 179–185.

Baylis J P, Cherrett J M and Ford J B 1986 A survey of the invertebrates feeding on living clover roots (*Trifolium repens* L.) using 32P as a radiotracer. Pedobiologia 29, 201–208.

Blackshaw R P 1984 The impact of low numbers of leatherjackets on grass yield. Grass For. Sci. 39, 339–343.

Blackshaw R A 1985 A preliminary comparison of some management options for reducing grass losses caused by leatherjackets in Northern Ireland. Ann Appl. Biol. 107, 279–285.

Bolger T and Curry J P 1980 Effects of cattle slurry on soil arthropods in grassland. Pedobiologia 20, 246–253.

Breymeyer A 1978 Analysis of the trophic structure of some grassland ecosystems. Pol. Ecol. Stud. 4, 55–128.

Breymeyer A I 1980 Trophic structure and relationships. *In* Grasslands, Systems Analysis and Man. Eds. A I Breymeyer and G M Van Dyne. Cambridge University Press, Cambridge, pp 799–819.

Burl S W, Hole F D and McCracken R J 1973 Soil Genesis and Classification, Iowa State University Press, Ames, Iowa, 360p.

Burn A J 1987 Cereal pest/predator interactions. *In* The Boxworth Project. 1986 Report. Ministry of Agriculture, Fisheries and Food, Tolworth, UK. pp 57–73.

Clarholm M 1984 Microbes as predators or prey. Heterotrophic, free-living protozoa: Neglected microorganisms with an important task in regulating bacterial populations. *In* Current Perspectives on Microbial Ecology. Eds. M J Klug and C A Reddy. A.S.M., Washington, pp 321–326.

Clarholm M 1985 Interactions of bacteria, protozoa and plants leading to mineralization of soil nitrogen. Soil Biol. Biochem. 17, 181–187.

Clements R O 1982 Some consequences of large and frequent pesticide applications to grassland. *In* Proc. 3rd Australas. Conf. Grassl. Invertebr. Ecol. Ed. K E Lee. South Australian Government Printer, Adelaide, pp 393–396.

Coulson J C and Whittaker J B 1978 E£cology of moorland animals. *In* The Ecology of some British Moors and Montane Grasslands. Eds. O W Heal and D F Perkins. Springer, Berlin, pp 52–93.

Crosby T K and Pottinger R P 1979 (Eds.) Proc. 2nd Australas. Conf. Grassl. Invertebr. Ecol. Government Printer, Wellington, 294 p.

Curry J P 1969 The qualitative and quantitative composition of the fauna of an old grassland site at Celbridge, Co. Kildare. Soil Biol. Biochem. 1, 219–227.

Curry J P 1976 Some effects of animal manures on earthworms in grassland. Pedobiologia 16, 425–438.

Curry J P 1987a The invertebrate fauna of grassland and its influence on productivity. I. The composition of the fauna. Grass For. Sci. 42, 103–120.

Curry J P 1987b The invertebrate fauna of grassland and its influence on productivity. II. Factors affecting the abundance and composition of the fauna. Grass For. Sci. 42, 197–212.

Curry J P 1987c The invertebrate fauna of grassland and its influence on productivity. III. Effects on soil fertility and plant growth. Grass For. Sci. 42, 325–341.

Curry J P and Bolger T 1985 Growth, reproduction and litter and soil consumption by *Lumbricus terrestris* L. in reclaimed peat. Soil Biol. Biochem. 16, 253–257.

Curry J P and Boyle K E 1987 Growth rates, establishment and effects on herbage yield of introduced earthworms in grassland on reclaimed cutover peat. Biol. Fertil. Soils 3, 95–98.

Curry J P and Cotton D C F 1983 Earthworms and land reclamation. *In* Earthworm Ecology. Ed. J E Satchell. Chapman and Hall, London, pp 215–228.

Czerwinski A, Jakubczyk H and Nowak E 1974 Analysis of a sheep pasture ecosystem in the Pieniny Mountains (the Carpathians). XII. The effect of earthworms on the pasture soil. Ekol. Pol. 22, 635–650.

Davidson R L 1979 Effects of root feeding on foliage yield. *In*

Proc. 2nd Australas. Conf. Grassl. Invertebr. Ecol. Eds. T K Crosby and R P Pottinger. Government Printer, Wellington, pp 117–120.

Davidson R L, Shackley A, Wolfe, V J and Donelan M J 1979 Anomalous increases in pasture yield after use of insecticides on soil. *In* Proc. 2nd Australas. Conf. Grassl. Invertebr. Ecol. Eds. T K Crosby and R P Pottinger. Government Printer, Wellington, pp 30–32.

Davidson R L, Wiseman J R and Wolfe V J 1970 A systems approach to pasture scarab problems in Australia. *In* Proc. 11th Intern. Grassl. Congr. Ed. M J T Normam, pp. 681–684.

Dickinson N M 1983 Decomposition of grass litter in a successional grassland. Pedobiologia 25, 117–126.

Drift, van der J and Witkamp M 1960 The significance of the breakdown of oak litter by *Enoicyla pusilla*. Archiv. Neerl. Zool. 13, 489–492.

East R, King P D and Watson R N 1981 Population studies of grass grub (*Costelytra zealandica*) and black beetle (*Heteronychus arator*) (Coleoptera: Scarabaeidae). N.Z.J. Ecol. 4, 56–64.

Ferrar P 1973 CSIRO dung beetle project. Wool Technol. Sheep Breed. 20, 73–75.

Gibbs G W 1976 The role of insects in natural terrestrial ecosystems. N.Z. Entomol. 6, 113–121.

Harding D J L and Stuttard R A 1974 Microarthropods. *In* Biology of Plant Litter Decomposition Vol. 2. Eds. C H Dickinson and G J F Pugh. Academic Press, London, pp 489–532.

Hartenstein R 1982 Soil macroinvertebrates, aldehyde oxidase, catalase, cellulase and peroxidase. Soil Biol. Biochem. 14, 387–391.

Hartenstein R 1986 Earthworm biotechnology and global biogeochemistry. Adv. Ecol. Res. 15, 379–409.

Hassal M and Sutton S L 1978 The role of isopods as decomposers in a dune grassland ecosystem. Scient. Proc. Roy. Dubl. Soc. 6A, 235–245.

Heal O W and Dighton J 1985 Resource quality and trophic structure in the soil system. *In* Ecological Interactions in Soil. Plants, Microbes and Animals. Eds. A H Fitter, D Atkinson, D J Read and M B Usher. Blackwell, Oxford, pp 339–354.

Heal O W and MacLean S F 1975 Comparative productivity in ecosystems: Secondary productivity. *In* Unifying Concepts in Ecology. Eds. W H Van Dobben and R H Lowe. Junk, The Hague, pp 89–109.

Henderson I F and Clements R O 1977 Grass growth in different parts of England in relation to invertebrate numbers and pesticide treatment. J. Br. Grassl. Soc. 32, 89–93.

Henderson I F and Clements R O 1979 Differential susceptibility to pest damage in agricultural grasses. J. Agric. Sci (Cambridge) 73, 465–472.

Henderson I F and Clements R O 1981 The effect of insecticide treatment on the establishment and growth of Italian ryegrass under different sowing conditions. Grass For. Sci. 35, 235–241.

Holter P 1979 Effect of dung-beetles (*Aphodius* spp.) and earthworms on the disappearance of cattle dung. Oikos 32, 393–402.

Hoogerkamp M, Rogäar H and Eijsackers H J P 1983 Effect of earthworms on grassland on recently reclaimed polder soils in the Netherlands. *In* Earthworm Ecology. Ed. J E Satchell. Chapman and Hall, London, pp 85–105.

Hopp H and Slater C S 1948 Influence of earthworms on soil productivity. Soil Sci. 66, 421–428.

Hughes R D, Tyndale-Biscoe M and Walker J 1978 Effect of introduced dung beetles (Coleoptera: Scarabaeidae) on the breeding and abundance of the Australian bushfly, *Musca vetustissima* Walker (Diptera: Muscidae). Bull. Entomol. Res. 68, 361–372.

Humphreys W F 1979 Production and respiration in animal populations. J. Anim. Ecol. 48, 427–453.

Hutchinson K J and King K L 1980 The effects of sheep stocking level on invertebrate abundance, biomass and energy utilization in a temperate, sown grassland. J. Appl. Ecol. 17, 369–387.

Hutchinson K J and King K L 1982 Invertebrates and nutrient cycling. *In* Proc. 3rd Australas. Conf. Grassl. Invertebr. Ecol. Ed. K E Lee. South Australian Government Printer, Adelaide, pp 331–338.

Hunt H W, Coleman D C, Ingham E R, Elliott E T, Moore J C, Rose S L, Reid C P P and Morley C R 1987 The detrital food web in a shortgrass prairie. Biol. Fert. Sol. 3, 57–68.

Ingham E R, Trofymow J A, Ames R N, Hunt H W, Morley C R, Moore J C and Coleman D C 1986. Trophic interactions and nitrogen cycling in a semi-arid grassland soil. 1. Seasonal dynamics of the natural populations, their interactions and effects on nitrogen cycling. J. Appl. Ecol. 23, 597–614.

Kain W M 1979 Pest management systems for control of pasture insects in New Zealand. *In* Proc. 2nd Australas. Conf. Grassl. Invertebr. Ecol. Eds. T K Crosby and R P Pottinger. Government Printer, Wellington, pp 172–179.

Keogh R G 1979 Lumbricid earthworm activities and nutrient cycling in pasture ecosystems. *In* Proc. 2nd Australas. Conf. Grassl. Invertebr. Ecol. Eds. T K Crosby and R P Pottinger. Government Printer, Wellington, pp 49–51.

King K L and Hutchinson K J 1976 The effects of sheep stocking intensity on the abundance and distribution of mesofauna in pastures. J. Appl. Ecol. 13, 41–55.

Kubiena W L 1953 The Soils of Europe. Murby, London.

Lacy H 1977 Putting new life in wormless soil. N.Z. Farmer 98, 20–22.

Lee K E 1982 (Ed.) Proc. 3rd Australas. Conf. Grassl. Invertebr. Ecol. Government Printer, Adelaide, 402p.

Lee K E 1983 The influence of earthworms and termites on soil nitrogen cycling. *In* New Trends in Soil Biology. Proc. 8th Intern. Coll. Soil Zoology, Louvain-la-Neuve (Belgium). Eds. Ph Lebrun, H M André, A De Meds, C Grégoire-Wibo and G Wauthy. Imprimeur Dieu-Brichart, Ottignies-Louvain-la-Neuve, pp 35–48.

Lutman J 1977 The role of slugs in an Agrostis-Festuca grassland. *In* Production Ecology of British Moors and Montane Grassland. Eds. O W Heal and D F Perkins. Ecol. Stud. 27, pp 332–347. Springer-Verlag, Berlin.

Luxton M 1983 Studies on the invertebrate fauna of New Zealand peat soils. V. Pasture soils on Kaipaki peat. Pedobiologia 25, 135–148.

Macfadyen A 1963 The contribution of the microfauna to total soil metabolism. Soil Organisms. Eds. J Doeksen and J van der Drift. North Holland Publishing Co., Amsterdam, pp 1–17.

Mackay A D, Syers J K, Springett J A and Gregg P E F 1982 Plant availability of phosphorus in superphosphate and a phosphate rock as influenced by earthworms. Soil Biol.

Biochem. 14, 281–287.

Mansell G P, Syers J K and Gregg P E H 1981 Plant availability of phosphorous in dead herbage ingested by surface-casting earthworms. Soil Biol. Biochem. 13, 163–167.

Mattson W J and Addy N D 1975 Phytophagous insects as regulators of forest primary production. Science (Washington) 190, 515–522.

Mitchell J 1973 A Model of Food Consumption by Three Grasshopper Species as Determined by Differential Feeding Trials. Ph.D. Thesis, Colorado State University, Fort Collins, Colorado.

McNeill S and Lawton J H 1970 Annual production and respiration in animal populations. Nature (London) 225, 472–474.

Nielsen C O 1962 Carbohydrases in soil and litter invertebrates. Oikos 13, 200–215.

Nielsen R L 1965 Presence of plant growth substances in earthworms demonstrated by paper chromatography and the Went pea test. Nature (London) 208, 113–114.

O'Connor F B 1967 The Enchytraeidae. *In* Soil Biology. Eds. A Burgess and F Raw. Academic Press, London, pp 213–257.

Olechowicz E 1976 The role of coprophagous dipterans in a mountain pasture ecosystem. Ekol. Pol. 24, 125–165.

Paustian K, Andrén, O, Boström U, Clarholm M, Hansson A-C, Johansson G, Lagerlöf J, Lindberg T, Pettersson R, Rosswall T, Schnürer J, Sohlenius B and Steen E 1987. Carbon and nitrogen budgets of four agroecosystems with annual and perennial crops, with and without N fertilization. *In* Theoretical Analysis of C and N cycling in Soil. Dissertation by K Paustian, Report 30, Swedish University of Agricultural Sciences, Uppsala.

Persson T and Lohm U 1977 Energetical significance of the annelids and arthropods in a Swedish grassland soil. Ecol. Bull. (Stockholm) 23, 1–211.

Petal J 1978 The role of ants in ecosystems. *In* Production Ecology of Ants and Termites. Ed. M V Brian. Cambridge University Press, London, pp 293–325.

Petal J 1980 Ant populations, their regulation and effect on soil in meadows. Ekol. Pol. 28, 297–326.

Petersen H and Luxton M 1982 A comparative analysis of soil fauna populations and their role in decomposition processes. Oikos 39, 287–388.

Petrusewicz K 1967 Concepts in studies on the secondary productivity of terrestrial ecosystems. *In* Secondary Productivity of Terrestrial Ecosystems, Vol. 1. Ed. K Petrusewicz. Polish Academy of Sciences, Warsaw, pp 17–49.

Plumb R T 1978 Invertebrates as vectors of grass viruses. Scient. Proc. Roy. Dubl. Soc. 6A, 343–350.

Raatikainen M 1967 Bionomics, enemies and population dynamics of *Javesella pellucida* (F.) (Hom. Delphacidae). Ann. Agric. Fenn. 6, 1–149.

Reichle D E 1968 Relation of body size to food intake, oxygen consumption and trace element metabolism in forest floor arthropods. Ecology 49, 538–542.

Ridsdill Smith T J 1977 Effects of root-feeding by scarabaeid larvae on growth of perennial ryegrass plants. J. Appl. Ecol. 14, 73–80.

Rodell C F 1977 A grasshopper model for a grassland ecosystem. Ecology 58, 227–245.

Rodell C F 1978 Simulation of grasshopper populations in a grassland ecosystem. *In* Grassland Simulation Model, Ecological Studies 26. Ed. G S Innis. Springer-Verlag, New York, pp. 127–153.

Rogäar H and Boswinkel J A 1978 Some soil morphological effects of earthworm activity; field data and X-ray radiography. Neth. J. Agric. Sci. 26, 145–160.

Rosswall T and Paustian K 1984 Cycling of nitrogen in modern agricultural systems. Plant and Soil 76, 3–21.

Salt G, Hollick F S J, Raw F and Brian M W 1948 The arthropod population of pasture soil. J. Anim. Ecol. 17, 139–150.

Satchell J E 1963 Nitrogen turnover by a woodland population of *Lumbricus terrestris*. *In* Soil Organisms. Eds. J Doeksen and J van der Drift. North Holland Publishing Company, Amsterdam, pp 60–66.

Schowalter T D 1981 Insect herbivore relationship to the state of the host plant: Biotic regulation of ecosystem nutrient cycling through eological succession. Oikos 37, 126–130.

Seastedt T R 1984 The role of microarthropods in decomposition and mineralization processes. Annu. Rev. Entomol. 29, 25–46.

Sharpley A N, Syers J K and Springett J A 1979 Effect of surface-casting earthworms on the transport of phosphorus and nitrogen in surface runoff from pasture. Soil Biol. Biochem. 11, 459–462.

Sinclair A R E 1975 The resource limitation of trophic levels in tropical grassland ecosystems. J. Anim. Ecol. 44, 497–520.

Sohlenius B 1980 Abundance, biomass and contribution to energy flow by soil nematodes in terrestrial ecosystems. Oikos 34, 186–194.

Springett J A 1983 Effect of five species of earthworm on some soil properties. J. Appl. Ecol. 20, 865–872.

Springett J A and Syers J K 1979 Effect of earthworm casts on ryegrass seedlings. *In* Proc. 2nd Australas. Conf. Grassl. Invertebr. Ecol. Eds. T K Crosby and R P Pottinger. Government Printer, Wellington, pp 47–49.

Standen V 1973 The production and respiration of an enchytraeid population in blanket bog. J. Anim. Ecol. 42, 219–245.

Stockdill S M J 1966 The effect of earthworms on pastures. Proc N.Z. Ecol. Soc. 13, 68–75.

Stockdill S M J 1982 Effects of introduced earthworms on the productivity of New Zealand pastures. Pedobiologia 24, 29–35.

Stockdill S M J and Cossens G G 1966 The role of earthworms in pasture production and moisture conservation. Proc. N.Z. Grassld. Assoc. 28, 168–183.

Stout J and Heal O W 1967 Protozoa. *In* Soil Biology. Eds. A Burgess and F Raw. Academic Press, London, pp 149–195.

Syers J K and Springett J A 1983 Earthworm ecology in grassland soils. *In* Earthworm Ecology. Ed. J E Satchell. Chapman and Hall, London, pp 67–83.

Valiela I 1974 Composition, food webs and population limitations in dung arthropod communities during invasion and succession. Am. Midl. Nat. 92, 370–385.

Van Hook R I Jr. 1971 Energy and nutrient dynamics of spider and orthopteran populations in a grassland ecosystem. Ecol. Monogr. 41, 1–26.

Van Rhee J A 1965 Earthworm activity and plant growth in artificial cultures. Plant and Soil 22, 45–48.

Vickerman G P 1987 Invertebrate fauna of cereal fields. *In* The Boxworth Project, 1986 Report. Ministry of Agriculture,

Fisheries and Food, Tolworth, U.K., pp 74–88.

Waters R A S 1951 Earthworms and the fertility of pasture. Proc. N.Z. Grassl. Assoc. 13, 168–175.

White E G 1974 Grazing pressures of grasshoppers in an alpine tussock grassland. N.Z. J. Agric. Res. 17, 357–372.

White E G 1979 Modelling the interactive effects of insects and stock animals on herbage production. *In* Proc. 2nd Australas.

Conf. Grassl. Invertebr. Ecol. Eds. T K Crosby and R P Pottinger. Government Printer, Wellington, pp 240–242.

Wiegert R G and Evans F C 1967 Investigations of secondary productivity in grasslands. *In* Secondary Productivity of Terrestrial Ecosystems, Vol. 2. Ed. K Petrusewicz. Polish Academy of Sciences, Warsaw, pp 499–518.

M. Clarholm and L. Bergström (Eds.), Ecology of arable land, 185–189.

Role of soil animals in C and N mineralisation

T. PERSSON

*Dept of Ecology and Environmental Research, Swedish University of Agricultural Sciences, Box 7072,
S-750 07 Uppsala, Sweden*

Key words: C mineralisation, microarthropods, microcosm, N mineralisation, soil animals, soil
arthropods

Abstract

Addition of single species of soil animals to animal-free microcosms often increases total heterotrophic
respiration, but sometimes additions of microarthropods have been reported not to increase or even decrease
CO_2 evolution rates. Most studies indicate that addition of soil animals increases net N mineralisation. In
a study with F/H layer materials from a spruce stand in central Sweden kept at two temperatures (5 and
15°C) and three moisture levels (15, 30 and 60% of WHC), addition of a mixed fauna of soil arthropods,
mainly microarthropods, could not be shown to change the CO_2 evolution rates in comparison with materials
where arthropods were absent. However, addition of the arthropods significantly increased net N mineralisation for each of the temperature and moisture combinations. The increase due to the arthropods was
dependent on soil temperature but not on soil moisture. Because the total net N mineralisation decreased
with decreasing soil moisture, the soil arthropods had a much larger relative effect on net N mineralisation
under dry than under moist conditions. It is concluded that soil arthropods are important in maintaining
net N mineralisation under dry conditions when the microflora is largely inactive. The microbial/faunal
release of mineral N is discussed in relation to the C:N of the substrate.

Animal contribution to C mineralisation

Many attempts in quantifying the role of soil
animals in organic matter decomposition have concentrated on measurements of respiratory metabolism. Soil animals are reported to contribute 1–5%
to total heterotrophic respiration in coniferous
forest soils (Huhta and Koskenniemi, 1975; Persson et al., 1980), 3–13% in deciduous forest soils
(Axelsson et al., 1984; Ellenberg et al., 1986; Kitazawa, 1971; Reichle et al., 1977; Satchell, 1971)
and 5–25% in mesic grasslands (Coulson and Whittaker, 1978, Persson and Lohm, 1977). Soil animals
generally contribute less to energy turnover in dry
grasslands than in mesic ones (Coleman et al., 1980;
Zlotin and Khodashova, 1980). The low figures in
coniferous forest soils reflect a situation where fungivores, bacterivores and carnivores are dominant
components of the soil fauna, while litter and root
feeders are scarce (Persson et al., 1980). In deciduous forests, litter feeders such as earthworms
are more abundant, and in grasslands the abundant
occurrence of both litter feeders and root-feeding
arthropods (see e.g. Bevan, 1962; Persson and
Lohm, 1977; Salt and Hollick, 1944) may explain
the very high figures of respiration.

Animal influence on C mineralisation

The figures of animal contribution to total
heterotrophic respiration do not reveal much about
how the presence of the animals affects the microbial respiration or the total level of heterotrophic
respiration. Therefore, many authors have studied
whether additions or exclusions of soil animals
stimulate or depress total metabolic activity in
soils.

As regards litter feeders, additions of animals
seem to stimulate C mineralisation and litter de-

composition. Standen (1978) found that addition of enchytraeid worms or tipulid larvae or both increased the weight loss of *Eriophorum* leaf litter in litter-bags. This was mainly due to a stimulation of the microbial activity. According to Hanlon and Anderson (1980), addition of the isopod *Oniscus* to oak leaf litter increased microbial CO_2 evolution when the animals were added at low densities but depressed microbial CO_2 evolution when added at high densities. The authors suggested that litter comminution, which exposed new surfaces for microbial attack, was the main factor in enhancing microbial activity, but obviously too much grazing counteracted this effect.

Additions of bacterial feeders, such as bacterial-feeding nematodes and amoebae, usually increase $\bar{C}O_2$ evolution in soil microcosms with bacteria (Coleman *et al.*, 1978; Woods *et al.*, 1982).

Additions of fungal feeders, mainly belonging to Collembola and Acari, seem to cause more diverse effects. A majority of studies indicates increased decay losses when microarthropods are added (Seastedt, 1984). Hanlon and Anderson (1979) reported that moderate grazing by the collembolan *Folsomia candida* on the fungus *Coriolus versicolor* stimulated evolution rates of microbial CO_2 whereas intensive grazing had the opposite effect. On the other hand, Andrén and Schnürer (1985) found no effects on CO_2 evolution rate, litter mass loss or microbial biomass after addition of different numbers of the collembolan *Folsomia fimetaria* to barley straw litter with a natural soil microflora. It appeared that this collembolan consumed bacteria and protozoa as well as fungi.

Addition of a diverse soil fauna (Collembola, Acari, Enchytraeidae and Nematoda) to birch litter, birch litter plus raw humus of a coniferous forest and raw humus alone was shown to increase CO_2 evolution by 32, 23 and 15%, respectively, in comparison with the same substrates without soil fauna (Setälä *et al.*, 1988). It was concluded that the animals enhanced the activity of soil microorganisms and, thereby, accelerated decomposition.

Animal influence on N mineralisation

Even if microbial grazers might have no or small effects on total CO_2 evolution rate, it has been suggested that they can increase nitrogen mineralisation (Persson, 1983). Data on increased nitrogen mineralisation have been presented by, for example, Woods *et al.*, (1982), who found that presence of the bacterial-grazing amoeba *Acanthamoeba polyphaga* in microcosms with bacteria increased the release of inorganic N in comparison with microcosms with bacteria alone. Clarholm (1985) reported that wheat plants grown in sterilized soil inoculated with soil bacteria plus protozoans contained 75% more nitrogen than plants grown in soil without protozoans. Griffiths (1986) found that both the ciliate protozoan *Colpoda steinii* and the nematode *Rhabditis* sp. increased the mineralisation of ammonium when added separately or together to cultures with the bacterium *Pseudomonas fluorescens.*, Thus, there are many evidences that the presence of microfauna (protozoans and nematodes) increases net nitrogen mineralisation.

Ineson *et al.* (1982) reported that presence of the collembolan *Folsomia candida* increased the leaching of ammonium and nitrate in oak leaf litter microcosms, and Anderson *et al.* (1983) found that presence of the collembolans *Tomocerus minor* and *Orchesella villosa*, the millipedes *Polydesmus angustus, Iulus scandinavicus* and *Glomeris marginata* and the lumbricid *Lumbricus rubellus* increased the leaching of ammonium and/or nitrate from incubated F-layer materials from an oak forest. On the other hand, presence of enchytraeids had no significant effect on nitrogen leaching in the same material (Anderson *et al.*, 1983).

A microcosm experiment

All of these studies on nitrogen mineralisation used one or a few animal species as model organisms. I was interested in how a diverse soil fauna might influence the C and N mineralisation, and I was particularly interested in the effects of soil arthropods, because these animals live in the air-filled pores of the soil and are less affected by drought than the fauna dependent on free soil water. My hypothesis was that presence of soil arthropods should increase C and N mineralisation and that their relative contribution to this mineralisation should be greater under dry conditions, when the microfloral activity is depressed, than under wet conditions, when the microflora is active.

To test this hypothesis, F/H material (pH 4.3, C:N 32) was taken from a 25-year-old stand of Norway spruce at Jädraås in central Sweden, some hundred meters from the pine stand described by Persson *et al.* (1980). The material was placed in Tullgren funnels for five days to extract and/or kill all soil arthropods and, furthermore, sieved, put into 50 cm^2 containers, rewetted to about 75% of the water holding capacity (WHC) and inoculated with a humus suspension from fresh mor humus to restore bacterial, fungal, protozoan and nematode populations. Half of the number of containers received soil arthropods by putting them under Tullgren funnels with 200 cm^2 fresh humus blocks, assuming that the larger area would compensate for inefficient extraction to obtain realistic animal densities in the containers. After the 5-day extraction, the material in the containers was dried to 75, 150 and 300% water content, corresponding to 15, 30 and 60% WHC, and placed for 110 days in rooms with 5 and 15°C constant temperature. Consequently, the experimental design was a 3-factor analysis with two temperatures, three moisture levels and presence/absence of arthropods, all in ten replicates.

The CO_2 evolution rates were highly dependent on temperature and soil moisture, but there were no clear differences between materials with arthropods present or absent. Consequently, there was no support for the hypothesis that the arthropods should increase total CO_2 evolution under dry conditions (T. Persson, unpubl.).

At the end of the incubation, exchangeable inorganic nitrogen was extracted by 0.5 M KCl. Ammonium made up more than 99% of all inorganic nitrogen extracted, and very small amounts of nitrate were found. Significantly ($p < 0.05$) more inorganic N had accumulated at 15°C than at 5°C, and significantly more inorganic N had accumulated for each increase in soil moisture content (Fig. 1).

Presence of soil arthropods, predominantly collembolans and mites, significantly ($p < 0.05$) increased the amount of inorganic N for all temperature and moisture conditions in relation to materials without soil arthropods. The increase due to the presence of soil arthropods was dependent

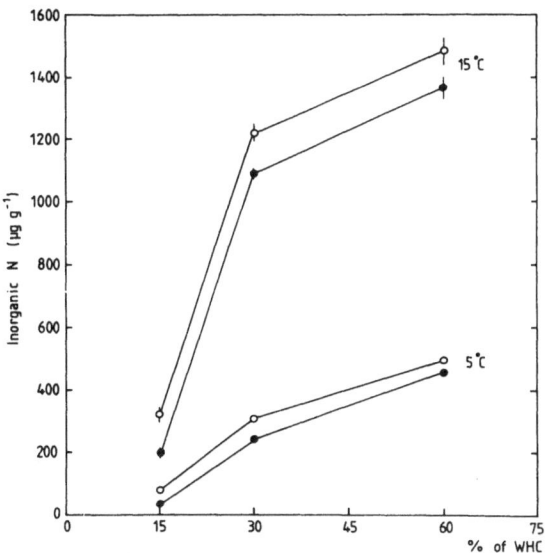

Fig. 1. Influence of soil temperature and soil moisture (% of WHC) on mineralised nitrogen after incubation for 110 days. Open rings indicate presence and filled rings indicate absence of soil arthropods. Bars indicate one S.E. when wider than one ring. Inorganic N was 51 ± 2 µg N g^{-1} at the start of the incubation.

on temperature, being about 50 µg g^{-1} at 5°C and 120 µg g^{-1} at 15°C (Table 1), *i.e.* corresponding to a Q_{10} of 2.4 which is close to the Q_{10} value of 2.6 given for oxygen consumption by collembolans (Persson and Lohm, 1977). The increase in net N mineralisation was, thus, well correlated with the overall increase in metabolic activity.

Soil moisture did not seem to have any influence on the arthropod contribution to net N mineralisation, being about the same for the WHC levels used for each temperature (Table 1).

Because net N mineralisation by the soil arthropods was almost unchanged and that by the whole soil organism community was decreased by decreasing soil moisture levels, the relative influence of the soil arthropods on nitrogen mineralisation increased with decreasing soil moisture (Table 1). At 5°C and 15% of WHC, the presence of arthropods even turned a net immobilisation into a net mineralisation and, therefore, the percentage given in Table 1 is higher than 100.

A realistic estimate of the temperature and moisture conditions at Jädraås, from where the soil was taken, is that the mean soil temperature during

Table 1. Accumulation of inorganic N (final value minus start value) after 110 days with soil arthropods present (P) or absent (A). Relative contribution (%) of soil arthropods to total net N mineralisation also indicated

Temperature (°C)	Moisture (% of WHC)	P ($\mu g \, N \, g^{-1}$)	A ($\mu g \, N \, g^{-1}$)	P-A ($\mu g \, N \, g^{-1}$)	$\frac{(P-A)}{P} \cdot 100$ (%)
5	15	27	−18	45	167
5	30	252	187	65	26
5	60	442	402	40	9
15	15	267	147	120	45
15	30	1156	1038	118	10
15	60	1433	1314	119	8

the growing season is slightly below 10°C and the mean moisture content about 30% of WHC in the humus layer (Jansson, 1977). According to the figures in Table 1, one might expect that the arthropods should contribute about 10–26% to total net N mineralisation in the F/H layer during this period. If the contribution by the microfauna and the enchytraeids, also being abundant in this type of soil, is taken into consideration, a total faunal contribution to net N mineralisation might be in the order of 30–50%, *i.e.*, very close to the estimates based on efficiency quotients and C:N ratios by Persson (1983).

A reason for the relatively great influence of the soil arthropods on net N mineralisation in a soil where their influence on C mineralisation is small is probably the high C:N ratio (32) in the microbial substrate. Bacteria and fungi have a C:N ratio of about 5 and, therefore, they must concentrate N from the substrate by a factor of 6–7 in the present study. This meant that fairly small amounts of N were probably mineralised per g of microorganism. Soil animals also have a C:N ratio of about 5 and, consequently, fungal and bacterial feeding soil animals should have no need to concentrate N and should excrete a great part of the N assimilated (Persson, 1983).

In conclusion, microbial feeding soil animals play an important role in promoting nitrogen mineralisation, especially in soil layers with high C:N ratios, where the microorganisms have a low release rate of nitrogen. Furthermore, soil arthropods seem to play a role in maintaining net N mineralisation under dry conditions, when the microflora and the water-dependent microfauna are largely inactive.

Acknowledgements

I thank Birgitta Vegerfors for statistical advice and Nigel Rollison for linguistic comments. Olof Andrén and Marianne Clarholm gave valuable comments to the manuscript. Grants were received from the Swedish Natural Science Research Council.

References

Anderson J M, Ineson P and Huish S A 1983 Nitrogen and cation mobilization by soil fauna feeding on leaf litter and soil organic matter from deciduous woodlands. Soil Biol. Biochem. 15, 463–467.

Andrén O and Schnürer J 1985 Barley straw decomposition with varied levels of microbial grazing by *Folsomia fimetaria* (L.) (Collembola, Isotomidae). Oecologia (Berlin) 68, 57–62.

Axelsson B, Lohm U and Persson T 1984 Enchytraeids, lumbricids and soil arthropods in a northern deciduous woodland – a quantitative study. Holarct. Ecol. 7, 91–103.

Bevan W J 1962 Observations on damage to grassland in East Yorkshire by larvae of the common leaf weevil, *Phyllobius pyri* L. and notes on its biology. J. Brit. Grassl. Soc. 17, 194–197.

Clarholm M 1985 Interactions of bacteria, protozoa and plants leading to mineralization of soil nitrogen. Soil Biol. Biochem. 17, 181–187.

Coleman D C, Anderson R V, Cole C V, Elliott E T, Woods L and Campion M K 1978 Trophic interactions in soils as they affect energy and nutrient dynamics. IV. Flows of metabolic and biomass carbon. Microbial Ecology 4, 373–380.

Coleman D C, Sasson A, Breymeyer A I, Dash M C, Dommergues Y, Hunt H W, Paul E A, Schaefer R, Ulehlová B and Zlotin R I 1980 Decomposer subsystem. *In* Grasslands, Systems Analysis and Man. Eds. A I Breymeyer and G M Van Dyne . International Biological Programme 19, pp. 609–655. Cambridge University Press, Cambridge.

Coleman D C, Reid C P P and Cole C V 1983 Biological strategies of nutrient cycling in soil systems. Adv. Ecol. Res. 13, 1–55.

Coulson J C and Whittaker J B 1978 Ecology of moorland animals. *In* Production Ecology of British Moors and Montane Grasslands. Eds. O W Heal and D F Perkins. Ecol. Stud. 27, 52–93. Springer-Verlag, Berlin.

Ellenberg H, Mayer R and Schauermann J 1986 Ökosystemforschung – Ergebnisse des Sollingprojekts 1966–1986. Ulmer, Stuttgart, 507 pp

Griffiths B S 1986 Mineralization of nitrogen and phosphorus by mixed cultures of the ciliate protozoan *Colpoda steinii*, the nematode *Rhabditis* sp. and the bacterium *Pseudomonas fluorescens*. Soil Biol. Biochem. 18, 637–641.

Hanlon R D G and Anderson J M 1979 The effects of collembola grazing on microbial activity in decomposing leaf litter. Oecologia (Berl.) 38, 93–99.

Hanlon R D G and Anderson J M 1980 Influence of macroarthropod feeding activities on microflora in decomposing oak leaves. Soil Biol. Biochem. 12, 255–261.

Huhta V and Koskenniemi A 1975 Numbers, biomass and community respiration of soil invertebrates in spruce forests at two latitudes in Finland. Ann. Zool. Fennici 12, 164–182.

Ineson P, Leonard M A and Anderson J M 1982 Effect of collembolan grazing upon nitrogen and cation leaching from decomposing leaf litter. Soil Biol. Biochem. 14, 601–605.

Jansson P E 1977 Soil properties at Ivantjärnsheden. Swed. Conif. For. Proj., Int. Rep. 54, 66 pp.

Kitazawa Y 1971 Biological regionality of the soil fauna and its function in forest ecosystem types. *In* Productivity of Forest Ecosystems. Ed. P Duvigneaud. Proc. Brussels Symp. 1969, pp 485–498. Unesco, Paris.

Persson T 1983 Influence of soil animals on nitrogen mineralisation in a northern Scots pine forest. *In* New Trends in Soil Biology. Eds. Ph Lebrun *et al*. Proc. 8th Int. Coll. Soil Zool., pp. 117–126.

Persson T and Lohm U 1977 Energetical significance of the annelids and arthropods in a Swedish grassland soil. Ecol. Bull. (Stockholm) 23, 1–211.

Persson, T, Bååth E, Clarholm M, Lundkvist H, Söderström B E and Sohlenius B 1980 Trophic structure, biomass dynamics and carbon metabolism of soil organisms in a Scots pine forest. *In* Structure and Function of Northern Coniferous Forests – An Ecosystem Study. Ed. T. Persson. Ecol. Bull. (Stockholm) 32, 419–459.

Reichle D E 1977 The role of soil invertebrates in nutrient cycling. *In* Soil Organisms as Components of Ecosystems. Eds U Lohm and T Persson. Proc. 6th Int. Coll. Soil Zool., Ecol. Bull. (Stockholm) 25, 145–156.

Salt G and Hollick F S J 1944 Studies of wireworm populations. I. A census of wireworms in pasture. Ann. Appl. Biol. 31, 52–64.

Satchell J E 1971 Feasibility study of an energy budget for Meathop Wood. *In* Productivity of Forest Ecosystems. Ed. P. Duvigneaud. Proc. Brussels Symp. 1969, pp 619–630.

Seastedt T R 1984 The role of microarthropods in decomposition and mineralization processes. Ann. Rev. Entomol. 29, 25–46.

Setälä H, Haimi J and Huhta V 1988 A microcosm study on the respiration and weight loss in birch litter and raw humus as influenced by soil fauna. Biol. Fertil. Soils 5, 282–287.

Standen V 1978 The influence of soil fauna on decomposition by micro-organisms in blanket bog litter. J. Anim. Ecol. 47, 25–38.

Woods L E, Cole C V, Elliott E T, Anderson R V and Coleman D C 1982 Nitrogen transformations in soil as affected by bacterial-microfaunal interactions. Soil Biol. Biochem. 14, 93–98.

Zlotin R I and Khodashova K S 1980 The role of animals in biological cycling of forest-steppe ecosystems. Dowden, Hutchinson & Ross, Stroudsburg, 221 p.

M. Clarholm and L. Bergström (Eds.), Ecology of arable land, 191–203.
© 1989 Kluwer Academic Publishers.

Inferring trophic transfers from pulse-dynamics in detrital food webs

H. W. HUNT, E. T. ELLIOTT and D. E. WALTER
Natural Resource Ecology Laboratory, Colorado State University, Fort Collins, CO 80523, USA

Key words: food web, microarthropods, microbes, nematodes, protozoans, soil

Abstract

In semiarid ecosystems, decomposers are active during numerous short periods following rainfall events, and most inactive in the intervening dry periods. Many studies concern season-long dynamics of decomposer populations, but less is known of the short-term dynamics during wet periods. These short-term dynamics may provide the key to understanding interactions between microbes and fauna.

The dynamics of populations in the detrital food web were followed after wetting large intact soil cores that had been removed from native shortgrass steppe, winter wheat, and fallow plots. The cores were sampled over a ten day period for bacteria, fungi, protozoa, and various functional groups of microarthropods and nematodes. The native sod had appreciably greater biomass of fungi, nematodes and microarthropods than did the cultivated plots, but there was no difference in bacteria or protozoans. The observed dynamics after wetting were different in two experiments which differed in temperature, soil water level, and the initial sizes of the populations. These results were interpreted in relation to a model of the structure of the detrital food web, and estimates were made of the rates of trophic transfers in the web. Consumption by protozoa was great enough for them to account for bacterial turnover, but consumption by fungivorous nematodes and microarthropods appeared to be too small to account for fungal turnover.

Progress in understanding the dynamics of detrital food webs requires a better definition of the functional groups of soil organisms, their resources, predators and population parameters, and the effects of soil structure and water content on trophic relationships.

Introduction

Soil fauna are thought to be important for decomposition and mineralization (Coleman, 1985; Moore *et al.*, 1988). However, some studies indicate that the fauna have no effect on nitrogen mineralization (Bååth *et al.*, 1981; Rosswall *et al.*, 1977), and Andrén and Paustian (1987) were able to account for virtually all the dynamics of barley straw decomposition without reference to fauna. Hunt *et al.* (1987) estimated that fauna are directly responsible for 37% of nitrogen mineralization in a shortgrass steppe, but their analysis was based on a steady state assumption, and did not address the dynamic aspects of interactions between fauna and microflora.

Most field studies of decomposer communities have followed the dynamics of populations over periods of months or years, using a sampling interval of weeks or months (Clarholm *et al.*, 1981; Elliott *et al.*, 1984; Ingham *et al.*, 1986). In contrast, many laboratory studies have used sampling intervals of days (Giesy, 1980). Laboratory studies using soils and organisms from shortgrass steppe ecosystems have revealed dramatic short term dynamics attributed to trophic interactions (Cole *et al.*, 1978; Hunt *et al.*, 1984), but conditions differed appreciably from the field, and it is not certain whether the results can be extrapolated to the field.

The shortgrass steppe of North America is a water limited system, with most precipitation falling as summer thundershowers (Smith, 1972). Carbon dioxide evolution from the soil indicates alternating active and inactive periods associated with wetting and drying (Hunt, 1977). Intense trophic interactions and dramatic population dynamics may occur during these wet periods and a pulse of microbial production is transferred to higher trophic levels. The nature of interactions among populations may be more readily apparent from studies of pulse-dynamics during the wet periods than from longer term studies in which sampling dates are chosen without respect to wetting events. We are unaware of any field studies of the short-term dynamics of the detrital food web, including all the major faunal groups, after a rainfall. Schnürer *et al.* (1986) followed the short-term dynamics of microbes, protozoans, and nematodes in a barley field after a single wetting event, but did not estimate microarthropod populations. Steinberger *et al.* (1984) and Parker *et al.* (1984) studied the dynamics of all the major groups during several rainfall events in a desert, but the sampling interval (six days) was probably too long to detect the early microbial dynamics.

It is not a simple matter to estimate the rates of trophic transfers from observations of biomass dynamics. A general problem is that secondary production might be passed along continuously from prey to predators, and not all appear as changes in prey biomass. However, predator production may give an indirect indication of prey production, which illustrates the importance of studying all components of the system simultaneously. In the hypothetical case of a system at steady state there are no dynamics even though trophic transfers are continuous, and therefore trophic transfers cannot be estimated from dynamics. Theoretically, it is impossible to estimate unique transfers from biomass dynamics without ancillary information on the rate of at least one transfer in the system. Carbon dioxide evolution is a suitable transfer that is easily measured. Parker *et al.* (1984) used CO_2 evolution along with a steady state assumption to estimate microbial production, and Clarholm (1981) used CO_2 evolution in a qualitative way to infer high bacterial production during a period not covered by biomass samples. Simulation modeling has been used to derive estimates of trophic transfers from information on biomass dynamics and CO_2 evolution in simplified microcosms (Hunt *et al.*, 1984), but apparently no one has done this for the whole detrital food web.

The objective of this study was to follow the dynamics of the major components of the detrital food web through a single wetting event under natural conditions, with emphasis on interactions between predator and prey populations. To avoid the logistic difficulties of controlling precipitation on the experimental area, and of taking frequent samples from a distant field site, intact soil cores were removed from the field, held in a greenhouse to dry, and subjected to a single wetting event. We developed a simple model to derive estimates of trophic transfers based on CO_2 evolution and biomass dynamics.

Materials and methods

Intact soil cores were taken from native shortgrass steppe, and from the planted and fallow phases of wheat plots at the University of Nebraska High Plains Agricultural Laboratory near Sidney, Nebraska. The soil is a Duroc loam, a fine silty, mixed mesic Pachic Haplustoll (Fenster and Peterson, 1979). The shortgrass steppe is dominated by blue grama (*Bouteloua gracilis*) and buffalo grass (*Buchloe dactyloides*). Wheat plots were established in 1970 when the sod was broken, and were planted to winter wheat (*Triticum aestivum*) in alternate years under a no-tillage management system (Fenster and Peterson, 1979).

Steel cylinders 32 cm in diameter and 46 cm long were driven into the ground and removed with intact soil on May 1, 1985. Shoots in both the wheat and the sod were only a few cm high and were not damaged in the process. The cylinders were transported to a greenhouse and held without watering until the first experiment started on May 13. The field design included four replicate cylinders from each of three treatments (native sod, no-till fallow, and no-till wheat crop) in each of three randomized blocks, for a total of 36 cylinders. The cylinders were split into two 'greenhouse' blocks in a balanced design. On May 13 the 18 cylinders in the first greenhouse block (E1) were watered with 2.25 cm of deionized water, which brought the top 9–10 cm to field capacity. The second greenhouse block (E2) was not watered until May 27, and the same amount of water wet the soil to only 7–8 cm because the soil was drier then. Except for the date of initiation of the two experiments, all aspects of experimental design and sampling procedures were identical. Each cylinder was sampled immediately before watering and at one, two, four, and ten days after watering.

Subsamples for all variables except for microarthropods were based on an homogenized composite of four small cores, 7 cm long by 2.2 cm diameter, from each cylinder. These cores were placed in plastic bags to prevent drying and returned immediately to the laboratory for processing. Holes left by the cores were lined with plastic bags which were filled with sand, tied off, and trimmed immediately after sampling. Area sample totaled about 14% of the cylinder area at the time of the last sample.

Bacteria were estimated by a direct count of FITC stained cells collected on 0.2 μm Millipore filters (Babiuk and Paul, 1970; Hobbie *et al.*, 1977) from 0.5 ml of a 1/100 dilution of soil suspension. Numbers were converted to biomass assuming an average cell size of 0.6 μm^3 and using the conversion factor of van Veen and Paul (1979). Fungal length was estimated by direct counts (Ingham and Klein, 1984), and an estimate of total fungal mass including empty hyphae was derived assuming a hyphal diameter of 3 μm and a ratio of 0.33 g dry wt. per cm^3 live volume (van Veen and Paul, 1979). Carbon dioxide evolution was not followed *in situ* in the cylinders, but the controls for the fumigation biomass method (Elliott *et al.*, 1984) provided a basis for an estimate of CO_2 evolution. Soil was assayed for water content, and soil temperature at 3.5 cm was monitored for most of the experiment using max-min thermometers.

Protozoans were enumerated using the most probable number method (Singh, 1946), and biomass estimates were based on cell sizes from Frey *et al.* (1985). Nematodes from Baermann extracts were counted and numbers were corrected for a 33.5% extraction efficiency (D. W. Freckman, unpublished data). Nematode trophic categories were assigned following Freckman *et al.* (1979), and individual sizes were based on observations (R.E. Ingham, unpublished data) from a shortgrass steppe site in northeastern Colorado.

Samples for microarthropods consisted of two 4.8 cm diameter cores to 10 cm depth. Each core was divided into 0–5 and 5–10 cm depth increments, and each depth increment was individually extracted using modified Merchant-Crossley high-gradient extractors (Leetham, 1975) over ethanol

for 7 days. Counts were corrected for extraction efficiency which was determined by a heptane flotation procedure (Walter *et al.*, 1987). Tullgren funnels were used to extract live animals into jars with moist plaster of Paris floors for use in feeding observations. Microarthropod samples were sorted into morphospecies, and slide mounted subsamples of Collembola, Mesostigmata, Oribatida, Astigmata, Prostigmata (excluding Tydeidae and Pygmephoridae), Symphyla, and Coleoptera were used to identify both immature and adult stages. Tydeidae, Pygmephoridae, and Pseudococcidae could not be sorted into species and were counted as morphotypes. Rare taxa such as pauropods, centipedes, pseudoscorpions, and a variety of insects were also counted as morphotypes. Trophic categorizations were determined by three methods: (1) direct observation of feeding by live animals; (2) oil immersion inspection of gut contents; and (3) literature records for the same or closely related species. Feeding observations are detailed in Walter *et al.* (1986), Walter (1987), Walter *et al.* (1987), and Walter *et al.* (1988). Herbivores and predators foraging aboveground are not considered in this paper.

Biomass of microarthropods was estimated using regressions derived from Edwards (1967) and average lengths calculated from slide mounted specimens divided into adult and immature categories for each species or morphotype. Since Mesostigmata at Sidney were smaller than and in different taxa from those used by Edwards (1967), we adjusted his regression coefficient for the mesostigmatid mites using dry weights obtained from cultures of mites of each body type. Biomass of all groups is expressed as μg dry weight per g dry soil.

ANOVAs were performed for the various biomass estimates, water content and temperature. Since the design involves repeated measures, the Greenhouse-Geyser adjustment was applied to degrees of freedom for tests involving time. Treatment effects were considered significant at the $p < 0.05$ level. A log transform was applied to stabilize the variance of the dependent variable in the ANOVAs when necessary, but all the data are presented on a linear scale, and 95% confidence intervals were employed for evaluating individual differences. Analyses were carried out using the BMDP software package (Dixon, 1983).

Results

Soil temperature was monitored beginning on day six of the first experiment (Fig. 1). It was hotter on average during the second experiment (E2) than during first experiment (E1), and maximum tem-

peratures reached 39°C. The soil was drier at the start of E2 than at the start of the E1 (Fig. 2) because of an additional two weeks of drying. Both the wheat and native plants showed signs of water stress before the start of E2, but these symptoms did not reappear after watering. Faster evapotrans-

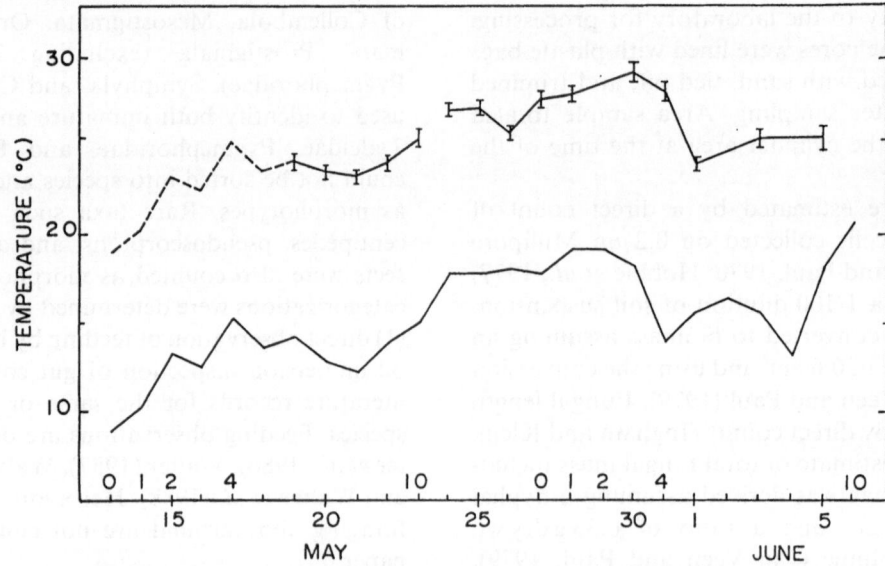

Fig. 1. Daily average air temperature for Fort Collins (lower line), and daily average soil temperature at 3.5 cm depth in the experimental units (upper line). Vertical bars are 95% confidence intervals. The dashed line segment is predicted soil temperature based on the observed correlation with air temperature. Sample dates for E1 (May 13–May 23) and E2 (May 27–June 6) are indicated above the x-axis.

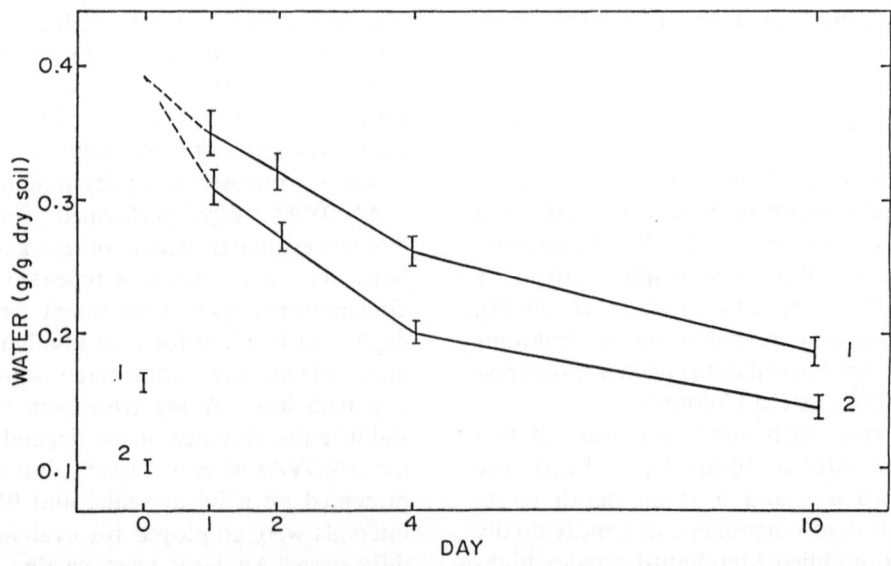

Fig. 2. Soil water before wetting and during drying at 0–7 cm in E1 (1) and E2 (2). Dashed lines are extrapolations of water content to a time immediately after wetting. Vertical bars are 95% confidence intervals.

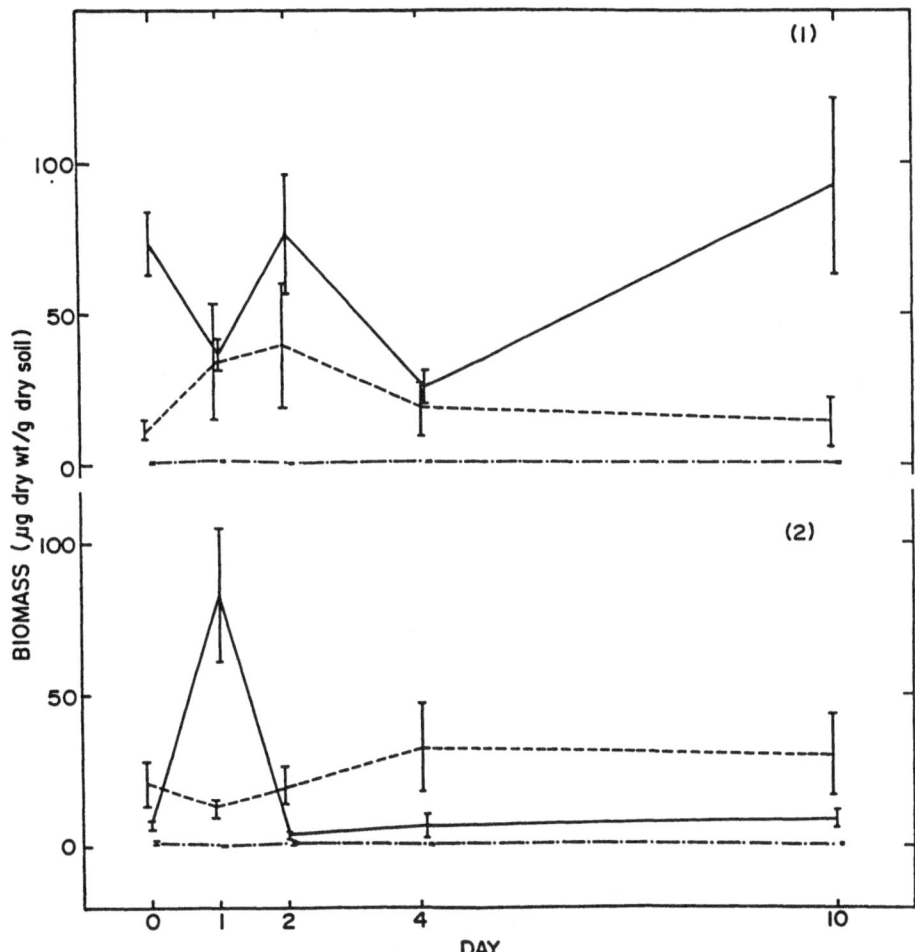

Fig. 3. Biomass of bacteria (solid line), protozoa (dashed), and bacterial feeding nematodes (dash-dot) in E1 (1) and E2 (2). Vertical bars are 95% confidence intervals.

piration at the start of E2 (Fig. 2) was probably related to higher temperatures.

Bacterial biomass did not differ among native sod, wheat or fallow, but biomass dynamics were very different in E1 and E2 (Fig. 3). In E1, the initial decrease in bacteria may be associated with the increase in protozoa. It is not clear why protozoa showed no increase until day four in E2. Bacterial biomass was much lower in E2 than in E1, lower even than that of protozoa. Bacterial feeding nematodes had appreciably lower biomass than protozoans, and showed no temporal dynamics.

Fungal biomass was about twice as great in native sod as in the wheat and fallow treatments. Since there was no interaction between treatment and time, the data have been averaged over treatment.

As was the case for bacteria, fungal dynamics differed in the two experiments, with a decrease in E1 (Fig. 4) and an increase in E2 (Fig. 5). In E1, the decrease in fungal biomass amounted to over 100 μg dry wt per g soil in ten days, yet the observed increases in fungal feeding microarthropods and nematodes (Fig. 4) were less than one μg apiece. In E2 (Fig. 5), fungal feeding nematodes decreased slightly while there was no change in fungal feeding microarthropods. The biomass of top predators was significantly greater in the sod (0.67 μg g^{-1}) than in the wheat (0.09) or fallow (0.16), and there were no temporal trends.

Carbon dioxide evolution over ten days in the controls of the fumigation biomass measurements did not vary significantly among systems or bet-

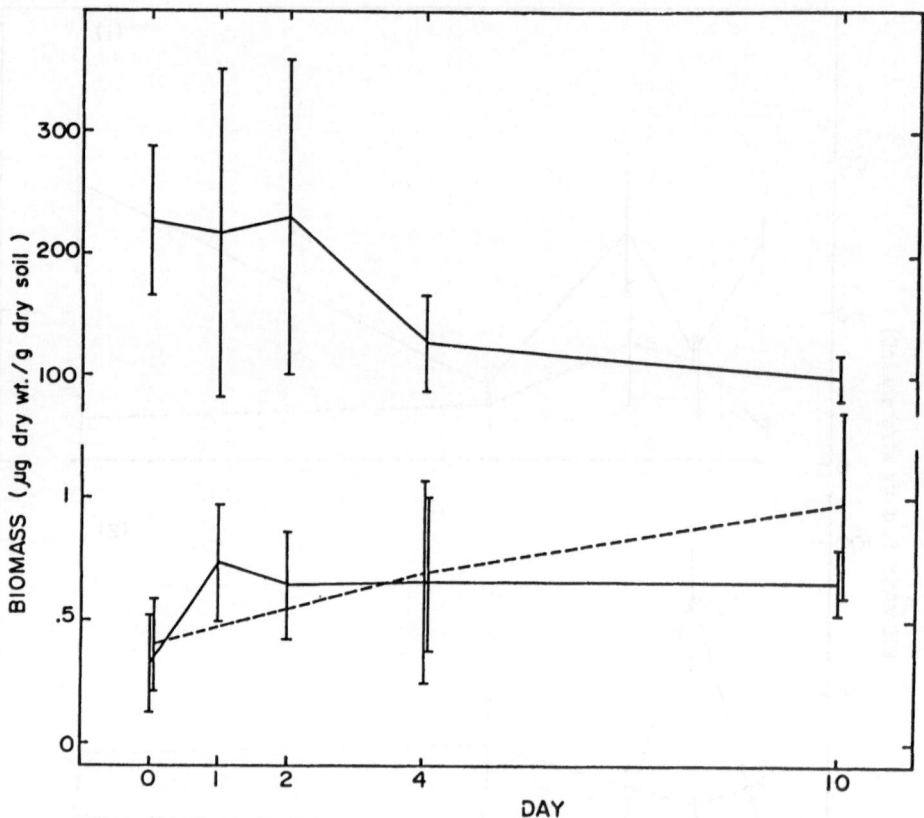

Fig. 4. Biomass of fungi (upper solid line), fungal feeding microarthropods (dashed line), and fungal feeding nematodes (lower solid line) in E1. Vertical bars are 95% confidence intervals.

ween experiments, but varied with date, from 13.0 $\mu g\,CO_2$-$C\,g^{-1}\,d^{-1}$ in samples taken before wetting, to 36.2, 41.3, 29.2, and 15.9 in samples taken after 1, 2, 4 and 10 days. In the following analysis we required an estimate of total CO_2 evolution in the cylinders. By assuming that the initial rate of carbon dioxide evolution from the fumigation controls was twice the average rate over the ten day incubation, that this initial rate was the same as the rate in the cylinders on the sample date, and by using linear interpolation between sample dates, we estimated a total of 538 $\mu g\,CO_2$-$C\,g^{-1}$ evolved over the course of the experiments.

Analysis

A simple mass balance model (Fig. 6) provides the basis for estimating the productivity of various functional groups in each of six data sets (fallow, wheat and sod in E1 and E2). The conceptual bases

of this model are similar to those of Heal and MacLean (1975). The model is not intended to simulate all the observed dynamics in Figs. 3–5, but it has been fitted to observe net changes in biomass, and has been constrained to use transfer rates compatible with the physiology and population attributes of the organisms. To allow for different transfer rates in the six data sets, the accumulative transfer T from resource to consumer in ten days ($\mu g\,g^{-1}$) is proportional to the biomass of both consumer C ($\mu g\,g^{-1}$) and resource R ($\mu g\,g^{-1}$):

$$T = k_1 \cdot C \cdot R, \qquad (1)$$

where k_1 is a rate constant. This is the simplest reasonable equation that incorporates effects of both resource and consumer level on the transfer rate. We do not have a measure of resource level for bacteria and fungi, so the transfer from substrate to microbes was assumed proportional to microbial biomass:

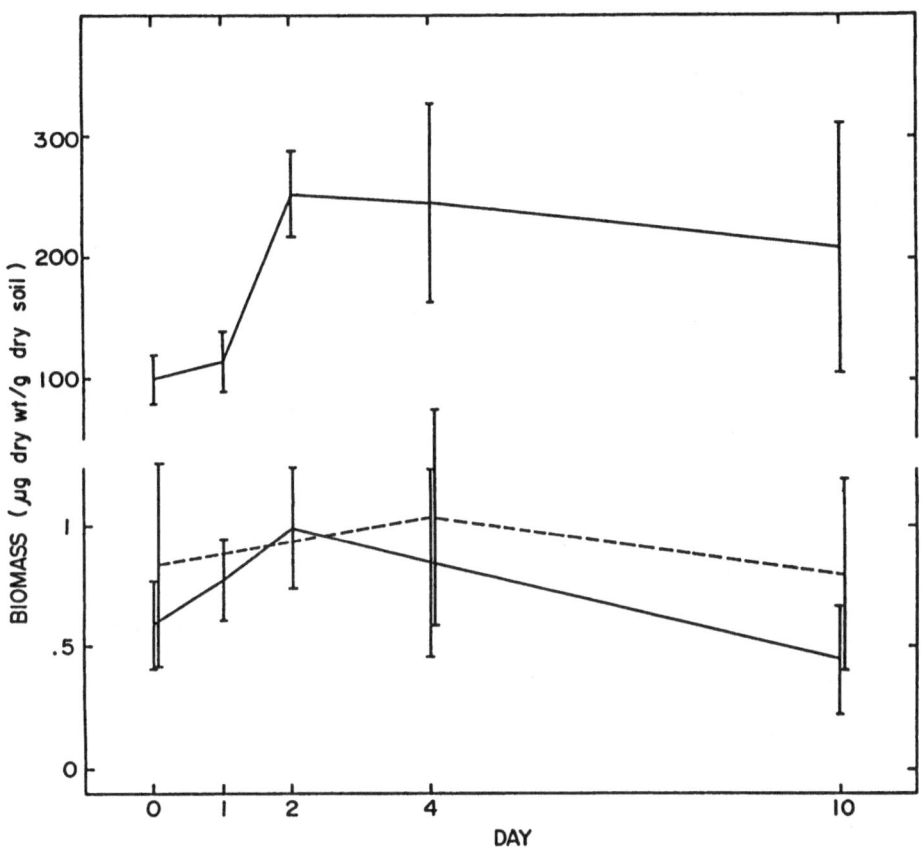

Fig. 5. Biomass of fungi (upper solid line), fungal feeding microarthropods (dashed line), and fungal feeding nematodes (lower solid line) in E2. Vertical bars are 95% confidence intervals.

$$T = k_2 \cdot F, \tag{2}$$

where F is biomass of bacteria or fungi, and k_2 is a rate constant.

The transfer from substrate to microbe was divided between a constant fraction Y to production and the remainder $(1.0 - Y)$ to carbon dioxide evolution. Consumption by fauna was divided among production, release of wastes, and respiration using parameters for the ratios of assimilation to consumption (A/C), and production to assimilation (P/A). We assumed the occurrence of death unrelated to consumption at a rate D proportional to biomass B:

$$D = k_3 \cdot B. \tag{3}$$

The model incorporates simultaneous inputs and outputs for every group. Change in biomass is assimilation minus losses to consumers minus other death. These changes were calculated in terms of the equations and parameters defined above, so that the model could be fitted to the data. Different sets of flow rates may yield the same changes in biomass, but these sets differ in the predicted CO_2 evolution. Thus, the CO_2 data are important for arriving at a reasonable set of transfer rates. Another source of information about transfers is the significant changes in biomass observed during the course of the experiments. For example, bacteria in E1 (Fig. 3) increased between days one and two and between days four and ten, by a total of $91.6 \, \mu g \, g^{-1}$. Therefore in the model, the product of yield Y and the transfer from substrate to bacteria can be no less than 91.6. Significant decreases in bacterial biomass in E1 total $87.6 \, \mu g \, g^{-1}$, which provides an estimate of the minimal loss to protozoa. Transfers in the model were constrained in this way to be compatible with summed significant increases and decreases for every functional group (Table 1). Finally, ratios of production to

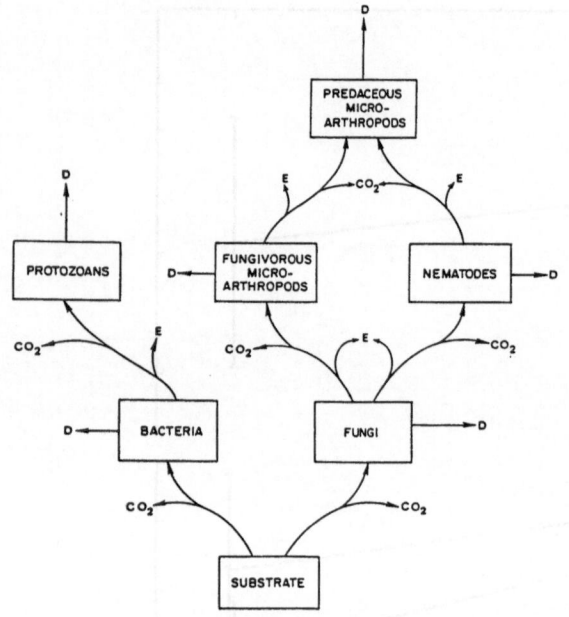

Fig. 6. Model for estimating components of secondary production in the food web. CO_2 is carbon dioxide generation, E is defecation, and D is death due to causes other than predation.

biomass in the model were required to be no greater than maximal values estimated for each functional group: bacteria ($9.9\,d^{-1}$, Hunt *et al.*, 1985); protozoans, based on amoebae, which make up most of

protozoan biomass ($0.5\,d^{-1}$, Hunt *et al.*, 1984); fungal feeding nematodes, assumed the same as bacterial feeding nematodes ($0.4\,d^{-1}$, Hunt *et al.*, 1984); fungal feeding microarthropods ($0.12\,d^{-1}$, Walter, unpublished data); and predaceous microarthropods ($0.24\,d^{-1}$, Walter, unpublished data).

We used the function minimizing algorithm of Powell (1965) to fit the model to the deduced CO_2 and observed biomass changes and to meet the above constraints on flow rates. To achieve a satisfactory fit, it was necessary to include non-predatory death of protozoans, predators, and fungi, but not of bacteria or fungivores. Also, some transfers were qualitatively different (required different parameter values) in E1 and E2, and others differed among fallow, wheat, and sod (Table 2). The resulting agreement between predicted and observed changes in biomass was good to excellent (Fig. 7). Table 3 gives the predicted transfer rates.

Discussion

Our results are significant for understanding the structure and dynamics of detrital food webs in

Table 1. Summed significant increases and decreases ($\mu g\,g^{-1}$) used to constrain transfer rates in food web model. Separate values for the six data sets were calculated only when the means for that functional group differed among data sets according to ANOVA

Functional group	E1			E2		
	Fallow	Wheat	Prairie	Fallow	Wheat	Prairie
Bacteria						
increases	91.6	91.6	91.6	81.3	81.3	81.3
decreases	87.6	87.6	87.6	79.5	79.5	79.5
Protozoa						
increases	27.6	27.6	27.6	20.2	20.2	20.2
decreases	0.0	0.0	0.0	0.0	0.0	0.0
Fungi						
increases	0.0	0.0	0.0	70.0	75.0	168.0
decreases	65.0	87.0	127.0	0.0	0.0	0.0
Fungivorous microarthropods						
increases	0.0	0.33	0.0	0.0	0.33	0.0
decreases	0.0	0.0	0.0	0.0	0.0	0.0
Fungivorous nematodes						
increases	0.0	0.0	0.34	0.0	0.0	0.0
decreases	0.0	0.0	0.0	0.0	0.0	0.54
Predators						
increases	0.0	0.0	0.0	0.0	0.0	0.0
decreases	0.0	0.0	0.0	0.0	0.0	0.0

Table 2. Parameter values for the food web model. Values in parentheses apply to all six data sets. A/C is the assimilation to consumption ratio, and P/A is production to assimilation

Transfer Parameter		E1			E2		
		Fallow	Wheat	Prairie	Fallow	Wheat	Prairie
Substrate to bacteria							
k2, eq. 2		9.2	9.2	9.2	27.0	27.0	27.0
yield	(0.23)						
Bacteria to protozoa							
k1, eq. 1		0.079	0.079	0.079	0.38	0.38	0.38
A/C	(0.5)						
P/A	(0.5)						
Protozoa death							
k3, eq. 3	(1.0)						
Substrate to fungi							
k2, eq. 2		10.1	10.1	4.4	10.1	10.1	4.4
yield	(0.23)						
Fungi to microarthropods							
k1, eq. 1		0.91	6.8	0.67	2.4	4.4	0.15
A/C	(0.3)						
P/A	(0.4)						
Fungi to nematodes							
k1, eq. 1		3.3	4.8	2.1	1.9	1.1	0.36
A/C	(0.3)						
P/A	(0.4)						
Fungal death							
k3, eq. 3		2.8	3.3	1.6	1.7	1.8	0.28
Microarthropods to predators							
k1, eq. 1	(1.3)						
A/C	(0.8)						
P/A	(0.4)						
Nematodes to predators							
k1, eq. 1	(4.9)						
A/C	(0.3)						
P/A	(0.4)						
Predator death							
k3, eq. 3	(0.22)						

Table 3. Transfer rates ($\mu g\,g^{-1}\,d^{-1}$) predicted by the model using parameter values in Table 2

Transfer	E1			E2		
	Fallow	Wheat	Prairie	Fallow	Wheat	Prairie
Substrate to bacteria	56.1	56.1	56.1	58.6	58.6	58.6
Bacteria to protozoa	11.1	11.1	11.1	13.4	13.4	13.4
Protozoan death*	2.5	2.5	2.5	2.4	2.4	2.4
Substrate to fungi	80.0	89.1	80.5	82.0	84.0	82.3
Fungi to microarthropods	0.049	0.32	0.75	0.085	0.31	0.79
Fungi to nematodes	0.63	0.31	2.00	0.14	0.048	1.70
Fungal death*	22.5	28.1	28.6	13.5	14.7	5.10
Microarthropods to predators	0.0052	0.0039	0.068	0.0052	0.0039	0.068
Nematodes to predators	0.060	0.016	0.20	0.041	0.0097	0.22
Predator death*	0.0036	0.0020	0.015	0.0036	0.0020	0.015

* Non-predatory death

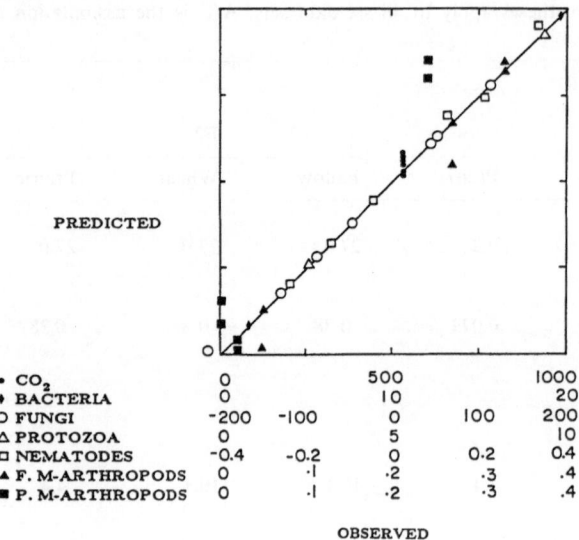

• CO₂	0		500		1000
♦ BACTERIA	0		10		20
○ FUNGI	-200	-100	0	100	200
△ PROTOZOA	0		5		10
□ NEMATODES	-0.4	-0.2	0	0.2	0.4
▲ F. M-ARTHROPODS	0	·1	·2	·3	·4
■ P. M-ARTHROPODS	0	·1	·2	·3	·4

OBSERVED

Fig. 7. Agreement between predicted and observed net changes in biomass ($\mu g\,g^{-1}$) and CO_2 evolution ($\mu g\,CO_2\text{-}C\,g^{-1}$). For each variable, the scales are the same for predicted and observed. The last two groups are fungivorous microarthropods and predaceous microarthropods.

dryland ecosystems, and the role of fauna in regulating microbial processes.

Average biomass levels in the sod compare fairly well with those reported by Hunt *et al.* (1987) for a shortgrass steppe site (CPER) in the same region, with several exceptions. The ratios of biomass in the present study to that at the CPER are as follows: bacteria 0.13, fungi 2.0, protozoa 1.6, nematodes 0.47, microarthropods 1.24. The distribution of trophic categories within the nematodes at Sidney (24% bacterial feeders, 8% fungal feeders, 59% root feeders, and 9% omnivore/predators) is very different from that at the CPER (76%, 5%, 3%, and 14%, respectively). The dominance of protozoa over bacterial feeding nematodes at Sidney is consistent with observations in pine forest (Clarholm, 1981), but is the reverse of previous findings at the CPER. The CPER is a more arid site than Sideny, and has a lower level of organic matter and less well developed soil structure. Furthermore, the CPER data are season-long averages, whereas the data from Sidney are for spring only. Thus there are several possible reasons for the differences between studies.

There were dramatic differences between experiments E1 and E2 in bacterial, fungal, and protozoan dynamics, despite the fact that the same per-

turbation was applied to two sets of cylinders taken from the field at the same time. In contrast, there was little difference in the faunal functional groups, although some individual taxa within the groups differed between experiments. The longer growth period and more severe drying in E2 could have resulted in a larger pool of labile plant organic matter than in E1, and this pool would support the more rapid growth of both bacteria and fungi observed the first two days after wetting (Fig. 3–5). However, this explanation fails to account for the similar bacterial and fungal dynamics in the fallow and in the two vegetated treatments. An accumulation of dead microbes and microbial products as the soil dried and a release from viable microbes upon wetting (Kieft *et al.*, 1987) could have occurred in fallow as well as wheat and sod treatments, and this would account both for the lower microbial biomass at the start of E2 and for the more rapid growth after wetting. It is surprising that relatively minor differences in conditions in the two experiments led to such different results. Pulse-dynamics after a wetting event may hold the key to understanding interactions between components of the detrital food web, but more work is necessary to characterize the variety of possible dynamics. Processes occurring during the dry periods, such as death of organisms and continued activity of Acarina must be accounted for to understand season-long dynamics.

The failure of the model to account for the data with the same parameter values across data sets (Table 2) is not surprising. Different temperature and water levels in the two experiments, and differences among fallow, wheat and sod in soil physical characteristics and substrate supply must affect the specific rates of organism activity. A more mechanistic model including effects of environment on organism activity might be useful for interpreting the differences among data sets.

The estimates of transfer rates through the food web based on the model (Table 3) are greater than indicated by direct examination of the data (Table 1) because the model allows for prey productivity to be passed continuously along to predators, without necessarily appearing as prey biomass. In spite of this model property, fungivorous nematodes and microarthropods accounted for only 8% of total losses from fungi (range of 2–33%, calculated from Table 3). This suggests that the majority of fungal turnover is related to normal evacuation of cytoplasm-filled hyphae (Paustian and Schnürer, 1987), to death from heat or water stress, or possibly to damage inflicted to hyphae during feeding by fungivores, rather than to consumption *per se*. One can make the same case without relying on the model results, by noting that the maximal ecological growth efficiency (production/consumption) for fungivores, based on maximal production/biomass ratios (above) and minimal losses from fungi (Table 1), average only 0.033 in E1. This number is well below most of the values for P/C of soil fauna listed by Luxton (1982), and therefore fungivore consumption cannot account for all fungal production under the conditions of our experiment. Little is known about mortality of fungi damaged but not consumed by fungivores. Information is needed on the rate of disappearance of killed and evacuated hyphae through decomposition. In contrast to the situation with fungivores, protozoans easily have the capacity to account for bacterial production.

Our method for estimating transfer rates could be profitably applied to the analysis of data from other detrital food webs (*e.g.*, Parker *et al.*, 1984; Steinberger *et al.*, 1984), or from any system in which it is difficult to make direct measurements of feeding rates. The approach requires good information on CO_2 evolution and biomass dynamics. It would be preferable to measure CO_2 evolution directly instead of estimating it indirectly as done in the present study. The protozoan and bacterial dynamics reported here are given in units of biomass, but actually we determined numbers and estimated biomass using average individual sizes from the literature. Since individual sizes may vary (Clarholm, 1981; Clarholm and Rosswall, 1980), our estimated biomass dynamics could be misleading. There is a great need for practical methods for estimating individual size of protozoans in the soil.

According to our method of analysis, population changes are inadequate as indicators of secondary production. For example, fungal production in the fallow treatment of E1, based on observed population increases (Table 1), is zero, while our estimate is 18.4 $\mu g\,g^{-1}d^{-1}$ [80.0 (uptake by fungi, Table 3) multiplied by 0.23 (yield, Table 2)]. To test this kind of prediction, independent information is needed about rate of production of populations. For example, the egg-ratio method, based on the

fraction of gravid females, has been used to estimate production in zooplankton (Threlkeld, 1979). Counts of egg-carrying females can be collected for microarthropods (Luxton, 1981) and nematodes (Sohlenius and Wasilewska, 1984), but apparently this kind of information has never been used to estimate production of soil fauna.

Besides reproductive status and population age distributions, the physiological condition of individuals may give an indication of activity and growth rate. Faster growing bacteria usually have larger cell sizes, and cell size distribution of field populations can be determined (Clarholm and Rosswall, 1980). The fraction of cryptobiotic individuals is an obvious indicator of activity which can be determined for protozoa (Ingham *et al.*, 1986; Parker *et al.*, 1984) and nematodes (Whitford *et al.*, 1981). Some of these indices of activity are strictly appropriate only for species, but they could prove useful for detecting large differences between functional groups in rate of production.

Acknowledgements

D. C. Coleman participated in planning the research. D. W. Freckman advised on methods for nematodes, and J. C. Moore and P. Chapman advised on the statistical design. R. E. Ingham, E. R. Ingham, J. C. Moore, and J. Heil aided in the early stages of data collection. E. Davis, D. Reuss, M. McMillan, and B. Hudgens did most of the laboratory analyses, and M. Fowler was responsible for data management. J. C. Moore reviewed the manuscript and made helpful suggestions regarding interpretation of results. Research supported by NSF Grant No. BSR-8418049.

References

Andrén O and Paustian K 1987 Barley straw decomposition in the field: A comparison of models. Ecology 68, 1190–1200.
Bååth E, Lohm U, Lundgren B, Rosswall T, Söderström B and Sohlenius B 1981 Impact of microbial-feeding animals on total soil activity and nitrogen dynamics: A soil microcosm experiment. Oikos 37, 257–264.
Babiuk L A and Paul E A 1970 The use of fluorescein isothiocyanate in the determination of bacterial biomass in grassland soil. Can. J. Microbiol. 16, 57–62.
Clarholm M 1981 Protozoan grazing of bacteria in soil: Impact and importance. Microb. Ecol. 7, 343–350.

Clarholm M, Popovic B, Rosswall T, Söderström B, Sohlenius B, Staff H, and Wiren A 1981 Biological aspects of nitrogen mineralization in humus from a pine forest podsol incubated under different moisture and temperature conditions. Oikos 37, 137–145.

Clarholm M and Rosswall T 1980 Biomass and turnover of bacteria in a forest soil and a peat. Soil Biol. Biochem. 12, 49–57.

Cole C V, Elliott E T, Hunt H W and Coleman D C 1978 Trophic interactions in soil as they affect energy and nutrient dynamics. V. Phosphorus transformations. Microb. Ecol. 4, 381–387.

Coleman D C 1985 Through a ped darkly: An ecological assessment of root-soil-microbial-faunal interactions. *In* Ecological Interactions in Soil. Ed. A H Fitter. Blackwell Scientific Publications, Oxford, England.

Dixon W J 1983 BMDP Statistical Software, 1985 Printing. Univerity of California Press, Los Angeles. 734 p.

Edwards C A 1967 Relationships between weights, volumes and numbers of soil animals. *In* Progress in Soil Biology. Eds. O Graff and J E Satchell. pp 585–594. North-Holland Publishing, Amsterdam, The Netherlands.

Elliott E T, Horton K, Moore J C, Coleman D C and Cole C V 1984 Mineralization dynamics in fallow dryland wheat plots, Colorado. Plant and Soil 76, 149–155.

Fenster C R and Peterson G A 1979 Effects of no-tillage as compared to conventional tillage in a wheat-fallow system. Research Bulletin 289. Agricultural Experiment Station. University of Nebraska-Lincoln.

Freckman D W, Duncan D A and Larson J R 1979 Nematode density and biomass in an annual grassland ecosystem. J. Range Manage. 32, 418–422.

Freckman D W, Hunt H W, Ingham R E and Fowler M. A comparison of Baermann and sugar floatation methods for quantitative estimates of nematodes in soil. (in prep.).

Frey J S, McClellan J F, Ingham E R and Coleman D C 1985 Filter-out-grazers (FOG): A filtration experiment for separating protozoan grazers in soil. Biol. Fertil. Soils 1, 73–79.

Giesy J P Jr. Ed. 1980 Microcosms in Ecological Research. U.S. Department of Energy. Springfield, Virginia, 1110 p.

Heal O W and MacLean Jr. S F 1975 Comparative productivity in ecosystems – secondary productivity. *In* Report of the Plenary Sessions of the First International Congress of Ecology. Eds. W H van Dobben and R H Lowe-McConnell. pp 89–198. Dr W. Junk, The Hague.

Hobbie J E, Daley R J and Jasper S 1977 Use of nuclepore filters for counting bacteria by fluorescence microscopy. Appl. Environ. Microbiol. 33, 1225–1228.

Hunt H W 1977 A simulation model for decomposition in grasslands. Ecology 58, 469–484.

Hunt H W, Cole C V and Elliott E T 1985 Models for growth of bacteria inoculated into sterilized soil. Soil Sci. 139, 156–165.

Hunt H W, Coleman D C, Cole C V, Ingham R E, Elliott E T and Woods L E 1984 Simulation model of a food web with bacteria, amoebae, and nematodes in soil. *In* Current Perspectives in Microbial Ecology. Eds. M J Klug and C A Reddy. American Society of Microbiologists, Washington, DC.

Hunt H W, Coleman D C, Ingham E R, Ingham R E, Elliott E

T, Moore J C, Rose S L, Reid C P P and Morley C R 1987 The detrital food web in shortgrass prairie. Biol. Fertil. Soils 3, 57–68.

Ingham E R and Klein D A 1984 Soil Fungi: Measurement of hyphal length. Soil Biol. Biochem. 16, 279–280.

Ingham E R, Trofymow J A, Ames R N, Hunt H W, Morley C R, Moore J C and Coleman D C 1986 Trophic interactions and nitrogen cycling in a semi-arid grassland soil. I. Seasonal dynamics of the natural populations, their interactions and effects on nitrogen cycling. J. Appl. Ecol. 23, 597–614.

Kieft T L, Soroker E and Firestone M K 1987 Microbial response to a rapid increase in water potential when dry soil is wetted. Soil Biol. Biochem. 19, 119–126.

Leetham J W 1975 A summary of field collecting and laboratory processing equipment and procedures for sampling arthropods at Pawnee Site. Grassland Biome, US IBP Tech. Rep. No. 284. Colorado State Univ., Ft. Collins.

Luxton M 1981 Studies on the oribatid mites of a Danish beech wood soil. IV. Developmental biology. Pedobiologia 21, 312–340.

Luxton M 1982 Quantitative utilization of energy by the soil fauna. Oikos 39, 342–354.

Moore J C, Walter D E and Hunt H W 1988 Arthropod regulation of micro- and mesobiota in below-ground detrital food webs. Annu. Rev. Entomol. 33, 419–439.

Parker L W, Freckman D W, Steinberger Y, Driggers L and Whitford W G 1984 Effects of simulated rainfall and litter quantities on desert soil biota: Soil respiration, microflora, and protozoa. Pedobiologia 27, 185–195.

Paustian K and Schnürer J 1987 Fungal growth-response to carbon and nitrogen limitation: A theoretical-model. Soil Biol. Biochem. 19, 613–620.

Powell M J D 1965 A method for minimizing a sum of squares of nonlinear functions without calculating derivatives. Comput. J. 7, 303–307.

Rosswall T, Lohm U and Sohlenius B 1977 Development d'un microcosme pour l'etude de la mineralization et de l'absorption radiculaire de l'azote dans l'humus d'une foret de coniferes (*Pinus sylvestris* L.). Lejeunia 84, 1–15.

Schnürer J, Clarholm M, Boström S and Rosswall T 1986 Effects of moisture on soil microorganisms and nematodes: A field experiment. Microb. Ecol. 12, 217–230.

Singh B N 1946 A method of estimating the numbers of soil protozoa, especially amoebae, based on their differential feeding on bacteria. Ann. Appl. Biol. 33, 112–119.

Smith F M 1972 Growth response of blue grama to thunderstorm rainfall. US IBP Tech. Rep. No. 157. p. 31. Colorado State Univ., Ft. Collins.

Sohlenius B and Wasilewska L 1984 Influence of irrigation and fertilization on the nematode community in a Swedish pine forest soil. J. Appl. Ecol. 21, 327–342.

Steinberger Y, Freckman D W, Parker L W and Whitford W G 1984 Effects of simulated rainfall and litter quantities on desert soil biota: Nematodes and microarthropods. Pedobiologia 26, 267–274.

Threlkeld S T 1979 Estimating cladoceran birth rates: The importance of egg mortality and the egg age distribution. Limnol. Oceanogr. 24, 601–612.

van Veen J A and Paul E A 1979 Conversion of biovolume

measurements of soil organisms, grown under various moisture tensions, to biomass and their nutrient content. Appl. Environ. Microbiol. 37, 686–692.

Walter D E 1987 Trophic behavior of "mycophagous" microarthropods. Ecology 68, 226–229.

Walter D E, Hudgens R A and Freckman D W 1986 Consumption of nematodes by fungivorous mites, *Tyrophagus* spp. (Acarina: Astigmata: Acaridae). Oecologia 70, 357–361.

Walter D E, Hunt H W and Elliott E T 1987 The influence of prey type on the reproduction and development of predatory soil mites. Pedobiologia 30, 419–424.

Walter D E, Kethley J and Moore J C 1987 A heptane flotation method for recovering microarthropods from semiarid soils, with comparisons to the Merchant-Crossley high-gradient extraction method and estimates of microarthropod biomass. Pedobiologia 30, 221–232.

Walter D E, Hunt H W and Elliott E T 1988 Guilds or functional groups? An analysis of predatory arthropods from a short-grass steppe soil. Pedobiologia 31, 247–260.

Whitford W G, Freckman D W, Elkins N Z, Parker L W, Parmalee R, Phillips J and Tucker S 1981 Diurnal migration and responses to simulated rainfall in desert soil microarthropods and nematodes. Soil Biol. Biochem. 13, 417–425.

M. Clarholm and L. Bergström (Eds.), Ecology of arable land, 205–216.

Can population and process ecology be combined to understand nutrient cycling?

O. W. HEAL[1], T. V. CALLAGHAN[2] and K. CHAPMAN[2]
*Institute of Terrestrial Ecology, [1]Bush Estate, Penicuik, Midlothian EH 24 0QB, UK and [2]Merlewood
Research Station, Grange-over-Sands, Cumbria LA11 6JU, UK*

Key words: decomposition, growth strategies, nutrient cycling, population ecology, process ecology

Abstract

Research on nutrient dynamics concentrates on direct or indirect measurement of processes with little
definition of populations involved, mainly because of technical difficulties and lack of interdisciplinary
understanding. As the young science of nutrient dynamics evolves and management of species and systems
becomes more sophisticated, there is a need for the science to draw on both population and process ecology.

How? The paper explores three ways in which knowledge of the two disciplines can be combined: First,
how the growth strategies of three perennial plants exploit and modify the spatial distribution of nutrients.
Second, a population model of *Carex bigelowii* is used to calculate nutrient acquisition, retention and release,
illustrating the potentially wider application of standard population techniques. Third, studies on litter
decomposition, including interactions between litter types, show that understanding of processes is incomplete and potentially misleading without analysis of population responses.

Introduction

Analysis of the transfer and transformation of
nutrients within ecosystems, subsystems and populations is a critical aspect of ecology. Essentially
nutrients control the growth of organisms, and the
dynamics and composition of communities. Control of nutrient circulation is also one of the main
aims in the management of crops and systems.
Lack of control results in sub-optimal production
and in pollution, particularly through disruption of
circulation within a terrestrial population (or
system) and loss to freshwater systems or through
infertility due to the sequestration of nutrients by
long-lived plants.

Approaches to the study of nutrient cycling are
many and varied—the description of the amount
distributed in components of an ecosystem, with
transfers estimated from known population
characteristics; measurement of plant or microbial
uptake (with or without experimental manipulation); chemical analysis of nutrient fractions; direct
measurement of transfers between spatially seg-
regated components, *e.g.* lysimetry; transformation
by microflora; and occasionally the direct analysis
of movement of labelled nutrients within the
system. These are readily recognised techniques
and show a trend of emphasis from population to
process methods. Thus research in nutrient cycling
has used techniques which directly or indirectly
measure the processes. Yet despite the fact that the
dynamics of nutrients are largely mediated by
organism populations, models of nutrient circulation usually omit or have minimal representation of
organisms (Fig. 1). Species variations and, more
importantly, the principles and theories of population regulation or succession are rarely expressed.

Why is population ecology so poorly reflected in
nutrient cycling research? There seem to be three
main limitations: (1) The key area of nutrient
control is in the soil and it is here that population
ecology (microbial, faunal and root) is weakest
because of the complexity of the soil system and the
lack of discriminatory techniques. (2) Most population ecologists are not interested in the resulting
processes. Conversely those concerned with

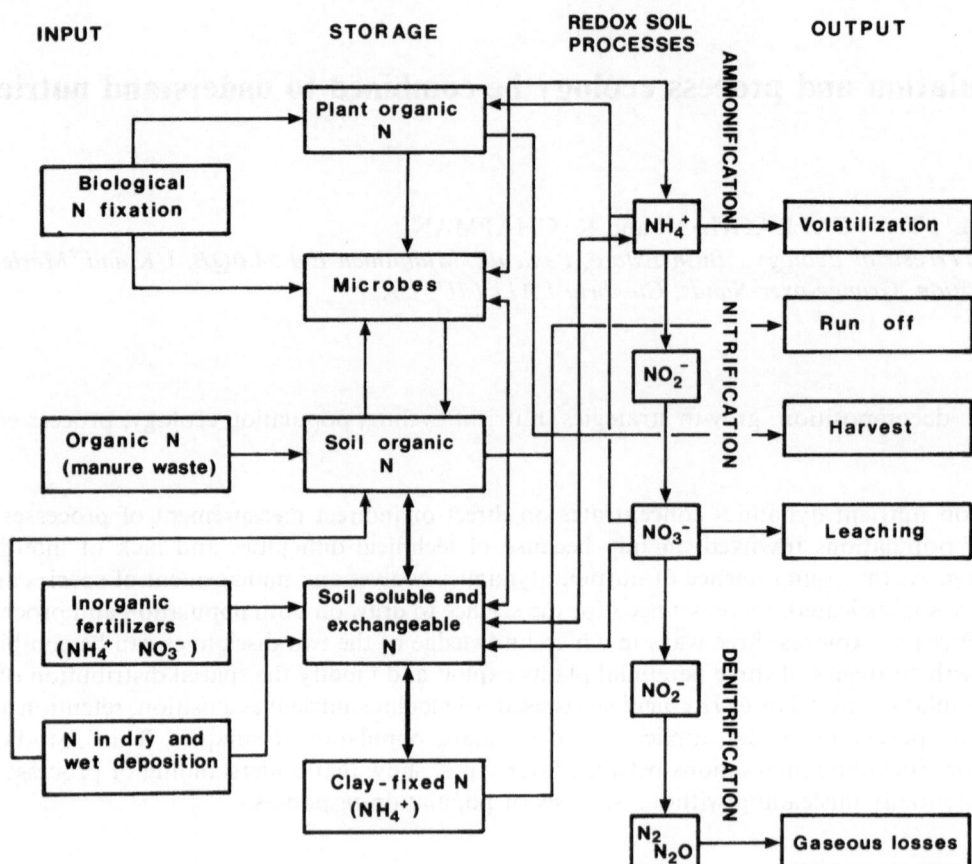

INPUT | STORAGE | REDOX SOIL PROCESSES | OUTPUT

Fig. 1. The basic inputs, storages, transfers and outputs of nitrogen in terrestrial ecosystems (from Bolin and Arrhenius, 1977).

nutrients are not usually conversant in population ecology, *i.e.* there is segregation of disciplines in precisely the area in which the subject requires integration. (3) Nutrient cycling research at the ecosystem level is in its infancy, a descriptive approach has dominated initially. Current and future questions in nutrient dynamics will require scientists to overcome these three limitations. For example intensive agriculture and forestry have simplified soil faunal and microbial populations and the need to introduce or reintroduce organisms is recognised (Anderson *et al.*, 1985; Huhta, 1979; Springett, 1985; Swift, 1984). An increasing range of plant and animal species and of mixtures of species are being used as crops, the selection of species requiring an understanding of nutrient requirements and interaction between individuals and species (Swift, 1984). Basic research has clearly shown that phases in population dynamics and interactions between species result in accelerated or retarded nutrient release and the search for general principles is beginning to appear (Anderson *et al.*, 1985; Coleman 1985). Thus there is a need and an opportunity to relate population ecology to the processes of nutrient cycling.

How? That is not so easy, and there are no quick solutions. In the following sections we simply explore a few inter-related ideas and examples of (1) how organism characteristics can control the circulation of nutrients within ecosystems (2) how population techniques may help in analysing nutrient processes (3) how process studies have overlooked the capacity of organisms to interact and thus modify nutrient processes.

Organism characteristics and the control of nutrient dynamics

A conventional approach to the analysis of nutrient dynamics is represented by the flow

Table 1. The response of plants to variation in the supply of nutrient resources — a general framework for analysis

Resource characteristics	Organism requirements		
	Acquisition	Retention and release	Quality of transformed resource
Temporal availability			
Short	Opportunistic, with fast growth and fibrous roots	Temporary sequestration and pulsed release by plants with periodic growth, annuals and biannuals	Good
Long	Competitive, with large, persistent roots, *e.g.* tap roots	Long term sequestration in perennial but steady and profligate release, *e.g.* deciduous habit	Good
Spatial availability			
Heterogeneous	Foraging rhizomes and stolons with many small root systems along axis	Indefinite growth with subsidised growing points, movement of nutrients and steady but patchy release	Poor
Homogeneous	Long-lived, large spreading roots	Non-clonal plants with no transfer of nutrients between generations: mainly trees	Intermediate
Quantity			
Small	Large weight allocation to roots; high uptake efficiency	Slow growing, low tissue concentrations in impoverished communities	Poor
Large	Small roots with low uptake efficiencies	Fast growing, high tissue concentrations, profligate release in species rich communities	Good

diagram (Fig. 1) which can be translated into a mathematical model. The focus of attention is the nutrient element. A rarely used alternative approach is to consider the organisms and examine how they respond to and control the nutrient resource. Basically (1) Nutrient resources and their availability to organisms vary in time, space, quantity and quality. (2) Organisms respond to variation in the resource by adopting different strategies in the acquisition, retention and utilisation of the nutrient resource. The selection concept of strategies is equally applicable to plants, animals and microbes, although it is recognised that nutrients are only one of the environmental factors to which the organisms respond. (3) The rate at which nutrients are transformed or transferred within a system is the product of 1 or 2, *i.e.* the net effect which is measured in process studies.

The characteristics of both the nutrient resource and the organism can be constructed as an interaction matrix (Table 1). The presentation is neither rigorous nor exhaustive; it is simply a checklist to focus attention on the dimensions of the problem. It is apparent that a range of strategies may be adopted to suit a particular nutrient supply; a concept which relates closely to the general considerations of Grime (1979) who stated that plants exhibited three primary strategies, each to varying degrees, associated with tolerance to competition, stress and disturbance. The approach is explored further by examining the strategies adopted by three plant species and the consequences for nutrient circulation.

Cannibalistic habit

Eriophorum vaginatum thrives in oligotrophic peat environments where large quantities of nutrients are stored in the peat but remain largely unavailable because of low decomposition rates. The nutrient resource is spatially homogenous relative to the size of the plant and temporal variation is determined mainly by the seasonal temperature fluctuations of northern latitudes. *E. vaginatum* has adopted a 'cannibalistic' strategy. It colonises its habitat by the growth of seedlings, but subsequent growth is vertical, rather than lateral. The production of intravaginal tillers with little rhizome extension results in the formation of tussocks which may survive for over 100 years (Mark *et al.*, 1985).

Tillers are only short lived. Death and decomposition of the leaves, leaf bases and roots of the older tiller generations occurs at the base of the tussock as it slowly expands. Consequently, tussock plants have twice as deep an organic horizon as plants rooted between tussocks (Chapin *et al.*, 1979). Concentrations of N, P and K are much higher in this organic soil than in the soil below the tussock and, at the end of the growing season, 75% of the *E. vaginatum* roots of younger tiller generations are found in the older and elevated portion of the tussock (Chapin *et al.*, 1979). Root density (9 cm root per cubic cm of soil) is sufficiently high for the roots to exploit the whole of the tussock volume each year.

Thus the strategy by which a tussock develops over a century is based on nutrient conservation accomplished by high internal nutrient transfer from older to younger generations of tillers (Chapin *et al.*, 1979; Defoliart *et al.*, 1988; Smith and Forest, 1978). The tussock growth form also ameliorates the microclimate of *E. vaginatum* tillers and the tussock soils. These soils are 6–8°C warmer than soils between tussocks and experience a 5–10% longer growing season (Chapin *et al.*, 1979).

The consequences for nutrient dynamics of the cannibalistic strategy and tussock growth form are (a) increasing spatial heterogeneity within the system as nutrients are concentrated by the closed nutrient cycle within the slowly expanding tussock, and (b) a smoothing of the temporal availability of nutrients held within a tussock. The strategy is, however, poorly adapted to respond to marked temporal variation in nutrient supply, such as pulses of nutrient availability at snow melt, and to disturbance. The tussock may be regarded as a stable long-lived microcosm and other growth forms, such as dwarf shrubs, cushion plants and tropical trees possess analogous strategies.

Nutrient foraging habit

Rhizomatous and stoloniferous plants adopt nutrient foraging habits of two basic types. In one, nutrient-rich patches are located but not exhaustively exploited as the plant continues to grow out of them. Weak competitors, such as *Lycopodium annotinum* with its small dichotomously branched roots, belong to this type, the mobility of which seems to be related to escape from competitors rather than a requirement to utilise nutrient-rich patches. The second group, contains plants which both locate and exploit patches of resources, *e.g. Glechoma hederacea* (Slade and Hutchings, 1987) and *Carex bigelowii* (Kershaw, 1962).

Lycopodium annotinum. In the spatially and temporally heterogeneous environment of stoney tundra, *L. annotinum* adopts a foraging strategy with an extensive, deterministic growth pattern which allows the retranslocation of nutrients from older senescing tissues to young growing points. Retranslocation is particularly effective with approximately 64% of phosphorus, 90% of potassium and 63% of nitrogen recycled (Callaghan 1980). Because of this effective nutrient conservatism, *L. annotinum* can survive with less than 5% of its dry weight invested in roots.

The rhizomatous and stoloniferous habits enable individuals to explore the heterogeneous environment, by subsidising growing points when they are in unsafe microsites (Callaghan, 1980; Callaghan, 1988; Headley *et al.*, 1988a, b), and by capitalising on those which find pockets of nutrients by increasing productivity. Such a strategy results in the spatial redistribution of nutrients, with growth in one place dependent upon nutrient uptake elsewhere.

Even the slow growing horizontal branches or 'stolons' of *L. annotinum* have an annual extension growth of 63 mm and its youngest annual increments contain means of 0.15, 0.05 and 0.36 mg of N, P and K respectively (compared with 0.27, 0.048 and 0.22 mg of N, P and K respectively in 15 year

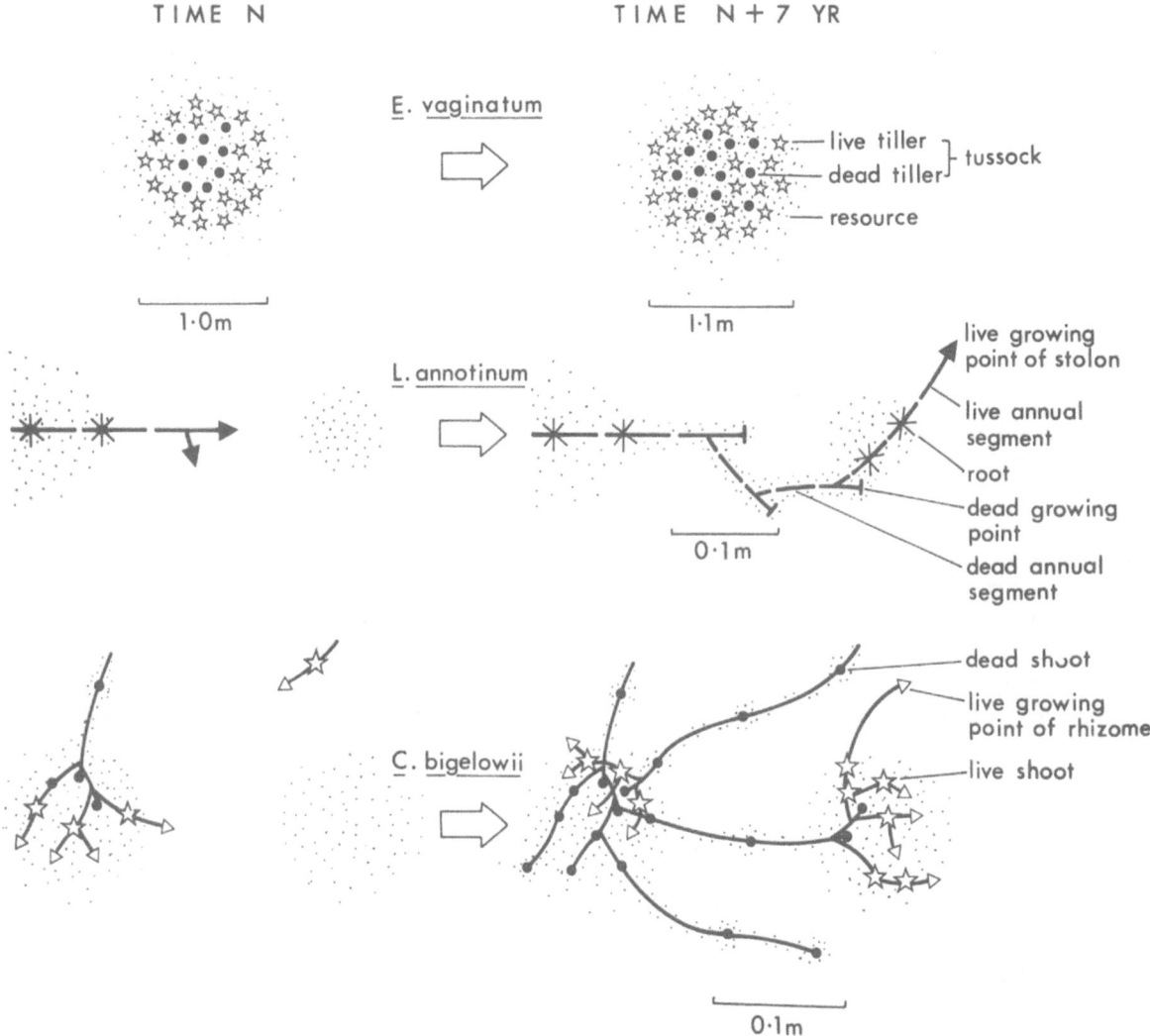

TIME N   TIME N + 7 YR

E. vaginatum

— live tiller
— dead tiller ⎤ tussock
— resource

1·0m   1·1m

L. annotinum

live growing
point of stolon

live annual
segment

root

dead growing
point

dead annual
segment

0·1m

C. bigelowii

— dead shoot
— live growing
point of rhizome
— live shoot

0·1m

Fig. 2. Change in spatial distribution of nutrients over time with 3 plant species of different growth strategies: *Eriophorum vaginatum* — cannibalistic (based on Fetcher and Shaver, 1982); *Lycopodium annotinum* — deterministic foraging (based on Callaghan *et al.*, 1986); *Carex bigelowii* — opportunistic foraging (based on Kershaw, 1962). Hatching indicates nutrient concentration.

old segments). Assuming that all the N, P and K in an annual segment of the stolon are taken up by its root, then after 5 years 10, 16 and 58% of the original N, P and K will have been moved 31 cm to the most recent annual increment, even though subsequent roots may have been produced.

Headley *et al.* (1989a) calculated that the maximum distance between a root and a growing point of a stolon of *L. annotinum* was 1 m and this represents at least 16 year's growth. In such a situation, 5.12 mg N, 1.22 mg P and 6.19 mg K would be moved away from the point of uptake while

0.15 mg N, 0.05 mg P and 0.36 mg K would be moved 1 m away over the 16 year period. These are likely to be overestimates because annual increments do not decrease as the frequency of rooting decreases and the age-specific nutrient content remains stable as rooting frequency decreases. However, the example illustrates (i) the extent of nutrient movement in a very slow growing species, (ii) the extent of retranslocation, and (iii) the long periods over which the nutrients may be sequestered. If rooting is impossible for 1 m, the growing point of that stolon of *L. annotinum* will die. Upon decomposition the nutrients contained in this axis will then become available for new individuals foraging in the same direction and more unfavourable ground may be traversed; this 'over-topping' is an important mechanism in primary colonisation of inhospitable surfaces such as rocks (Callaghan *et al.*, 1986; Svensson and Callaghan, 1988).

Carex bigelowii. With its capacity to tiller, and its rhizomatous growth form, *C. bigelowii* has a foraging strategy which is opportunistic rather than deterministic as with *L. annotinum*. Exploratory tillers of *C. bigelowii* invade space in a pattern controlled by branching angles, amount of extension growth, and numbers and positions of new tiller generations produced. When a pocket of nutrients is located, the tillers proliferate, increase their productivity and form 'nodes' in rhizome systems (Kershaw, 1962). Subsequent development of tillers may be relatively independent from their parents. This foraging results in much less translocation than in *L. annotinum* and it tends to increase spatial heterogeneity. Opportunistic species such as *C. bigelowii* possess an additional advantage over determinstic species: if young tiller generations die in nutrient poor areas, or through disturbance, reserve buds on old, senescing tillers which normally die, break dormancy and draw directly on the nutrients of the senescing structures (Jonsdottir and Callaghan, 1988).

The extent to which the growth characteristics and nutrient contents of these three species influence nutrient dynamics and spatial pattern is indicated in Fig. 2. The species illustrate how the growth characteristics of perennials have evolved to exploit the limited nutrient resource of their environment. They emphasise the capacity of plants to exploit environmental heterogeneity and to reduce (*Lycopodium, Carex*) or enhance (*Eriophorum*) that heterogeneity. All three species appear to have developed methods of retaining a high proportion of the nutrients which they acquire and are selected for particular niches. The mechanisms of separation of the nutrient resource which is captured by different species within a community are varied, *e.g. Vaccinium vitis-idaea* and *V. uliginosum* have complementary rooting depths and possibly time of uptake (Karlsson 1982); *Arrhenatherum* has a short period of rapid nutrient uptake complementing the longer period of slow uptake by *Festuca* (Rorison *et al.*, 1983). Thus there is a tendency for individual species to exploit spatial and temporal heterogeneity, but the extent to which they differentially exploit individual elements or states of elements is largely unknown — with the exception of nitrate and ammonium utilising species, but these are usually from different sites and little is known about differential utilisation within a site.

The illustrations for plant species have parallels in faunal and microbial population ecology for example in the concepts of optimal foraging strategies, niche separation and competitive saprophytic ability. A few studies relate nutrient dynamics to organism populations, *e.g.* the calculation of throughput of nitrogen in *Lumbricus terrestris* (Satchell, 1963), or the measurements of net nutrient release through fauna-microflora interactions (Anderson *et al.*, 1985; Clarholm, 1985; Coleman, 1985) or estimation of nutrient flux by calculations based on microbial population fluctuations (Ausmus *et al.*, 1976). These types of study have rarely examined experimentally or theoretically, the nutrient consequences or nutrient efficiencies of different strategies, *e.g.* the effect of the territorial burrowing *Lumbricus terrestris* as distinct from the foraging *Lumbricus castanaea*, or of different termite species. In microbial ecology the influence of different growth characteristics on nutrient dynamics is increasingly recognised, *e.g.* Swift (1984), Heal and Ineson (1984), based on the extensive literature of physiology, resource use and microbial succession. The relevance to process ecology is in understanding the effects of management practices which selectively or non selectively remove elements of the fauna or microflora; in the selection of complementary plant species for mixed cropping systems and in improving the synchrony of nutrient dynamics in production and decomposition subsystems (Anderson and Swift, 1983). It is possible that the adoption of different nutrient acquisition and retention strategies results in the higher and more consistent nutrient content of fauna and bacteria compared to higher plants and fungi. This is an open question.

The logical consequence of adopting a strategy which conserves nutrients within the individual or population is a tendency towards high nutrient use efficiency, *i.e.* higher dry matter production per unit nutrient assimilated (Chapin and Tryon, 1983; Vitousek and Reiners, 1975) and towards closed nutrient cycles (Herrera *et al.*, 1978). The closed nutrient cycling hypothesis in tropical forests results from nutrient limitation in an environment in which light, temperature and moisture tend to be non-limiting. In contrast in taiga and tundra it is possible that comparable nutrient conservatism, including the cushion or tussock habit, is in response to limited growth due to low temperatures and short growing seasons. In this case nutrients are conserved not because of an absolute limitation but because of the time or energy penalty which is incurred in their uptake.

Population techniques in the analysis of nutrient turnover

A major advantage of methods employed by population biologists is the use of statistical tools to model discontinuous variables such as birth and death — and hence nutrient turnover. Population growth rate can be calculated from probability transition matrices, which contain age- or stage-specific probabilities of survival and fecundity, and vectors of age class distributions. Callaghan (1976) describes how the following matrix was constructed.

A population model of the sedge Carex bigelowii

A transition probability matrix for *Carex bigelowii*, a long-lived rhizomatous sedge with an opportunistic foraging strategy, is presented in Table 2. The first 4 rows and 4 columns contain probabilities related to easily recognisable tillers seen in the field, *i.e.*, they have rhizomes below ground and shoots with green leaves above ground (Callaghan, 1976). The remaining columns and rows relate to old tillers which have lost their shoots, but, surprisingly, remain alive and grow for some years. As almost no information is available on the subsequent pattern of survival and death of these rhizomes, each tiller which survives for 3 years, is assumed to survive for 7 more years (Assumption 1) in the following calculations. An alternative assumption could be a constant rate of mortality.

The values in the model are age-specific probabilities of survival and age-specific factors of fecundity. For example, the value 0.829 (Table 2) is the probability with which a 1 year old tiller will survive to become a 2 year old tiller and the value of 1 in column '3 years' and row '4 years' (Table 2) signifies that 100% of the 3 year old tillers survive to become 4 year old tillers. No 10 year old tillers survive as all the survival probabilities are zero. The value 1.39 in the first row of the model represents the number of new tillers formed by a one year old tiller.

To model the changes in the tiller populations, the matrix is multiplied by a vector of current age class distribution to generate a new one for the first transition. Subsequent distributions are predicted by iteratively multiplying the matrix by each successive vector (Callaghan, 1976). As no data are available for the age class distributions of the older, below ground, tillers, they are assumed to be absent at the start of the population transitions, and to accumulate with time.

Table 2. Transition probability matrix (M) of survival (mortality = 1-survival) and fecundity of *Carex bigelowii* and vector (V) of current age class distribution (tillers/m²). The inset square presents measured probabilities for tillers with green shoots while the remaining assumed probabilities are for old tillers remaining just as rhizomes

Age class (years)	M Age class (years)								V
	0	1	2	3	4⁺		10	11	
0	0.393	1.39	0.04	0	0	...	0	0	27.18
1	0.453	0	0	0	0	...	0	0	16.14
2	0	0.829	0	0	0	...	0	0	22.08
3	0	0	0.175	0	0	...	0	0	3.4
4	0	0	0	1	0	...	0	0	0
10	0	0	0	0	0	...	0	0	0
11	0	0	0	0	0	...	0	0	0

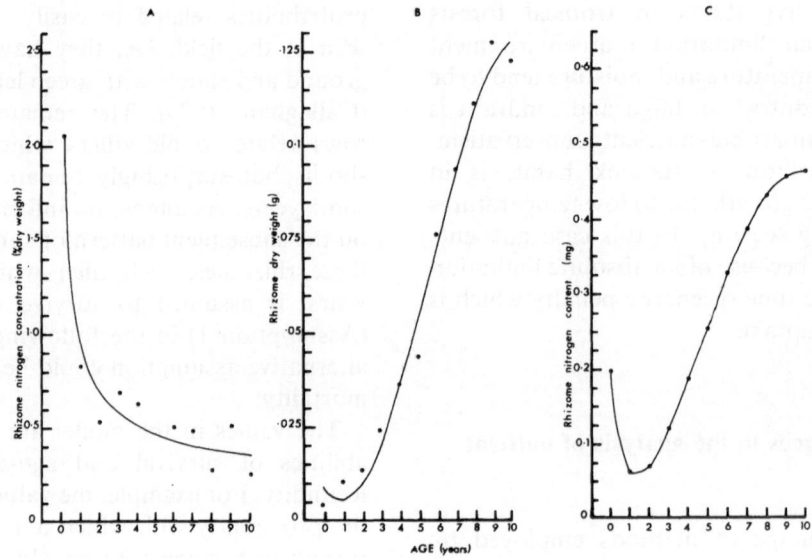

Fig. 3. Age-specific nitrogen concentrations in rhizomes of *Carex bigelowii* (A) age-specific dry weight of rhizomes (B) age-specific rhizome nitrogen content (C) rhizome nitrogen content calculated from A and B. Regression for A: $y = 0.8607X^{-0.381}$, B: $y = 0.0101 - 0.00892X + 0.00506X^2 - 0.000299X^3$ ($r^2 = 0.987$, $n = 11$).

Application of the model to nutrient dynamics

Nutrient flux through, and within, the population of *C. bigelowii* tillers can be calculated from the size of the tiller populations and the age-specific nutrient content of the tillers. From the age-specific nitrogen concentrations of rhizomes of *C. bigelowii* tillers (Fig. 3), inferences can be made about the quality of particular age classes of tiller as substrate for decomposers. In the example described below, nitrogen concentrations are only available for rhizomes and not whole tillers. The following is intended as a description of a methodological approach rather than a presentation of primary data.

The nitrogen content of rhizomes (Fig. 3C) can be calculated by multiplying the age-specific nitrogen concentrations (Fig. 3A) by the age-specific dry weight (Fig. 3B), each predicted from a regression equation to smooth out sample variability of the data presented in Callaghan (1976). It can be seen that the youngest age class has a relatively high N content (Fig. 3C), despite its small dry weight (Fig. 3B); because of its high N concentration (Fig. 3A).

Nitrogen uptake and retention per m² by tillers can be estimated by multiplying the age-specific N contents of rhizomes (Fig. 3C) by the numbers of tillers of given age classes predicted by the tran-

sition probability matrix (Table 2). Similarly, N loss can be estimated by multiplying age-specific N contents by the numbers of tiller dying in each age class. It is important to note that two heroic assumptions are made in addition to Assumption 1: *i.e.*, there is no difference in the N concentration of tillers dying and surviving in a particular age class (Assumption 2) and there is no retranslocation of N out of senescing tillers (Assumption 3). All three assumptions are unlikely to hold in practice, but quantitative evidence is lacking.

The model indicates that after a small initial decrease, there is a gradual increase, in the number of tillers in the population over a 10 year period (Table 3). Most of the tillers (just less than one half) are in the youngest age class. The amount of N held within the population of tillers over a 10 year period increases at a faster rate than the number of tillers (Table 3). Most of the N is initially locked up in the youngest age class and, subsequently, in the old rhizomes.

Even though there is an increase in the amount of N held in the population with time, considerable quantities of N are released each year through the death of tillers (Table 3). Indeed up to one half of the N held in living rhizomes is released to the decomposition cycle each year via tiller death. The amount of N lost is consistently greater (up to 20 times) than the extra N incorporated in living tillers

Table 3. Nitrogen balance of a population of *Carex bigelowii* tillers calculated over a 10 year period with values from Table 2 and Fig. 3

	Year 1	Year 10
Number of living tillers m^{-2}	68.8	87.3
Number of dead tillers m^{-2}	35.9	37.4
N content of living tillers mg m^{-2}	13.37	18.23
N content of dead tillers mg m^{-2}	6.73	6.28
N increment in living tillers mg m^{-2} y^{-1}	0.36	0.39
N loss in death of tillers mg m^{-2} y^{-1}	6.73	6.28
N turnover through tiller population mg m^{-2} y^{-1}	7.09	6.67

due to population increase (Table 3). The N released through tiller death is mainly in the youngest and oldest age classes. However, the N concentrations of the two age classes vary considerably, by up to 6 × (Fig. 3A). Thus, much of the nitrogen which is released is at high concentration and should be a good resource for decomposition. On the other hand, however, a substantial component of the released N will be in the form of large old rhizomes with low N concentrations (Fig. 3A). These will be poor resource, and as expected, are persistent in the field.

A critique of the method

Disadvantages 1. The 3 assumptions are all likely to be incorrect, and the N release is over-estimated, particularly as retranslocation is almost certain to occur, thereby reducing N concentrations within dying individuals within an age class relative to the survivors. 2. The example has not presented data on organs, such as leaves, stems and flowers, as data are not available. 3. The population model is simplistic and contains no feed-back mechanisms to relate, for example, population growth rate to tiller density, *i.e.* each transition element is static.

Advantages 1. The method provides a useful framework, which has highlighted gaps in our knowledge, and can be used to integrate the data filling these gaps. 2. The method shows how an apparently stable population, with an increasing N demand, is, in fact, a very dynamic system with relatively large quantities of N recycling.

The importance of population interactions in process ecology

Enhanced nutrient mobilisation by soil microflora through the direct or indirect effect of faunal activity is well recognised (Coleman, 1985; Visser, 1985). Fauna-microflora interactions, as well as interactions between different micro-organism-groups are particularly amenable to laboratory experiments. In the field, experiments involving exclusion, elimination or addition have had some success in quantifying the functional role of selected taxa in decomposition and nutrient mineralisation.

The general mechanisms of interaction between species are reasonably well understood, but they do not yet allow quantitative prediction. Although Anderson *et al.* (1985) show a general relationship between faunal biomass and microbial mineralisa-

Table 4. Nitrogen and phosphorus leachate (kg ha^{-1} y^{-1}) from surface organic matter (L and F horizons) in tree mixtures compared with that expected from the two species in monocultures (Chapman, 1986)

	Expected	Observed	Difference (%)
Inorganic-N			
Picea-Pinus	29	42	+13 (+44)
Picea-Alnus	34	25	− 9 (−26)
Picea-Quercus	20	12	− 8 (−40)
Phosphate-P			
Picea-Pinus	1.9	3.0	+1.1 (+58)
Picea-Alnus	1.2	0.7	−0.5 (−42)
Picea-Quercus	0.9	0.0	−0.9 (−100)

tion of N, the precision of techniques currently available seems inadequate to define precisely the complex of biological interactions. Is it simply a question of technical precision? Are we searching for a tool which allows tracking of specific populations, *e.g.* genetic or isotopic markers? Or is the soil system fundamentally too complex for precise analytical solution? We tend to believe that the system is too complex. This is not a defeatist attitude. It simply means that in many situations we should concentrate on measurement of the processes which are the net result of population interactions. In so doing we take a 'black box' approach except where a dominant population of organisms can be analysed and manipulated with some precision, *e.g.* large fauna or there is a target population of particular concern, *e.g.* a pest or potential symbiont such as a mycorrhizal fungus.

The 'black box' approach is, however, dangerous. Even where the process can be easily measured information on the population characteristics can help in interpretation. One example is in decomposition of mixed litters. In decomposition research, most measurements have been made on plant litters of a single species but recent studies by Carlyle and Malcolm (1986) and Chapman *et al.* (1988) have shown significant interactions between populations of litters of different species. In the case of four species examined in NW England, mixtures of *Pinus sylvestris* and *Picea abies* litter showed an increase in respiration and mineralisation (measured as leachate) compared with that expected from measurements of the individual litters. In contrast two other mixtures (*Alnus glutinosa* with *P. abies*, and *Quercus robur* with *P. abies*) showed a decrease in activity (Table 4). The litter interactions are detectable in a field experiment, established in 1955, with replicated blocks planted in monoculture or two-species mixture. The litter interactions observed both in the field and in artificial laboratory mixtures (Chapman *et al.* 1988) correspond with increased or decreased earthworm biomass and with growth responses of the trees (Brown and Harrison, 1983).

There are two aspects of population interaction in those litters. At one level the litters themselves represent cohorts of each species which interact. A relatively simple deterministic experiment is required, using a large number of litter types, to define and quantify the generality of the interaction

effect and the chemical characteristics involved. This would provide a predictive but not a mechanistic model — the mechanisms of interaction are at a more detailed level of resultion, *i.e.* the organisms level. Here we are ignorant. There is the obvious Hypothesis (1) that enhanced rates of decomposition and mineralisation occur through translocation by filamentous microflora from nutrient rich to nutrient poor litters allowing more rapid use of carbon substrates and hence nutrient release. This could be tested experimentally but the evidence that mineralisation can be retarded in mixtures containing nutrient rich *Alnus* leaves argues against this hypothesis. Hypothesis (2) would invoke the presence of chemicals which are effective inhibitors of microbial activity in some litters but are used as substrates in others. An unlikely explanation and very difficult to test. Hypothesis (3) would argue that earthworms are the causal agent responding differentially to heterogeneity of food supply — but the laboratory experiments without earthworms produced the same response as field litters with earthworms, *i.e.* their observed population changes are responsive rather than causal. And so on!

The key point, apart from the important identification of litter interactions, is that the phenomenon detected by process measurements cannot be understood without recourse to population ecology. We can use the process knowledge empirically but its application is liable to be in error sometimes until we understand the population mechanisms.

Conclusions

Population and process ecology have developed along independent lines. Understanding of nutrient cycling has been the province of process ecologists who have largely ignored the relevance of developments in population ecology. Descriptive studies in nutrient cycling have provided reasonable simulations of major field patterns and the effects of variation in climatic and soil physico-chemical factors. Such models are often adequate for management requirements especially in stable monocultures. However, these models do not adequately express the changes in pattern of nutrient cycling of more complex successional systems, or the potential nutrient efficiency of differing species combinations. It is in these situations where biological flexibility needs to be expressed. Thus it is argued that the next stage in understanding of population dynamics will be dependent on the inclusion of more explicit information of populations.

In this paper we have illustrated an approach which builds on the understanding of growth strategies or characteristics of different species and interprets them in terms of nutrient dynamics. This approach is distinct from the widely used descriptive measurement of changes in the number or mass of organisms present. As illustrated, however inadequately, population models can be used to calculate the contribution of individual species to the distribution of nutrients in a system. These models, when given temporal and spatial dimensions, can be used to explore the complementarity of species and the potential of combinations of species to exploit the nutrient resources of a system. In the final example, that of nutrient release from interacting litters, it is only through further studies of component populations that the processes can be understood — and the capability for management extended beyond empiricism.

References

Anderson J M and Swift M J 1983 Decomposition in tropical forests. *In* Tropical Rain Forest: Ecology and Management. Eds. S L Sutton, T C Whitmore and A C Chadwick. pp 287–309. Blackwell Scientific Publ., Oxford.

Anderson J M, Huish S A, Ineson P, Leonard M A and Splatt P R 1985 Interactions of invertebrates, micro-organisms and tree roots in nitrogen and mineral element fluxes in deciduous woodland soils. *In* Ecological Interactions in Soil. Ed. A H Fitter, pp 377–392. Blackwell Scientific Publications, Oxford.

Ausmus B S, Edwards N T and Witkamp M 1976 Microbial immobilization of carbon, phosphorus and potassium: Implications for forest ecosystem processes. *In* The Role of Terrestrial and Aquatic Organisms in Decomposition Processes. Eds. J M Anderson and A Macfadyen. pp 397–416. Blackwell Scientific Publications, Oxford.

Bolin B and Arrhenius E 1977 Nitrogen – an essential life factor and a growing environmental hazard. Ambio 6, 96–105.

Brown A H F and Harrison A F 1983 Effects of tree mixtures on earthworm populations and nitrogen and phosphorus status in Norway spruce (*Picea abies*) stands. *In* New Trends in Soil Biology. Ed. P Lebrun. pp 101–108. Ottignies, Louvain-la-Neauve.

Callaghan T V 1976 Strategies of growth and population dynamics of tundra plants. 3. Growth and population

dynamics of *Carex bigelowii* in an alpine environment. Oikos 27, 402–413.

Callaghan T V 1980 Strategies of growth and population dynamics of tundra plants. 5. Age-related patterns of nutrient allocation in *Lycopodium annotinum* from Swedish Lapland. Oikos 35, 373–386.

Callaghan T V 1988 Physiological and demographic implications of modular construction in cold environments. *In* Plant Population Ecology. Eds. A J Davy, M J Hutchings and A R Watkinson, Blackwell Scientific Publications, Oxford. pp 111–135.

Callaghan T V, Headley A D, Svensson B M, Lixian L, Lee J A and Lindley D K 1986 Modular growth and function in the vascular cryptogram *Lycopodium annotinum*. Proc. R. Soc. Lond. B. 228, 195–206.

Carlyle J C and Malcolm D C 1986 Nitrogen availability beneath pure spruce and mixed larch + spruce stands growing on a deep peat. I. Net N mineralisation measured by field and laboratory incubations. Plant and Soil 93, 95–114.

Carlyle J C and Malcolm D C 1986 Nitrogen availability beneath pure spruce and mixed larch + spruce stands growing on a deep peat. II. A comparison of N availability as measured by plant uptake and long-term laboratory incubations. Plant and Soil 93, 115–122.

Chapin F S III, Van Cleve K and Chapin M C 1979 Soil temperature and nutrient cycling in the tussock growth form of *Eriophorum vaginatum*. J. Ecol. 67, 169–189.

Chapin F S III and Tryon P R 1983 Habitat and leaf habit as determinants of growth, nutrient absorption and nutrient use by Alaskan taiga forest species. Can. J. For. Res. 13, 818–826.

Chapman K, Whittaker J B and Heal O W 1988 Metabolic and faunal activity in litters of tree mixtures compared with pure stands. Agric. Ecosystems Environ. 24, 33–40.

Chapman K C 1986 Interactions between tree species: decomposition and nutrient release from litters. Unpub. Ph.D. thesis, University of Lancaster.

Clarholm M 1985 Possible roles for roots, bacterial, protozoa and fungi in supplying nitrogen to plants. *In* Ecological Interactions in Soil. Ed. A H Fitter. pp 355–365. Blackwell Scientific Publications, Oxford.

Coleman D C 1985 Through a ped darkly: An ecological assessment of root-soil-microbial-faunal interactions. *In* Ecological Interactions in soil. Ed. A H Fitter. pp 1–21. Blackwell Scientific Publications, Oxford.

Defoliart L S, Griffith M, Chapin F S III and Jonasson S 1988 Seasonal patterns of photosynthesis and nutrient storage in *Eriophorum vaginatum* L. on Arctic sedge. Functional Ecology 2, 185–194.

Fetcher N and Shaver G R 1982 Growth and tillering patterns within tussocks of *Eriophorum vaginatum*. Holarctic Ecology 5, 180–186.

Grime J P 1979 Plant Strategies and Vegetation Processes. Wiley & Sons, New York.

Headley A D, Callaghan T V and Lee J A 1988a Water uptake and movement in the clonal plants *Lycopodium annotinum* L. and *Diphasiastrum complanatum* (L.) Holub. New Phytol. 110, 487–502.

Headley A D, Callaghan T V and Lee J A 1988b Phosphate and nitrate movement in the evergreen clonal plants *Lycopodium annotinum* L. and *Diphasiastrum complanatum* (L.) Holub.

New Phytol. 110, 487–495.

Heal O W and Ineson P 1984 Carbon and energy flow in terrestrial ecosystems—relevance to microflora. *In* Current Perspectives in Microbial Ecology. Ed. M J Klug and C A Reddy. pp 394–404. American Society for Microbiology, Washington.

Herrera R, Jordan C F, Klinge H and Medina E 1978 Amazonian ecosystems: Their structure and functioning with particular emphasis on nutrients. Interciencia 3, 223–231.

Huhta V 1979 Effects of liming and deciduous litter on earthworms (Lumbricidae) populations of a spruce forest, with an inoculation experiment on *Allolobophora caligonosa*. Pedobiologia 19, 340–345.

Jonsdottir I S and Callaghan T V 1988 Interrelationships between different generations of interconnected tillers of *Carex bigelowii*. Oikos. 52, 120–128.

Karlsson S 1982 Ecology of a deciduous and evergreen dwarf shrub: *Vaccinium uliginosum* and *Vaccinium vitis-idaea* in subarctic Fennoscandia. PhD. Thesis, University of Lund, Sweden.

Kershaw K A 1962 Quantitative ecological studies from Landmannahellir, Iceland. II. The rhizome behaviour of *Carex bigelowii* and *Calamagrostis neglecta*. J. Ecol. 50, 171–179.

Mark A F, Fetcher N, Shaver G R and Chapin F S III 1985 Estimated ages of mature tussocks of *Eriophorum vaginatum* along a latitudinal gradient in central Alaska, USA. Arct. Alp. Res. 17, pp 1–5.

Rorison I H, Peterkin J H and Clarkson D T 1983 Nitrogen source, temperature and the growth of herbaceous plants. *In* Nitrogen as an Ecological Factor. Eds. J A Lee, S McNeill and I H Rorison. pp 189–209. Blackwell Scientific Publications, Oxford.

Satchell J E 1963 Nitrogen turnover by a woodland population of *Lumbricus terrestris*. *In* Soil Organisms. Eds. J Doeksen and J van der Drift. pp 60–66. North Holland Publishing Company, Amsterdam.

Slade A J and Hutchings M J 1987 The effects of nutrient availability on foraging in the clonal herb *Glechoma hederacea*. J. Ecol. 75, 639–650.

Smith R A H and Forrest G I 1978 Field estimates of primary production. *In* Production Ecology of British Moors and Montane Grasslands. Eds. O W Heal and D F Perkins. pp 17–37. Springer Verlag, Berlin.

Springett J A 1985 Effect of introducing *Allolobophora longa* Ude on root distribution and some soil properties on New Zealand pastures. *In* Ecological Interactions in Soil. Ed. A H Fitter. pp 399–405. Blackwell Scientific Publications, Oxford.

Svensson B M and Callaghan T V 1988 Apical dominance and the simulation of metapopulation dynamics in *Lycopodium annotinum*. Oikos 51, 331–342.

Swift M J 1984 Microbial diversity and decomposer niches. *In* Current Perspectives in Microbial Ecology. Eds. M J Klug and C A Reddy. pp 8–16. American Society for Microbiology, Washington.

Visser S 1985 Role of soil invertebrates in determining the composition of soil microbial communities. *In* Ecological Interactions in Soil. Ed. A H Fitter. pp 297–317. Blackwell Scientific Publications, Oxford.

Vitousek P M and Reiners W A 1975 Ecosystem succession and nutrient retention: A hypothesis. Bioscience 25, 376–381.

M. Clarholm and L. Bergström (Eds.), Ecology of arable land, 217–240.

Perspectives on measurement of denitrification in the field including recommended protocols for acetylene based methods

JAMES M. TIEDJE, STEPHEN SIMKINS[1] and PETER M. GROFFMAN[2]
Dept. of Crop and Soil Sciences and of Microbiology and Public Health, Michigan State University, E. Lansing, MI 48824, USA. Present addresses: [1]Dept. Plant and Soil Sciences, Univ. of Massachusetts, Amherst, MA 01003, USA and [2]Dept. of Natural Resources Science, Univ. of Rhode Island, Kingston, RI 02881, USA

Key words: acetylene inhibition, denitrifiers, geostatistics, kriging, nitrogen, nitrogen-15, nutrient cycling, soil cores, terrestrial ecosystems

Abstract

Of the biogeochemical processes, denitrification has perhaps been the most difficult to study in the field because of the inability to measure the product of the process. The last decade of research, however, has provided both acetylene and ^{15}N based methods as well as undisturbed soil core and *in situ* soil cover sampling approaches to implementing these methods. All of these methods, if used appropriately, give comparable results. Thus, we now have several methods, each with advantages for particular sites or objectives, that accurately measure denitrification in nature. Because of the general usefulness of the acetylene methods, updated protocols for the following three methods are given: gas-phase recirculation soil cores; static soil cores; and the denitrifying enzyme assay also known as the phase 1 assay. Despite the availability of these and other methods, denitrification budgets remain difficult to accurately establish in most environments because of the high spatial and temporal variability inherent in denitrification. Appropriate analysis of those data includes a distribution analysis of the data, and if highly skewed as is typically the case, the most accurate method to estimate the mean and the population variance is the UMVUE method (uniformly minimum variance unbiased estimator). Geostatistical methods have also been employed to improve spatial and temporal estimates of denitrification. These have occasionally been successful for spatial analysis but in the attempt described here for temporal analysis the approach was not useful.

Discussions of the importance of denitrification have always focused on quantifying the process and whether particular measured quantities are judged to be a significant amount of nitrogen. A second line of evidence discussed here is the extant genetic record that results from natural selection. These analysis lead to the conclusion that strong selection for denitrification must currently be occurring, which implies that the process is of general significance in soils.

Introduction

Denitrification is the process by which nitrogenous oxides, principally NO_3^- and NO_2^-, are reduced to dinitrogen gases, N_2 and N_2O. Most denitrification is carried out by respiratory denitrifiers that gain energy by coupling N-oxide reduction to electron transport phosphorylation (Tiedje, 1988). Nearly all respiratory denitrifiers prefer to use O_2 as their electron acceptor and will reduce N-oxides only when O_2 is not available. Denitrification is reasoned to be an important biogeochemical process since it appears to balance nitrogen fixation by recycling fixed nitrogen to the atmosphere. Most of the research on denitrification has been by agricultural scientists in an effort to understand and hopefully minimize loss of the nutrient that most often limits crop growth. Further reasons for in-

218   *Tiedje* et al.

terest in denitrification are: in waste treatment it can remove excess nitrate; it can decrease nitrate contamination of groundwaters, it affects atmospheric composition through the production and consumption of N_2O and thus has impact on climate; and it can produce toxic intermediates, NO and NO_2^-, the latter which can lead to carcinogenic nitrosamines (Tiedje, 1988).

The first need to study any process is an assay method. This has been the major obstacle in the study of denitrification and is why progress in quantifying this process has not been extensive despite its discovery over a century ago. There are three reasons why methodology has been such a limitation to the study of denitrification. (1) The best assay method for a process is to measure its product, but the Earth's atmosphere is 80% N_2 which rules out the use of this approach except in certain sealed microbial or biochemical laboratory studies. An alternative approach is to measure substrate disappearance, which is usually unsatisfactory for denitrification because of the diverse sources and fates of nitrate. (2) Given the above, the next best approach is to use a radioactive isotope, but for nitrogen there are none convenient for frequent use. Hence alternative methods must be used and these require more assumptions or are less direct. (3) Of the biogeochemical processes, denitrification is the most dynamic with several environmental regulators *e.g.* oxygen, nitrate, carbon (Tiedje, 1988). The complex variation in and interactions of these regulators results in considerable variability of denitrification over time and space (Burton *et al.*, 1984; Folorunso and Rolston, 1984; Parkin *et al.*, 1985; Parkin *et al.*, 1987). This makes quantitation of denitrification imprecise, costly, and makes conclusions difficult to establish. To overcome these obstacles more effort has been required than for other biogeochemical processes to arrive at reasonable and reliable methods for study of denitrification in the field. We believe these methods now exist and that it is time to focus on studying the ecology of the process. As this chapter was derived from a talk which was to emphasize findings from our laboratory, this chapter reflects this perspective as well. Furthermore, we summarize recommended protocols for those acetylene methods for which we have considerable experience as this information is not available elsewhere.

How important is denitrification: the perspective from natural selection?

Discussions of the importance of denitrification always focus on quantifying the process and whether particular measured quantities are judged to be a significant amount of nitrogen. There is another line of evidence, however, that should be considered in evaluating the importance of denitrification. This evidence is based on whether the trait has left a significant impact, *i.e.* has it been selected, as evidenced in the genetic record of the extant natural communities. Evolution by natural selection is based on the principle that species, and their physiological processes, were derived from variants in the progeny of previous generations that were more fit for their environment, and thus reproduced more extensively. If we look at the selection and thus the historical importance of this process in the soil ecosystem. This evidence is as follows:

1. Among the soil biogeochemical processes, denitrification appears to be second only to respiration in the number of gene copies present. We base this statement on the fact that denitrifiers make up 1 to 5% of the culturable soil microbial population (Tiedje, 1988), which makes denitrifier genes more numerous than genes for nitrogen fixation, nitrification and even cellulose decomposition for example. Since it is costly to organisms to maintain unneeded gene sequences (and denitrification requires many genes), this large number of gene copies should not have been maintained unless the trait is of sufficient and frequent value to the populations that contain the genes. It is hard to envision that brief, sporadic denitrification events or low levels of denitrification, as is usually envisioned for most soils (Aulakh *et al.*, 1982; Duxbury *et al.*, 1986; Mosier *et al.*, 1986; Myrold, 1988; Rolston *et al.*, 1982; Sexstone *et al.*, 1985; Terry *et al.*, 1986), are enough to maintain this large quantity of denitrifying genes in the indigenous communities. The implication is that the essential use of the denitrification process is more extensive than this.

2. Denitrification is widely distributed among procaryotes, and is even found in archebacteria. In fact it is now easier to list those bacterial groups in which denitrification is not present (Tiedje, 1988). It is hard to imagine how a process could have such

a wide phylogenetic spread without it being a process under strong selection. The fact that denitrification is found in both the eubacteria and archebacteria (Tiedje, 1988) together with what is believed to be it's relatively recent evolutionary origin (Betlach, 1982) suggests that there must have been lateral transfer of these genetic sequences among genera in soil communities. Such a spread only becomes recognized and established if there is strong selection.

3. The key denitrifying enzyme, nitrite reductase, appears to have evolved twice since both Cu- and heme cd_1-based nitrite reductases are common in soil denitrifiers (Coyne *et al.*, ms submitted). The niche for denitrification must have been substantial for two separate enzyme systems to have evolved to carry out the same physiological reaction. The heme based enzyme may have evolved more recently but even if so, the Cu based enzyme is still readily found in soil denitrifiers today.

4. The denitrifying nitrite reductase, and particularly the heme cd_1-type, seem to be highly conserved structures. Of the more than 100 soil denitrifiers that we have surveyed by Western immunoblots, the immunological specificity and molecular weight suggest conserved protein structures (Coyne *et al.*, ms submitted). Nitrous oxide reductase and nitrate reductase appear to be even more highly conserved (Korner, 1987; Michalski and Nicholas, 1988). Proteins that carry out processes vital to cells tend to be highly conserved. Thus denitrification would seem to be an important process to microbial cells, and thus in ecosystems it must have been significant.

The above arguments all suggest that denitrification is a valuable process to microbial communities. It has long been known that denitrification is coupled to electron transport phosphorylation (ETP) and provides about 60% of the energy (ATP) of oxygen respiration (Koike and Hattori, 1975). This energy yield, while less than respiration, is considerably more than provided by fermentation. Thus, in anaerobic environments in which nitrate is present, organisms with the denitrifying capacity would clearly benefit. However, it is hard to imagine that nitrate-rich anaerobic niches are sufficiently frequent in time and space to account for the advantage of denitrification as evidenced by the genetic record.

One explanation might be that previous conditions on Earth were more conducive to denitrification than current conditions and that the present genetic record is historical and doesn't reflect current conditions, *i.e.* that denitrifying genes are currently being shed. However, there are two arguments against this explanation. First, the majority of denitrifiers freshly isolated from Nature readily loose part or all of their denitrifying capacity during cultivation on laboratory media (Abd-el-Malek *et al.*, 1974; Gamble *et al.*, 1977). In addition, in at least one case, denitrifier genes appear to be plasmid-borne (Romermann and Friedrich, 1985). The location of denitrifier genes is not yet well studied, but when it is we would expect the plasmid location to not be rare. Thus, many denitrification gene sequences seem not to be stably maintained and should be readily lost if they were not continually being selected for in the progeny that retained them. Hence, current conditions must be selecting for denitrification. Second, gene sequences even with a low negative selection coefficient, *e.g.*, such as for gene maintenance, should be lost in 100 to 1000 generations. Even if generations in soil are few, this negative selection should be apparent in 10 to 100 years. There is little evidence that soil conditions important to denitrification have changed significantly in recorded history to suggest that conditions conducive to denitrification have changed the selective pressure.

In support of the argument that current conditions are still selecting for denitrifiers, we found that the ratio of denitrification enzyme activity/microbial biomass C was consistently higher in poorly drained soils than in well drained soils, and in fine-textured soils than in course-textured soils (Groffman and Tiedje, 1989b). These data suggest that in environments more conducive to denitrification, selection favors denitrifiers relative to other soil microorganisms.

The conclusion from the above analysis is that the denitrification process is still significant enough to be selecting and thus maintaining diverse and extensive populations of denitrifiers in soil. In order to provide the energy benefit to the populations to explain this result, denitrification would seem to be more common than reported in most soil studies. One explanation may be that most agricultural studies are biased toward measuring denitrification only during the cropping season when soil aeration is high and nitrate and carbon are low. The evidence to explain the apparent dilemma does not yet exist, but new nucleic acid and antibody probe methods and gene sequencing provide important tools to help resolve this issue.

Methods to measure denitrification

A number of different methods have been used to measure denitrification in the field. These include the acetylene and ^{15}N methods discussed here, nitrate/chloride ratios, nitrate disappearance, nitrogen balance, N gas production in sealed chambers, non-random isotope distribution, N production calculated from soil gas gradients, and micrometeorological methods (for recent reviews of these methods see Hauck, 1986 and Smith, 1988). The acetylene and ^{15}N methods are the most reliable and widely useful and these are further discussed here.

Acetylene inhibition methods

The ability of acetylene to cause denitrifiers to accumulate N_2O from NO_3^- was first noted by Fedorova *et al.*, (1973) and its use for denitrification assays was demonstrated in pure cultures in 1976 independently by Balderston *et al.* (1976) and Yoshinari and Knowles (1976). These contributions are major milestones in denitrification measurement and have led to an explosion of denitrification studies as well as to an improved understanding of the process. Some of the advantages of the acetylene method have been reviewed by Duxbury (1986). The major advantage is the tremendous improvement in sensitivity over previous methods: detection limits of 0.5 ng N/g soil·day (core)(Parkin and Tiedje, 1984) or 1 g/ha·day (field covers)(Duxbury, 1986). Other important advantages include (i) the use of the natural nitrate substrate pool, (ii) the large number of samples that may be assayed so that the spatial and temporal distributions can be analyzed and appropriate statistical analysis applied (Parkin *et al.*, 1987 (iii) the relatively low cost of the method (and especially for the analytical equipment) compared to the ^{15}N and

^{13}N methods, and (iv) the versatility of the method allowing lab, field, and remote site studies. It is these advantages that have made denitrification more widely studied and by a larger number of research groups.

There are some pitfalls with the acetylene method that the user must be aware of so that they can be avoided. These have been discussed by Keeney (1986); Rolston (1986) and Tiedje (1988). They are: (i) acetylene affects on other processes such as nitrification, sulfur cycling, and methanogenesis, (ii) acetylene inhibition can fail because not enough acetylene is present, which is particularly important when organic matter is high and/or nitrate concentrations are low. The acetylene can be biodegraded; the latter occurs only after the population is enriched by exposure to acetylene for approximately one week (Terry and Duxbury, 1985), (iii) contaminants in the acetylene may affect denitrifiers, and (iv) the dispersal of the acetylene, the recovery of N_2O, and the significant water solubility of N_2O are all important physical aspects that can lead to inaccurate results. Of these pitfalls the only one which cannot be overcome with appropriate care and design is the acetylene inhibition of nitrification. Fortunately this is an important consideration only for samples in which the nitrate concentration is very low. We have never found this to be a problem in agricultural soils but we have found it to be a problem in soils from unfertilized, natural ecosystems. In these latter cases, we have still found the acetylene method to be useful (*e.g.* Robert and Tiedje, 1984) when the gas-phase recirculation core (described below) is used.

The initial studies of denitrification on soil using this method were done with mixed, sieved or slurried soils. For measurement of natural rates, however, it is important to preserve the natural soil structure. Disruption of structure often stimulates denitrification by providing a new supply of carbon (Myrold and Tiedje, 1985; Parkin and Tiedje, 1984; Sexstone *et al.*, 1988), but in sandy soils it can cause a decrease in denitrification due to increased exposure of microsites to oxygen (Parkin and Tiedje, 1984). Thus, it is now generally accepted that maintaining the natural soil structure is required for measurement of natural rates of denitrification. The methods now used and described in this chapter accommodate this requirement.

Isotope methods

The use of ^{15}N in N cycling studies became popular during the 1960's, with denitrification being calculated by difference, i.e. denitrification was equated to the amount of ^{15}N unaccounted for at the end of the experiment. Since unaccounted for ^{15}N actually represents the sum of all experimental errors (Rolston *et al.*, 1979), as well as denitrification, the 'difference' method is not very accurate for denitrification studies. Although the first techniques for quantifying denitrification by direct measurement of $^{15}N_2$ were reported in the 1950's (Hauck *et al.*, 1958; Nommik, 1956), direct ^{15}N field measurements of denitrification did not begin until the mid to late 1970's (Rolston *et al.*, 1976). Early direct measurements, which had low sensitivity, have been improved upon (Siegel *et al.*, 1982), and fluxes as low as $50\,g\,N\cdot ha^{-1}\,d^{-1}$ can now be measured (Duxbury, 1986). Very sensitive measurements of denitrification can be obtained using ^{13}N ($> 10^6$ orders of magnitude more sensitive than ^{15}N, Tiedje *et al.*, 1981), but there have been few studies using this isotope due to its short half-life.

The ^{15}N difference methods were developed for assessment of fertilizer N losses, and not for denitrification measurement. To calculate denitrification, it is necessary to account for leaching and volatilization losses, for ^{15}N that is immobilized, and for unlabeled N that is mineralized during the course of the experiment. The errors associated with leaching, mineralization, and immobilization measurement are often greater than denitrification, greatly reducing the effectiveness of the difference method for calculating denitrification N losses.

Using ^{15}N isotope dilution techniques in combination with mathematical modeling allows for calculation of the gross rates of different N cycle processes, permitting more precise measurements of processes than those produced by simple mass balance (Juma and Paul, 1981). Myrold and Tiedje (1986) used mathematical modeling and nonlinear parameter estimation, along with ^{15}N, to simultaneously calculate estimates of denitrification, mineralization, immobilization and nitrification. Although rates could be determined for all the N cycle processes (Table 1), denitrification was the most poorly estimated because of the low 'sensitivity' of the denitrification parameter, due to the low rate of denitrification activity relative to other N cycle processes under the conditions of this study. Nonetheless, this is the only approach that allows quantitation of all the rates of the N cycle processes as they naturally interact.

The most important recent advancement in ^{15}N, field-based measurements has been the direct measurement of denitrification by the $^{30}N_2$ mass. The basis for this method is the sensitivity provided by measurement of the $^{30}N_2/(^{28}N_2 + ^{29}N_2)$ ratio since this ratio at natural abundance is low (Duxbury, 1986; Siegel *et al.*, 1982; Smith, 1988). While superior to the difference method, the $^{30}N_2$ method also has disadvantages. A primary problem is the necessary enhancement of the NO_3^- pool by the ^{15}N addition, which increases denitrification rates unless the *in situ* NO_3^- pool is high enough to make rates independent of NO_3 concentration. A second problem is determing the source of the ^{15}N gas produced. The methods of Rolston *et al.* (1976) and Siegel *et al.* (1982) require a uniform distribution of ^{15}N in the soil NO_3^- pool. Since the NO_3^- pool is dynamic, with production and consumption occurring in diverse soil microenvironments, it is unlikely that uniform ^{15}N distribution can be maintained over moderate to long time periods. Heinemeyer *et al.* (1988) in a phytotron study comparing the $^{30}N_2$

Table 1. Simultaneous estimation of several N cycle rates using isotope dilution, modeling and non-linear parameter estimation

Process/Pool	First order rate constants (day^{-1})		N fluxes (mg N kg^{-1} day^{-1})	
	Clay loam	Sandy loam	Clay loam	Sandy loam
Mineralization	0.0104	0.016	1.5	0.56
Immobilization	0.108	1.27	0.1–3.0	0.7–4.2
Nitrification	1.23	1.39	1–34	0.7–4.6
Denitrification	0.0132	0.001	0.54–0.96	0.04–0.015
Active N fraction (mg N g^{-1} soil)	145	35		

From Myrold and Tiedje, 1986.

method with the ^{15}N balance method, found close agreement between the methods for up to 30 days but losses by ^{15}N balance were higher over longer time periods.

A second ^{15}N method that directly measures *in situ* denitrification is the isotope dilution of the product pool. In this case the sample atmosphere is switched to $^{15}N_2$ and any denitrifier product ($^{14}N_2$) dilutes the ^{15}N enrichment of the $^{15}N_2$ (Limmer *et al.*, 1982). This method avoids problems with substrate alteration and uniform label distribution, but it is difficult to remove all $^{14}N_2$ from a sample without either extensive alteration of its physical structure, or holding the sample for extended time periods.

The high sensitivity of ^{13}N based methods makes it possible to minimize substrate enhancement while measuring denitrification (Smith *et al.*, 1978; Tiedje *et al.*, 1981). However, the short half-life (9.6 minutes) and the requirement of access to a cyclotron or Van de Graaff accelerator to produce the isotope, has limited application of these methods to only a few laboratories (reviewed in Tiedje *et al.*, 1981). It is impossible to do field studies with ^{13}N, and its use is best reserved for questions that cannot be addressed by other techniques. Therefore, it will not be discussed in this chapter.

Comparison of methods

Acetylene vs ^{15}N

One of the most important aspects of any method for measurement of a biochemical process is whether the natural substrate concentration and distribution is altered in any way. Methods that rely on surrogate substrates, *e.g.* acetylene in place of N_2 for the nitrogenase assay or any isotope used as substrate, are always subject to criticism since it is difficult to ensure in complex matrices like soil that the concentration of the surrogate achieves and maintains the natural concentration at all microsites. Therefore, methods that employ the natural substrate avoid this potential criticism. Thus a major advantage of the acetylene method over the ^{15}N method is that it uses the natural substrate pool and measures a direct result of that process. With the N-isotopes (both ^{15}N and ^{13}N), achieving uniform labeling of natural substrate

pools without altering natural concentrations has been particularly difficult if not impossible.

The few studies that have directly compared C_2H_2 and ^{15}N methods are summarized in Table 2. Rolston *et al.* (1982) and Mosier *et al.* (1986) conducted very similar field studies, comparing $^{15}N_2$ production to N_2O production in C_2H_2 treated soils during three irrigation events. Although both studies concluded that there were not important differences between the two methods, estimates of N loss were consistently higher with the C_2H_2 than with the ^{15}N method. This difference may have arisen because multiple treatments of the plots with C_2H_2 caused enrichment of C_2H_2 degraders, which could have increased the rates if available C was limiting denitrification (Terry and Duxbury, 1985; Topp and Germon, 1986; Yeomans and Beauchamp, 1982). This explanation of differences is supported by the fact that in the study by (Mosier *et al.*, 1986), the most marked difference between methods was observed in the last of three irrigation cycles, when the development of C_2H_2 degrading denitrifiers would be most likely.

Parkin *et al.* (1985) also compared the two methods but compared an C_2H_2 based soil core recirculation method with the ^{15}N difference method. Again there were no statistically significant differences between methods, but the N losses were higher when measured by the ^{15}N difference method than by the C_2H_2 method in both soils studied. Higher soil moisture in ^{15}N microplots than in soil outside of the microplots probably enhanced the ^{15}N losses. Furthermore, the difficulty in maintaining uniformly labeled NO_3^- pools, make measurement of total denitrification by ^{15}N imprecise. Parkin *et al.* (1985) observed that rates measured in microplots by ^{15}N were normally distributed while the C_2H_2 core measurements were log-normally distributed (and usually are, Parkin *et al.*, 1988). While normally distributed data are easier to present and synthesize than log-normally distributed data, the advantage in this case is small, since total variance does not appear to be any lower with the ^{15}N method than with the C_2H_2 method.

Since the acetylene and ^{15}N methods gave similar estimates of N loss, we believe that both methods, when used correctly, are equally acceptable quantitative measures of denitrification in nature. Both methods have different advantages and disadvantages that ought to be evaluated for any situation.

Table 2. Summary of field studies in which the acetylene inhibition and ^{15}N methods have been compared for measurement of N losses by denitrification

Study	Condition	Total denitrification		Comments	Authors
		Acetylene Method	^{15}N Method		
I	3 irrig/wk	4.3	4.1	*In situ* cover method: $^{15}NO_3^-$ added	Rolston *et al.*, 1982
	1 irrig/wk	3.4	3.2	Acetylene pumped into soil	
	1 irrig/2wk	2.7	1.9		
II	Clay loam	3.27	10.1	Recirculation cores for acetylene method and ^{15}N balance in *in situ*	Parkin *et al.*, 1985
	Sandy loam	1.71	2.65	cylinder for ^{15}N method	
III	June irrig	2.15	1.86	*In situ* cover method: $^{15}NH_4^+$ added	Mosier *et al.*, 1986
	Early July irrig	1.00	0.75	Acetylene allowed to diffused into soil	
	Late July–Aug. irrig	8.96	6.44[a]		

[a] Significantly different ($P < 0.1$), the rest of the comparisons are not significantly different (statistical comparisons were not done in study I). The single instance of significant difference is caused by one outlier (n = 4), suggesting a log normal distribution. If analyzed by log normal statistics this significant difference may disappear.
irrig = irrigation.

In choosing a method the specific questions to be addressed, the characteristics of the site to be studied, the availability of required equipment and the cost are all important considerations. Comparison studies, while valuable, are frustrating due to the difficulties in demonstrating significant differences due to the high spatial and temporal variability of this process. Since both methods are sound, it is more important to move forward and investigate underlying principles of the denitrification process than to dwell on denitrification methodology. The methods are not now the problem — it is the dynamic nature of the process with its many regulators that cause the problem in budget quantitation. A 'better' method is not the route to understanding the process nor to integrating true variation.

Cores vs in situ *covers*

The need to add C_2H_2 to soil in a controlled atmosphere provided the stimulus for the use of extracted soil cores in denitrification research. The use of cores can be problematic since the coring process may disturb the soil system, and create effects on denitrification rates that are difficult to interpret. This concern led to the development of chamber methods for measuring denitrification. These methods involve placing covers over the soil surface and either measuring the accumulation of

N_2O in the air space of the box or sweeping air through the box and analyzing N_2O in the exit air stream (Jury *et al.*, 1982). Chamber methods have been used either with or without C_2H_2, and several approaches have been developed for introducing C_2H_2 into in-field chambers (Burton *et al.*, 1984; Hallmark and Terry, 1986; Ryden *et al.*, 1979). The main advantage of chamber methods is that they allow for in-field measurement of actual fluxes of N gases from soil to the atmosphere.

There are several problems with chamber methods however, that derive from the fact that physical effects inhibit diffusion and cause emission of N gases from the soil surface to be divorced from biological production of those gases. Jury *et al.* (1982) reported that several weeks of monitoring may be required to accurately assess production of N gases associated with a particular rainfall or irrigation event. Soil temperature, which strongly controls N_2O solubility and diffusion, and which varies diurnally, has a strong effect on emissions measured by chamber methods (Blackmer *et al.*, 1982). Diffusion problems can be easily overcome with cores however, either with forced air flow (See: Gas-phase recirculation core below), or by thorough mixing of the air space of the soil core (See: Static core below).

While core methods can disturb the soil environment, chamber methods also affect soil physical conditions. Disturbance effects from driving cylinders into soil can increase rates of gas emission

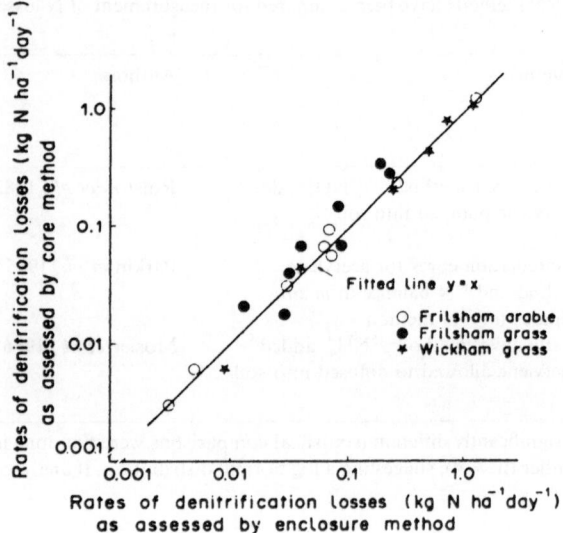

Fig. 1. Comparison of rates of denitrification measured using a cover method with those measured by a static core method on three sites. Reproduced from Ryden *et al.*, Soil Biol. Biochem. 1987, 19, 753–757, by permission.

significantly (Matthias *et al.*, 1980). This effect may be caused by release of gases physically trapped in soil spaces (Goodroad and Keeney, 1984). If so, waiting for some period between insertion of the chamber and sealing should alleviate this problem (Matson and Vitousek, 1987). Chambers can significantly increase soil temperatures (Matthias *et al.*, 1980), especially if long (> 1 hour) incubations are used. A final problem with chambers occurs if gas concentrations become sufficiently high to inhibit diffusion of gases out of the soil. Using short incubation times or flow through chambers minimize this effect.

Burton and Beauchamp (1984) compared two extracted core techniques with two *in situ* chamber methods, but high variability in rates complicated comparisons between the methods. In a comprehensive comparison of cores and chambers, Ryden *et al.* (1987) found a very strong relationship (slope not significantly different from 1) between denitrification rates in cores versus chambers, over a wide range of denitrification rates (Fig. 1). In very wet soils, they found that cores were superior to chambers due to the difficulty of introducing C_2H_2 into, and the slow diffusion of N_2O out of these soils. An additional advantage of cores is that it is

possible to run numerous core incubations, cheaply and quickly, while chamber measurements can be expensive and time consuming, limiting the number of replicates and/or sites that can be analyzed.

Core and chamber methods can differ in the method of extrapolating point measurements to larger areas. While chamber measurements must be extrapolated using surface area, core measurements can be extrapolated using either surface area or soil weight. For surface area extrapolation, the flux measurement made from under the area of the chamber is extrapolated to a hectare or square meter basis and represents the activity for that area over the entire soil profile. Values are extrapolated by weight by calculating flux on a per gram of soil basis and extrapolating to an areal basis using bulk density values for the soil under study. Extrapolation by weight allows for evaluation of the contribution of different soil depths and is useful for making comparisons between sites on a per unit weight basis, negating the effects of bulk density.

In summary, when compared directly, cores and chambers provided equivalent measurements of denitrification. While chambers may provide more accurate measurements of instantaneous flux of N gases from the soil to the atmosphere, cores appear to give more direct estimates of N gas production by biological processes. Total N gas production is needed for N budget questions important to ecosystem production and water quality; therefore the core measurements are best for this parameter. Nitrous oxide flux is needed for atmospheric chemistry questions; the chamber methods are most appropriate in this case.

Considerations and protocol for acetylene based methods

Gas phase recirculation core

Background. This method was originally described by Parkin *et al.* (1984) and was developed in our lab from the original concept of Kaspar (1984) and Kaspar and Tiedje (1980). It is based on the principle that acetylene distribution and N_2O recovery from intact soil cores can be more quickly and accurately achieved by introducing mass flow through soil macropores, and that denitrification rates can be more accurately measured in a sealed, repeatedly sampled system. In this method, a membrane pump recycles soil gas plus acetylene between the soil core and the gas chromatograph sampling loop. The increase in N_2O is continuously measured and the denitrification rate is obtained within 2 hours.

The advantages of the method are: (i) the natural soil structure and thus microsites with their carbon, nitrate and oxygen concentration are preserved, (ii) the assay is the most rapid (less than 2 hours) of any that maintains natural soil structure, (iii) because the production of N_2O is continuously measured, and linearity established, there is less uncertainty about whether the observed rate is influenced by limitations in gas diffusion, (iv) the analytical error is very low (CV < 10%), (v) cores of larger diameter are more accurately assayed than with the static core method (vi) samples with low nitrate concentrations can be analyzed, since a decline in N_2O accumulation rate is readily apparent when the nitrate concentration becomes rate limiting, and (vii) this system is the most convenient for experiments where the same core is reused for determining the effects of other treatments on denitrification. Examples of the latter include addition of water with or without nitrate (Groffman and Tiedje, 1989a; Robertson and Tiedje, 1984), the denitrification hysteresis during wetting versus drying cycles (Groffman and Tiedje, 1988), effect of different oxygen or acetylene concentrations (Parkin and Tiedje, 1984; Robertson and Tiedje, 1987), the effect of air filled porosity (Sexstone *et al.*, 1988), and measurement of the maximum denitrification rate under anaerobic conditions (argon plus acetylene as the recycled gas).

The maximum (anaerobic) denitrification rate is effectively used as the last measurement on a core, so that the previously measured rates on that core can be reference to this standard condition as a means of maximizing the treatment effect and minimizing the influence of variation among replicate cores (Myrold and Tiedje, 1985; Parkin and Tiedje, 1984; Sexstone *et al.*, 1988).

The disadvantages of the recirculation method include the following. Since gas is pumped through the soil, this slight increase in pressure could break water films extending oxygen to more microsites than occurs naturally; however we have not been able to demonstrate this to be more than a theoretical problem (Parkin *et al.*, 1984). Other disadvantages are that the number of cores that can be analyzed is too limited for some purposes (30 cores per day on the system described below), the method will not work for clay soils that are wet, and the equipment required, especially if automated, is complex and moderately expensive and cannot be used at remote sites.

Although not now the most popular acetylene-based method, the recirculation method remains the method of choice when advantages ii through vii (above) are important.

Recommended protocol. The method remains largely as described in Parkin *et al.* (1984). Subsequent improvements have been in automation, valving and plumbing, temperature control of the cores, and in data handling. The diagram of one loop in our current system which has the capacity to simultaneously measure eight cores is shown in Fig. 2. Soil cores of 4.7 cm diameter × 10 to 20 cm length are taken in a plastic liner which fits inside a steel coring device which has a sharpened, tapered, cutting tip. The soil core should fit snuggly in the plastic liner to minimize edge flow of the pumped gas. The metal core is driven into the soil by a hand operated slide hammer. Cores compacted more than 5% are discarded. After collection, the plastic cores are removed from the driver, capped on both ends with butyl rubber stoppers, and placed on ice

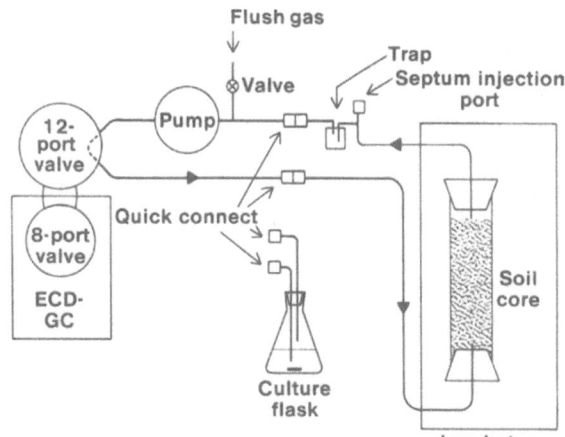

Fig. 2. Diagram of one of the eight parallel loops in our recirculation system to measure denitrification. This system can be easily shifted from core to flask samples by the quick connect fittings. The system can also be easily flushed to remove acetylene or change O_2 content by opening the valve for the flush gas and disconnecting the top quick connect. Acetylene or other gases are added through the septum injection port.

for transport to the laboratory. Cores have been stored at 4°C for up to 19 days without significantly affecting the denitrification rate (Parkin *et al.*, 1984). This feature is often important for large field studies and efficient use of the analytical equipment.

The analytical portion of our system consists of two gas chromatographs each equipped with two ^{63}Ni electron capture detectors and four analytical columns. An 8-port valve (Valco, column switch/ backflush to vent) mounted in the GC oven connects two analytical columns to each detector. Thus four analytical columns in each of two gas chromatographs allow eight soil cores to be simultaneously analyzed. At alternating intervals, computer controlled valves switch the two analytical columns between analysis of the sample and backflush. The backflush prevents acetylene and water from reaching the detector. The analytical columns are 1.8 m × 0.32 cm o.d. stainless steel packed with Porapak Q. They are operated at 55°C with a carrier gas flow of 15 ml/min and a backflush of 30 ml/ min. The carrier gas is 95% argon and 5% methane and the detector temperature is 300°C. The column conditions are adjusted as needed to optimize the separation of N_2O and CO_2 since the latter can interfere with quantitation of N_2O. We have found that the relative sensitivity of electron capture detectors for N_2O *vs.* CO_2 varies with manufacturer and that it is important to establish whether a particular instrument is adequate for N_2O analysis. Sample peaks are integrated on computing integrators (Hewlett Packard 3390A) with the data accumulated on a microbuffer and then transferred daily to data files in a personal computer. Detection limits on our system are 0.1 ng N_2O-N/ml and 3.0 μg CO_2/ml (Robertson and Tiedje, 1984).

The recirculation system cycles the gas phase between the soil core and a 0.5 ml sampling loop by means of a membrane pump (Neptune Dyna, Scientific Products, McGraw Park, IL) (Fig. 2). The pumps are operated at full speed which recycles the soil atmosphere at 200 to 300 ml/min for most cores. Swagelock quick connects are used to detach the soil core portion if attachment to another type of sample is desired, *e.g.* a culture flask. Serum vials (6 ml) are used as water and soil particle traps prior to the membrane pump. The tubing used throughout is 1/8″ stainless steel, which

is important if a reactive gas like NO is to be measured. The sample loop is connected to the gas chromatograph by means of a 12-port valve (Valco, External sample) which connects to two recirculation loops. Both valves are stainless steel with 1/8 inch fittings and are actuated by pneumatic controllers. The valves are controlled by solenoids which are activated from a program stored in a small computer and operated through an external events actuator board. This system allows cores to be analyzed unattended.

To measure the denitrification rate the stored cores are first warmed to the desired temperature and then mounted in the recirculation system (Fig. 2). We currently house the cores in an incubator (with heating and cooling capability) at the *in situ* soil temperature. We have also used room temperature and corrected rates to the *in situ* temperature using a Q_{10} of 2 for denitrification (Rolston *et al.*, 1984; Knowles, 1981). The oxygen concentration of the recirculating gas can be adjusted to that measured *in situ* for the soil macropores (*e.g.* ranged from 14 to 18% in one of our studies), although this is usually not necessary since the sensitivity of denitrification rates to oxygen concentration in this range is low (Parkin and Tiedje, 1984). The acetylene concentration is recommended to be 20% (20 kPa) to insure effective inhibition of N_2O reduction even in soils low in nitrate. At 10 min intervals the computer controlled valves shunt 0.5 ml of the recirculating gas to the analytical columns. N_2O measurements should be made until a linear pattern of N_2O accumulation (constant denitrification rate) is seen. This usually occurs within 15 to 30 min for coarse textured soils and 1 to 2 h for fine textured soils.

Denitrification rates are calculated by multiplying the N_2O concentration in the gas phase by the volume of the gas in the recirculation-core system, correcting for the N_2O dissolved in the aqueous phase, and dividing by the dry weight of the soil. The gas volume is determined by a pressure transducer after injecting a known volume of air into the system (Parkin *et al.*, 1984). The core water content is determined gravimetrically, and the dissolved N_2O is then calculated from the Bunsen relationship (Tiedje, 1982). The detection limit for denitrification rate is 100 ng-N·m^{-2}h^{-1} or 24 mg-N·ha^{-1}day^{-1} (Robertson and Tiedje, 1984).

Table 3. Comparison of variability of N cycle processes and soil parameters in 0.5 ha old-field on sandy loam sampled in fall

	Coefficient of variation[a]
N mineralization	58
Nitrification	70
Denitrification	275
CO_2 production	61
Moisture	52
pH	5

[a] number of samples was 301.
From Robertson *et al.*, 1988.

Static core

Background. The static core is similar to the recirculation core in concept except that the gas phase is static during incubation. This static system has two important advantages over the recirculation system: it offers the capacity to obtain measurements on an even larger number of cores, *e.g.* 200 per day instead of 30 per day on the recirculation system, and the analytical system is less complex and also allows work at remote field sites. The increased sample capacity is particularly important because of the high temporal and spatial variability of denitrification. This is easily illustrated by comparing the coefficients of variation for denitrification to other processes or parameters measured at a study site (Table 3). Given this situation, the preference in field studies at least, is for high sample capacity. The larger sample capacity provided by the static core system has also allowed for the first time, proper statistical evaluation of field denitrification rates (Parkin *et al.*, 1988, and p. 232).

Table 4. Summary of the characteristics and development of the static core method for measurement of field rates of denitrification

Authors	Core size and seal	Incubation conditions	Acetylene concentration	Sampling times	Precautions to achieve gas distribution	No. of replicate cores per site or treatment
Aulakh *et al.*, 1982	6.0 × 15 cm in glass jar	In shade at field temp.	5 kPa	24 h	Al core had slits to foster gas exchange with jar atmosphere	4
Burton and Beauchamp, 1984	5.0 × 10 cm, sealed	*In situ*	1 kPa	0, 2, 4 h	Double wall cylinder with holes to allow acetylene to enter soil from interwall reservoir	4
Groffman, 1985	2.0 × 8 cm, septa on ends	In shade outside	10 kPa	0 and 6 h	Loose fit of soil core in tube	12
Robertson *et al.*, 1987	2.2 × 20 cm septa on ends	Lab, 20–22°C	5 ml (10– 15 kPa)	0 and 24 h[a]	Pumping with 50 ml syringe at 0 time and before each sampling	20
Klemedtsson, 1986 and Svensson *et al.*, 1985	3.2 × 10 cm inside larger sealed core	Lab, 15°C	10 kPa	5 and 15 h	Plastic core had holes punched in sides to foster gas exchange	15
Parkin *et al.*, 1987	1.7–20 cm[c] × 16 cm	Lab, 24–26°C	10 kPa	3, 6 and 18 h	Loose fit of soil core in tube, and automated pumping with syringe[d]	36
Ryden *et al.*, 1987	3.4 × 10 cm inside jar	*In situ*	5 kPa	0 and 24 h	No core support used; texture and roots maintained core integrity	5
Groffman and Tiedje, 1989a and Rice *et al.*, 1988	2.2 × 15 cm, septa on ends	Lab, 22°C[b]	10 kPa	2 and 6 h	Pumping with 30 ml syringe	20
Myrold, 1988	2.5 × 20 cm, septa on ends	*In situ*	10 ml	0 and 24 h	Pumping with 50 ml syringe at 0 time and before each sampling	10

[a] Random subset of 6 cores sampled at 4 h intervals to confirm linearity of response over the 24 h period.
[b] Rates corrected to *in situ* temperature using Q_{10} of 2 (Rolston *et al.*, 1984).
[c] Core diameter of > 4.2 cm recommended from this study to be the optimum for yielding the most reliable estimates of natural denitrification rate (Parkin *et al.*, 1987).
[d] Cores were taken with steel tubes and the intact soil core was then transferred to more loosely fitting plastic tubes for incubation.

A major limitation of the static core system is that the acetylene and N_2O gas distribution is not as efficient. Consequently, other accomodations must be made in the design to improve gas distribution. Table 4 summarizes the features used by various investigators as this static core method has evolved. Two points are important to minimizing the gas distribution problem. First, the core size is usually small both to aid gas distribution and to facilitate driving and handling of larger numbers of cores, and second, all designs (Table 4) have some feature, *e.g.* syringe pumping or exposed core sides, to improve gas distribution. The two other features in the evolution of the static core method are also notable: the number of replicate cores per treatment has increased and the length of assay period has become shorter and more well defined. Both improvements are important to making the static core a more reliable and accurate denitrification assay method.

Recommended protocol. The features of the more recent method (Table 4) are sufficiently similar that any design following these general characteristics is probably acceptable. Development of the high sample capacity static core methods were initiated in the early 1980's by the Swedish group for the 'Ecology of Arable Land Project', and in our laboratory by Phil Robertson and Tim Parkin. Subsequently, Parkin has independently developed a more automated, high capacity protocol (Parkin, 1985; Parkin *et al.*, 1988; Parkin *et al.*, 1987) to support his statistical work.

For use of the static core there are certain features and precautions that warrant discussion. Soil cores have most commonly been taken in a plastic tube housed inside of a steel tube that is driven into the soil. Parkin, however, has recently collected the soil in a steel tube sampler and then transferred the soil core to a slightly larger diameter plastic tube for incubation (Parkin *et al.*, 1987). The loose fit allows acetylene and N_2O to mix more readily along the walls of the core reducing the length of the gas diffusion path into soil. This approach will probably only work in soils that maintain structural integrity through the transfer. While cores of 2 to 4 cm diameter have been used (Table 4), Parkin reports that 10 to 15 kg of soil (> 4.2 cm diameter cores), gives the most reliable results because this sample size was necessary to reasonably sample the

'hot spots' of denitrification (Parkin *et al.*, 1987). However, cores of this size are more difficult to drive into soil and make it heavier to handle large core numbers. Thus, these practical considerations may outweight the slight increase in accuracy afforded by the larger core size.

Cores can be incubated in place, but most commonly are incubated in the laboratory. Both are acceptable methods, but if the core is transported to the laboratory they should be kept stored on ice during transit and held at 4°C. Before incubation, the cores are brought to the desired temperature, degassed if significant N_2O has accumulated to reduce the sensitivity of the measurement, sealed, and 10% (10 kPa) of acetylene added.

Acetylene from a cylinder should be scrubbed through a sulfuric acid train to remove the contaminating acetone (Tiedje, 1982). Acetylene generated from the reaction of carbide rock with water has no acetone and is free of other contaminants that might interfere with the denitrification assay (Hyman and Arp, 1987) (contaminants are produced from the water-carbide reaction but these are not substrates for denitrifiers nor inhibitors of denitrification at the concentrations produced). The most convenient way to foster acetylene distribution throughout soil is to create mass flow by alternately reducing and increasing pressure in the soil pore space, which can be accomplished by pumping with a large syringe. This pumping should be done immediately after the acetylene is added and prior to each gas sampling. Gas samples are taken by disposable plastic syringes and transferred to evacuated gas vials. We use 3-ml, preevacuated Venoject™ vials (Terumo Scientific, N.J.) but any vial used should first be checked for background N_2O and any other contaminants that interfere with the N_2O analysis. Samples of 4 ml are recommended for injection into this vial of ~ 3.3 ml capacity. This volume will not pop the stoppers and the vials can be stored for several months if sealed with silicon. If stored, vials should be surveyed to verify pressurization (*i.e.*, no leakage) prior to analysis. To be certain of the correct quantitation, a series of N_2O standards can be prepared and stored in the same manner. Satisfactory internal standards are not available; we have tried helium but it is too insensitive to detection by electron capture, requiring excessive dilution of the sample.

The N_2O (and CO_2) is analyzed by gas chromatography as described for the recirculation system. Because the static core system can lead to a larger number of samples (thousands), gas vial analysis becomes the rate limiting experimental step. At least three automated systems have been built for this purpose (Klemedtsson, 1986; Parkin, 1985; Robertson and Tiedje, 1985). We are also aware of one report where a commercial autosampler system was used to sample 1 ml vials in a denitrification study (Lowrance and Smittle, 1988).

Particularly important in the static core method is the time of sampling. We now recommend not to sample at 0 time but to use a later time to establish the initial point, after the acetylene is better distributed. If only two sampling times are used, shorter intervals are recommended, *e.g.* 2 and 6 h to insure that nitrate does not become rate limiting. If large sample analysis capacity is available, three sampling points are recommended to insure linearity.

To determine the N_2O produced, the total gas volume of the core and its water content are needed. The pore volume can be estimated from the total known volume of the core and the bulk density of the soil, which is estimated by measuring the length of the soil core and its dry weight. The moisture content is determined gravimetrically. The dissolved N_2O is calculated by using the Bunsen coefficient (Tiedje, 1982). An alternative used by Parkin is to inject a large volume of acetylene into the core prior to incubation and then to measure the pressure increase by means of a pressure transducer (Parkin *et al.*, 1987). This determines if there are leaks as well as allowing calculation of the total pore volume. The gas phase is then mixed by syringe and the excess pressure vented to atmospheric presure prior to the start of incubation. After incubation, the moisture content and soil dry weight are determined gravimetrically.

High spatial and temporal variability in denitrification rates necessitates that careful consideration be given to sampling strategy. Two ways of dealing with the variability problem are; i) taking a large number of samples and ii) accommodating the temporal and spatial variations characteristic of the site in the sampling design.

Taking a large number of samples requires using a static core technique. With this technique up to 200 cores can be dug, incubated, and headspace

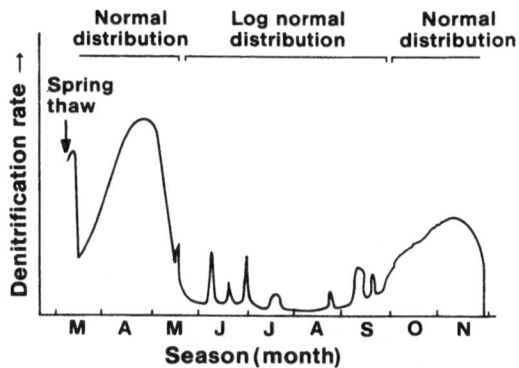

Fig. 3. A generalized diagram of the seasonal pattern of denitrification rates for the northern temperate soils. Peaks of activity in the summer are due to rainfall or to pockets of decaying organic matter. The frequency distribution of rates for these periods is also shown (Groffman and Tiedje, 1989a).

samples collected in one day. With a gas chromatograph equipped with two detectors and two column switching/backflushing valves, 30 samples an hour (200 per day) can be analyzed by manual injection. With an autosampler, even more analyses can be done. We completed weekly sampling of this type for 10 sites over an 80 km² area for several months (Groffman and Tiedje, 1989a).

Considering temporal patterns of denitrification can aid in designing sampling strategies. For example, denitrification activity in temperate ecosystems often occurs during brief periods of high soil wetness and low plant activity in early spring and fall (Goodroad and Keeney, 1984; Groffman and Tiedje, 1989a; Myrold, 1988; Schmidt *et al.*, 1988). From our experience, a generalized seasonal pattern of denitrification in our region is illustrated by Fig. 3. Such a pattern may be used as a model for a sampling strategy in which the periods of greater activity are more intensively sampled and less active periods, *e.g.* the summer, are infrequently sampled. The seasonal pattern of variability (*e.g.* Fig. 3) also provides guidance in the number of samples needed.

As a minimum guideline we recommend taking 20 cores per site and 12 to 20 samplings per year planned to encompass periods of higher activity.

Denitrifying enzyme activity (Phase 1)

Background. The denitrifying enzyme assay (DEA), also known as the phase 1 assay, measures the concentration of functional denitrifying enzymes in a sample at the time of sample collection (Smith and Tiedje, 1979; Tiedje, 1982). This assay does not measure the denitrifying activity of the natural sample, but the denitrifying enzyme concentration of that sample does reflect the environmental history of that site. This assay has been used in a comparative manner to characterize samples and to study experimental treatment effects on denitrification, but it has not been used to provide information on field denitrification rates. Recently, however, two new lines of evidence suggest that this assay may also be useful in field studies of denitrification.

In one study (Groffman and Tiedje, 1989b), found that DEA was strongly correlated with the measured annual denitrification N loss in forest soils of southern Michigan. Furthermore, this assay was found to correlate with soil texture and drainage characteristics of catenas and could be used to predict the denitrification N loss of these sites. Since this assay is more easily done than core assays, a larger number of sites can be sampled, perhaps improving large scale estimates of denitrification. The relationship of DEA to denitrification at this larger temporal and spatial scale may be revealing the effects of selection discussed earlier (p. 218). If so, this provides a rationale for why the DEA may be predictive of natural denitrification losses.

In a second study, Parkin and Robinson (1989) have used the phase 1 assay (DEA) in a stochastic model along with respiration rate to predict denitrification frequency distribution and mean rates. Their approach was based on the fact that a highly variable process, like denitrification, probably cannot be explained by a deterministic model. One of their stochastic models accurately predicted the frequency distribution as well as the mean denitrification rates. This study also illustrates how DEA offers potential for estimation of field denitrification rates.

Recommended protocol. This protocol is based on the phase 1 assay described by Smith and Tiedje (1979) and Tiedje (1982). The method described here includes some further improvements in components and in convenience.

The principle of the method is based on optimizing all requirements for enzymatic activity-saturation with nitrate, an electron donor, no oxygen, and no diffusion limitation — so that the rate of N_2O production is proportional to denitrifying enzyme content. The method can also be used to test whether one of the substrates is limiting by not adding that substrate to the assay.

Soil (25 g) is placed in a 125 ml-Erlenmeyer flask containing 25 ml of a solution of 1 mM glucose, 1 mM KNO_3, and 1 g/l of chloramphenicol. Chloramphenicol blocks protein synthesis, thus extending the period of linear N_2O accumulation. The flasks are capped with gas impermeable stoppers and made anaerobic by alternately flushing with argon and evacuating 4 times. Purified acetylene is added to the flask to achieve a final concentration of 10% (10 kPa) in the gas phase. Higher concentrations of acetylene should be used if the organic content and biological activity are unusually high (Yeomans and Beauchamp, 1978; Kaspar *et al.*, 1981). The soil slurries are incubated on a rotary shaker. Three replicates are recommended.

The headspace gas is sampled by syringe and the N_2O measured by gas chromatography as described above. At least four determinations should be made during the incubation period to establish linearity. The recommended incubation period is 1 h and should not go beyond 2 h. The dissolved N_2O which is substantial in this case, should be corrected for by using the Bunsen relationship (Tiedje, 1982). Samples can be stored in evacuated glass vials if they cannot be analyzed directly. In mixed, homogeneous, anaerobic soil, the coefficient of variation should be 5 to 15%. Higher variation may indicate incomplete anaerobiosis or natural patchiness of denitrifiers in soil.

In situ *soil cover*

Since *in situ* soil cover techniques are not extensively used in our laboratory, we will not recommend a protocol for their use. We will, however, review the different approaches to adding C_2H_2 to soil and provide some considerations for the development and use of cover methods.

The simplest approach to adding C_2H_2 to a soil cover system is to introduce C_2H_2 into the headspace of the cover. With this approach, the time required for C_2H_2 to diffuse throughout soil can be considerable, and the area of soil that will have

C_2H_2 concentrations sufficient to inhibit N_2O reduction is unknown. As discussed earlier, it is necessary to minimize the time that the cover is in place to avoid temperature and diffusion problems under the covers. Passive diffusion of C_2H_2 into soil can be enhanced by using a hollow 'double-wall' chamber design (Burton and Beauchamp, 1984), or by perforated tubes inserted into the soil which are connected to an above ground manifold through which acetylene flows (McConnaughey and Duxbury, 1986). The advantage of adding acetylene by diffusion is that no aeration changes are induced since mass flow of soil gas is avoided.

Ryden *et al.* (1979) and Ryden and Dawson (1982) developed a procedure where C_2H_2 is introduced into soil by radial diffusion from probes inserted into the soil. With this system, C_2H_2 concentrations required to inhibit N_2O reduction are established within 15 to 30 minutes, and denitrification rates can be measured over a 1 to 2 hour period. Air is continuously swept through the chamber and accumulated N_2O is trapped on molecular sieve, avoiding diffusion problems caused by N_2O buildup in the chamber. The main drawback with this system is the time and expense required to set-up the chambers in the field, which limits the numbers of replicates that can be run.

A third technique for introducing C_2H_2 into soil involves adding C_2H_2 saturated water to field chambers (Hallmark and Terry, 1985; Terry *et al.*, 1986). The major drawback of this technique is that the moisture addition decreases the oxygen status of the soil and thus increases denitrification, but it can be used to approximate irrigated soils or to simulate rainfall events. An additional problem is that C_2H_2 concentrations sufficient to inhibit N_2O reduction may not be maintained as the soil dries. Hallmark and Terry (1985) recommend using both C_2H_2 saturated water and radial diffusion to introduce C_2H_2 into irrigated soils.

At certain locations, most notably sites with acid soils, N_2O may be the natural terminal product of denitrification, eliminating the need for introducing C_2H_2 into soil. Spatial and temporal patterns of denitrification were characterized at an acid soil in Michigan using this approach (Christensen and Tiedje, 1988). The drawback to this approach rests on the uncertainty of whether N_2O reduction is uniformly inhibited and whether respiratory denitrification is the principal source of the N_2O

(Robertson and Tiedje, 1987). While bulk soil pH and N_2O reductase activity may be low at a site, microsites of high pH and denitrification activity associated with decomposing plant material may have significant N_2O reduction.

Data analysis

Analysis of skewed data

Many of the difficulties inherent in denitrification research arise after measurements of denitrification rates have been made and appropriate methods to analyze the data are sought. These complexities in the analysis of denitrification data occur principally as a result of the tendency of the frequency distribution of rates to be much better approximated by the lognormal than by the normal distribution. Denitrification rates have been observed to be lognormally distributed when measured using surface chambers (Duxbury and McConnaughey, 1986; Folorunso and Rolston, 1984), in intact soil cores (Parkin *et al.*, 1985), or in anaerobically incubated soil slurries (Parkin *et al.*, 1987). Members of the denitrification group within the project 'The Ecology of Arable Land' have also observed that rates measured in intact soil cores are

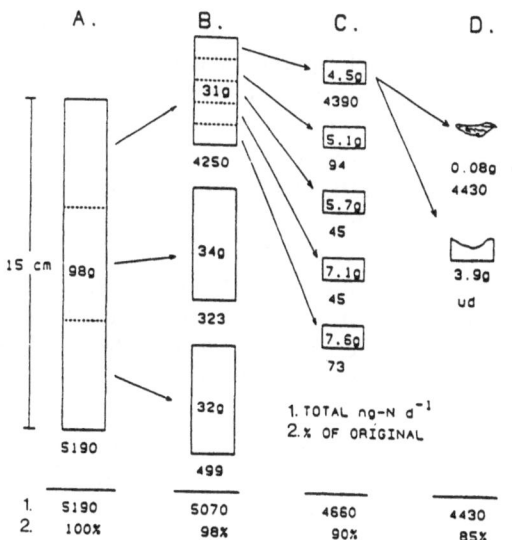

Fig. 4. Results from segmentation of a soil core to localize the active denitrifying sites. Reproduced from Parkin, 1987, Soil Sci. Soc. Am. J. 51, 1194–1199, by permission.

almost invariably better described by the lognormal than the normal distribution.

While the high variability of denitrification has long been recognized, and the skewed frequency distributions are now being realized, the underlying basis for these patterns had not been investigated until Parkin (1987) demonstrated that decaying particulate organic matter created 'hot spots' of denitrification. In one example (Fig. 4), he found that 85% of the denitrification activity of a 98 g soil core was found in a 0.08 g piece of decaying plant leaf. Such 'hot spots' would likely be non homogeneously dispersed, and thus would give rise to the observed lognormal frequency distributions (Parkin, 1987). The generalized seasonal pattern of denitrification (Fig. 3) consists of spring and fall periods when moisture, and probably carbon, are more plentiful (and uniform), and the frequency distribution is observed to be normal (Groffman and Tiedje, 1989a). However, in the summer particularly moisture is much more limited and the organic matter 'hot spots' are the only sites sufficiently depleted in O_2 to allow denitrification, thereby resulting in lognormal distributions for this period (Groffman and Tiedje, 1989a).

The possibility that rates of denitrification may be lognormally distributed requires that the first steps in analyzing denitrification rate data center on the testing of the rates for statistically significant departures from normality. The Kolmogorov-Smirnov test for goodness of fit is one tool for the identification of significant departures from normality. High positive skewness of the distribution of a set of denitrification rates is a strong indication that those rates may be closer to lognormal than normal. The statistical significance of suspiciously high coefficients of skewness can be determined through reference to tabulated values for this coefficient that are statistically significant indications of departures from normality for different sample sizes and at different levels of probability (Pearson and Hartley, 1958; Zar, 1974).

If denitrification data are found not to be normally distributed, the next step appropriate for their analysis would be to apply the same tests for normality to the logarithms of the rates. If the logarithms of the rates are found to be well approximated by the normal distribution, *i.e.*, if the rates are lognormally distributed, then the logarithmically transformed rates can be safely used for popular parametric statistical tests, such as analyses of variance and covariance or regression analyses, that provide accurate results only for normally distributed data. Thus, reliable evaluation of the significance of differences in denitrification rates between experimental treatments or of the significance of trends in denitrification rates is possible using familiar statistical tests even when those rates are lognormally distributed.

Best method to estimate mean and variance of denitrification rates. A surprisingly challenging problem is the accurate estimation of the true mean rate of denitrification from a limited number of rate measurements. Folorunso and Rolston (1984) have calculated that denitrification rates would have to be measured in more than 4000 samples to be able to calculate a value for the mean denitrification rate that was within 10% of the true mean rate for a 3×36 m experimental plot with highly variable, lognormally distributed rates. Parkin *et al.* (1988) have evaluated three different methods for calculating the mean from limited numbers of lognormally distributed measurements. The first of these methods, the arithmetic average, gave relatively efficient and unbiased estimates of the population mean. A maximum likelihood method based on a transformation of the mean and variance of the logarithms of the original measurements gave estimates for the mean that not only were less accurate than the arithmetic average but were also biased overestimates by as much as 73%. The problems of inaccuracy and bias with the maximum likelihood method were exacerbated by decreasing sample size or increasing skewness of the frequency distribution. A third method evaluated by Parkin *et al.* (1988) is a uniformly minimum variance unbiased estimator (UMVUE) that incorporates mathematical expressions that correct for the bias inherent in the maximum likelihood method. The UMVUE method was found to be the most accurate of the three techniques for the estimation of the mean of a lognormal population based on a limited number of samples. The superiority of the UMVUE method was most evident for small sample sizes ($n < 30$) from highly skewed distributions, and its use was recommended for such samples because the accuracy of the UMVUE method is sufficiently greater than that of the arithmetic average to justify the more elaborate calculations that it requires.

Parkin *et al.* (1988) also reviewed three methods for the estimation of population variance analogous to the three methods for mean estimation. They found, again, that the maximum likelihood estimate of population variance was the least accurate of the three methods for small sample sizes. The familiar formula for variance (the sum of the squared departures from the sample mean divided by one less than the sample size) was the least accurate of the three methods for larger sample sizes ($20 < n < 100$). The UMVUE method was the most accurate estimator of population variance for all sample sizes and different degrees of skewness examined by Parkin *et al.* (1988), and they recommended its use for all log-normal samples except for those of small size ($n < 20$) drawn from distributions of low skewness, *e.g.*, with a coefficient of skewness less than 2.

Geostatistical methods

Given the extreme variability of denitrification

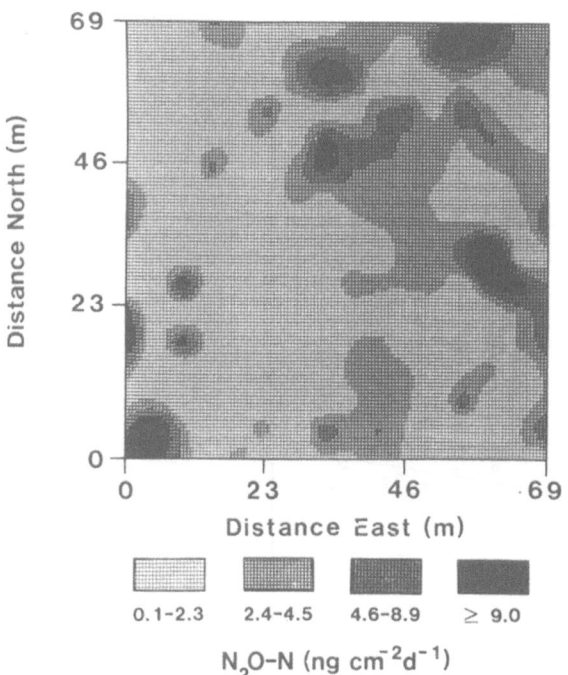

Fig. 5. Isopleth for denitrification across site derived by punctual kriging. The patterns show the variability of this process in this 0.5 ha old field. Reproduced from Robertson *et al.*, 1988, Ecology 69, 1517–1524, by permission.

rates in the field, the integration of rates over space and time becomes a focus of concern. For this purpose, methods of data analysis taken from the body of theory known as geostatistics are gaining in frequency of use for denitrification research (Folorunso and Rolston, 1984; Folorunso and Rolston, 1985; Parkin *et al.*, 1987; Robertson *et al.*, 1988). One geostatistical technique, kriging, offers the ability to predict (with known estimation variance) denitrification rates at unsampled locations through optimal interpolation of rates at sampled locations.

For rates predicted by kriging to be more accurate estimates than the simple average of all the measured rates, the rates of denitrification must exhibit autocorrelation: a tendency for the variance of the rates at locations close to one another to be less than the variance of rates at widely separated points. Robertson *et al.* (1988) found that denitrification rates in their study exhibited marked spatial autocorrelation, and consequently they were able to use kriging to predict the rates over their 69×69 m study site at 1-m intervals as shown in Fig. 5. Denitrification rates do not always appear to be autocorrelated. Folorunso and Rolston (1984) found evidence of spatial autocorrelation of denitrification rates in only one out of twelve transects examined.

Attempts at temporal kriging

Previous efforts to apply geostatistical methods to denitrification research have focused on the use of kriging for the prediction of rates of denitrification at unsampled locations in space. In this section, geostatistical methods such as kriging are applied to the prediction of denitrification rates at unsampled points (dates) in time. For the sake of brevity, the following discussion of temporal kriging assumes some familiarity with the general techniques and terminology of geostatistics. Good general introductions to geostatistics include Journel and Huijbregts, 1978; Vieira *et al.*, 1983; and Webster, 1985.

The massive data base on denitrification rates that has been assembled by members of the project 'The Ecology of Arable Land' is well suited for the evaluation of temporal kriging. The denitrification group measured denitrification rates in three dif-

234   *Tiedje* et al.

ferent crops (grass, barley, and lucerne) in two positions relative to the crop plants (within and between rows), and these measurements were carried out at relatively frequent intervals over two entire field seasons (1982 and 1983). We are, thus, in a position of being able to analyze these data to determine whether rates of denitrification are temporally autocorrelated in all of these different field treatments. If evidence of autocorrelation is observed, it then would be possible to examine the stability of the semivariograms from year to year and from crop to crop.

The inhibition of nitrous oxide reductase by acetylene as described by Klemedtsson (1986) was used for the measurement of the rates of denitrification in intact soil cores. For all of the following analyses, common logarithms of the rates of denitrification measured in individual cores were taken,

and then the logged rates were averaged. The rates were logged because members of the denitrification group of the project had determined that the rates were almost invariably better described by the log-normal than the normal distribution. These daily logged rates were averaged separately for the three different crops and the two different row-positions.

As an initial evaluation of the extent to which temporal autocorrelation was present in the (averaged logged) rates of denitrification in these studies, semivariograms were constructed for each of the field treatments for both years when full field-season data were collected. No evidence of temporal autocorrelation was evident in the semivariograms for denitrification rates in the barley or lucerne treatments in either year whether between or within crop rows; the semivariograms were flat showing a pure nugget effect (Journel and

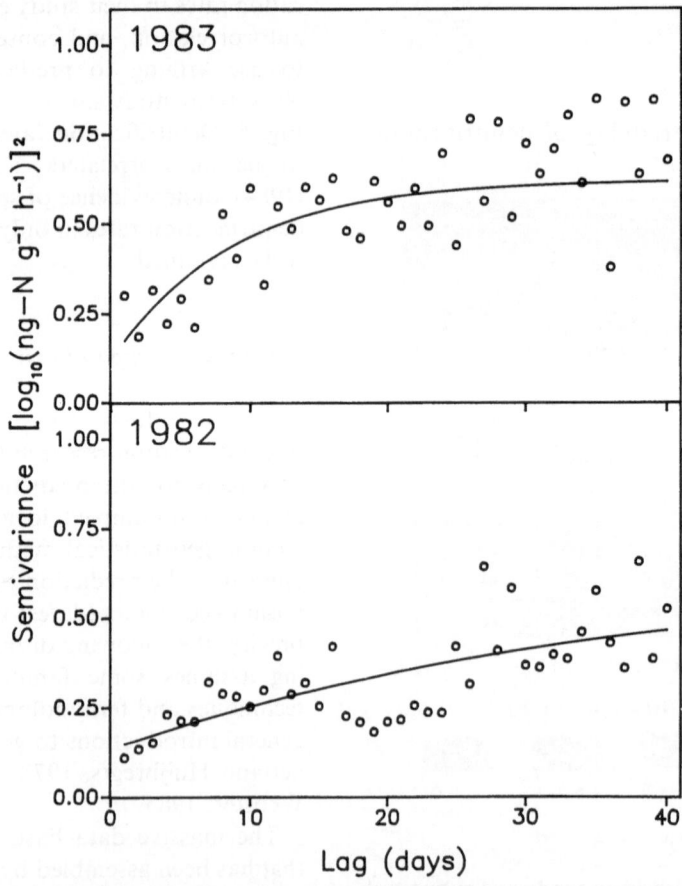

Fig. 6. Semivariograms constructed from logarithmically transformed denitrification rates measured within rows of grass over the course of two field seasons. The smooth curves correspond to the exponential model for semivariograms fit to the experimental values for semivariance by weighted nonlinear regression.

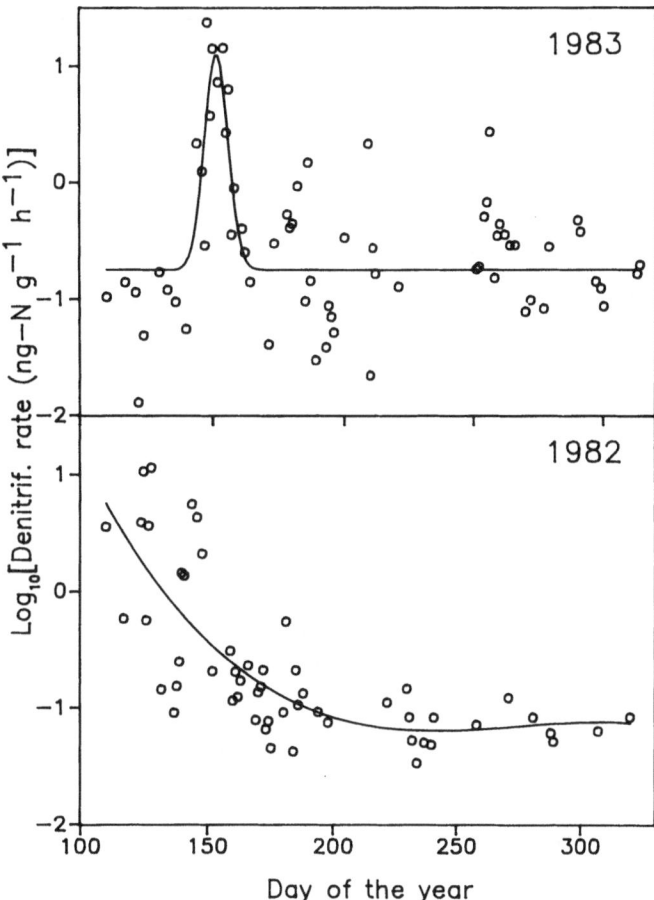

Fig. 7. Trends in logarithmically transformed denitrification rates measured within rows of grass over the course of two field seasons. The trends were removed from the data for subsequent analyses using the curves shown fit to the data.

Huijbregts, 1978; Webster, 1985). Rates of denitrification measured between rows of grass also failed to show evidence of temporal autocorrelation. In contrast, temporal autocorrelation did appear to be present in denitrification rates measured within rows of grass during both 1982 and 1983. Semivariograms calculated from these rates are shown in Fig. 6. The smooth curves shown are fits of the exponential model (Journel and Huijbregts, 1978; Webster, 1985) to the experimental values for semivariance. The models were fit using weighted nonlinear regression in which the weights corresponded to the number of pairs of data used at each value for lag. The observed tendency for semivariance to increase with increasing lag implies that measurements of denitrification made close together in time tended to be more similar to one

another than were measurements separated by a long period of time.

To determine whether trends in the denitrification rates within rows of grass were responsible for the observed autocorrelation in the rates, the data were analyzed by curvilinear regression to determine whether denitrification rate varied through either the 1982 or 1983 field season as a simple linear or polynomial function of time. A significant ($P < 0.001$) fit of a cubic polynomial to the logarithms of the rates of denitrification was found for the data collected in 1982 but not for data from 1983. The lower half of Fig. 7 shows the polynomial of best fit and the daily averages of the logarithms of the denitrification rates measured during the 1982 field season. Sixty percent of the variation in the logarthmically transformed rates

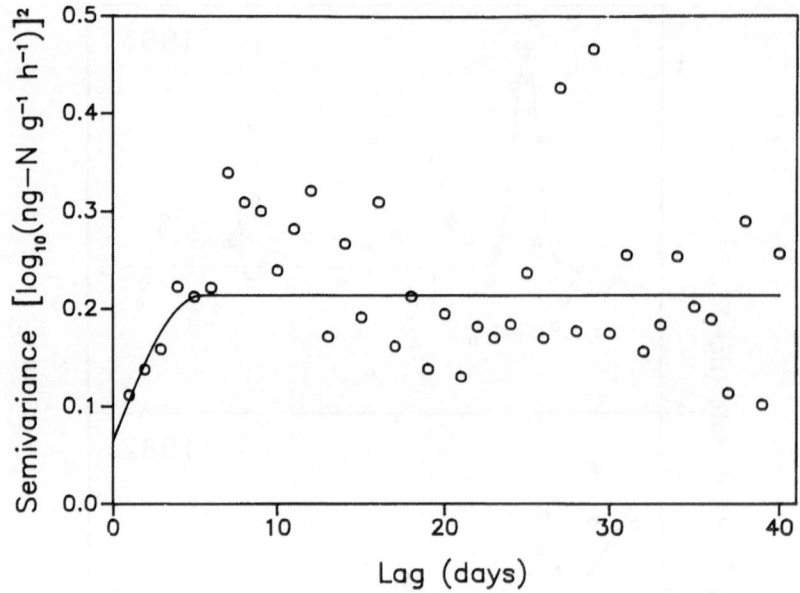

Fig. 8. Semivariogram constructed from the residuals left by a fit of a cubic polynomial function of time to logarithmically transformed denitrification rates measured within rows of grass during 1982. The smooth curve shown corresponds to the spherical model for semivariograms.

measured in 1982 could be attributed to the trend.

Although a simple polynomial failed to provide a statisticaly justifiable fit to the denitrification rates from 1983, a trend can be observed in these data in that a clear peak in activity occurs at about day 150. This peak followed an addition of 120 kg per ha of fertilizer N (Klemedtsson, 1986). Because the peak of denitrification activity could almost be predicted as a consequence of fertilizer addition, it seemed ill-advised to regard the data as stationary, *i.e.*, lacking a trend. Consequently, a model was arbitrarily chosen that had the right general shape to fit the peak, and this model was fit by nonlinear regression to the data. This bell-shaped model (the shape is the same as the normal distribution) is shown fit to the data from 1983 in the upper half of Fig. 7. The model explained a significant ($P < 0.001$) portion (53.9%) of the variability of the logarithmically transformed denitrification rates. Significant trends, thus, appeared to be present in the rates of denitrification measured within rows of grass in both years of the study.

To test for the presence of autocorrelation, semivariograms were constructed for the residuals left by the models used to account for the trends in the rates of denitrification within rows of grass. The semivariogram for the residuals of the model fit to

the 1983 data (not shown) was flat offering no indication of temporal autocorrelation. In fact, this semivariogram indicated that rates of denitrification measured one day apart tended to be more dissimilar to one another than rates measured at any more widely separated intervals up to one month. In contrast, the residuals left by the polynomial model used to remove the trend in denitrification rates in grass rows during 1982 appeared to be autocorrelated. A semivariogram constructed from this data appears in Fig. 8 with a fit of the spherical model (Webster, 1985). Two consequences of the removal of the trend in the 1982 data can be seen from a comparison of Fig. 8 with Fig. 6. First, the range over which autocorrelation appears to exist is greatly diminished by the removal of the trend. In addition, detrending has the expected effect of reducing the maximum value for semivariance achieved at the longest lags.

Autocorrelation appeared to be present in the denitrification rates measured between crop rows in grass leys during 1982 even when a trend in those data was removed with a cubic polynomial. The presence of autocorrelation offers an opportunity to use kriging to predict denitrification rates on unsampled dates with more accuracy than would be possible using only the polynomial model. It seemed of potential value to evaluate the accuracy of those predictions. Accordingly, kriging was used in a jackknifing procedure (Vieira *et al.*, 1983) to predict detrended values of denitrification for all the days when measured values were available. These predicted values were compared to the measured, detrended values, and it was determined that the kriged predictions could account for only 15.2% of the variation in denitrification rates after the polynomial trend had been removed. This reduction in unexplained variation is particularly unimpressive when it is compared to the variation explained (60%) by the polynomial trend in the 1982 data.

The inability of kriging to produce more accurate estimates for the 1982 study in the grass ley can be at least partially attributed to the sampling pattern. Although it is not readily apparent in Fig. 7, denitrification rates were measured within rows of grass during 1982 on many dates that were not within 3 days of another sampling date. Unfortunately, no autocorrelation appears to be present in the detrended rates of denitrification for rates measured 4 or more days apart (Fig. 8). Kriging would not be expected to provide good estimates for rates on dates separated by 4 or more days from the nearest sampling date because the kriging predictions are based on a weighted average of the measured rates in which more weight is given to rates measured on dates close enough to have autocorrelated rates. Thus, for dates separated by 4 or more days from the nearest date of sampling kriging generates a prediction that is nothing more than the unweighted average of all the data except for the rate (temporarily deleted for jackknifing) actually measured on the date for which the prediction is to be made.

The principal conclusion that can be drawn from our analyses of data on denitrification collected through time is that geostatistics does not appear to offer immediately useful tools for the analysis of this kind of data. Supporting this conclusion was our inability to find any evidence of temporal autocorrelation for rates measured in two (barley and lucerne) out of the three crops used by the denitrification group within the project 'Ecology of Arable Land'. Moreover, rates in grass leys that were measured between (rather than within) plant rows also failed to show any evidence of temporal autocorrelation. The rates measured within rows of grass showed superficial autocorrelation. For rates measured during 1983 this superficial autocorrelation was entirely attributable to a nonlinear trend resulting from fertilizer addition. During 1982, denitrification rates within grass rows also showed a strong trend, but there appeared to be evidence of autocorrelation in the rates that was not directly attributable to the trend. This autocorrelation permitted estimates to be generated using kriging that were more accurate than those offered by the detrending regression alone, but the improvement was very small in a quantitative sense. The modest utility of geostatistics in analyzing the data from only one of three crops, in only one of two sampling positions relative to plant rows, and in only one of two years does not encourage one to believe that geostatistical techniques will play a major role in future studies of the variation of denitrification rates measured through time.

Conclusions and recommendations

1. Acetylene and ^{15}N based methods as well as soil core and *in situ* cover sampling methods all give comparable estimates of field denitrification rates. Some of the methods are better suited for particular objectives or sites, but all have been proved to be sound methods for measurement of terrestrial denitrification.

2. The difficulty in quantifying denitrification lies not with the methods, as they accurately measure the process, it is due to the dynamic nature of the process that causes high temporal and spatial variability. Because of this, improvements in quantitation of denitrification are more likely to come from better approaches to analyze, model, and predict the variability than from further work on methodology. It's doubtful, however, that denitrification budget estimates will ever approach the accuracy of most other biogeochemical cycle measurements.

3. In studies of denitrification, perhaps we have too often asked an inappropriate question, namely 'How much nitrogen is lost by denitrification?'. While it is an important question, it can divert too many resources in one direction if that question is too difficult to answer and especially if the same approaches are repeatedly used. In our view, greater opportunities for the future lie in research

at scales other than the traditional field plot and on questions that have been relatively ignored. Particularly important are studies at both larger (landscape, regional) and smaller (microsites, organism, enzyme, gene) scales. We have illustrated examples of the concepts and approaches for work at some of these scales elsewhere (Groffman *et al.*, 1988). Important future opportunities also lie in understanding how denitrification is regulated at the molecular level and how these mechanisms are coupled to the environmental triggers at the microsite. On balance, we believe understanding of denitrification would benefit from more diversity in the questions asked and the scales studied.

Acknowledgements

We particularly thank all those denitrification researchers at Michigan State University who have helped in the evolution of the methods described here: in approximate chronological order they are: Mary Firestone, Scott Smith, Henry Kaspar, Tim Parkin, Alan Sexstone, David Myrold, Phil Robertson, Chuck Rice, and Søren Christensen. We also thank Leif Klemedtsson and Bo Svensson for the Arable Lands denitrification database that was used in the attempt at temporal kriging. The support for this research was provided by the U.S. National Science Foundation.

References

Abd-el-Malek Y, Hosny I and Emam N F 1974 Evaluation of media used for enumeration of denitrifying bacteria. Zentralbl. Bakteriol. Parasitenkd. Infektionski Hyg. 2 Abt. 415–421.

Aulakh M S, Rennie D A and Paul E A 1982 Gaseous nitrogen from cropped and summer-fallowed soils. Can. J. Soil Sci. 62, 187–196.

Balderston W L, Sherr B and Payne W J 1976 Blockage by acetylene of nitrous oxide reduction in *Pseudomonas perfectomarinus*. Appl. Environ. Microbiol. 31, 504–508.

Betlach M R 1982 Evolution of bacterial denitrification and denitrifier diversity. Antonie van Leeuwenhock J. Microbiol. 48, 585–607.

Blackmer A M, Robbins S G and Bremner J M 1982 Diurnal variability in rate of emission of nitrous oxide from soils. Soil Sci. Soc. Am. J. 46, 937–942.

Burton D L and Beauchamp E C 1984 Field techniques using the acetylene blockage of nitrous oxide reduction to measure denitrification. Can. J. Soil Sci. 64, 555–562.

Christensen S and Tiedje J M 1988 Denitrification in the field: Analysis of spatial and temporal variability. *In* Nitrogen Efficiency in Agricultural Soils. Eds D S Jenkinson and K A Smith. pp 295–301. Elsevier Appl. Sci. London, N.Y., USA.

Duxbury J M and McCounnaughey P K 1986 Effect of fertilizer source on denitrification and nitrous oxide emissions in a maize-field. Soil Sci. Soc. Am. J. 50, 644–648.

Duxbury J M 1986 Advantages of the acetylene method for measuring denitrification. *In* Field Measurement of Dinitrogen Fixation and Denitrification. Eds. R D Hauck and R W Weaver, pp 73–91. Soil Science Society of America, Madison, Wisconsin, USA.

Federova R I, Milekhina E I and Ilkukhina N I 1973 Possiblity of using the 'gas-exchange' method to detect extraterrestrial life: Identification of nitrogen-fixing organisms. Izv. Akad. Nauk Arm. SSR Biol. Nauki 6, 797–806.

Folorunso O A and Rolston D E 1984 Spatial variability of field-measured denitrification gas fluxes. Soil Sci. Soc. Am. J. 48, 1214–1219.

Folorunso O A and Rolston D E 1985 Spatial and spectral relationships between field-measured denitrification gas fluxes and soil properties. Soil Sci. Soc. Am. J. 49, 1087–1093.

Gamble T N, Betlach M R and Tiedje J M 1977 Numerically dominant denitrifying bacteria from world soils. Appl. Environ. Microbiol. 33, 926–939.

Goodroad L L and Keeney D R 1984 Nitrous oxide emissions from soils during thawing. Can. J. Soil Sci. 64, 187–194.

Groffman P M 1985 Nitrification and denitrification in conventional and no-tillage soils. Soil Sci. Soc. Am. J. 49, 329–334.

Groffman P M and Tiedje J M 1988 Denitrification hysteresis during wetting and drying cycles in soil. Soil Sci. Soc. Am. J. 52 1626–1629.

Groffman P M and Tiedje J M 1989a Denitrification in north temperate forest soils: Spatial and temporal patterns at the landscape and seasonal scales. Soil Biol. Biochem. (Accepted).

Groffman P M and Tiedje J M 1989b Denitrification in north temperate forest soils: Relationships between denitrification and environmental parameters at the landscape scale. Soil Biol. Biochem. (Accepted).

Groffman P M, Tiedje J M, Robertson G P and Christensen S 1988 Denitrification at different temporal and geographical scales: Proximal and distal controls. *in* Adv. in Nitrogen Cycling in Agricultural Ecosystems. Ed. J R Wilson. pp 174–192. CAB International, Wallingford, U.K.

Hallmark S L and Terry R E 1985 Field measurement of denitrification in irrigated soils. Soil Sci. 140, 35–44.

Hauck R D 1986 Field measurement of denitrification — an overview. *In* Field Measurement of Dinitrogen Fixation and Denitrification. Eds. R D Hauck and R W Weaver. pp 59–72. Soil Science Society of America, Madison, Wisconsin, USA.

Hauck R D, Melsted S W and Yankwich P E 1958 Use of N-isotope distribution in nitrogen gas in the study of denitrification. Soil Sci. 86, 287–291.

Heinemeyer O, Haider K and Mosier A 1988 Phytotron studies to compare nitrogen losses from corn planted soil by the 15-N balance or direct dinitrogen and nitrous oxide measurements. Biol. Fert. Soils 6, 73–77.

Hyman M R and Arp D J 1987 Quantification and removal of some contaminating gases from acetylene used to study gas-utilizing enzymes and microorganisms. Appl. Environ. Microbiol. 53, 298–303.

Journel A G and Huijbregts C J 1978 Mining Geostatistics. Academic Press, New York.

Juma N G and Paul E A 1981 Use of tracers and computer simulation techniques to assess mineralization and immobilization of soil nitrogen. *In* Simulation of Nitrogen Behavior of Soil-Plant Systems. Eds. M J Frissel and J A Van Veen. pp 145–154. Center for Agricultural Publishing and Documentation, Wageningen.

Jury W A, Letey J and Collins T 1982 Analysis of chamber methods used for measuring nitrous oxide production in the field. Soil Sci. Soc. Am. J. 46, 250–255.

Kaspar H F 1984 A simple method for the measurement of N_2O and CO_2 flux rates across undisturbed soil surfaces. N.Z. J. Sci. 27, 243–246.

Kaspar H F and Tiedje J M 1980 Response of electron-capture detector to hydrogen, oxygen, nitrogen, carbon dioxide, nitric oxide and nitrous oxide. J. Chromatography 193, 142–147.

Kaspar H F, Tiedje J M and Firestone R B 1981 Denitrification and dissimilatory nitrate reduction to ammonium in digested sludge. Can. J. Microbiol. 27, 878–885.

Keeney D R 1986 Critique of the acetylene blockage technique for field measurement of denitrification. *In* Field Measurement of Dinitrogen Fixation and Denitrification. Eds. R D Hauck and R W Weaver. pp 103–115. Soil Science Society of America, Madison, Wisconsin, USA.

Knowles R 1981 Denitrification. *In* Soil Biochemistry, Vol. 5, p 240. Eds. E A Paul and J N Ladd. Marcel Dekker, Inc. New York.

Koike I and Hattori A 1975 Growth yield of a denitrifying bacterium *Pseudomonas denitrificans* under aerobic and denitrifying conditions. J. Gen. Microbiol. 88, 1–10.

Korner H, Runzke K, Dohler K and Zumft W G 1987 Immunochemical patterns of distribution of nitrous oxide reductase and nitrite reductase (cytochrome cd_1) among denitrifying pseudomonads. Arch. Microbiol. 148, 20–24.

Klemedtsson L 1986 Denitrification in arable soil with special emphasis on the influence of plant roots. Report 32, Dept. of Microbiology. Swedish University of Agricultural Sciences, Uppsala, Sweden.

Klemedtsson L, Svensson B H and Rosswal T 1977 The use of acetylene inhibition of nitrous oxide reductase in quantifying denitrification in soils. Swedish J. Agric. Res. 7, 179–185.

Limmer A W, Steele K W and Wilson A T 1982 Direct field measurement of N_2 and N_2O evolution from soil. J. Soil Sci. 33, 499–507.

Lowrance R and Smittle D 1988 Nitrogen cycling in a multiple-crop vegetable production system. J. Environ. Qual. 17, 158–162.

McConnaughey P K and Duxbury J M 1986 Introduction of acetylene into soil for measurement of denitrification. Soil Sci. Soc. Am. J. 50, 260–263.

Matson P A and Vitousek P M 1987 Cross-system comparisons of soil nitrogen transformations and nitrous oxide flux in tropical forest ecosystems. Global Biogeochemical Cycles 1, 163–170.

Matthias A D, Blackmer A M and Bremner J M 1980 A simple chamber technique for field measurement of emissions of nitrous oxide from soils. J. Environ. Qual. 9, 251–256.

Michalski W P and Nicholas D J D 1988 Immunological patterns of distribution of bacterial denitrifying enzymes. Phytochemistry 27, 2451–2456.

Mosier A R, Guenzi W D and Schweizer E E 1986 Field denitrification estimation by nitrogen-15 and acetylene inhibition techniques. Soil Sci. Soc. Am. J. 50, 831–833.

Mosier A R, Guenzi W D and Schweizer E E 1986 Soil losses of dinitrogen and nitrous oxide from irrigated crops in north eastern Colorado. Soil. Sci. Soc. Am. J. 50, 344–348.

Myrold D D 1988 Denitrification in ryegrass and winter wheat cropping systems of western Oregon. Soil Sci. Soc. Am. J. 52, 412–415.

Myrold D D and Tiedje J M 1986 Simultaneous estimation of several nitrogen cycle rates using ^{15}N: Theory and application. Soil Biol. Biochem. 6, 559–568.

Myrold D D and Tiedje J M 1985 Diffusional constraints on denitrification. Soil Sci. Soc. Am. J. 49, 652–657.

Nommik H 1956 Investigations on denitrification in soil. Acta Agric. Scan. 6, 195–228.

Parkin T B 1985 Automated analysis of nitrous oxide. Soil Sci. Soc. Am. J. 49, 273–276.

Parkin T B 1987 Soil microsites as a source of denitrification variability. Soil Sci. Am. J. 51, 1194–1199.

Parkin T B, Kaspar H F, Sexstone A J and Tiedje J M 1984 A gas-flow soil core method to measure field denitrification rates. Soil Biol. Biochem. 16, 323–330.

Parkin T B, Meisinger J J, Chester S T, Starr J L and Robinson J A 1988 Evaluation of statistical estimation methods for lognormally distributed variables. Soil Sci. Soc. Am. J. 52, 323–329.

Parkin T B and Robinson J A 1989 Stochastic models of soil denitrification. Appl. Environ. Microbiol 55, 72–77.

Parkin T B, Sexstone A J and Tiedje J M 1985 Comparison of field denitrification rates determined by acetylene-based soil core and nitrogen-15 methods. Soil Sci. Soc. Am. J. 49, 94–99.

Parkin T B, Starr J L and Meisinger J J 1987 Influence of sample size on measurement of soil denitrification. Soil Sci. Soc. Am. J. 51, 1492–1501.

Parkin T B and Tiedje J M 1984 Application of a soil core method to investigate the effect of oxygen concentration on denitrification. Soil Biol. Biochem. 4, 331–334.

Pearson E S and Hartley H O 1958 Biometrika Tables for Statisticians, Volume 1. Cambridge Univ. Press, London.

Rice C W, Sierzega P E, Tiedje J M and Jacobs L W 1988 Stimulated denitrification in the microenvironment of a biodegradable organic waste injected into soil. Soil Sci. Soc. Am. J. 52, 102–108.

Robertson G P, Huston M A, Evans F C and Tiedje J M 1988 Spatial variability in a successional plant community: Patterns of nitrogen mineralization, nitrification, and denitrification. Ecology 69, 1517–1524.

Robertson G P and Tiedje J M 1987 Nitrous oxide sources in aerobic soils. Nitrification, denitrification and other biological processes. Soil Biol. Biochem. 19, 187–193.

Robertson G P and Tiedje J M 1985 An automated technique for sampling the contents of stoppered gas-collection vials. Plant and Soil 83, 453–457.

Robertson G P and Tiedje J M 1984 Denitrification and nitrous oxide production in successional and old-growth Michigan forests. Soil Sci. Soc. Am. J. 383–389.

Robertson G P, Vitousek P M, Matson P A and Tiedje J M 1987 Denitrification in a clearcut Loblolly pine (*Pinus taeda* L) plantation in the southeastern U.S. Plant and Soil 97, 119–129.

Rolston D E 1986 Limitations of the acetylene blockage technique for field measurement of denitrification. *In* Field Measurement of Dinitrogen Fixation and Denitrification. Eds. R D Hauck and R W Weaver. pp 93–101. Soil Science Society of America, Madison, Wisconsin, USA.

Rolston D E, Broadbent F E and Goldhamer D A 1979 Field measurements of denitrification. II. Mass balance and sampling uncertainty. Soil Sci. Am. J. 43, 703–708.

Rolston D E, Fried M and Goldhamer D A 1976 Denitrification measured directly from nitrogen and nitrous oxide gas fluxes. Soil Sci. Soc. Am. J. 40, 259–266.

Rolston D E, Hoffman D L and Toy D W 1978 Field measurement of denitrification. I. Flux of N_2 and N_2O. Soil Sci. Soc. Am. J. 42, 863–869.

Rolston D E, Rao P S C, Davidson J M and Jessup R E 1984 Simulation of denitrification losses of nitrate fertilizer applied to uncropped, cropped and manure-amended field plots. Soil Sci. 137, 270–279.

Rolston D E, Sharpley A N, Toy D W and Broadbent F E 1982 Field measurement of denitrification. III. Rates during irrigation cycles. Soil Sci. Soc. Am. J. 46, 289–296.

Romermann D and Friedrich B 1985 Denitrification by *Alcaligenes entrophus* is plasmid dependent. J. Bacteriol. 162, 852–854.

Ryden J C, Lund L J, Letey J and Focht D D 1979 Direct measurement of denitrification loss from soils. II. Development and application of field methods. Soil Sci. Soc. Am. J. 43, 110–118.

Ryden J C and Dawson K P 1982 Evaluation of the acetylene-inhibition technique for the measurement of denitrification in grassland soils. J. Sci. Food Agric. 33, 1197–1206.

Ryden J C, Skinner J H and Nixon D J 1987 Soil core incubation system for the field measurement of denitrification using acetylene-inhibition. Soil Biol. Biochem. 19, 753–757.

Schmidt J, Seiler W and Conrad R 1988 Emission of nitrous oxide from temperate forest soils into the atmosphere. J. Atmos. Chem. 6, 95–115.

Sexstone A J, Parkin T B and Tiedje J M 1988 Denitrification response to soil wetting in aggregated and unaggregated soil. Soil Biol. Biochem. 20, 767–769.

Sexstone A J, Parkin T B and Tiedje J M 1985 Temporal response of soil denitrification rates to rainfall and irrigation. Soil Sci. Soc. Am. J. 48, 99–103.

Siegel R S, Hauck R D and Kurtz L T 1982 Determination of $^{30}N_2$ and application to measurement of N_2 evolution during denitrification. Soil Sci. Soc. Am. J. 46, 68–74.

Smith C J 1988 Denitrification in the field. *In* Advances in Nitrogen Cycling in Agricultural Ecosystems. Ed J R Wilson. pp 387–398. C.A.B. International, Wallingford, U.K.

Smith M S, Firestone M K and Tiedje J M 1978 The acetylene inhibition method for short-term measurement of soil denitrification and its evaluation using nitrogen-13. Soil Sci. Soc. Am. J. 42, 611–615.

Smith M S and Tiedje J M 1979 Phases of denitrification following oxygen depletion in soil. Soil Biol. Biochem. 11, 261–267.

Svensson B H, Klemedtsson L and Rosswall T 1985 Preliminary field denitrification studies of nitrate-fertilized and nitrogen-fixing crops. *In* Denitrification and the Nitrogen Cycle. Ed. H L Golterman. pp 157–169. NATO Conference Series I: Ecology Vol. 9. Plenum Press, London.

Terry R E and Duxbury J M 1985 Acetylene decomposition in soils. Soil Sci. Soc. Am. J. 49, 90–94.

Terry R E, Jellen E N and Breakwell D P 1986 Effect of irrigation and acetylene exposure on field denitrification measurements. Soil Sci. Soc. Am. J. 50, 115–120.

Tiedje J M 1982 Denitrification. *In* Methods of Soil Analysis, Park 2nd ed. Ed. A L Page. Agronomy Monogr. 9, 1011–1026. Amer. Soc. Agron., Madison, Wisc.

Tiedje J M 1988 Ecology of denitrification and dissimilatory nitrate reduction to ammonium. *In* Biology of Anaerobic Microorganisms. Ed. A J B Zehnder. pp 179–244. John Wiley & Sons, New York.

Tiedje J M, Firestone R B, Firestone M K, Betlach M R, Kaspar H F and Sørenson J 1981 Use of ^{13}N in studies of denitrification. *In* Short-lived radionuclides in Chemistry and Biology. Eds. J W Root and K A Krohn. pp 295–315. American Chemical Society.

Tiedje J M, Sexstone A J, Parkin T B and Revsbech N P 1984 Anaerobic processes in soil. Plant and Soil 76, 197–212.

Topp E and Germon J C 1986 Acetylene metabolism and stimulation of denitrification in an agricultural soil. Appl. Environ. Microbiol. 52, 802–806.

Vieira S R, Hatfield J L, Nielsen D R and Biggar J W 1983 Geostatistical theory and application to variability of some agronomical properties. Hilgardia 51, 1–75.

Webster 1985 Quantitative spatial analysis of soil in the field. Adv. Soil Sci. 3, 1–70.

Yeomans J C and Beauchamp E G 1982 Acetylene as a possible substrate in the denitrification process. Can. J. Soil Sci. 62, 139–146.

Yeomans J C and Beauchamp E G 1978 Limited inhibition of nitrous oxide reduction in soil in the presence of acetylene. Soil Biol. Biochem. 10, 517–519.

Yoshinari T and Knowles R 1976 Acetylene inhibition of nitrous oxide reduction by denitrifying bacteria. Biochem. Biophys. Res. Commun. 69, 705–710.

Zar J H 1974 Biostatistical Analysis. Prentice-Hall, Inc., Englewood Cliffs, N. J.

M. Clarholm and L. Bergström (Eds.), Ecology of arable land, 241–246.
© 1989 Kluwer Academic Publishers.

Impact of agricultural landscape structure on cycling of inorganic nutrients

LECH RYSZKOWSKI and ALINA BARTOSZEWICZ
Institute of Agrobiology and Forestry, Polish Academy of Sciences, PL-60-809 Poznan, Swieczewskiego 19,
Poland and Department of Soil Sciences, Agricultural Academy, PL-60-623 Poznan, Mazowiecka 42,
Poland

Key words: biogeochemical barrier, groundwater, landscape, nutrient leaching, shelterbelt

Abstract

Increased application of mineral fertilizers resulted in an increase in leaching of nitrogen and other elements from the investigated arable soils. Differences in the depth of the groundwater table were ascertained in order to determine the directions of subsurface water flows influenced by gravity. Biogeochemical barriers chosen for study included shelterbelt, forests and meadows situated so that the groundwater outflow from neighbouring cultivated fields passed under them. The analyses of element concentrations in subsurface water flows showed that shelterbelts, forests and meadows had a strong influence on the chemistry of water draining from the arable fields. The most pronounced effects concerned nitrate. All direct and indirect evidence indicates that by manipulating the plant cover in agricultural landscapes the chemistry of subsurface and surface water can be greatly influenced.

Introduction

In agricultural watersheds large amounts of migrating nutrients are usually leached out from cultivated soils. There is no doubt that the increase in application rates of commercial fertilizer has resulted in elevated concentrations of nitrogen and other elements in waters situated in and around agricultural areas (*e.g.*, Brink, 1978; Duncan and Rzoska, 1980; Frissel, 1977). Furthermore, decomposition of soil organic matter due to tillage activities also contributes to elevated concentrations of elements in water bodies (Viets, 1971).

To decrease water pollution caused by agricultural activities it has been proposed that shelterbelts (planted rows of trees) or stretches of meadow intersecting an agricultural landscape be established to counteract wind and water erosion. Such measures are hypothized to decrease the rate of eutrophication of inland waters (Ryszkowski, 1974a). Studies to test this hypothesis were carried out as part of a project on energy flow and matter fluxes coordinated by the Institute of Agrobiology and Forestry in Poznan, Poland (Ryszkowski, 1974b; 1979; 1980).

Area description

The studied area is located about 40 km south of Poznan in the West Polish Lowland. The terrain consists of a rolling plain of slightly undulating moraine with many drainage basins. The differences in elevation between high- and low-lying areas are not more than a few metres. Light soils with high infiltration capacities are found on the higher parts of the area. Peat soils, having relatively high water-retention capacities, occur in small depressions. The depth to the groundwater table ranged from 0.5 to 5.0 m below the soil surface. The depth at any given spot fluctuated during the year depending on the specific conditions.

The climate in the area is one of the warmest in Poland with a mean annual air temperature of 8°C. The growing season, with air temperatures above 5°C, lasts 225 days. On average, it begins March 21 and ends October 30. Mean annual precipitation (1881–1985) amounts to 527 mm, which could be a factor limiting plant growth.

About 70% of the area is covered by cultivated fields. Meadows and pastures cover ca. 12%, while shelterbelts and forests cover ca. 14%. The rest of the region is composed of villages, roads, small lakes, channels and waterlogged areas. In general, the typical crop composition is ca. 50% cereals (rye, wheat, barley, and oats), ca. 25% row crops (beets, potatoes) including rape-seed, ca. 10% perennial fodder crops, and ca. 15% other crops.

Shelterbelts, *i.e.*, planted strips of woody plants which can consist of several parallel rows, are characteristic components of the Turew landscape. These shelterbelts consist of false acacia, poplars, oaks, pines, spruces, and a small number of other tree species.

Mechanisms of manipulating groundwater chemistry using biogeochemical barriers

By taking the estimated relationship between climatological characteristics and the growth stage of vegetation into account, as well as the solar energy used for evapotranspiration, the amount of water transpired by various types of plant cover has been estimated (Kedziora and Olejnik, in press; Olejnik, 1986; Tamulewicz and Wos, in press). Ryszkowski and Kedziora (1987) showed that a shelterbelt can transpire about 40% more water than a field of wheat (Table 1) when located in the same area. There are two main reasons for this phenomenon: First, there is a great difference in the plant cover structure between these two ecosystems. Trees have much longer roots than wheat plants; therefore the former are able to take up water from deeper layers of the soil profile com-

pared with the latter. Thus trees have better access to water. The second reason for the higher evapotranspiration of trees is the relatively higher wind speeds and turbulence to which the shelterbelt canopy is exposed. This results in greater vapour exchange over shelterbelts than over cultivated fields. The comparison between shelterbelts and meadows showed that, for the same reason, shelterbelts transpired about 23% more water than meadows (Table 1). Thus shelterbelts function as powerful 'environmental water pumps' and should thereby exert great influence on groundwater chemistry when groundwater is within direct or indirect (capillary water) reach of the tree root system. Meadows, which are usually located in terrain depressions, often along draining channels, have root systems reaching shallow groundwater reservoirs. Grass roots could thereby affect the chemistry of passing groundwater.

Ion exchange capacities of soils under shelterbelts or meadows are different from those in cultivated field soils, which could also lead to differential effects on passing groundwater. In this paper, however, no attempt was made to differentiate between the effects of these two mechanisms.

Control of groundwater chemistry by biogeochemical barriers

In addition to determining groundwater flow direction, information on infiltration rates of soil profiles was gathered. Watershed fragments were

Table 1. Evapotranspiration (mm) in various ecosystems in the Turew region during the growing season March 21–October 31. (After Ryszkowski and Kedziora, 1987)

Ecosystem	Shelterbelt	Meadow	Rapeseed	Beet	Wheat
Evapotranspiration	542	439	416	405	389

Table 2. Element concentrations (mg l^{-1}) in groundwater under cultivated field and part of a forest adjacent to the field, during August 1982–September 1986. n = 182 for cultivated field; n = 89 for forest. (After Bartoszewicz and Ryszkowski, *in press*)

Elements	Cultivated field		Forest	
	Mean	Range	Mean	Range
NO_3^--N	22.2	1.8–210	1.0	0.1–4.7
NH_4^+-N	2.5	0.3–10.8	2.0	0.4–9.3
PO_4^{3-}-P	0.21	<0.01–1.11	0.11	<0.01–1.03
K^+	8.1	1.6–34.0	4.2	1.1–2.7
Ca^{2+}	158	96–392	82	25–120
Mg^{2+}	15.8	0.5–36.0	8.6	2.4–14.1

chosen that contained shelterbelts, forests and meadows situated so that the groundwater outflow from neighbouring cultivated fields passed under them. The soils of the chosen plots are presented in detail by Margowski *et al.* (1976) and Bartoszewicz (1979; 1985). The depth of the groundwater table in the field and forest system varied from 3.5 to 5 m below the soil surface during the four years of study. In the field and shelterbelt system the groundwater table varied from 1.1. to 5.0 m, and in the field and meadow system the groundwater table was always shallow, varying from 0.6 to 2 m.

Observations in the field and the forest as well as in the shelterbelt and the meadow showed that the forest as well as the meadow had strong influence on the chemistry of water draining from the arable fields. The most pronounced effects concerned concentrations of nitrate (Table 2). Since nitrates are not adsorbed by soil colloids to any significant extent, these differences can mainly be attributed to the action of a complex set of biological factors determining nitrogen passage through afforested soils.

A considerable decrease in the nitrate concentration of groundwater in the forest was observed 50 m from the field. Mean nitrate concentrations only exceeded 5 mg N l^{-1} in the forest groundwater during periods when the mean concentration in the field groundwater was 35 and 55 mg N l^{-1}. Significant differentiation in nitrate concentration also occurred in groundwater of the shelterbelt (Table 3). During the study, the mean value of nitrate in the forest and shelterbelt groundwaters was 22 and 34-fold lower than that in neighbouring fields. However, exceptionally high leaching of nitrate from the arable soil also caused a stepwise increase of nitrate concentrations in groundwater of the forest and the shelterbelt (Fig. 1), confirming the connection between groundwaters of the cultivated fields and that of the shelterbelt or forest.

Among other elements, a proportionally similar concentration decrease occurred in calcium and magnesium as groundwater passed from field to forest or shelterbelt (Tables 2 and 3). The concentration levels of these elements in water flowing through the forest or shelterbelt were significantly lower than those flowing through neighbouring arable fields. Small amounts of phosphorus compounds were detected which contributed little to the overall ionic composition of the groundwater.

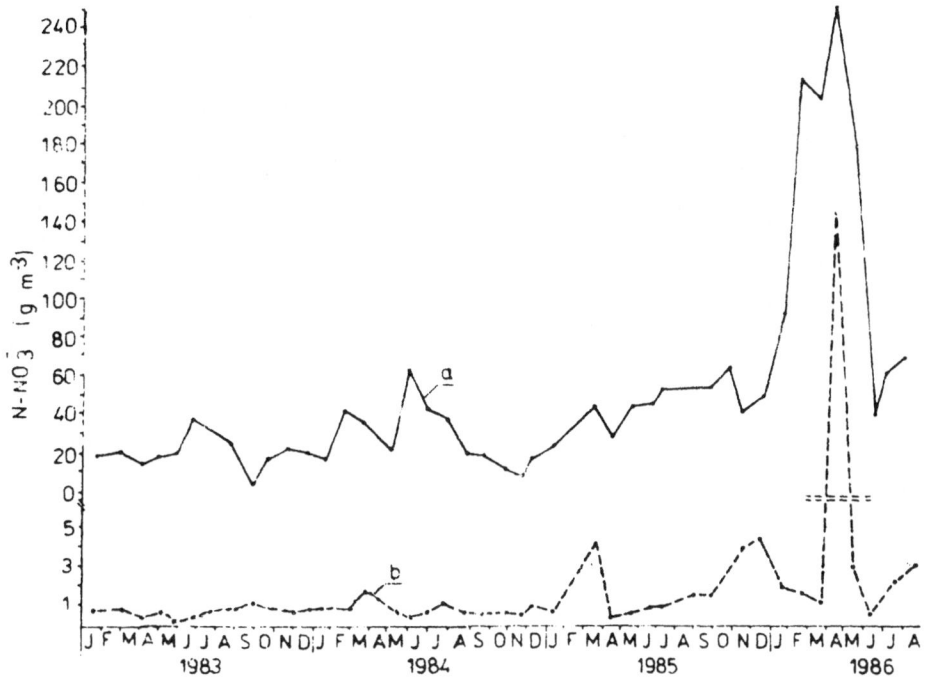

Fig. 1. Mean monthly concentrations of NO$_3$-N in groundwater: January 1983–August 1986. (After Bartoszewicz and Ryszkowski, *in press*) **a** – cultivated field; **b** – shelterbelt.

Table 3. Element concentrations $(mg\,l^{-1})$ in groundwater under cultivated field and shelterbelt, during August 1982–September 1986. n = 74 for cultivated field; n = 47 for shelterbelt. (After Bartoszewicz and Ryszkowski, *in press*)

Elements	Cultivated field		Shelterbelt	
	Mean	Range	Mean	Range
$NO_3^- $-N	37.6	2.4–248	1.1	0.1–4.3
$NH_4^+ $-N	2.1	0.6–10.0	4.5	0.1–23.5
$PO_4^{3-} $-P	0.08	<0.01–0.73	0.06	<0.01–0.55
K^+	5.6	2.5–17.0	10.7	2.6–49.1
Ca^{2+}	198	113–640	116	47–492
Mg^{2+}	41.1	17.0–72.1	18.4	0.5–74.4

Table 4. Concentrations $(mg\,l^{-1})$ of the different nitrogen forms in groundwaters and surface waters in large pine forest and arable land. (After Margowski and Bartoszewicz, 1976)

Form of nitrogen	Pine-forest		Arable land	
	Groundwaters	Stream waters	Groundwaters ditch waters	Drainage
$NO_3^- $-N	0.3	0.1	12.6	8.1
$NH_4^+ $-N	2.7	1.3	1.4	0.8
Organic-N	1.5	2.4	0.9	0.8
Total-N	4.5	3.8	14.9	9.8
Mineral-N in % of total-N	66.7	36.8	94.0	90.8

Nevertheless, phosphate concentrations were observed to decrease as water passed through the forest or the shelterbelt.

Conversely, changes in potassium and ammonium concentrations in groundwater passing under the cultivated fields and tree stands were observed. Mean concentrations of potassium and ammonium in water from the shelterbelt were twice as high as those in water from the neighbouring fields (Table 3). The differences with respect to nitrate and ammonium levels in groundwater between arable fields and forest-shelterbelts deserve attention. Nitrate predominates in water from arable fields, while groundwaters in forests and shelterbelts have relatively higher concentrations of ammonium. The nitrogen economy observed within shelterbelts or small forests situated among arable fields resembles that of large forest complexes (cf. Table 4). Thus, one can conclude that the differences in the rates of nitrification and denitrification processes characteristic of large forest ecosystems persist in small forests or shelterbelts surrounded by cultivated fields and cause increased concentrations of ammonium in groundwater passing under

Table 5. Mean concentrations of elements $(mg\,l^{-1})$ in groundwater from fields, meadows and drainage ditches. n = 57 for field; n = 68 for meadow 15–25 m from field; n = 51 for meadow 60–75 m from field; n = 33 for water in drainage ditch, both periods. (After Bartoszewicz and Ryszkowski, *in press*)

Elements	Groundwater			Water in drainage ditch	
	Field	Meadow		Period 1972–76	Period 1981–85
		15–25 From field	60–75 m From field		
$NO_3^- $-N	10.4	2.4	0.7	1.8	1.1
$NH_4^+ $-N	2.4	4.6	2.9	1.3	1.8
$PO_4^{3-} $-P	0.19	0.23	0.07	0.07	0.11
K^+	8.5	10.7	6.6	8.4	10.2
Ca^{2+}	162.3	262.0	195.1	80.0	128.1
Mg^{2+}	16.6	26.1	21.5	16.1	17.4

them. The enrichment of potassium in groundwater under shelterbelts may be associated with decomposition of litter.

In groundwater from fields and in groundwater passing under a stretch of meadow 80–90 meters wide the greatest changes in ion content occurred for nitrate (Table 5). Significant decreases in nitrate concentration in groundwater were observed 15–25 m into the meadow. At sampling points more distant from the field, the mean concentration of nitrate was 15 times lower than that in groundwater under the cultivated field. This was true during the entire study period. Changes in ammonium, phosphorus, and potassium concentrations in water were less obvious. In contrast, the groundwater concentrations of calcium and magnesium in the meadow were higher than those in the field.

Chemistry of water in open ditches

The concentrations of nitrate, ammonium, phosphate, and calcium were lower in ditch water than in groundwater of the cultivated field, while potassium and magnesium showed slightly higher concentrations in ditch water. The ditch canal, 4 m wide and 0.5 m deep, drains water from a lake situated 6000 m from the ditch. Along its 6000 m run, water in the ditch crosses a moor and collects sewage from several small villages. Thus the chemistry of the ditch water is influenced not only by the inflow of subsurface water from the studied landscape, but also from other sources. However, the nitrate concentration in ditch water was, on average, nine times less than that of groundwater in the cultivated field. Simultaneously, the nitrate concentration in ditch water was slightly higher than that in groundwater of the meadow collected 60–75 m from the field boundary. This difference is most likely the combined result of nitrification in surface water and pollution by sewage. It should be noted that the concentrations of the six investigated ions in ditch water have not changed during the last 14 years, during which time the application rates of artificial fertilizers in the region have been practically the same.

Despite all the factors influencing ion concentrations in ditch water, there is little doubt that meadows enhance 'self-purification' processes in running waters.

Chemistry of drainage water

Another example of the impact of plant cover on drainage water chemistry in agricultural watersheds is provided by comparative analysis of the chemistry of stream outflow from watersheds having various percentages of arable land. Results of such analyses do not have as direct bearing on the problem as do the studies of forests, shelterbelts and meadows, reported above, where the changes in water passing directly under a given plant cover were observed. Nevertheless, the information can be useful when addressing questions related to water pollution. In thirteen watersheds classified as either high, moderate or low intensity agricultural areas, Borowiec *et al.* (1978) observed leaching of ions from a typical hectare of the catchment area for two years. Stress criteria were based on plant structure (arable land, forest and grassland) of the watershed, as well as tillage intensity. Despite the overlap between some of the estimates, the differences between the three stress categories were well defined (Table 6). In low intensity watersheds, which contained an average of 30% arable land, 50% of the calcium, and 54% of magnesium were leached in comparison with corresponding losses from the high intensity watersheds. Thus as the area of land under cultivation increases there is a corresponding increase in leaching of elements.

Conclusions

The results of this investigation indicate that considerable amounts of mineral components can be leached from light soils. In addition to causing decreases in agricultural production, leaching also decreases the purity of agroecosystem waters. High concentrations of nitrate are particularly dangerous. The mean concentration of nitrate within the studied watershed was about $13\,mg\,N\,l^{-1}$ during the study (1972–1976). Recently (1981–1985), this value exceeded $20\,mg\,N\,l^{-1}$.

It was found that small patches of forest or meadow interspersed within a predominantly agricultural area had a strong influence on the matter cycling of biogenic components over the entire area. As water from cultivated fields passed under shelterbelts or meadows the element concentrations were significantly reduced. The most pronounced

Table 6. Influences of plant cover on ion leaching. (After Borowiec *et al.*, 1978)

Degree of agricultural stress	Water-sheds	Areas in % of			Outflow of elements during two years ($g\,m^{-2}$)				
		Arable fields	Forests	Grass lands	$NO_3^- $-N	PO_4^{3-}-P	K^+	Ca^{2+}	Mg^{2+}
High	1	53	12	35	1.6	0.03	3.3	38	2.8
	2	62	8	28	1.4	0.04	3.0	35	3.3
	3	53	32	13	1.1	0.04	2.6	31	2.3
	4	51	21	27	0.7	0.02	2.9	32	2.7
	Mean	55	18	26	1.2	0.03	2.4	34	2.8
Moderate	1	60	20	20	1.2	0.02	1.3	36	2.7
	2	44	36	19	0.7	0.02	1.7	24	1.9
	3	49	32	18	0.6	0.02	1.5	6	2.0
	4	45	43	12	0.6	0.02	1.5	26	2.2
	5	40	36	23	0.6	0.02	1.7	22	1.6
	Mean	48	33	18	0.7	0.02	1.5	26	2.1
Low	1	38	44	17	0.6	0.02	1.6	23	1.8
	2	29	45	26	0.5	0.01	1.0	26	1.4
	3	21	65	14	0.4	0.01	0.9	23	1.6
	4	32	47	21	0.4	0.01	0.9	15	1.2
	Mean	30	50	19	0.5	0.01	1.1	22	1.5

changes were observed in nitrate concentration.

All of the direct and indirect findings in this study point in the same direction – *i.e.*, that by manipulating the plant cover on agricultural landscapes the chemistry of subsurface and surface waters can be influenced substantially.

Acknowledgements

This study was carried out within the project 'Natural bases for protection and management of agricultural landscapes' (CPBP 04.10.03) supported by the Ministry of Education in Poland.

References

Bartoszewicz A 1979 Mineralization of ground and surface waters in different soils. Roczniki Akademii Rolniczej Poznan 91, 1–53. (*In Polish*).

Bartoszewicz A 1985 Concentration of some ions in ground water of arable soils. Roczniki Adademii Rolniczej Poznan 127, 19–31. (*In Polish*).

Bartoszewicz A and Ryszkowski 1989 Influence of shelterbelts and meadows on the chemistry of ground water. *In* Dynamics of an Agricultural Landscape. Ed. L Ryskowski. Springer Verlag, New York (*in press*).

Borowiec S, Skrzynski T and Kucharska T 1978 Migration der mineralischen Bestandteile aus den Boden der Nizina Szczecinska. Polish Scientific Publishers, Warszawa. 68 p. (*In Polish*).

Brink N 1978 Nitrogen leaching from arable land. Ekohydrologi 2, 31–39. Swedish University of Agricultural Sciences, Uppsala. (*In Swedish*)

Duncan N and Rzoska J Eds. 1980 Land Use Impacts on Lake and Reservoir Ecosystems. Facultas-Verlag, Wien. 294 p.

Frissel M J Ed. 1977 Cycling of mineral nutrients in agricultural ecosystems. Agro-ecosystems 4, 1–354.

Kedziora A and Olejnik J 1989 Heat balance structure of agroecosystems. *In* Dynamics of an Agricultural Landscape. Ed. L Ryszkowski. Springer-Verlag, New York (*in press*).

Margowski Z and Bartoszewicz A 1976 Leaching of basic nutrients into ground waters. Nawozenie a ertrofizacja wod. Wyzsza Szkola Inzynierska Zielona Gora. (*In Polish*).

Margowski Z, Bartoszewicz A and Siwinski A 1976 Soil formed from boulder loam containing sand in the upper layers of the Koscian Plain. Pol. Ecol. Stud. 2, 5–13.

Olejnik J 1986 Influence of the vegetation on turbulence fluxes of heat and vapour in boundary layer. PhD thesis. Department of Agrometeorology of Agricultural Academy in Poznan, Poland.

Ryszkowski L 1974a Matter cycling in agroceanoses. Zeszyty Problemowe Postepow Nauk Rolniczych. 155, 19–38. (*In Polish*).

Ryszkowski L Ed. 1974b Ecological effects of intensive agriculture. Polish Scientific Publishers, Warszawa. 84 p.

Ryszkowski L 1979 Croplands *In* Grassland Ecosystems of the World. Ed. R T Coupland. pp 301–331. Cambridge University Press, Cambridge.

Ryszkowski L 1980 Introduction to the volume: Research on agricultural landscape. Pol. Ecol. Stud. 6, 105–111.

Ryszkowski L and Kedziora A 1987 Impact of agricultural landscape structure on energy flow and water cycling. Landscape Ecology 1, 85–94.

Tamulawicz J and Wos A 1989 Radiation characteristics of the landscape. *In* Dynamics of an Agricultural Landscape. Ed. L Ryszkowski. Springer Verlag, New York (*in press*).

Viets F G 1971 Water quality in relation to farm use of fertilizer. BioScience 21, 460–467.

M. Clarholm and L. Bergström (Eds.), Ecology of arable land, 247–251.

The movement of nutrients across heterogeneous landscapes

PAUL G. RISSER

Department of Biology, University of New Mexico, Albuquerque, NM 87131, USA

Key words: biogeochemical cycles, ecosystem, landscape ecology

Abstract

Over the past two decades, nutrient budgets have been constructed for a number of specific ecosystems ranging from croplands to forests. However, relatively few studies have focused on the transfers between and among ecosystem types. These transfers and their dynamics depend on properties of the landscape units themselves as well as various mechanisms that influence inter-ecosystem transfers driven by such as atmospheric, hydrological and biological conditions. The movements of nutrients, especially nitrogen, and phosphorous, are subject to several processes such as absorption, adsorption, transformation and immobilization. Thus, the particular pathway of transfer across the landscape and the rates of movement depend on both transfer mechanisms and ecosystem properties. This paper reviews the current understanding of nutrient flows across the landscape, suggests some generalities, and presents recommendations for future research.

Introduction

During the past two decades, considerable progress has been made in understanding biogeochemical cycles. Much of this research effort has been devoted to describing the behavior of watersheds, especially as these have been subjected to various management techniques. More recently, attention has been focused on the pattern in which landscape units are positioned within the landscape. Of particular importance are the topographical position and the juxtapositions of landscape units. These landscape settings are crucial to understanding the ways in which watersheds operate in general, but especially at the land-water interface (Bormann and Likens, 1981; Decamps, 1984; Verry and Timmons, 1982). This emphasis on the positioning of landscape unit within a watershed has led to an appreciation of the importance of ecological processes at a variety of temporal and especially spatial scales ranging from the landscape (Forman and Godron, 1986; Risser *et al.*, 1984; Urban *et al.*, 1987) to smaller patches (Loucks, 1985; Pickett and White, 1985). Despite this focus on both biogeo-

chemical cycles and landscape dynamics, there have been relatively few attempts to quantify the behavior of nutrient dynamics across landscapes. Thus, the purposes of this paper are to focus specifically on the atmospheric, hydrological and biological pathways by which nutrients move across the heterogeneous landscape, to describe known examples where landscape characteristics affect the behavior of these nutrient transfers, and to postulate potentially valuable applications of the results of this research.

Evidence of landscape-nutrient dynamic interactions

By far the most conspicuous examples of vegetation controls of nutrient dynamics at the landscape level have involved the role of bottomland forests and wetlands (Brinson *et al.*, 1984; Elder, 1985; Karr and Schlosser, 1978; Richardson, 1985). Both wetlands and bottomland or riparian forests capture nutrients, organic matter, and sediments. These materials are differentially captured by each vegetation type. As will be shown below, specific losses or chemical transformations frequently occur within these landscape units.

Nutrients travel across the landscape from one ecosystem unit to another. Understanding the processes of transport is essential for building simplified or comprehensive views of landscape nutrient dynamics. For example, much of the phosphorous moving across the landscape does so attached to sediment particles. Thus, the degree to which a vegetation zone within the landscape retains sediment is an indicator of its nutrient (phosphorous) trapping capacity. This capacity of vegetation to trap sediments, and therefore nutrients, depends upon several variables (Karr and Schlosser, 1978); surface water levels and their periodicities relative to the height of the vegetation; width and slope of the vegetated area; morphological characteristics of the individual plants and physiognomic or structural characteristics of the vegetation; size distribution of the incoming sediment particles; rate of water flow through the vegetation zone; and characteristics of the slope above the vegetated area. This trapping efficiency has been measured in several instances and the acquired information is perhaps adequate to construct a preliminary, even generally applicable, regression model of these processes. This simple model could then be tested under various situations where relatively routine data are available describing these components of the landscape, *e.g.*, water quality data and vegetation measurements. If these simple models were combined with geographic information system data bases containing the appropriate spatially-oriented data, projections could be made of the ways in which alterations of the landscape would affect the nutrient retention capacity of the landscape as a whole.

Although the distribution pattern of vegetation across the landscape clearly affects the transport of sediment and nutrients, the soil characteristics are also important. Schimel *et al.* (1985b) compared the soil organic matter dynamics on paired rangeland and cropland toposequences in North Dakota, USA. Prolonged agricultural practices had caused soil organic matter losses from the croplands, but the amounts of these losses were significantly different among the three soils: sandstone, shale, and siltstone toposequences of montmorillonitic, typic haploboralls and argiboralls.

Table 1. Proportional loss of organic carbon after 44 years of cultivation from three soil toposequences in eastern Colorado, USA (Adapted from Schimel *et al.*, 1984)

Toposequence position	Loss of organic carbon (%)		
	Sandstone	Siltstone	Shale
Summit	− 54	− 45	− 45
Shoulder	− 40	− 49	− 53
Upper backslope	− 56	− 39	− 24*
Lower backslope	− 18*	− 49	− 24*
Footslope	− 34	− 37	− 24
Average weighted by area	− 34	− 46	− 35

* Depositional soil

The sandstone soils were relatively resistant to organic matter loss because they were able to maintain apparently stable organic matter concentrations despite erosion. As compared to the rangeland soils, the shale cropland soils were resistant to erosional effects because the clayey soil retained higher amounts of organic matter (Table 1). However, the siltstone sequence was unstable with larger losses of organic matter because much of the organic matter in the soil profile was near the soil's surface and the soil had a low resistance to erosion. The loss of organic matter from the landscape was a function of soil type, topographic position, and the amount of the watershed composed of each soil. Thus, a landscape characterized by a number of different soil types and topographic positions will demonstrate a heterogeneous pattern of organic matter production, retention, and organic loss to erosional processes.

This study indicates the importance of considering the landscape heterogeneity when measuring the behavior of landscape units, even larger more widespread watersheds. Total watershed carbon and soil dynamics would depend upon the amount of each soil type. Though the outputs of the entire watershed were not calculated and attributed to the outputs of each soil type, such an analysis could be constructed. An example of such a study is the detailed evaluation of four subwatersheds of the Little River in the Georgia (USA) Coastal Plain from 1979 to 1981, where Lowrance *et al.* (1984a, b, c; 1985) measured the inputs and outputs of nitrogen, phosphorous, potassium, calcium, magnesium and chlorine. These watersheds were composed of 40, 36, 54 and 50% agricultural row crop and pasture vegetation types. Fertilizer inputs exceeded precipitation inputs of all elements, and except for chlorine, there were greater total inputs than outputs from the watersheds. The two watersheds with more agricultural land had higher streamloads of nitrogen, potassium, calcium and chlorine, with NO_3-N streamloads higher by 1.5 to 4.4 times that of the watersheds with lesser amounts of agricultural land. Rowcrops were the primary source of NO_3-N while pastures contributed more NH_4-N (Lowrance *et al.*, 1983). Though there were differences among landscape units, several general properties can be described: the vegetation demonstrated a large buffering capacity for nitrogen and phosphorous, less of a buffering capacity in terms of potassium, calcium and magnesium, and little effect on chlorine. It was also notable that about 56 kg ha^{-1} yr^{-1} nitrogen was either retained within the watersheds or lost to gaseous emissions. Or stated differently, from these watershed studies between 45 and 55% of the nitrogen inputs were unaccounted for or lost from the uplands of these watersheds.

These investigations of broad landscape flows have indicated that various materials flow at differential flow rates, and mass balance calculations have indicated that significant losses of nitrogen occur but these losses depend upon land use patterns. Another example demonstrating the importance of landscape heterogeneity at a smaller, more subtle scale is a study by Schimel *et al.* (1986) on the volatilization of nitrogen from what appeared to be a relatively simple grazed shortgrass steppe in Colorado, USA. The pattern of soil and forage properties within such a grassland influences cattle behavior, and as a consequence, the loss of nitrogen through the deposition pattern of urine (Senft *et al.*, 1985). Within this particular landscape, the upper slopes were characterized by coarser textured soils than those on the lower slopes. Loss of nitrogen was higher from the coarse upland sites (0.016 g N m^{-2} yr^{-1}) than from the finer textured lowland sites (where nitrogen loss was negligible), although the acutal mechanisms of ammonium retention as a function of soil characteristics could not be determined from the results of this study. However, it is clear that even in this apparently uniform landscape of a shortgrass steppe on rolling topography, it would be an error to assume that nitrogen volatilization loss would be

the same throughout the landscape because losses from lowland and upland sites are so different. Indeed, such an assumption would significantly overestimate the volatile loss of nitrogen from this landscape because of the lower loss rates from the lowland sites. Other differences between upland and lowland sites have been found in N_2O flux and the processes of mineralization and nitrification (Schimel *et al.*, 1985a). The results indicate that landscapes may be quite heterogeneous and as a result, different landscape units affect the flows of nutrients across the landscape.

These three studies have been described in some detail because they represent an important direction in our understanding of landscape-level biogeochemical processes. In each case, the landscape heterogeneity has influenced the behavior of nutrient dynamics within small parts of a watershed or landscape. In some cases, the differences have been dramatic, thus indicating the amount of information lost by simply invoking the average behavior at the level of the watershed, whether this watershed is a relatively small but seemingly uniform grassland landscape (Colorado, USA), or whether it consists of many square kilometers of mixed agricultural landscape in the North American southeastern Coastal Plain.

As indicated at the beginning of this discussion, focusing on bottomland and riparian forests is a reasonable and obvious approach for evaluating the effect of landscape heterogeneity on nutrient flows. However, the previously described studies indicate that landscapes usually contain a variety of units so it is necessary to understand how each unit behaves. The last point in this discussion will finalize the contention that just knowing how each unit behaves is an insufficient approach to landscape-level biogeochemical processes. Indeed, it is necessary to understand the interactions among these units.

In an important study of the nutrient dynamics within a heterogeneous landscape, Peterjohn and Correll (1984) considered the flows of nitrogen, phosphorous and carbon. The 16.3 ha basin located in Maryland, (eastern seaboard of the U.S.) consisted of 10.4 ha of corn cropland and 5.9 ha of hedgerows and riparian forest dominated by sweetgum (*Liquidambar stryciflua*) and red maple (*Acer rubrum*). Nutrient samples were collected from both surface and subsurface flows through the

Table 2. Percent change in concentration of selected nutrients across riparian forest within an agricultural watershed in Maryland, eastern U.S. (Peterjohn and Correll, 1984)

Nutrient	Meters within forest		
	0[a]	19	50
Nitrate	100	40	21
Ammonium	100	30	26
Organic-N	100	23	15
Total-P	100	29	19

[a] Uphill edge of riparian forest

soil; also, measurements or calculations were made of nutrient uptake by plants in the cropland and the riparian forest. On the basis of mass balance, the watershed as a whole captured most of the nutrient inputs. As might be expected, the nutrient retention capability of the native vegetation and its associated soil system was greater than that of the corn field. In fact, the riparian forest trapped considerable amounts of material in the surface runoff (expressed as kg per ha of forest): 4.1 particulates; 11.0 particulate organic-N; 0.83 dissolved ammonium-N; 2.7 nitrate-N; and 3.0 total particulate-P. In addition, the forest removed 45 kg per ha of nitrate-N in subsurface flows.

A transect of samples aligned through the riparian forest enabled a measurement of the rate of nutrient capture across the forest. Most (60–75%) of the nutrients were captured within the first 19 m of the forest (Table 2). The dominant pathway of total-N loss from the forest was through the subsurface pathway; total phosphorous loss was evenly divided between subsurface and surface losses (Table 3). Only about one-third of the nitrogen loss in the subsurface flow could be accounted for by uptake in the vegetation. Significant amounts of denitrification therefore may occur in the soils of the riparian forest (Peterjohn and Correll, 1984).

This study re-emphasizes the importance of landscape pattern in understanding the behavior of heterogeneous watersheds. But even more significant is the realization that simply summing or aggregating similar landscape units for the purpose of predicting watershed behavior will not be significant. Indeed, the configuration and juxtaposition patterns of landscape units will affect the collective behavior of the units within a landscape.

Conclusions

For the past two decades, the primary previous focus of research on nutrient dynamics has been on developing budgets for individual nutrients and on understanding the soil and vegetation porcesses within single and presumably uniform vegetation types. These studies have permitted the understanding of many of the basic processes controlling the status and transformations of nutrients such as nitrogen and phosphorous. Furthermore, the biotic and abiotic controls of these processes are reasonably well understood in most of the major types of ecosystems.

In this paper, several selected recent studies have been summarized which indicate the importance of the processes which control the movements across heterogeneous landscapes rather than just within landscape units. This approach represents a potentially very rich area of research since it mandates an extension of our understanding about single ecosystems to the spatial and temporal scales that define landscapes and eventually to even broader and more heterogeneous regions. Nutrient flows are controlled by a number of biotic and abiotic processes. The studies described in this paper indicate that there are some similarities in results which will permit generalization to other

Table 3. Inputs and outputs (%) of nitrogen and phosphorous measured for a riparian forest in Maryland, eastern U.S. (Peterjohn and Correll, 1984)

Transfer pathway	Nitrogen		Phosphorous	
	Input	Output	Input	Output
Precipitation	17	0	4	0
Surface	22	25	94	59
Groundwater	61	75	2	41
% Retention		98		80

ecological situations, for example, phosphorous attached to sediment particles can perhaps be modelled with reasonably simple sediment transfer models which have already been developed and are in use in agricultural systems. On the other hand, two of these studies indicate nitrogen losses approaching 50% while traversing upland and riparian forested areas. The processes by which these losses occur are not completely understood, and as a result, more empirical research and trial process models (*e.g.*, denitrification, microbial immobilization, decomposition of small and large organic debris) will necessarily precede simple landscape-level model applications.

The clearly necessary objective of extending nutrient flows and dynamics to the landscape level will eventually require models of nutrient flows described from specific field measurements. At the moment more investigations will be required for evaluating the less-well understood processes, especially those that depend upon landscape heterogeneity and configuration patterns of landscape units. However, the papers discussed here indicate that preliminary models could be constructed of some landscapes at the present time. These models could take one of several forms. For example, landscape units could be described in terms of nutrient inputs and outputs, then the topographic sequences of these units could be aggregated to the landscape level while still preserving the juxtapositional relationships of the units. Alternatively, landscape flows could be measured and modelled as a landscape process where the characteristics of each landscape unit affect the total flow. As these approaches mature, there will be a whole new set of principles to extend our understanding of nutrient dynamics into different spatial and temporal scales. And just as importantly, it will offer descriptive models which will be of great use to those who have the responsibility of managing the heterogeneous landscapes of the World.

References

Brinson M M, Bradshaw M D and Kane E S 1984 Nutrient assimilative capacity of an alluvial floodplain swamp. J. Appl. Ecol. 21, 1041–1057.

Bormann F H and Likens G E 1981 Pattern and Process in a Forested Ecosystem. Springer-Verlag, New York.

Decamps H 1984 Towards a landscape ecology of river valleys. *In* Trends in Ecological Research in the 1980s. Eds. J H Cooley and F B Golley. pp 163–178. Plenum Press, New York, 344 p.

Elder J F 1985 Nitrogen and phosphorous speciation and flux in a large Florida river wetland system. Water Resource Res. Bull. 21, 724–732.

Forman R T T and Godron M 1986 Landscape Ecology. John Wiley and Sons, New York. 619 p.

Karr J R and Schlosser I J 1978 Water resources and the land-water interface. Science 201, 229–234.

Loucks O L 1985 Looking for surprise in managing stressed ecosystems. BioScience 35, 428–432.

Lowrance R, Todd R and Asmussen L 1983 Waterborne nutrient budgets for the riparian zone of an agricultural watershed. Agriculture, Ecosystems and Environment 10, 371–384.

Lowrance R, Todd R and Asmussen L 1984a Nutrient cycling in an agricultural watershed. I. Phreatic movement. J. Environ. Qual. 13, 22–27.

Lowrance R, Todd R and Asmussen L 1984b Nutrient cycling in an agricultural watershed. II. Streamflow and artificial drainage. J. Environ. Qual. 13, 27–32.

Lowrance R, Todd R, Fail J Jr., Leonard R and Asmussen L 1984c Riparian forests as nutrient filters in agricultural watersheds. BioScience 34, 374–377.

Lowrance R R, Leonard R A, Asmussen L E and Todd R L 1985 Nutrient budgets for agricultural watersheds in the southeastern coastal plain. Ecology 66, 287–296.

Peterjohn W T and Correll D L 1984 Nutrient dynamics in an agricultural watershed: Observations on the role of a riparian forest. Ecology 65, 1466–1475.

Pickett S T A and White P S 1985 The Ecology of Natural Disturbance and Patch Dynamics. Academic Press, New York. 472 p.

Richardson C J 1985 Mechanisms controlling phosphorous retention capacity in freshwater wetlands. Science 228, 1424.

Risser P G, Karr J R and Forman R T T 1984 Landscape Ecology: Directions and Approaches. Illinois Natural History Survey, Special Publication Number 2, Champaign. 18 p.

Schimel D, Stillwell M A and Woodmansee R G 1985a Biogeochemistry of C, N, and P in a soil catena in the shortgrass steppe. Ecology 66, 276–282.

Schimel D D, Coleman D C and Horton K A 1985b Soil organic matter dynamics in paired rangeland and cropland toposequences in North Dakota. Geoderma 36, 201–214.

Schimel D S, Parton W J, Adamsen F J, Woodmansee R G, Senft R L and Stillwell M A 1986 The role of cattle in the volatile loss of nitrogen from a shortgrass steppe. Biogeochemistry 2, 29–52.

Senft R L, Rittenhouse L R and Woodmansee R G 1985 Factors influencing selection of resting sites by cattle on shortgrass steppe. J. Range Manag. 38, 295–299.

Urban D L, O'Neill R V and Shugart H H Jr. 1987 Landscape ecology. BioScience 37, 119–127.

Verry F S and Timmons D R 1982 Waterborne nutrient flow through an upland-peatland watershed in Minnesota. Ecology 63, 1456–1457.

M. Clarholm and L. Bergström (Eds.), Ecology of arable land, 253–259.

Development, validation and applications of simulation models for agroecosystems: Problems and perspectives

P.S.C. RAO, J.W. JONES and G. KIDDER
Institute of Food and Agricultural Sciences, University of Florida, Gainesville, FL 32611, USA

Key words: model development, model validation, model applications, spatial modelling, uncertainty
analysis

Abstract

Several key issues relevant to the development, validation, and applications of simulation models for soil-crop systems are discussed. Development of models with modular components providing multiple options for conceptual representation of the system components is recommended. Combining simulation models with the concepts of artificial intelligence will facilitate the development of user-interactive interfaces which permit the user to customize the model, based on expert guidance, for a specific application. Most models do not account for the spatial and temporal variabilities in input parameters. Uncertainty in model predictions resulting from such variations in the input parameters needs to be accounted for. Minimum data sets required for model development and validation as well as objective criteria for assessing model performance need to be identified. The application of crop-soil simulation models to estimate the probable success of a specific crop production management recommendation (*i.e.*, risk analysis) and evaluating the regional variations in crop performance using spatial modelling techniques are discussed.

Introduction

It is widely recognized that the vast complexity and diversity of agroecosystems makes it difficult to understand thoroughly the intricate interactions among the various system components and the factors controlling system processes. Considerable effort is usually required to investigate experimentally even the most important processes and factors. Given the limited understanding of the agroecosystems, and other biological systems as well, it has been argued that development and testing of simulation models may be a futile exercise. On the other hand, it is also argued that attempts to develop simulation models contribute to organization of existing knowledge and will provide guidance for further research by revealing gaps in the knowledge base on the system of interest. As a result of the efforts of a large number of scientists working independently in several countries, and more recently

in coordinated efforts, the early skepticism on the utility of simulation models has now given way to a more optimistic outlook. During the past decade, a number of simulation models have been developed for predicting the performance of several crops under a variety of management practices and for a broad range of environmental conditions.

A critical review of the available crop-soil system models or a synthesis of the published data on crop responses under various climatological conditions and management treatments is beyond the scope of this paper. Rather, we will discuss some key issues that need to be addressed in further development and testing of simulation models and their application for management of agroecosystems. Our experiences in developing and testing simulation models for water and nitrogen dynamics in arable lands and coupling these to models for crop growth and farm economics are used here as a basis for such a discussion.

254   *Rao* et al.

Development of simulation models

A number of simulation models have been developed and tested for predicting nitrogen dynamics in the crop root zone. These models have been reviewed, classified, and organized in a number of different ways (*e.g.*, Frissel and van Veen, 1981; Tanji, 1982). The available models vary considerably in their level of complexity and conceptual completeness (Rao *et al.*, 1982).

Model development strategy

The spatial and temporal scales at which the different components or processes are modeled vary considerably among models and often even within a given model. The intended use of the model usually governs the level of complexity at which a given component of the system is modeled. Since models have been put to multiple uses (*e.g.*, research *vs.* management) and because different model developers conceptualize the system or system components in various ways, a number of models have been developed.

In essentially all of the models currently available one of the two approaches have been utilized in model formulation: (1) a comprehensive model is developed that accommodates all possible applications and uses of the model; or (2) several versions of the model are developed, each for a different application. For example, in developing a simulation model for studying water and fertilizer management options for a rice crop, it is necessary to first decide if there will be separate model versions for upland and flooded rice or a single version of the model. A single comprehensive model will need to handle the significant differences in the physical and biochemical processes occurring within the root zone of these cropping systems. On the other hand, different versions of the model would only account for specific processes occurring in upland or flooded rice. Adopting the first approach to model development generally requires the writing of an extremely large and inefficient computer code, and the model input parameter requirements are likely to be large. Choosing the second approach leads to a simulation model of somewhat limited utility and requires the model users to be familiar with the various versions of the same crop-soil model.

Buttler and Rhea (1987) identified the following three reasons as being the most important as to why simulation models are not more widely used: (1) inflexibility of computer codes to easily accomodate various applications of the model; (2) poor documentation of the model; and (3) lack of 'user-friendly' interfaces. An attractive alternative is to develop a modular computer code which offers the model user multiple options for modelling a given system process. An efficient, front-end, user-friendly interface can be written such that a potential user can, in fact, customize his own model for a specific application by selecting among the various submodel modules that are available as a part of the computer code. For example, a model for a rice crop then would contain two separate submodels for predicting transport and transformations of the nitrogen species within the root zone; one for the upland crop and the other for the flooded case. Each of these submodels, in turn, can contain other modules which permit the user to select among various levels of complexity of describing a specific process. Each of the submodels would have to provide a list of input and output parameters as well as the rules or criteria by which the user can make a decision in choosing among the submodels. It is feasible to include modules that allow for calculating the values for model input parameters either on the basis of available data bases or by using empirical estimation methods (*e.g.*, regression equations). The modularity of the crop-soil models necessitates that the input and output formats of all submodels of a given system component or process be compatible.

Development of simulation models as described above has just begun to be attempted. Uschold *et al.* (1985) described the development of a front-end, user-interactive interface for formulating ecological models. Buttler and Rhea (1987) presented a general purpose simulation model for describing water flow in the soil-plant-atmosphere continuum. The approach they used in model development was to 'create a program that would provide the user with more than one option to simulate a particular process, the choice being dependent on available input data, desired level of complexity, and objectives.' For example, their model provides the user with several options for computing evapotranspiration and for predicting water flow. They are extending this modelling approach for simulating nitrogen transport and transformation in the crop root zone.

On-going efforts at the University of Florida are exploring the feasibility of using artificial intelligence concepts in organizing submodels with various levels of complexity and matching them to the user's intended application and the availability of input and output data. The combining of simulation models with the concepts of aritifical intelligence (*e.g.*, expert systems) has the potential for increasing the use of models with expert guidance as decision aids in the management of agroecosystems.

Representation of uncertainty

With a very few exceptions, most of the simulation models currently in use can be termed deterministic, *i.e.*, the model parameters are considered to be single-valued. Given the spatial and temporal variability of soils, crops and climate, each of the model parameters is, in fact, represented by a population with a unique frequency distribution. Addiscott and Wagenet (1985) state that the key distinction between the two types of modelling approaches is that the deterministic models 'presume that a system or a process operates such that the occurrence of a given set of events leads to a uniquely-definable outcome,' whereas the stochastic models 'presuppose that the outcome will be uncertain and are structured to account for this uncertainty.' Thus, the deterministic models utilize only a single realization of each parameter value (usually the mean value) and produce a single realization (not necessarily the mean) of the output. Such models then fail to account for the uncertainty (*e.g.*, the confidence intervals about the mean) of the model outputs which result from the variability in the model inputs. This becomes an especially important consideration in comparing model outputs with measured data in model testing and validation efforts. While the application of scaling concepts (Tanji *et al.*, 1981; Wagenet and Rao, 1983) and Monte Carlo techniques (Amoozegar-Fard *et al.*, 1982; Carsel *et al.*, 1988; Persaud *et al.*, 1985) have been evaluated for certain components of the system (*e.g.*, water and solute transport submodels), such techniques have been used only to a limited extent in developing agroecosystem simula-

tion models. A number of geostatistical techniques useful for dealing with the spatial and temporal variability encountered in agroecosytems have been reviewed by Nielsen and Alemi (1989).

Specific problems with flooded soils

In contrast to the large number of models available for describing nitrogen dynamics in upland soils (either irrigated or rainfed), models for flooded soils are almost nonexistent. A lack of interest in modelling such systems is surprising given that flooded rice is one of the major staple food crops of the world. Since in flooded soils (*e.g.*, those planted to paddy rice) saturated, steady water flow conditions prevail, simple submodels can be used to represent water and solute transport in the root zone. However, submodels for the biochemical processes controlling nitrogen and carbon transformations may need to be fairly complex. For example, oxidized regions form around rice roots growing in an otherwise anoxic, flooded soil. The dynamics of the evolution of such oxidized rhizocylinders and their impact on the gaseous nitrogen losses of ammonium fertilizers via denitrification are not well understood. Somewhat similar problems are also encountered in quantifying the enhancement of denitrification losses as a result of increased amounts of available organic carbon near the plant roots (Klemedtsson, pers. comm.) as well as in predicting denitrification rates in well-drained aerobic soils that may contain anoxic microsites (Leffelaar, 1987).

Nitrogen simulation models for flooded soils are currently being developed and tested (Rao *et al.*, 1984; 1987). Such models need to be coupled to crop growth models for rice. With only minor modifications, these models have also been used to describe nitrogen dynamics in shallow aquatic systems (Reddy *et al.*, 1988).

Validation of models

The validation of a model involves comparison of the model simulations against measured data. Thus, the model input parameters characterizing the soil-crop system must be known, and experimental data to compare with the model outputs must be available.

Minimum data sets

Depending on the level of complexity, each simulation model will require a minimum amount of input and output data without which model validation cannot be completed. Following Nix (1984) we refer to these data as the 'minimum data set.' He identified the following hierarchy of minimum data sets needed for model development and testing: (1) The absolute minimum data required for a simple analysis of the interactions between crop genotype and the environment as well as to compare the crop performance at widely-spaced sites or seasons are designated as Level 0; (2) Data collected to provide a basis for development and testing of process-based models of crop growth, development, and yield are designated as Level 1; and (3) Data needed to develop and test specific process models (*e.g.*, photosynthesis, nutrient uptake) and to couple them to formulate crop growth models are designated as Level 2.

It is evident that the effort and resources needed to collect the required data rapidly increase with the level of the minimum data set. In recognition of such resource constraints on most field studies, Nix (1984) also suggested that, 'Since it is neither necessary nor practical to model a crop system or subsystem in more detail than required for explanation and useful prediction, it should be obvious that our major interests will be served by minimum data sets at Level 0 and 1. Provided that the whole system is monitored at Level 0, there is no reason why a particular system should not be monitored at higher levels (1 or 2) if it is of particular interest.'

While some of the model input parameters can be 'estimated' by calibrating the model to measured data, independent estimates of all the model input parameters are desirable. Researchers conducting field studies on crop yield responses to water and nutrient management do not usually collect the required minimum data sets, even at Level 0. Of those that do, sufficient auxiliary data are seldom collected to meet the Level 1 requirements. Thus, an unfortunate limitation is that data collected in the vast majority of field trials cannot be used for model development and verification. It has been argued that the primary objective of most field trials is *not* to gather data suitable for model validation, but to develop management guidelines for crop production. However, with little extra effort the complimentary data satisfying both needs can be collected. The minimum data sets and the appropriate experimental protocols for collecting them need to be identified through coordinated efforts of inter-disciplinary teams of modelers and experimental scientists.

Criteria for model performance

In order to merit the confidence of a user, a simulation model must be able to adequately predict the system responses under a variety of conditions (*e.g.*, crop yield in a number of climatic zones or for different management conditions). This requires that quantitative and statistically-valid criteria be established for judging the goodness-of-fit of the model predictions to measured data. Most model validation efforts involve a qualitative and subjective judgement of the success of the model.

A common technique used for assessing the performance of a simulation model is to plot the measured and predicted responses (*e.g.*, crop yields; nutrient storage in the root zone) under a wide range of experimental conditions and using linear regression techniques to check for a 1:1 correspondence (*i.e.*, a unit slope and zero intercept). The statistical validity of this technique may be questioned on the grounds that both the measured data and the predicted outcomes are subject to error. The sum of the squared deviations between the measured and predicted values has also been used as an indicator of the goodness-of-fit. A better statistical technique is to test for the equivalence of the measured and predicted statistical distributions of a given output (*e.g.*, crop yield). While Monte Carlo simulation techniques may be used to generate the frequency distribution of the model output, experimental measurements are usually insufficient for valid determination of the statistical frequency distribution of the measured data.

Consideration of spatial variability

Rao and Wagenet (1986) noted that the total variability in field measurements of pesticide residue concentrations consists of two parts: intrinsic variability and extrinsic variability. The first may be a random component and is attributed to the variability introduced by the inherent variability in soil properties. Extrinsic variability is introduced as a result of non-random patterns caused by management practices such as row cropping, drip irrigation, and banded applications of pesticides. Such effects are also important in field studies on nitrogen dynamics in the crop root zone. Saka (1984) reported that the changes in the inorganic nitrogen concentration distributions in the soil profile within a corn field were highly variable, primarily as a result of banded fertilizer application (side-dressing). Appropriate soil sample protocols (*e.g.*, numbers and locations of samples) to account for such variability are not always used in field studies, casting doubt on the statistical validity of the data. Data for the crop parameters (such as yield, N uptake, *etc.*) are less likely to be subject to the errors introduced by the extrinsic variability because usually the soil volume explored by the crop is much larger than that sampled for nutrient content analysis.

Applications of models

The initial stimulus for the development of a model is usually the desire to organize the available knowledge base in order to *understand* the interactions among the system processes and to *predict* the likely system responses. As the confidence in the simulation model grows after a period of testing and validation, it is used for providing guidance in *management*.

Risk analysis

The traditional approach for providing crop production management guidance to farmers has been to use yield-response data gathered from repeated field trials. By conducting such field trials over a sufficiently broad range of soils, climate, and management practices, it is hoped that the recommendations made for a given crop are valid. However, the variations in the observed crop yield responses resulting from a number of site-specific factors (*e.g.*, differences in soils or year-to-year variations in the rainfall patterns) make it difficult to make consistent and reliable recommendations to the farmers.

The traditional empirical approach based on data from field trials has another shortcoming. It fails to provide a measure of the *risk* associated with a particular management recommendation. For example, the probability that a specific management practice will in fact produce an optimum crop yield or maximize net profit is seldom provided by the traditional approach. The impact of the seasonal variations in rainfall patterns on crop yield at a given site or the crop response under different irrigation and fertilizer application schedules are also not evaluated in a systematic manner. The use of simulation models in developing management strategies may overcome such limitations imposed by the traditional approach. Crop-soil model simulation may provide guidance in crop production management in a more time- and cost-effective manner than do field trials.

Monte Carlo techniques and crop simulation models have been used to perform risk analysis and to determine the profitability of installing irrigation systems for soybean production (Boggess and Amerling, 1983; Monson *et al.*, 1986). These authors have also applied these techniques to forecast the expected variations in soybean yield responses as a consequence of the within-season decisions to irrigate the crop. Jones *et al.* (1983) utilized Monte Carlo techniques and a simulation model for pesticide behavior in the crop root zone to assess the likelihood of groundwater contamination as influenced by the timing of pesticide application and year-to-year variations in weather.

Spatial modelling

The spatial patterns in crop yields over an entire geographic area (*e.g.*, a watershed) are also of interest. For example, can we identify specific regions of lower (or higher) crop yields if a uniform management scheme were to be imposed over an entire geographic area? Can we identify portion of a geographical area where leaching losses are expected to be the highest and might contribute disproportionately towards groundwater contamination?

The use of spatial modelling techniques to address such questions is receiving increasing attention (Khan and Liang, 1985; Burrough, 1986; Hoshikawa and Matsuda, 1987; Uehara and Yost, 1987; Torssell, 1989). Spatial modelling techniques can be employed to predict crop performance over an entire area, given the spatial patterns in soil properties, climate, and other model input parameters. Such spatial data bases can be digitized and retrieved using a number of commercial Geographic Information Systems (GIS), and combined with specific crop-soil simulation models to display the areal patterns in crop yield or the leaching potentials for a given management. Khan *et al.* (1986) coupled a GIS system developed at the University of Hawaii and simple indices of pesticide leaching to identify areas within a watershed that are likely to have high potential for groundwater contamination.

With the increasing availability of fairly inexpensive commercial GIS software designed for personal computers, the prospects for coupling crop-soil models with GIS software and the applications of spatial modelling are likely to rise. While the spatial data needed for this purpose already exist in a number of maps, they have not been digitized and cannot be readily accessed by the GIS software. This situation needs to be remedied.

References

Addiscott T A and Wagenet R J 1985 Concepts of solute leaching in soils: A review of modelling approaches. J. Soil Sci. 36, 411–424.

Amoozegar-Fard A, Nielsen D R and Warrick A W 1982 Soil solute concentration distributions for spatially-varying pore-water velocities and apparent diffusion coefficients. Soil Sci. Soc. Am. J. 46, 3–9.

Boggess W G and Amerling C B 1983 A bioeconomics simulation of irrigation investments. Southern J. Agric. Econ. 16, 85–91.

Burrough P A 1986 Principles of Geographic Information Systems for Land Resource Assessment. Monographs on Soil and Resource Survey, No. 12, Oxford Sci. Publ., Clarendon Press, Oxford, UK.

Buttler I and Rhea S J 1987 General purpose simulation model of water flow in the soil-plant-atmosphere continuum. Appl. Agric. Res. 2, 230–234.

Carsel R F, Parrish R S, Jones R L, Hansen J L and Lamb R L 1988 Characterizing the uncertainty of pesticide leaching in agricultural soils. J. Contam. Hydrol. (*In press*).

Frissel M J and van Veen J A Eds. 1981 Simulation of nitrogen behaviour in soil-plant systems. Centre for Agric. Publ. and Documentation, Pudoc, The Netherlands.

Hoshikawa K and Matsuda M 1987 Estimation and verification of the regional potential productivity. II. Study on the estimation of the mass-production in a rural area. Trans. Japan Soc. Irrig. Drainage and Reclamation Eng. 128, 11–21. (*In Japanese*).

Jones R L, Rao P S C and Hornsby A G 1983 Fate of aldicarb in Florida citrus soils. 2. Model evaluation. *In* Characterization and Monitoring of the Unsaturated (Vadose) Zone. Eds. D Nielsen and M Curl. pp 959–978. National Water Well Assoc., Dublin, OH.

Khan M A and Liang T 1985 A nutrient requirement mapping system for new crop introduction. Agric. Systems 16, 25–53.

Khan M A, Liang T, Rao P S C and Green R E 1986 Use of a geographic information system to assess the potential for groundwater contamination with pesticides. Trans. Amer. Geophys. Union 67, 278.

Leffelaar P A 1987 Dynamics of partial anaerobiosis, denitrification, and water in soil: Experiments and simulation. Doctoral Thesis, Agricultural University Wageningen, Wageningen, The Netherlands, 117 p.

Monson M, Boggess W G and Jones J W 1986 A user's guide to IRRIGATE: An irrigation decision aid. Florida Coop. Ext. Serv. Circular 720, Univ. of Florida, Gainesville, FL.

Nielsen D R and Alemi M H 1989 Statistical opportunities for analyzing spatial and temporal heterogeneity of field soils. Plant and Soil 115, 285–296.

Nix H A 1984 Minimum data sets for agrotechnology transfer. *In* Proc. Intern. Symp. on 'Minimum Data Sets for Agrotechnology Transfer.' pp. 181–188 Intern. Crops Res. Inst. for Semi-Arid Tropics (ICRISAT), Hyderabad, India.

Persaud N, Giraldez J V and Chang A C 1985 Monte-Carlo simulation of non-interacting solute transport in a spatially heterogeneous soil. Soil Sci. Soc. Am. J. 49, 5652–568.

Rao P S C and Wagenet R J 1986 Spatial variability of herbicides in field soils: Methods for data analysis and consequences. Weed Sci. 33(Suppl. 2), 18–24.

Rao P S C, Jessup R E and Hornsby A G 1982 Simulation of nitrogen in agroecosystems: Criteria for model selection and use. Plant and Soil 67, 35–43.

Rao P S C, Jessup R E and Reddy K R 1984 Simulation of nitrogen dynamics in a flooded soil. Soil Sci. 138, 54–62.

Rao P S C, Jessup R E and Reddy K R 1987 Nitrogen dynamics in the root zone of flooded rice: Development of a simulation model NRICE. Agronomy Abstracts, p. 191, Am. Soc. Agron., Madison, WI. (*In manuscript*).

Reddy K R, Jessup R E and Rao P S C 1988 Nitrogen dynamics in an eutrophic lake sediment. Hydrobiologia (*In press*).

Saka A R 1984 Nitrogen movement, retention, and uptake in the corn (*Zea mays* L.) root zone as influenced by cultivation and water management. Ph.D. Dissertation, University of Florida, Gainesville, FL, 178 p.

Tanji K K 1982 Modelling of the soil nitrogen cycle. *In* Nitrogen in Agricultural Soils, Ed. F J Stevenson, Agronomy Monograph 22, pp 721–772, Am. Soc. Agron., Madison, WI.

Tanji K K, Mehran M and Gupta S K 1981 Water and nitrogen fluxes in the root zone of irrigated maize. *In* Simulation of Nitrogen Behaviour in Soil-Plant Systems, Eds. M J Frissel and J A van Veen. pp 51–66. Center for Agric. Publ. and Documentation, Pudoc, Wageningen, The Netherlands.

Torsell B W R 1989 The application to agriculture of predictive plant production models based on regional experimental data. *In* Ecology of Arable Land – Perspectives and Challenges. Eds. M Clarholm and L Bergström. pp 1–7. Kluwer Academic Publishers, Dordrecht, The Netherlands.

Uehara G and Yost R S 1987 Spatial and temporal agroecosystem variability. *In* Future Developments in Soil Science

Research, Eds. L L Boersma *et al.* pp 43–57. Soil Sci. Soc. Am., Madison, WI.

Uschold M, Harding N, Muetzelfedt R and Bundy A 1985 An intelligent front-end for ecological modelling. *In* Adv. in Artificial Intelligence, Ed. T O'Shea. pp 13–22. Elsevier Sc.

Publ., Amsterdam, The Netherlands.

Wagenet R J and Rao B K 1983 Description of nitrogen movement in the presence of spatially-variable soil hydraulic properties. Agric. Water Mangmt. 6, 227–242.

M. Clarholm and L. Bergström (Eds.), Ecology of arable land, 261–272.

Statistical opportunities for analyzing spatial and temporal heterogeneity of field soils

D.R. NIELSEN and MOHAMMAD H. ALEMI
Department of Land, Air and Water Resources, University of California, Davis, CA 95616, USA

Key words: coherency, geostatistics, kriging, state-space, stochastic, time-series analysis

Abstract

Statistical techniques for analyzing data in the agricultural sciences have traditionally followed the pioneering efforts of R.A. Fisher who assumed that observations in the field were independent and identically distributed. Such techniques, proven useful in the past and still being used today for comparing the merits of different management practices or different treatments, are presently giving way to additional methods that are based upon observations being spatially or temporally correlated. It is physically more sensible to expect soil attributes to be correlated when they are measured at adjacent points in space or time. Spatially repetitious patterns of soil attributes for physical and biological processes occurring at distances of a few molecules to those of kilometers are also expected. Opportunities to use geostatistics, time series analyses, state-space models, spectral analyses of variance, lagged regression models and other alternative techniques for analyzing multidimensional random fields are available to enhance the understanding of agro-ecosystems. Approaches to modeling and fitting data using stochastic partial differential equations and scaling techniques also help reveal the underlying processes occurring in field soils. Inclusion of these alternatives in the development of an agro-ecological technology leads to improved sampling designs to better entire management units, rather than ascertaining the impact of particular, sometimes arbitrary treatments applied to a set of small plots using analysis of variance methods.

Introduction

Although agricultural scientists have conducted experiments throughout the world at an accelerated pace during the last half century, they have been less than successful explaining and managing biological processes that occur at different scales of space and time. The soil chemist, for example understands the kinetics of chemical reactions within a homogeneous soil sample but cannot usually extend the results to a larger, more heterogeneous rhizosphere. Likewise, the microbiologist cannot easily account for shifts in microbial populations within the rhizosphere when the crop canopy experiences nonuniform, larger scale microclimate perturbations. And many agronomists or plant scientists never "trust their luck" with only one year's data, but rely on fertilizer trials made repetitiously during a period of several years. Too often they have found that the results from one year are not "typical" or that the response of the crop to the treatments was biased owing to a variety of local conditions not included in the experimental plan.

Because of the educational opportunities and research techniques usually made available to agricultural or biological scientists, most field experiments conducted by them consist of relatively small plots receiving replicated treatments arranged in a randomized block or similarly related statistical design. The size of the plot is usually selected on the basis of the area needed for a "good" measurement of crop yield, the area perceived to effectively establish a treatment, and the area considered adequate for obtaining samples or that area required for various cultural practices. The total collage of plots is usually squeezed into a sufficiently small area to hopefully guarantee the entire experiment is performed on a uniform site representative of a larger management unit. The investigators have the belief that an analysis of variance of the result stemming from their semi-empirical treatments will allow the development of superior management alternatives for agricultural production for sites of similar characteristics. Such alternatives are indeed developed with whole fields being treated or managed the same, but seldom are quantitative observations made without resorting to the use of additional treated plots to assess the appropriateness or impact of the management alternative.

The above experimental scenarios stem from the pioneering work of R.A. Fisher. Recognizing that the properties and attributes of our natural resources are heterogeneous and vary from one location to another, he developed, in the absence of computers, the concept of analysis of variance which quantified the impacts of treatments within prescribed limits of probability. In a sence, he was able to ignore where soils vary from location to location. The benefits of his pioneering statistical efforts have been enormous, but today's complex environmental problems need to be alternatively approached using different, more global methods. We must be able to monitor fields or regions at particular locations and times to ascertain the behavior of our natural resources integrated over different scales of space and time and to be able to ascertain the value of a plant or soil attribute at unsampled locations within prescribed limits of accuracy and precision.

The objective here is to briefly introduce statistical concepts that require and take advantage of experimental observations manifesting spatial and/or temporal dependence and to provide a few examples of the many available methods having potential relevance to microbial ecology of arable lands. We give here a perhaps overly simplistic presentation of the theory and its potential practice. A secondary objective is to suggest the advantages of stochastic equations and models over the more common deterministic formulations normally used in agricultural and environmental sciences.

Because of the spatially continuous nature of the landscape, soil formation processes and climatic zones, it is not unreasonable to expect that soil measurements taken in close proximity to each other would be correlated. It is also reasonable to expect spatial or temporal variations in one or more ecological attributes to be related to varia-

tions of other parameters or properties indigenous to the landscape. Using statistical techniques that rely on observations made at different scales being spatially or temporally dependent allows greater opportunity to explore and understand biological and physical processes occurring in arable lands.

An approach for examining spatial structure is time series analysis, a methodology used to study temporally and spatially correlated data. Both time domain and frequency domain analysis are treated in time series methodologies (Shumway, 1988). Time domain analysis expresses the observation variable in terms of a linear function of its past values and a random error, and is used for estimating equally spaced univariate and multivariate processes. This analysis, known as autoregressive modeling (Box and Jenkins, 1970), may not easily be used to estimate missing observations. Another analysis, known as state-space modeling introduced by Kalman (1960) and Kalman and Bucy (1961), yields two linear equations called observation and state equations which describe the observed series. Two different kinds of errors are recognized. The first is the observation error which might be due to measurement error or disturbances observed in the data not related to the variable under study. The second is the state error caused by the noise inherent in the variable under study. Frequency domain analysis assigns the variations of the observation variable to combinations of sine and cosine waves. Spectral analysis examines the series to find those frequencies that contribute the most to the total variance assuming an expected value or mean for the observation variable.

An approach utilizing the spatial structure of a data set is kriging which is an estimation technique for regionalized variables that exhibit significant spatial autocorrelation (see, for example Ripley, 1981). The theory of regionalized variables was developed by Matheron (1963) and has been applied in the field of mining in estimation of ore reserves and more recently has been applied to agronomic and soil water properties (Vieira *et al.*, 1983). Applications of this theory are now popularly known as geostatistics. The kriging estimate is a weighted sum of the observed data in which the weights are determined by using the spatial distribution of the observations. In addition to kriging Journl and Huijbregts (1978) developed the geostatistical cokriging technique of estimating a variable that is undersampled by considering its spatial correlation with other variables that are more densely sampled (David, 1977). Cokriging was used by Vieira *et al.* (1983) to estimate agronomic properties. Many other applications of regionalized variable analyses or geostatistics are available and continue to be developed. We only introduce them here and begin with the definition of spatial dependence.

Quantification of spatial dependence

Consider a set of observations $Z(x_1)$, $Z(x_2)$, . . . , $Z(x_n)$ at locations x_1, x_2, . . . x_n where each location defines a point in 1-, 2-, or 3-dimensional space. The correlogram $r(h)$ of the observations is estimated by

$$r(h) = \left[\frac{\sigma^{-2}}{n(h) - 1} \right] \sum_{i=1}^{n(h)} \left[Z(x_i) - \bar{Z} \right]$$
$$\times \left[Z(x_i + h) - \bar{Z} \right] \qquad [1]$$

where $n(h)$ is the number of pairs of observation points a distance h apart, $\bar{Z}$ the mean and σ^2 the variance. The correlogram has possible values from -1 to 1. Some idealized correlograms are given in Figure 1. In Figure 1A the value of $r(0)$ is unity while $r(h)$ gradually decreases to zero as h increases. Figure 2A shows autocorrelation of a nematode population calculated from 625 data points separated by a distance of 6.5 m (Alemi *et al.*, in manuscript). For large separation distances no correlation or spatial dependence exists. For a set of observations in 1 dimension the integral scale l defined by

$$l = \int_0^\infty r(h) \, dh \qquad [2]$$

gives a measure of the distance h beyond which the pairs of observations are considered independent. The correlation length λ defined by

$$r = \exp(-h/\lambda) \qquad [3]$$

gives still another measure beyond which the observations are considered independent. Numerically, the value of λ is that of h when r equals $1/e$.

Figure 1B is a correlogram manifesting a set of observations that are not correlated. In other words, for all separation distances, the observations are independent or may be said to be purely random. Figure 1C illustrates a set of observations whose values vary somewhat sinusoidally in space. The correlogram of an ideally sinusoidal set of observations would be a cosine curve having the same period as the observations and an amplitude of unity. In general, the correlogram shows fluctuations with the same kinds of periods as the original observations, except that all the fluctuations have been put in phase so that they reach a maximum value at $h = 0$. Figure 2B shows correlograms of water flow and its salt concentration measured monthly in a stream during a 6-year period. The autocorrelation functions indicate that both variables have strong cyclic behavior with a period of 12 lags (a yearly cycle).

Another measure of spatial dependence is given by the variogram $\gamma(h)$ estimated by

$$\gamma(h) = \frac{1}{2n(h)} \sum_{i=1}^{n(h)} [Z(x_i + h) - Z(x_i)]^2. \qquad [4]$$

Figure 3 provides an illustrative example of the several potential features of a variogram. Although $\gamma(0) = 0$, $\gamma(h)$ will appear to remain nonzero in many cases as h approaches zero as shown by the value C_0 commonly called the nuggest. The value $h = a$, called the range, is the maximum separation distance for which pairs of observations remain

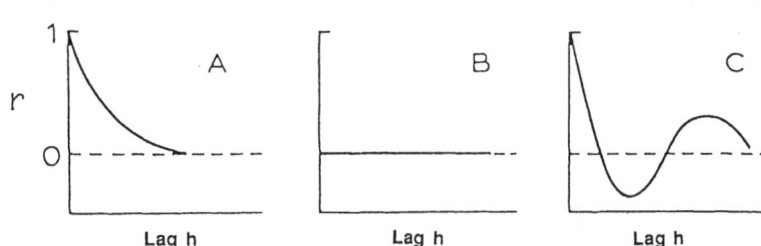

Fig. 1. Idealized correlograms. **A** is well-behaved and decreases monotonically to zero, **B** is for spatially independent or random observations. **C** is for observations that manifest repetitions, cyclic variations.

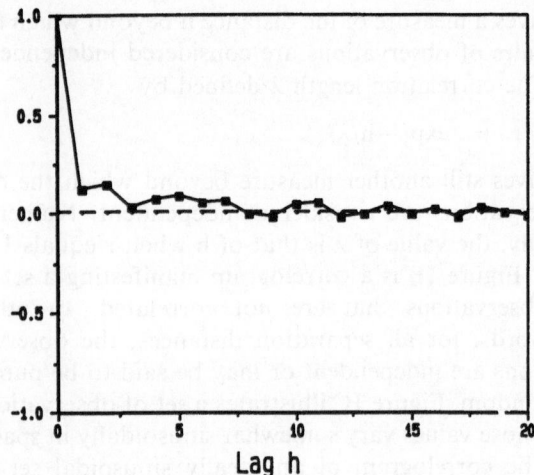

Fig. 2A. Autocorrelation function for observations of *Paratrichodorus* populations.

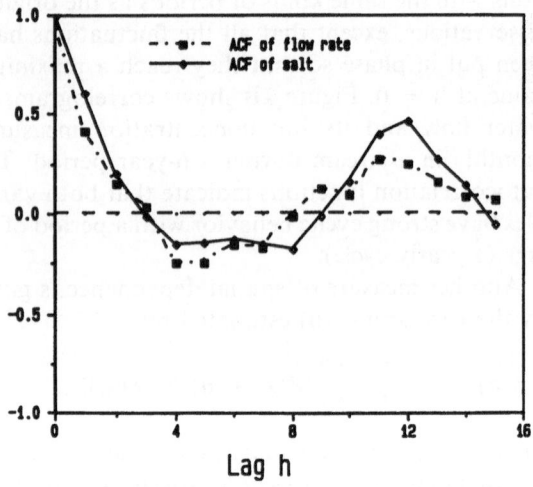

Fig. 2B. Autocorrelation functions of water flow and salt concentration.

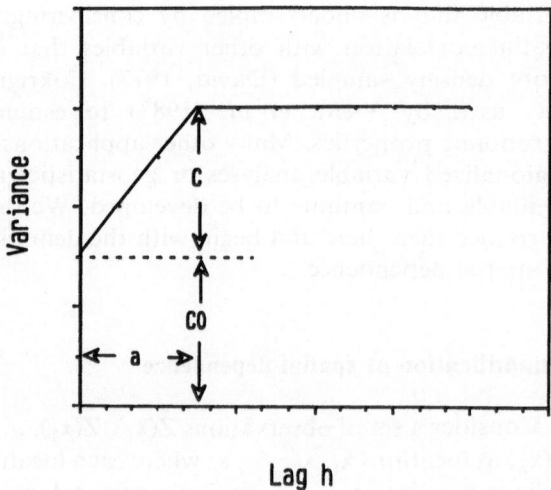

Fig. 3. Idealized variogram with range a, sill (C + C_0) and nugget C_0.

correlated. For greater separation distances γ remains equal to C + C_0 and is called, the sill. Some variograms do not manifest one or more of the features C_0, C or a. The use of the variogram provides an opportunity for interpolation and mapping schemes, for identifying new observation locations to improve estimates for a total region or subregion, and to ascertain the impact of the size of observation on the scale of the field problem being studied as well as to develop efficient sampling strategies.

We assume that the reader is relatively unfami-

liar with the concepts of time series and geostatistical analysis of data. Hence in the following section these methods are described briefly with real world examples.

Autoregressive integrated moving average (ARIMA) models

A stationary process Z_s is said to be autoregressive of order p, AR(p) if

$$Z_s = \sum_{k=1}^{p} \Phi_k Z_{s-k} + \omega_s \qquad s = 1,2,\ldots \qquad [5]$$

where ω_s is a white noise (independent and identically distributed normal variable) and Φ_k are coefficients to be estimated. The moving average of order q is defined as

$$Z_s = \omega_s - \sum_{k=1}^{q} \theta_k \omega_{s-k} \qquad s = 1,2,\ldots \qquad [6]$$

where ω_s is again white noise and θ_k are coefficients to be estimated. The mixed autoregressive process of order p and moving average of order q, ARMA (p, q) is defined as

$$Z_s = \sum_{k=1}^{p} \Phi_k Z_{s-k} - \sum_{k=1}^{q} \theta_k \omega_{s-k} + \omega_s. \qquad [7]$$

ARMA modeling can be applied to a nonstationary series, if the nonstationarity is removed for

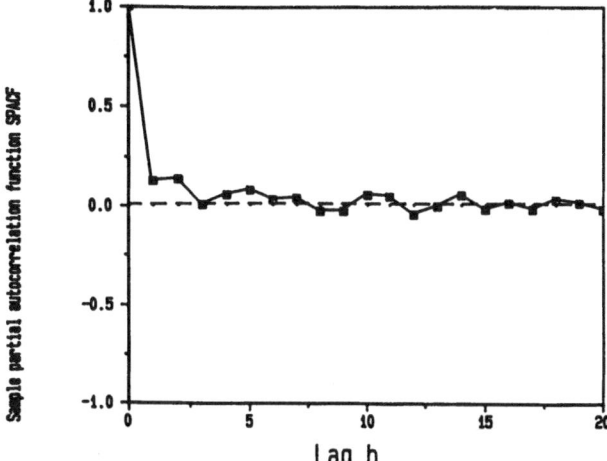

Fig. 4. Sample partial autocorrelation of *Paratrichodorus* series.

example by differencing the original series. If a difference of order d is applied, and an autoregressive model of order p and moving average of order q is used, the model is called autoregressive integrated moving average of orders p, d, and q or ARIMA (p, d, q)

The first step in ARIMA modeling is model identification. Autocorrelation and partial autocorrelation functions are the main tools in detection of the nonstationarity and model identification. For a complete description of the procedures for model identification and testing the adequacy of the model see Shumway (1988). The sample partial autocorrelation (SPAC) of a series of nematode

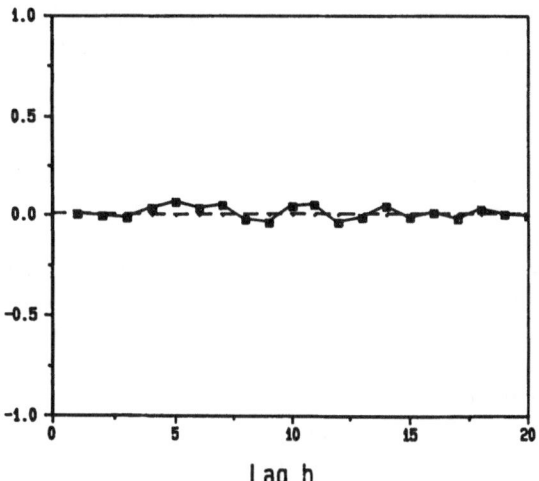

Fig. 5. Autocorrelation function of residuals of the autoregressive model.

(*Paratrichodorus*) populations given in Figure 4 indicates that the SPAC is significant at lags one and two. Thus an AR(2) seems to be the appropriate model. The ACF of residuals (differences between the observed and estimated values in the series) is given in Figure 5 with no significant peak (ACF less than $1.96 (S)^{-0.5}$ where S is the length of the series). Hence, the AR(2) is an adequate model to describe the nematode series and can be written as

$$Z_s = 0.1068 Z_{s-1} + 0.1324 Z_{s-2} + \omega_s. \quad [8]$$

We note that the nematode population at location s can be described in terms of its magnitude at the two previous locations (s−1) and (s−2).

State-space models

State-space models have been used to describe data in many different scientific endeavors. The multivariate version of the model is defined by the observation equation

$$\mathbf{Z}_s = \mathbf{A}_s \boldsymbol{\xi}_s + v_s \quad s = 1,2,3,\ldots.,S \quad [9]$$

where the p-dimensional vector $\mathbf{Z}_s = (Z_{s1},\ldots, Z_{sp})$ is represented as a linear combination of unobserved q-dimensional vector $\boldsymbol{\xi}_s = (\xi_{s1},\ldots,\xi_{sq})$, using a pxq design matrix $\mathbf{A}_s$. The additive noise vector v_s is a pxl vector of measurement error related to the observation vector $\mathbf{Z}_s$. It is assumed that v_s are zero mean independent random vectors with covariance matrix $\mathbf{R}$. The unobserved or so called state vector $\boldsymbol{\xi}_s$ satisifies the state equation

$$\boldsymbol{\xi}_s = \mathbf{B} \boldsymbol{\xi}_{s-1} + \omega_s, \quad s = 1,2,\ldots,S \quad [10]$$

where $\mathbf{B}$ is a q × q transition matrix and ω_s is the state noise vector independent of v_s with mean zero and covariance matrix $\mathbf{Q}$. The initial value of $\boldsymbol{\xi}_s$ is assumed to be $\boldsymbol{\xi}_0$ with mean vector $\boldsymbol{\mu}$ and covariance matrix $\boldsymbol{\Sigma}$.

In state-space modeling $\boldsymbol{\mu}, \boldsymbol{\Sigma}, \mathbf{B}, \mathbf{Q}$, and $\mathbf{R}$ are estimated from the observed series $\mathbf{Z}_s$, s = 1,2,..,S by an iterative procedure using Kalman filtering and expectation maximization (EM) algorithm given in Shumway and Stoffer (1982) and Shumway (1988). An example of application of a state-space model is given in Figure 6 where a series of cotton yield observations and that of a nematode population were used to estimate the cotton yield

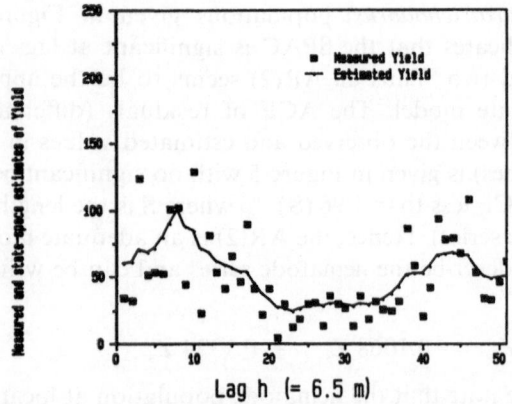

Fig. 6. Measured and state-space estimates of cotton yield.

series. The observation equation is

$$\mathbf{Z}_s = \mathbf{A}_s \, \boldsymbol{\xi}_s + \boldsymbol{\nu}_s. \qquad [11]$$

The state equation for the bivariate process using *Paratrichodorus* (P) and cotton yield (Y) series is

$$\begin{Bmatrix} P \\ Y \end{Bmatrix}_s = \begin{bmatrix} 0.92 & 0.23 \\ 0.01 & 0.94 \end{bmatrix} \begin{Bmatrix} P \\ Y \end{Bmatrix}_{s-1} + \begin{Bmatrix} \omega_P \\ \omega_Y \end{Bmatrix} \qquad [12]$$

with variance ν equal to 683 and 165, and those of ω equal to 4277 and 944, respectively. Using the standard deviation of variances, the magnitude of measurement and process errors can be calculated. The solid line given in the figure stems from equation [12] and tends to "smooth" the cotton yield data on the basis that there is a spatial correlation between cotton yield and nematode infestation. Our example has involved only two observation variables — cotton yield and number of nematodes. State-space modeling easily handles several variables, does not require long series of observations, and is an excellent method to identify which local field properties or attributes tend to be related to the behavior of an observation variable of primary interest. Noting that the solid line in Figure 6 does not indicate a constant yield of cotton across the field implies that additional properties or attributes affecting crop yield are not accounted for in the bivariate state space model. The use of a multivariate model would potentially allow the identification of other properties or attributes that are responsible for the manner in which the cotton yield varies across the field.

Spectral analysis

Arable land is often cultivated with animals or machines in repeatable, cyclic patterns during one or more growing seasons. In addition, irrigations, fertilizers, agrochemicals, and crop residues are usually applied on or into the soil in cyclic patterns having various periodicities or frequencies depending upon the spacing between plant rows or the method of cultivation or irrigation. Under such conditions observations $Z(x_n)$ made in the field will yield a correlogram r(h) that can be used to identify those portions of the spatial variance structure that are associated with sinusoidal patterns of specific periods or frequencies. The identification is made from a spectral analysis by calculating the power spectra from the correlogram in one or more directions. For one direction, the power spectra f(v) is

$$f(v) = 2 \int_0^\infty r(h) \, \cos(2\pi v h) \, dh \qquad [13]$$

where v is the frequency of the periodic observations. A graph of f(v) reveals relative maxima at those values of v for which the observations manifest a sinusoidal behavior.

Soil surface temperature observations of a fallow, 6-ha field measured with an infrared thermometer on a 36 by 64 regular grid 2 days after a 2 cm rain are examined here to illustrate the use of spectral analysis (Bazza, 1985). Figures 7 and 8 show 3-dimensional plots of all 2304 observations viewed from the east and north, respectively. The most striking feature of the data is that variability is not the same across the field. An obvious feature is that variability in the E-W direction is completely stationary as compared with that in the other direction. The correlogram of the observations for the E-W direction indicated spatial independence while that for the N-S direction revealed a periodic behavior. The graph of f(v) for the N-S direction (Fig. 9) shows a relative maximum of a frequency of about 0.06 corresponding to a period of about 18 lags. Figure 10 of average values of temperature of all 36 columns shows this periodic behavior about a mean of 6.6°C along the N-S direction. Also given in the figure is a graph of the salinity of the water used to irrigate the field for cotton production during the previous summer. Notice that there appears to be an inverse relation between the soil tem-

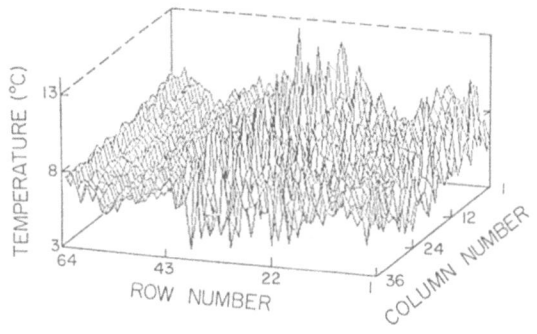

Fig. 7. Soil surface temperature observations viewed from the east.

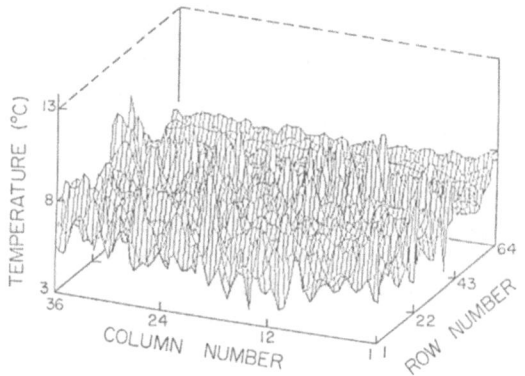

Fig. 8. Soil surface temperature observations viewed from the north.

perature and the salinity level of the irrigation water. The cross spectrum for two series Y_s and X_s can be written as

$$f_{xy}(v) = C_{xy}(v) - i \, q_{xy}(v) \qquad [14]$$

where $f_{xy}(v)$ is cross spectrum between two series, $C_{xy}(v)$ and $q_{xy}(v)$ are called cospectrum and quad-

spectrum, respectively and i is the imaginary number. The cross-spectrum is the frequency dependent measure of the covariance between two series Y_s and X_s. The dependence between the two series Y_s and X_s with frequency can be explained by the coherence function between the two series. The coherence, defined by

$$\gamma_{yx}^2(v) = |f_{xy}(v)|^2/f_x(v) \, f_y(v) \qquad [15]$$

is the frequency dependent correlation between two series. If Y_s is an exact linear version of X_s then coherence becomes equal to unity at all frequencies.

After first calculating the cross-correlogram between the temperature versus salinity data already given as average values in Figure 10 in a manner analogous to equation [1] only for two variables. The cross-spectrum $f_{xy}(v)$ comes from equation [14]. The strong negative peak shown in Figure 11 indicates that the erratic small differences in temperature values across the field and salinity are spatially related at a frequency of about 0.06. At that frequency, the magnitude of the coherence was nearly unity and indicates that the two series, temperature and salinity, were almost perfectly related. The reason behind the cyclic temperature behavior is probably due to the salinity causing lower soil water potential, lower evaporation and reduced hydraulic conductivity tending to prevent redistribution of soil water.

The concept of time stability of a spatially measured series as defined by Vachaud *et al.* (1985) has recently been expanded to include general linear transformations in time and to account for the occurrence of spatial scale dependency (Kachanoski and de Jong, 1988). Vachaud defined time dependency as the time invariant association between spatial location and classical statistical parametric

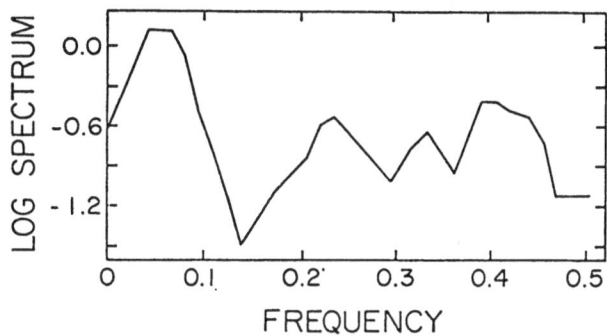

Fig. 9. Spectrum of surface soil temperture.

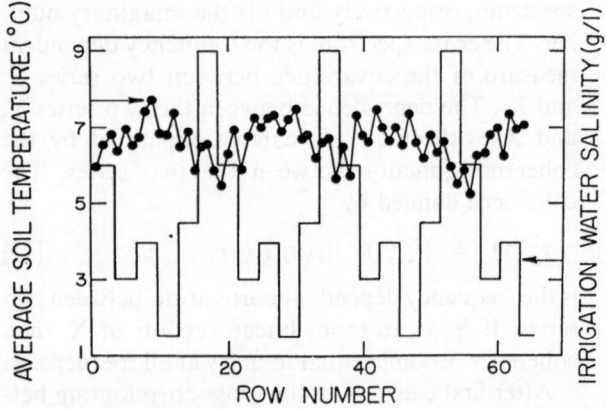

Fig. 10. Row averages of surface soil temperature and salt content of irrigation water.

values. For example, a given location in a field will tend to have a soil water content that remains in the same ranked order with those of other locations. In other words, a location that is relatively dry at one time compared to others will remain relatively dry at other times while a wet location will remain wet.

Let D represent the spatial domain of interest, j represent a horizontal spatial location vector in D, and $S_t(j)$ represent the cumulative soil water storage to depth L at location j and time t. If $s_t(j)$ is the realization of the variable $S_t(j)$ at location j and time t, the relative deviation from the expected or mean soil water storage $E[S_t(j)]$ at the same time across the spatial domain D is given by (Vachaud *et al.*, 1985):

$$\delta_t(j) = \frac{S_t(j) - E[S_t(j)]}{E[S_t(j)]}. \qquad [16]$$

According to Vachaud *et al.*, a small temporal variation in $\delta_t(j)$ (*i.e.* time independence) is an indication of time stability, thus

$$\delta_{t_1}(j) \simeq \delta_{t_2}(j) \qquad [17]$$

where t_1 and t_2 represent two different measurement times. Substitution of equation [16] into equation [17] and taking the variables $S_{t_1}(j)$ and $S_{t_2}(j)$ in lieu of the observations $s_{t_1}(j)$ and $s_{t_2}(j)$ respectively, gives:

$$S_{t_2}(j) \simeq \frac{E[S_{t_2}(j)]}{E[S_{t_1}(j)]} S_{t_1}(j). \qquad [18]$$

Thus time stability as defined in equation [18] implies a linear relationship between soil water storage at time t_1 and t_2 across all spatial locations.

Since spatial variation is often scale dependent, it follows that time stability may also be scale dependent. For example, localized surface runoff may significantly alter the spatial variation of soil water storage on a small scale, but the changes may be insignificant compared to large scale variations. Other factors, such as climatic variations may affect time stability at large scales.

The persistence of a spatial pattern at different spatial scales was examined by Kachanoski and de Jong (1988) by expanding on the concept that time stability requires a linear relationship between successive time samplings. The presence of a significant linear relationship at different spatial scales is tested using spatial coherency analysis introduced by equation [15]. In this case, the coherency function estimates the proportion of the spatial variance of $S_{t_2}(j)$ which can be explained by the

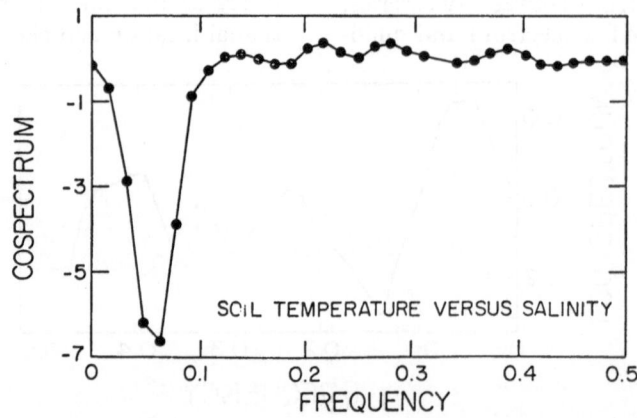

Fig. 11. Cross spectrum of surface soil temperature and salt content of irrigation water.

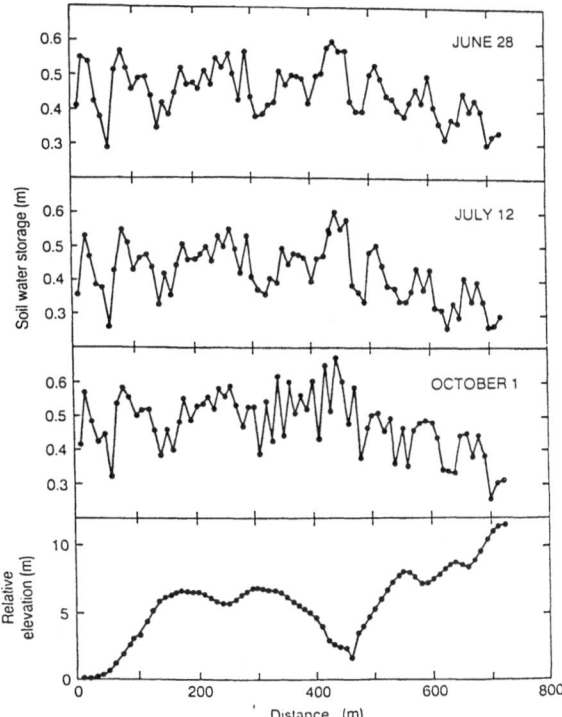

Fig. 12. Measured salt water storage and relative elevation for the study site.

spatial variance of $S_{t_i}(j)$, as a function of spatial scale. Thus, the coherency function is a direct measurement of time stability as a function of spatial scale. A more detailed discussion has been given by Kachanoski *et al.* (1985a) which in turn is a summary of the direct Fourier method of spectrum estimation (Brillinger, 1981).

The soil of the study area of Kachanoski and de Jong was a moderately fine textured, moderately calcareous Typic Haploboroll resting on a gently rolling to undulating topography. Neutron probe access tubes were installed in a fallow field every 10 m in a 720 m long transect in a north-south direction. Soil water measurements were taken at all locations with depth on June 28, July 12 and October 1. Elevation measurements were taken at each location and 10 m east and west of each location using a rod and surveying level. The grid of elevation readings around each soil water sampling location was used to estimate local surface curvature by a least squares surface analysis. No significant precipitation was recorded during the first two sampling dates, (June 28-July 12), thus this period is characterized by drying. The

next sampling period (July 12-October 1) had significant rainfall and represents a time of net soil water recharge. Power spectra of cumulative soil water storage observations were calculated for each of the three measurement dates. Coherency spectra were estimated for the June 28 and July 12, and July 12 and October 1 paired sampling dates.

Measurements of surface elevation and cumulative soil water storage are illustrated in Figure 12. As can be observed in the figure, the variations in soil water storage appear to persist between sampling data. In fact, a statistical analysis indicates that 94% of the observed spatial variability after the drying period measured July 12 was explained by the variability present on June 28.

The coherency spectra for the drying and recharge periods are shown in Figure 13. The coherency spectrum for the drying period, June 28-July 12, is greater than 0.90 for almost all spatial scales and indicates the presence of a higher linear correlation and thus temporal stability across all spatial scales. The coherency spectrum for the recharge period shows a high coherency for spatial scales greater than 50 m. For spatial scales less than approximately 40 m the coherency drops below the 95% confidence level for a significant linear relationship. The coherency is especially low for scales less than 30 m. The coherency clearly indicates that recharge has significantly altered the spatial pattern of soil water storage on a scale of less than 40 m, but has not affected the pattern at larger scales. Thus, time stability was present for spatial scales greater than 40 m but not for smaller scales. Large scale variations of soil water storage are no doubt related to elevation, especially in the prairie regions. However, smaller scale variations may be more affected by local conditions (curvature, gradient, *etc.*). The local variability of surface curvature resulted in variations of runoff and subsequent recharge which significantly altered the spatial pattern of soil water storage at a small scale while not affecting the large scale pattern.

Kachanoski *et al.* (1985b) have used coherency analysis to examine the effects of cultivation on the persistence of the spatial patterns of a number of soil properties. The change in spatial variance patterns as a function of scale were used to infer the scale of processes such as horizontal soil movement and mixing by tillage. In a similar manner, scale dependent time stability studies of soil water, avail-

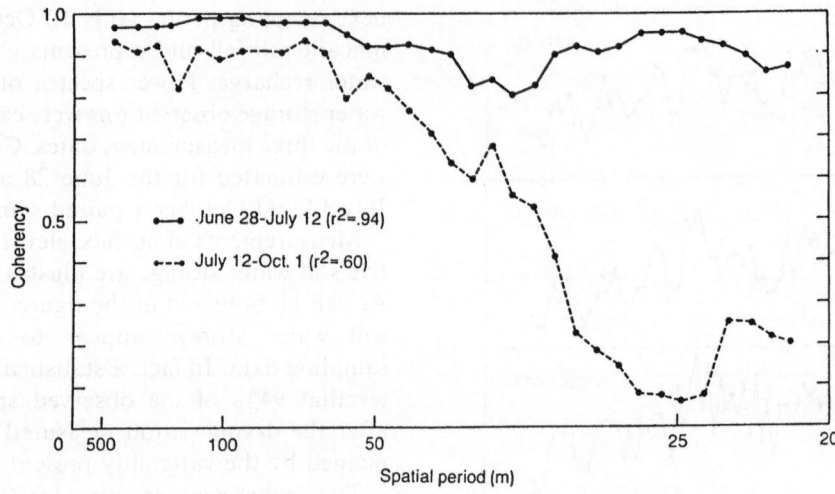

Fig. 13. Coherency spectra for the drying (June 28–July 12) and recharge (July 12–Oct. 1) periods.

able carbon and nutrient sources, using coherency analyses should be able to infer spatial scales of hydrologic microbial and crop community processes.

Geostatistics

An example of the variogram given in Figure 3 for the nematode populations measured by Alemi *et al.* (1988) on a 6.5 m grid across an agricultural field is given in Figure 14. The values of the range, sill and nugget are 32.5 m (5 lags), 8500 and 7000, respectively. We may use this variogram to krige or

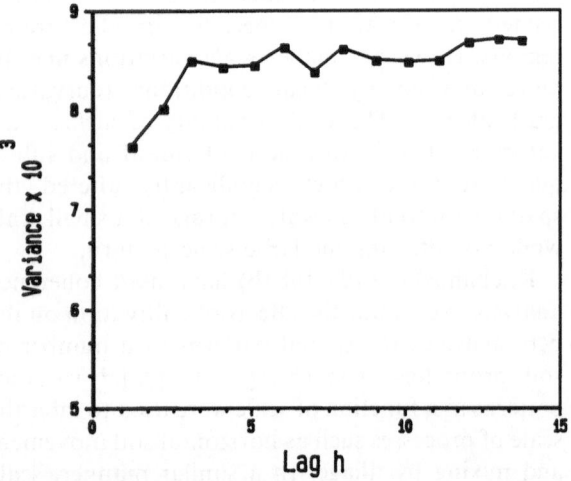

Fig. 14. Average semivariogram of *Paratrichodorus*.

estimate the nematode population at all locations across the agricultural field. The variogram can be used for estimating nematode values Z^* at locations x_0 where measurements have not been made. The procedures lead not only to the estimated values Z^*, but also an estimate of $var(Z - Z^*)$ which indicates the reliability of the estimate. "Error maps" can be prepared to reveal areas within the field where the variance is highest and presumably where additional sampling should be made.

In the kriging procedure, an estimate Z^* is assumed to be a linear function of the known values Z already measured at N other locations

$$Z^*(x_0) = \sum_{i=1}^{N} \lambda_i Z(x_i) \qquad [19]$$

by choosing the weight factors λ_i such that the expected value and variance of $[Z^*(x_0) - Z(x_0)]$ are zero and a minimum, respectively. The variogram showing the spatial variance structure of Z is used to estimate the values of λ_i. Cokriging is a similar procedure that takes advantage of the spatial structure of the covariance between two variables such as cotton yield and nematode. Still another procedure, block kriging, allows interpolation between averages of values centered at locations instead of individual observations for each location. Universal kriging provides for combination of a deterministic trend across the field to coexist with random variations while disjunctive kriging can be used to identify areas within a field having values expected to be above or below a threshold magnitude. Hence, there are many attractive kriging and cokriging analyses that can be used to develop and evaluate alternative sampling grids or locations relative to a specific management system.

The excellent comprehensive reference of Journel and Huijbregts (1978) on geostatistics and that of Warrick *et al.* (1986) applied to soil science provide essential details and conditions omitted from the above introductory presentation. Many additional statistical opportunities exist when one recognizes that spatial dependence also depends upon direction and that more than one variable can be treated simultaneously.

Stochastic equations

It is not possible to model a system relative to what is "a typical response to management practices" or what is "a typical pattern in ecology" without understanding the spatial and temporal variations for both small and large scales. "Models are being constructed, refined, elaborated, tinkered with and displayed with little or no effort to link them to the real world," Pielou (1981). The standard approach to field scale or plot scale problems has been to construct analytical or numerical solutions of the classical differential equations, assuming that the parameters take on known values which can be easily evaluated by a few observations. Although there have been major advances in numerical modeling techniques, standard modeling approaches are giving way to stochastic methods which incorporate the effects of natural or anthropogenically-induced variability. Stochastic models are more realistic representations of field situations than deterministic models and they provide better predictions owing to the probabilistic nature of the data incorporated. Because stochastic models can be examined in terms of the variability of the model parameters, data collection can be improved to emphasize those parameters having the greatest effects on the results.

During the past decade, stochastic analyses of groundwater (Bakr *et al.*, 1978) and unsaturated flow (*e.g.*, Yeh *et al.*, 1985) have been accelerating in number and kind of applications. Nielsen *et al.* (1987) recently reviewed conceptual alternatives for modeling water and solute transport in the vadose zone. Promising areas for further research in microbial ecology of arable lands await similar development.

References

Alemi M H, Ferris H and Nielsen D R 1989 Geostatistical analyses of the spatial and temporal variability of nematode populations (in manuscript).

Bakr A A, Gelkhar, L W, Gutjahr A L and MacMillan J R 1978 Stochastic analysis of spatial variability in subsurface flows. Water Resour. Research 14, 263–271.

Bazza M 1985 Modeling of Soil Surface Temperature by the Method of two-dimensional Spectral Analysis. Ph.D. Thesis, UC Davis.

Box G E P and Jenkins G M 1970 Time series analysis, forecasting, and control. Holden Day, Inc. San Francisco. CA.

Brillinger D R 1981 Time Series, Data Analysis and Theory. Holden Day Inc. San Fransisco, CA.

David M 1977 Geostatistical Ore Reserve Estimation. Centre D'Informatique Geologique, Fountainbleau, France.

Journel A G and Huijbregts Ch 1978 Mining Geostatistics. Academic Press, London.

Kachanoski R G and de Jong E 1988 Scale dependence and the temporal persistence of spatial patterns of soil water storage. Water Resour. Research 24, 85–91.

Kachanoski R G, Rolston D E and de Jong E 1985a Spatial and spectral relationships of soil properties and microtopography. I. Density and thickness of A horizon. Soil Sci. Soc. Am. J. 49, 804–812.

Kachanoski R G, Rolston D E and de Jong E 1985b Spatial variabilty of a cultivated soil as affected by past and present microtopography. Soil Sci. Soc. Am. J. 49, 1082–1089.

Kalman R E 1960 A new approach to linear filtering and prediction theory. Trans. ASME J. Basic Eng. 82, 35–45.

Kalman R E and Bucy R S 1961 New results in linear filtering and prediction theory. Trans. ASME J. Basic Eng. 83, 95–108.

Matheron G 1963 Principles of geostatistics. Econ. Geology 58, 1246–1266.

Nielsen D R, van Genuchten M Th and Jury W A 1987 Monitoring and analyzing water and solute transport in the vadose zone. Proceedings of International symposium on groundwater monitoring and management, 23–28 March 1987. Dresden, GDR.

Pielou E C 1981 The usefulness of ecological models: A stock-taking. Quarterly Review of Biology 56, 17–31.

Ripley B B 1981 Spatial Statistics. Wiley, New York.

Shumway R H 1988 Applied Statistical Time Series Analysis. Prentice-Hall, Englewood Cliffs, NJ.

Shumway R H and Stoffer D S 1982 An approach to time series smoothing and forecasting using EM algorithm. J. Time Series Analysis, 3, 253–264.

Vachaud G, Passerat De Silane A, Balabanis P and Vauclin M 1985 Temporal stability of spatially measured soil water probability density function. Soil Sci. Soc. Am. J. 49, 822–827.

Vieira S R, Hatfield J L, Nielsen D R and Biggar J W 1983 Geostatistical theory and application to variability of some agronomic properties. Hilgardia, 51(3).

Warrick A W, Myers D E and Nielsen D R 1986 Geostatistical methods applied to soil science. *In* Methods of Soil Analysis, Part I, Physical and Mineralogical Methods Ed. A Klute. pp 53–82. Am. Soc. Agronomy, Madison, WI.

Yeh T-C J, Gelhar L W and Gutjahr A L 1985 Stochastic analysis of unsaturated flow in heterogeneous soils. Water Resour. Research 21, 447–456.

M. Clarholm and L. Bergström (Eds.), Ecology of arable land, 273–279.

Models by decision makers and ecologists, can they be coupled?

M.J. FRISSEL and T.N. OLSTHOORN
National Institute of Public Health and Environmental Protection (RIVM), P.O. Box 1, 3720 BA Bilthoven, The Netherlands

Key words: ecology, environment, management, models, nitrogen, pollution

Abstract

Ecologists tend to improve their models by including more and more detail, while decision makers prefer models at a very high aggregation level with a minimum of detail. This seems to lead to a controversy between the two groups of modelers. Ecologists may wonder if their results will ever be used by policy makers. On the basis of experience gained by the development of management models, possible solutions to this problem are indicated. The principle of the Pollution Management Model of the Netherlands National Institute of Public Health and Environmental Protection (RIVM) is described. Existing knowledge was successfully coupled using a framework, containing a source term part which describes a particular action, *e.g.*, a release of a pollutant, a pathway part which describes the behaviour of the pollutant in the environment and a dose effect part which describes the harmful effects at a particular location. Advantages and disadvantages of this approach are discussed.

Introduction

Increasing interest has recently been shown by governmental authorities in the use of mathematical environment management decision models. These models take conventional types of knowledge into account, *e.g.*, the feeding habits of birds, the cycling of pollutants, plant nutrient cycles — as studied within the Swedish project on the Ecology of Arable Land (Persson and Rosswall, 1983). The use of such models is far from simple.

Ecological studies are not tailored to answer questions posed by decision makers. For example, finding that a particular pollutant is responsible for the extinction of a particular species would not help the decision maker to specify emission levels. Similarly, a biologist might be aware that a decrease in the number of snakes is related to an increase in the intensity of road traffic; nevertheless, if you asked him to predict the development of the snake population in related to the construction of a new highway, he would tell you to get lost.

In principle, a decision model can be generally described by the flow diagram shown in Fig. 1. The

diagram starts with the description of a particular action, in model jargon this action is the "source term". It is not always simple to describe an action in ecological terms, but once that has been accomplished the following two steps are usually easier to take because relatively more experience has been gained on how to use existing scientific knowledge. These steps may consider the behaviour of particular animals, plants or pollutants, *etc.*, in the environment. Reproduction and distribution terms, as well as transfer and uptake terms, are considered. In general physical processes, such as atmospheric and hydrological transport, are relatively well known. Chemical processes, such as reactions, adsorption and desorption, are somewhat less known, while knowledge on biological processes is mostly fragmentary. The next step describes the harmful effects as a function of the environmental parameter derived. The dose/effect relationship is usually poorly known. Often, experts do not agree on the main mechanism causing a particular harmful effect. As a consequence, criteria on admissible concentrations, *etc.*, are often ambiguous, rendering decision-making difficult. Another problem for

273

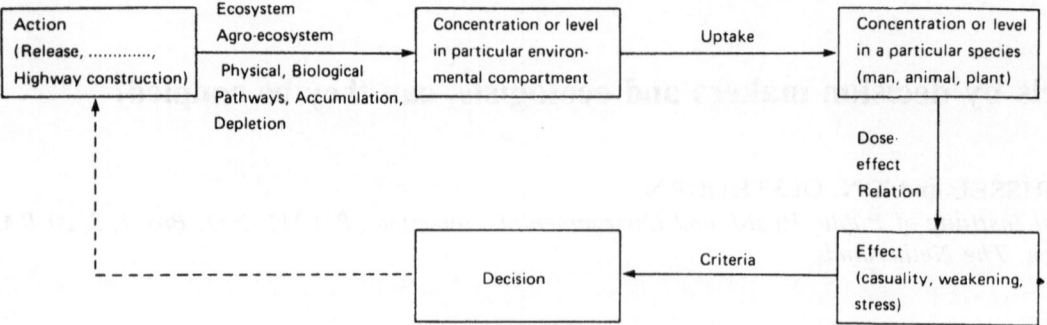

Fig. 1. A generalized decision model.

decision makers is caused by the need to include well described or hard criteria in decision models, while in ecology, criteria are soft, *i.e.*, rather vaguely described.

Hard and soft criteria

The difference between hard and soft criteria can be illustrated with one of the oldest types of models, *i.e.*, a map. A map fulfills the modern definition of a model perfectly. It is a mathematical presentation of reality, and details can be added or deleted, depending on the purpose. Many decisions that have shaped our society were made by using maps. Although it would be interesting to review history at this point, this publication will have to limit itself to an example based on two contemporary persons: one who wants to visit city centers in the most economic way and another who wants to take a tour of the more interesting sites. The criteria for the first person are hard and, consequently, the routing of such a trip can be computed with a proper program. Many organizations plan the delivery of goods in this way. In contrast, the tourist's wishes are not well defined. Perhaps the tourist wants to see plenty of scenery but has no firm idea about how much he wants to expend, in terms of money, time or effort, to see some additional scenery. He uses soft criteria and consequently, there is no program that can help him determine the optimal way to spend his day.

For the tourist this is no big deal, he is quite happy with his soft criteria, but for the environmentalist soft criteria is a real problem. Society has to invest capital, time and effort to obtain results;

how can these be evaluated? There is currently a trend among politicians to create hard criteria, *e.g.*, in the Western Alps a population of 100 eagles is wanted. The criteria how this hard decision can be realised is vague, however; consequently, the hardness offers no real benefit.

An advantage for radioecologists

One of the scientific disciplines that has received considerable attention during the last decade is radiation protection. Calculations of the number of possible casualties are quite normal in radiation protection. Consequently, there is a trend to approach other risks in the same way. Fig. 2 shows part of a radiation risk model.

Of all possible sources only three are considered: 1) the radon escaping from walls due to the presence of Ra-226 in almost all building materials, 2) cosmic rays and 3) radionuclides released by a nuclear power plant during normal operation. All three sources contribute to the radiation dose. However, they cannot simply be added together to obtain one total radiation dose because each type of radiation differs with respect to the intensity of its effects. Instead, each of the three doses is multiplied by a correction factor. In this way, the differences in effect intensity can be compensated for, and a total equivalent dose can be calculated, which can be related to expected effects (Fig. 2). Immediately it becomes clear what an advantage radiation protectionists have: they calculate a total equivalent dose, and because the relation between the equivalent dose and its effects is well documented, they can make reliable estimates of possible damage.

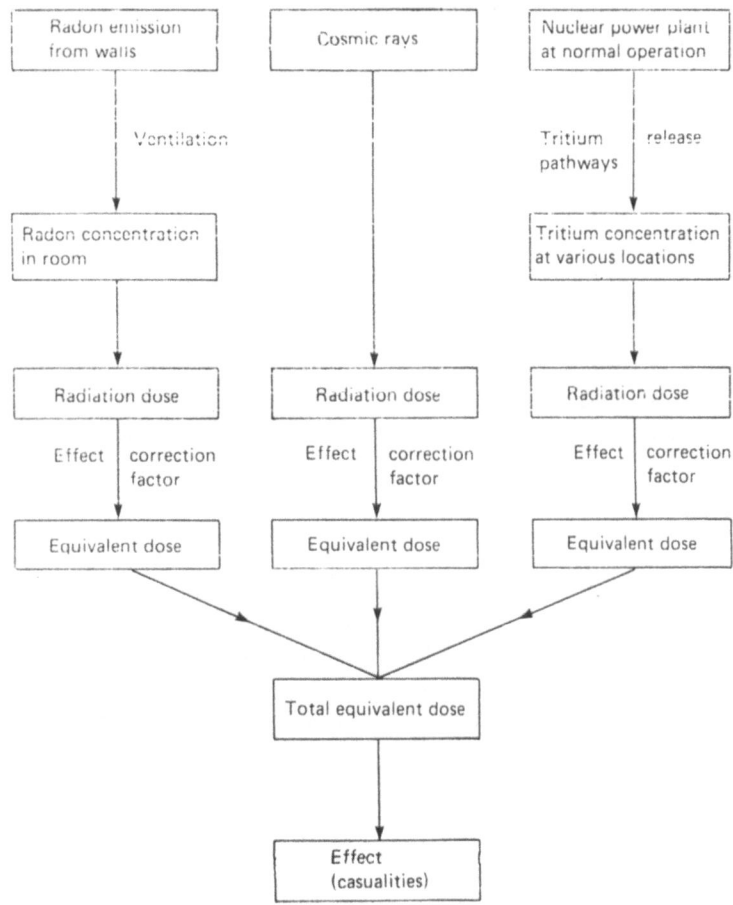

Fig. 2. An integral radiation dose model.

A problem for environmentalists

Environmentalists have no simple way to combine the effects of Cd, Zn, NO$_3$, pH, *etc*. Moreover, the problem is exacerbated where synergistic effects have to be considered. For example, the separately determined exposure thresholds for Cd and Zn may not apply when both are simultaneously present in a system. There are no simple solutions to this difficulty. In a particular ecosystem study it is possible to derive regression equations between environmental conditions and ecological status. It is also possible to determine the number of plant species per m^2 in a large number of locations. The same can be done for the presence of rare or unwanted species. The values can then be related statistically to other variables, such as groundwater level, nutrient states, or levels of trace elements,

pollutants, rocks, etc. The research capacity required for such projects is enormous and its applicability limited. Moreover, the regression equation does not elucidate the mechanism responsible for the changes occurring in ecological status. Therefore, such relations have been derived for special cases only. The authors are not aware of any project that has tried to integrate results of such studies.

Other decision-making possibilities are, in principle, not very different from the methods used by someone who wants to buy a washing machine with the help of data from a consumer's union test (Table 1). Experts are being asked to describe criteria, not in the form of mathematical equations, but in wordings, recommended levels, *etc*. Results of different scenarios are arranged in matrices and tested by comparing them with the opinions of

Table 1. A decision model for use when buying a washing machine

Type of Washer	Price, ECU's	Cleaning of Laundry	Suitable for PO$_4$-free detergents	Service availability	Number of washing programmes
Blue bird	300	+	−	+ +	2
Red diver	400	+	+	−	3
Yellow sparrow	50	+ +	+	+	4

+ = sufficient, + + = excellent, − = not sufficient.

these experts. Although this is indeed a subjective method, it can still be effective. One of the challenges of modern management is to develop ways to optimize the use of expert groups, including modelers and non-modelers.

An integrated model

Returning to the mathematical models, it should be mentioned that a characteristic of such models is their high degree of abstraction. It always surprises, and even irritates, the developers of traditional ecological models. A typical example of such a management model is the Integrated Model of the National Institute of Public Health and Environmental Protection in the Netherlands. The model itself consists of a framework only; it contains no data, nor does it describe processes (Fig. 3). The only task of the model is to link existing information. It consists of three parts: a source-term part, a pathway part and an effect part. One generalized example of a source-term part is shown in Fig. 4. Fourteen main categories of polluting sources are included: Chemistry, Agriculture, Traffic, *etc.* These main categories are split into 58 subcategories. For particular activities specific categories may be added, which makes the model rather powerful. Another strong feature of the model is that, because of its simplicity, it can be coupled to other

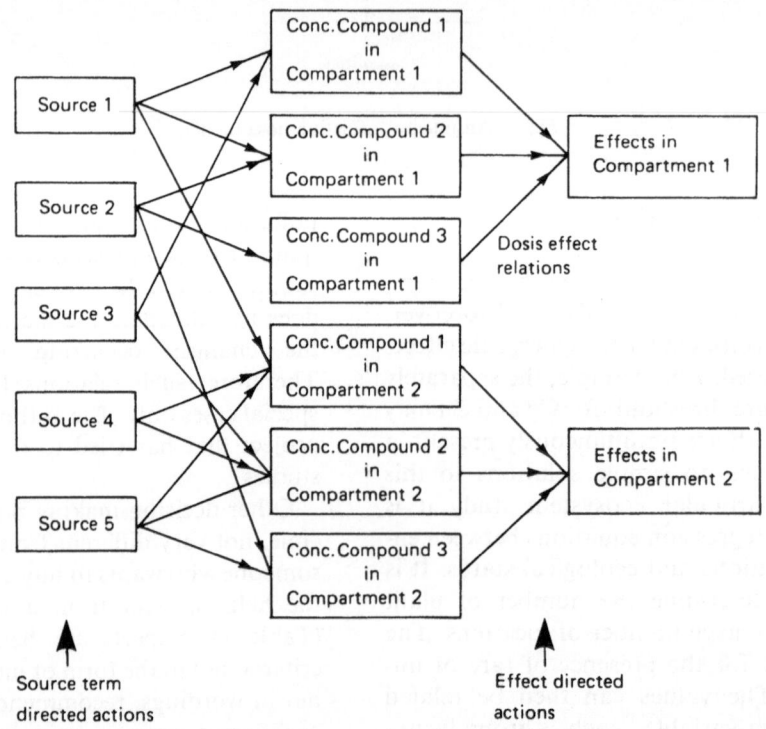

Fig. 3. Lay out of an integral model for the impact of releases of pollutants into the environment.

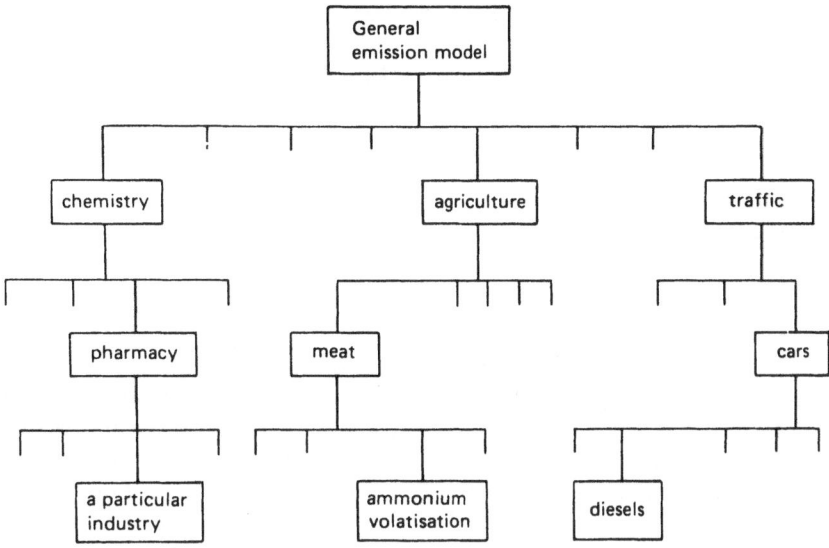

Fig. 4. An example of a source term model at a very high degree of abstraction.

prognoses, *e.g.*, data on the number of cars, split into different categories and on their average speeds, split into speed categories. With a set of conversion factors it is thus possible to immediately calculate the effect of a speed limit on the emissions of exhaust gases like SO_2 and NO_x. Coupling with the economic model at the Netherlands Central Planning Bureau is relatively simple because car number is one of the leading parameters in the model. In the same way coupling is possible with the energy model used by the National Energy Research Center as well as with the agricultural development model created at the Agricultural Economics Research Institute. The power of the source-term model is that it can handle large quantities of basic information, such as data files on emissions, production levels, prognoses, conversion factors, *etc*. When using the integrated model approach, these sets of basic information are neither generated nor stored in the model itself. Instead, the data are generated by internal or external specialists who also take responsibility for maintaining and updating the model parts and data files. The disadvantage of the model is, as previously mentioned, its simplicity. A simplistic model is necessary in order to take an entire country into account as well as a long time span. The problems are clearly demonstrated in the case of agricultural emissions.

Nitrogen inputs from agriculture, an example of a source-term part

The agricultural source term considers the input into the environment of NO_3 ions, NH_3, and phosphate, the last of which is not discussed here (Kuik, 1987). The only sources of NO_3 and NH_3 that the model takes into account are those stemming from fertilizers and animal excreta. The model demonstrates clearly which type of difficulties may occur. Input data are usually based on an extrapolation of the past. The predicted 7% reduction in the amount of agricultural area in Holland between 1985 and 2000 is the first complication. Presently, the Dutch agricultural area consists of approx. $2 \cdot 10^6$ ha: $1.2 \cdot 10^6$ ha grassland, $0.6 \cdot 10^6$ ha arable land and $0.1 \cdot 10^6$ horticultural land. The extent to which each of these categories of land usage will be affected by this reduction is unknown.

Between 1950 and 1985 the number of cows in the Netherlands increased from $2.7 \cdot 10^6$ to $5.2 \cdot 10^6$. Milk production increased during the same period from 3000 to 5300 kg per cow per year. Recently, the European Commission issued regulations designed to reduce the production of milk in the European Community. Thus, the revised prognosis is that there will be $5.0 \cdot 10^6$ cows by the year 2000. The value of this prediction is doubtful since the decrease is not based on ecological or economical restrictions, but on political ones. The number of pigs increased from $2 \cdot 10^6$ in 1950 to $11.6 \cdot 10^6$ in 1985. In this case another restriction emerged. In the Netherlands, pigs are nowadays mainly kept indoors in areas with sandy soils. The slurry type manure of these pigs is applied to these sandy soils. In some areas the intensive application of manure heavily contaminated the groundwater. Consequently, severe restrictions on the application of manure were issued. The expected number of pigs has therefore been reduced to 'only' $12.0 \cdot 10^6$ by the year 2000.

The distribution pattern of land use, *i.e.,* arable land, grassland and pig breeding, is uneven. To account for this the Netherlands is divided into 14 subareas, each with its own characteristic land use pattern. Thus the key problem now is to determine how many cattle will be present in each of the subareas in the future. But other complications also exist. For example, the present manure production levels are uncertain since conversion factors may differ by 20 to 30 percent. For the time being nitrogen production for cows and pigs is estimated to be 44 and 20 kg per year and animal respectively. These estimates translate into nitrogen production levels of 295 and 240 kg per ha, respectively, for the most contaminated subareas. The mean value, which also takes poultry into account (occupying only a very limited space), is close to these figures, *i.e.,* 270 kg per year per ha for the total Dutch agricultural area. Part of the nitrogen in the manure as well as in the fertilizers leaches into the groundwater. The equation for N leached from areas with an excess of manure in $kg\,ha^{-1}\,yr^{-1}$ is:

Leached N = 0.25 (N in manure) + 75. To calculate the amounts of N leached from fertilizers, Table 2 is used. It is a typical example of a compilation of data from several publications. The release of ammonium stems mainly from manure. The mean value for the Netherlands is $51\,kg\,N\,ha^{-1}\,yr^{-1}$.

The pathway part

An integrated model takes all source terms into account simultaneously, as is shown in Fig. 3. Whether or not the same pathway model can be used for several source terms depends on the type of pollutant and type of target. Even if some source terms can be combined, the task initially seems rather hopeless. The complexity becomes so overwhelming that it does not seem possible to model all pathways in a proper way.

Therefore, integral modelers do not bother with diffusion, mass flow mediated transport, bacterial growth, algae, amoebae or predators. This part of the task is left to specialists who develop pathway models and calculate relations between source term patterns and concentrations in the target components. These specialists make separate calculations of the relations between source term 1 and the concentrations in the compartments 1 to n. This approach has also the advantage that integral modelers can always use updated pathways, into which the most recent scientific information has been incorporated.

However, the method can only be applied to pre-considered pathways. If an unforeseen process

Table 2. Leaching of N (NO_3) as a function of N-fertilizer rate

| Application rate ($kg\,N\,ha^{-1}\,yr^{-1}$) | Leaching ($kg\,N\,ha^{-1}\,yr^{-1}$) | | | |
| | Arable land | | Grassland | |
	Sandy soil	Clay soil	Sandy soil	Clay soil
100	70	30	5.5	5.5
150	90	40	6	6
200	100	45	7.5	6
250	110	50	12	10
300	130	50	15	10
350	150	60	60	18
400	170	70	70	20
450	185	75	80	25
500	200	80	110	30

has to be considered, a new set of relations must be prepared. This is typical for a management model: it has not been developed for unexpected situations; instead it is used to determine the effects of reducing source terms or (partly) eliminating pathways, for example, by prohibiting consumption of particular products and by prohibiting crop production on contaminated soils.

The dose-effect part

The last step, involving dose effect relations, presents another difficulty. Pollutant levels that are just tolerable in one situation may be toxic in another. If the ecological situation is well described, expert experience may be sufficient to make the necessary modifications to the model. Because of the high abstraction level used by integrated models, however, they do not supply the information necessary for such expert decisions. The problem is particularly troublesome for soil, sediment, marine and biological compartments. Models dealing with air and fresh surface water do not suffer from this difficulty. Effects of atmospheric SO_2 and O_3 deposition on vegetation can therefore be modeled. The modeling of NO_x is difficult, because the relation between the atmospheric NO_x concentration and ecological damage is not known. Moreover, although almost everyone is convinced that there is a relation between the NO_x concentration in air and forest decline, there is no agreement on the mechanisms responsible. So far, five to six mechanisms have been proposed, but, in general, they are considered as being circumstantial or improbable (Cowling *et al.*, 1986). Yet, there exists consensus that certain combinations of mechanisms and conditions do in fact lead to forest decline. Since such opinions cannot be definitely modeled, requests by decision makers for hard ecological criteria cannot be satisfactorily fulfilled.

In conclusion

This paper's title, which was suggested by the Symposium Organizing Committee, poses the question as to whether ecologists' models and decision makers' models can be coupled. This question is actually somewhat misleading. Decision makers do not develop models themselves — they use models developed by ecologists. One of the remarkable differences is, however, that decision makers use source-term scenarios that are not based on ecological assumptions but on economic and social ones. This is indeed a necessity for decision makers. Ecologists should appreciate that decision makers are not just running after the facts, but are also trying to look ahead. Nevertheless, ecologists and decision makers do have one thing in common: both use their models as tools and not as representations of reality. There is also another important difference: models of decision makers have to follow national frontiers or other jurisdictional frontiers: thus lumped models cannot be modified with new ideas. In contrast, ecological models follows natural boundaries and can be expanded as far as imagination goes. Clearly ecological models should not apply solely to a single nation; instead, their application should be at the ecosystem — or even the global — level.

Acknowledgements

The authors are very grateful for the helpful comments received from Drs. J.A. van Veen and A.H.M. Bresser.

References

Cowling E, Krahl-Urban B and Schimanslag Chr 1986 Wissenschaftliche Hypothesen zur Erklärung der Ursachen Waldschäden. KFA, Jülich, FRG pp 120–125 (*In German*).

Kuik O 1987 Ammoniak emission 1980–2000. RIM-21. Institute for Environmental Studies, Free University, Amsterdam.

Persson J and Rosswall T 1983 Opportunities for research in agricultural ecosystems in Sweden. *In* Nutrient Cycling in Agricultural Ecosystems. Eds. R R Lowrance, R L Todd, L E Asmussen and R A Leonard. pp 61–71. The University of Georgia, College of Agriculture Experimental Station. Special Publication No. 23.

M. Clarholm and L. Bergström (Eds.), Ecology of arable land, 281–295.

The use of nitrogen fertiliser in agriculture. Where do we go practically and ecologically?

P. NEWBOULD

Macaulay Land Use Research Institute, Craigiebuckler, Aberdeen, Scotland

Key words: crop responses, environmental effects, future strategy of use, losses, nitrogen fertiliser, research needs

Abstract

Nitrogen fertilisers were produced in 72 countries in 1982, total world capacity being 99 mt of N, having been 50 mt in 1970. Consumption was 31.8 mt in 1970, rising to 60.3 mt in 1980 (Av. annual growth rate 7%). Forecasts suggest N use of 90 mt in 1990 rising to between 111–134 mt by the year 2000.

The large amount of N added to only some 11% of the earth's land surface as fertilisers, coupled with concurrent increases in biological N fixation, mainly by grain legumes, is bound to result in increases in the total N content of soils, waters, crop residues and municipal wastes. The need to use N to produce sufficient food and fibre for the 7 billion humans must be set against the need to maintain a 'good' and safe environment. Nitrate levels are increasing in both surface and ground water supplies. The amount of ammonia and oxides of nitrogen in the atmosphere produced by volatilisation and denitrification from soils and animal excreta is also rising. Such increases may have detrimental environmental effects to human health and to the ecology of downstream or 'polluted' non-agricultural ecosystems though the severity and extent of these effects requires verification. As yet, there is little hard evidence of direct damage to human health due to high levels of nitrate in diet or of NH_3 and NO_x in the atmosphere, but effects on natural and forest ecosystems in some areas are proven.

With this background, strategies are examined which should help to increase the efficiency with which N is utilised by crops and animals and so decrease losses of nitrogen from farmland.

These include the selection of optimum N fertiliser practices based on knowledge of plant requirements, soil N supply, and the use of carefully chosen times, methods and forms of N fertiliser application. Other technological approaches such as use of slow release fertilisers, chemicals that inhibit certain biological transfers of N in soils and amendments added to N fertilisers, to soils or to animal excreta to alter their chemical properties could be developed. Greater use of legumes and enhanced levels of N_2-fixation may also help to lessen the need for N fertiliser.

To achieve further improvement in the ways of using N in agriculture, more precise knowledge is needed of the dynamics of nitrogen turnover in soils, of translocation and assimilation in plants, and of interactive flows between soil, plants and animals, and the atmosphere. Only with full understanding of the many biological processes that affect N in ecosystems obtained by multidisciplinary research will it be possible to determine the guidelines for environmentally kind, socially acceptable and economically sound management of nitrogen utilisation in agriculture.

Introduction

The increased use of nitrogen fertiliser, coupled with the use of new responsive strains of arable crops and improved success in the control of pests, diseases and weeds has led to large increases in crop production in many developed countries. Nitrogen fertiliser has given farmers great flexibility in the management of their systems of production whether arable or livestock. Because of favourable cost/product price relationships it has increased also the profitability of many farm enterprises. However, successful and profitable use of fertiliser N has contributed to overproduction in the Western World and, at the same time, because nitrogen can 'leak' from soils and especially grazed pastures the greater overall use of nitrogen has given rise to concern amongst environmental and health specialists. Lobby pressure to reduce and in some cases ban the use of nitrogen fertiliser in agriculture has been building up (Thompson, 1985). This paper attempts to examine the topic of N fertiliser use and to provide a view of what farmers should be doing to stay in business and what scientists need to be studying to assist in this process. The paper briefly describes the present usage of N fertiliser in the world and reports on present projections for future use. The case for continued use of nitrogen in agriculture and possible ecological consequences are examined. Methods to improve the efficiency of use of nitrogen fertiliser and of the research needed to further reduce leaks from agricultural systems are also described. An alternative title for the paper (or

the challenge!!) could be 'how to continue using N fertiliser whilst controlling losses'? The topic is a huge one with a voluminous literature and this brief paper attempts to indicate key points ('sign posts') only but with reference to published sources where fuller discussion of particular aspects can be found. The following volumes in particular provide excellent and detailed coverage of all aspects of the topic (Aldrich, 1980; Hauck, 1984; Haynes, 1986; Stevenson, 1986).

Most soils contain large amounts of nitrogen usually as organic forms which are sometimes associated with the clay fraction. Each year some of this organic nitrogen is mineralised to ammonium-N (NH_4^+-N) and nitrate-N (NO_3^--N), forms which plants can take up through their roots. Approximately 3% of the total amount of organic N in a soil is mineralised each year (Henzell and Ross, 1973) which in the UK means about 80–100 kg/ha of plant available N. Rainfall or dry deposition may add 10–20 kg N/ha. Most arable crops and grass require from 100–400 kg N/ha to produce an economically viable yield. Farmers apply N fer-

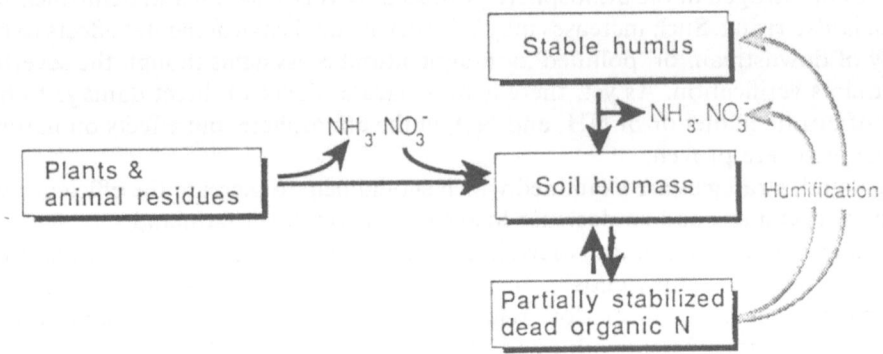

Fig. 1. The internal cycle of N in Soil (from Stevenson, 1986)

Table 1. Main processes affecting N in soil

Ammonification	Org. N	→	$NH_3^+ \underset{-H+}{\overset{H+}{\rightleftarrows}} NH_4^+$		
Nitrification	NH_4^+	→	NO_2^-	→	NO_3^-
Denitrification	NO_3^-	→	NO_x^-	→	N_2
Assimilation (Immobilisation)	NO_3^-	→	NH_3	→	Amino Acids, → Purines → Proteins, Nucleic Acids

Fixation of NH_4^+ by Clay Minerals

Fixation of NH_3 by Soil Organic Matter

Reactions of NO_2^- with Humic and Fulvic Acids

tiliser to provide the optimum amounts of N their crops require. Since the efficiency of fertiliser N use by plants is only about 50% farmers must add from 200–800 kg N/ha as fertiliser depending on type of crop, and soil N supply to provide sufficient for the economic production of a crop.

Most fertilisers contain nitrate or forms of N (e.g. urea and ammonium) which can be converted to nitrate. Nitrate is soluble in water and is not retained by exchange processes in soil. Thus when rain or irrigation water passes through the soil the dissolved nitrate can appear in drainage water or in the case of soils overlying porous strata, in the deep ground water. All soils whether fertilised or not contain nitrate which can appear in drainage and ground water. It is the appearance of nitrate in river water or in deep aquifers which is of major concern to environmentalists.

The key to what happens to nitrogen fertiliser lies in the internal cycle of N in the soil (Fig. 1). The residues of crops and animals are decomposed by means of soil fauna and the soil biomass to partially stabilised organic N and mineral N and then to stable humus. The main processes affecting N in soil are listed in Table 1 and many of these will be referred to throughout this paper. Less is known about the last three processes shown in the Table than the others.

Present and projected use

Nitrogen fertiliser is the most widely used of the three major plant nutrients N P K. It is produced mainly as NH_3 and is used either in this form or in a range of liquid and solid derivatives. The main solid N fertilisers with the percentage share of world production capacity in 1980 shown in paranthesis are urea (35%), ammonium nitrate (20%), ammonium sulphate (8%) and calcium ammonium nitrate (6%) (Lastigzon 1981). Urea is the most popular solid N fertiliser and urea and anhydrous ammonia are the dominant N products traded on the International Market (Stangel, 1984).

Nitrogen fertilisers were produced in 72 countries in 1982 with the total world capacity being 99 mt of N having been only 50 mt in 1970. Forty-five percent of the world's capacity to produce N lay within three countries with the following percentage capacity USSR (20.8), USA (16.2) and the Peoples Republic of China (8.5). Not all this capacity has been utilised with consumption being almost 32 mt in 1970 rising to 60.3 mt in 1980 at an average growth rate of 7% per year (Fig. 2). Estimates of the future use of N fertiliser depend on so many factors that they are extremely tenuous but the dotted curve on Figure 1 indicates a summation of projections made by UNIDO (1980), FAO (1981) and Stangel (1984). In the year 2000 it is estimated that world N demand might fall into the range 111 to 134 mt (UNIDO, 1980).

Since the use of N fertiliser is linked with the ability to pay for it there is great variation between the major world economic groupings in the use of N fertilisers on a per hectare basis. While the world average N use on arable land and permanent crops in 1980 was 42 kg/ha for the three major economic regimes, the amounts were 58 (Developed Market Economies), 66 (Centrally Planned Economies) and 18 kg N/ha for the Developing Market Econ-

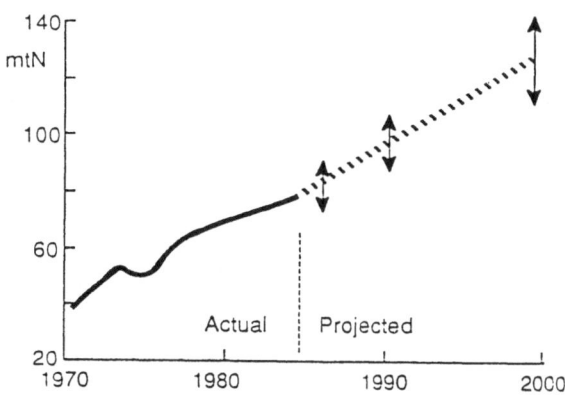

Fig. 2. Consumption of Nitrogen Fertiliser in the world. (UNIDO, 1980; FAO, 1981; Stangel, 1984).

omies respectively (FAO 1982). For individual countries in 1979 the range of application rates on average varied markedly from about 240 kg N/ha in The Netherlands, through 55 in the USA and 32 in the USSR to 13 kg N/ha in Brazil (FAO, 1981).

Within any one country there is great variability in application rate of N fertiliser between crops and therefore between regions where particular crops are concentrated. Using the United Kingdom as an example overall trends in the use of N, K_2O and P_2O_5 fertiliser are shown on Figure 3. Consumption of N in 1985 was 1.6 mt with a rising trend reflecting the world picture; by contrast after a period of little change in the use of K_2O and P_2O_5 the Figure illustrates a small tendency in recent years for use of these nutrients to increase also. No one in the United Kingdom is prepared to project a figure for N use in the year 2000. The Figure illustrates what would happen if the predicted trend for the world applied to UK. However, there are already signs of a down turn in the use of N fertiliser in the UK and the lobby pressure from environmentalists discussed later in this paper coupled with the need to lessen agricultural production in the EEC may result in there being little change in the use of N fertilisers from the present level. The variation in rates of N used on different crops in the UK is shown in Table 2. The Table also lists the levels of fertiliser recommended for these same range of crops by the Government Advisory Services. This data is appended to indicate that the human factor has to be considered when discussing the topic of fertiliser use since farmers tend to apply more than the recommended rates to their high value crops *e.g.* wheat and potatoes. Account must be taken of this type of perception and reaction when considering responses to fertilisers and attempts to describe guidelines for their efficient use.

Table 2. Current rates of fertiliser N in use in UK. Kg N ha^{-1} (Fertilisers Manufacturers Association 1987)

	Recommended by ADAS/SAC	Used 1985	
		Eng/Wales	Scotland
Winter wheat	180	192	215
Spring barley	100	102	98
Potatoes	130	198	173
Oil seed rape (Winter)	240	279	270
Grass Up to	375	131	130

Responses

Nitrogen is used as a fertiliser because it is essential for plant growth and because soils do not have sufficient N in available form to support the production levels that farmers and nations require. For example, a 10 tonne crop of winter wheat now quite commonly achieved by farmers in the UK will contain some 200 kg of N in the grain, 40 kg in the straw and a further 30 kg in the roots and stubble giving a total nitrogen requirement of 270 kg N per ha (Jenkinson, 1986). This crop may be grown on a soil containing 5000 kg organic N per ha of which 2–3% *i.e.* 100–150 kg N/ha is mineralised each year. Even at the higher level of available soil N, the 10 tonne winter wheat crop will require all the recommended application (Table 2) of 180 kg N fertiliser/ha on the assumption that from 30–40% of that added does not appear in the crop and is described as lost (see p8). Lidgate (1984) has estimated that about 51% of the UK average wheat yield from 1980 to 1983 was directly attributable to the use of N fertiliser. Similar considerations can be applied to other crops and to animal production.

Response curves of crop yield to N fertiliser can be constructed for all crops and for a range of soil types and climatic situations. Unfortunately they

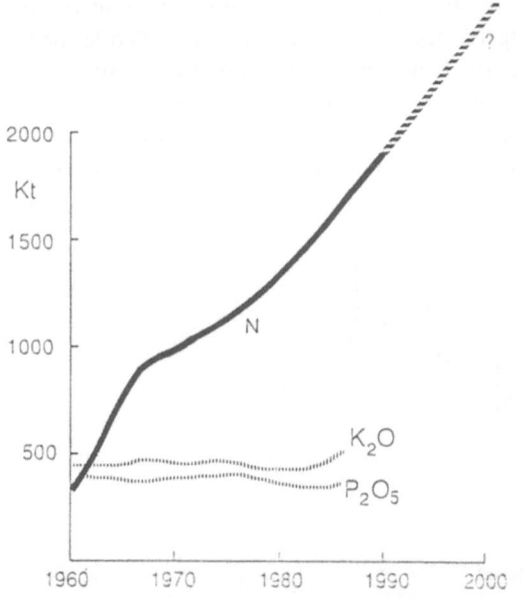

Fig. 3. Fertiliser use in the United Kingdom. (Royal Commission on Environmental Pollution, 1979; Gasser, 1982; Fertiliser Manufacturers Association, 1987).

are variable in their ability to predict the need for N fertiliser since there is no well accepted method for soil testing for available soil N (Goh and Haynes, 1986). A biological optimum level of N fertiliser can be described: above this level of fertiliser addition yields may decrease from that at the optimum and/or the quality of the product may be damaged by the accumulation of excessive amounts of nitrate or other N compounds in the crop. In practice the use of N fertilisers is decided by economic rather than biological considerations.

The profit and loss relationships in using a fertiliser to enhance crop production are illustrated by Figure 4 adapted from Jollans (1985). It can be seen that the farmer makes a loss without fertiliser and also if he uses too much fertiliser. In the case illustrated the maximum profit in relation to fertiliser use is achieved at a yield level somewhat less than the biological optimum. Considerations of this type are influenced to a great extent by national economics and relative price movements in the cost of fertilisers and crop products and in retail prices. In the UK between 1980 and 1986 the retail price index rose by 45%, the cost of livestock and crop products by 25% while the cost of fertilisers rose by only 5% (Fertiliser Manufacturers Association, 1987). However, the latter figure for August 1986 masks a decline of 16% in fertiliser price since summer 1985 primarily due to the large increase in imported urea at a cheap price. Overall since 1980 fertiliser prices have risen by less than inflation and less than the prices farmers have received for their product.

This discussion may give the impression that N fertiliser is the sole determinant of yield. In reality the achievement of yield potentials has been due to improved cultivars, improved soil physical and general chemical conditions, good moisture control, good management practices including the control of pests and diseases in addition to the supply of N. Stanford and Legg (1984) concluded that "additions of N fertilisers periodically throughout the growing season in irrigation water provides the best means of fertilization without injury to plants". A view supported by the work of Ingestad and Lund (1986) with trees.

These viewpoints may become important when attempts to devise more efficient ways to manage nitrogen fertilisers are discussed later in this paper (p.17).

Loss of N from agricultural soils

Nitrogen is lost from soil by leaching of nitrate, by denitrification to NO_x or N_2, by volatilisation of ammonia and by erosion of soil and/or soil organic matter. Leaching of nitrate from arable soils tends to take place in winter and early spring and except with heavily fertilised crops such as oil seed rape

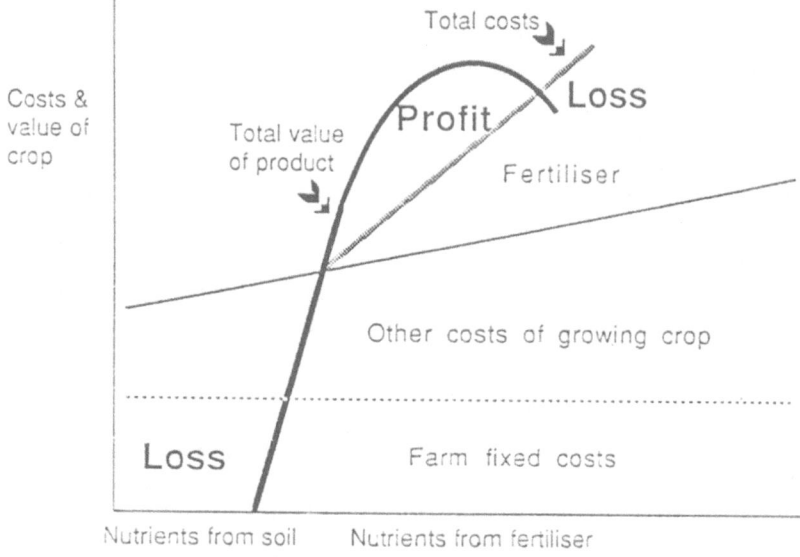

Fig. 4. Profit and loss in the use of N fertiliser in crop production. (after Jollans, 1985).

and potatoes the immediate source of the nitrogen is rarely unused fertiliser but is more often organic nitrogen. This is mineralised to nitrate in the autumn when soils are moistened by rain, with temperatures still relatively high and plant demand is low or has stopped altogether. Until recently it was believed that the major source of leached nitrate appearing in water supplies in East and South East England was from arable crops with grassland making little contribution (Hood, 1976; Royal Society, 1983). However recent work by Ryden *et al.*, (1984) indicates that grassland should be considered either as grazed or cut for conservation. Significant amounts of nitrogen (162 kg N per ha) were leached each year from grassland grazed by cattle to which 420 kg fertiliser N was added ha^{-1}yr^{-1}. This was 5.6 times greater than that leached below a comparable cut sward and exceeded the losses normally observed from arable land. On the basis of this data White (1984) concluded that 'further intensification of livestock farming in the UK was likely to exacerbate the problem of nitrate pollution of water supplies'. However, losses of nitrate from drainage from grassland in the UK which receives about 1 mt of fertiliser N each year is variable. For example Haigh and White (1986) examined losses of nitrogen in tile drains under a clay soil beneath grazed grassland and found the total amounts of N lost by leaching in 1983/84 were 17.5 and 48.7 kg N ha^{-1} in the two years representing 9 and 43% respectively of the N added. The wide variation between the two years was attributed to differences in the quantity and timing of N fertiliser applications, the dryness of the preceding summer and the duration and intensity of stocking.

Significant losses of nitrate by leaching also occur when old grassland is ploughed up and the mineralisation of the accumulated organic matter and humus is stimulated by cultivation and aeration with no live grass plants to absorb the nitrate. Nitrate can be leached to the field drains and so enter streams and rivers or in some freely draining soils it can by-pass the drainage system and enter the ground water which supplies deep aquifers often used as a source of drinking water.

Denitrification, occurs when anaerobic conditions are produced in soils in the presence of substrate nitrate, a source of carbohydrate and the appropriate micro-organisms. Nitrogen can be lost rapidly by denitrification when soils are warm and wet, conditions which can apply when fertilisers are applied in spring prior to heavy rain storms. This mechanism of loss of nitrogen either returns N_2 to the atmosphere or adds to the burden of nitrous oxide which as will be described later is considered to be a hazard to the protective ozone layer of the upper atmosphere.

Volatilisation of ammonia can occur in arable farming when large amounts of farmyard manure are used but this loss pathway is mainly of importance from feed lots, from intensively grazed grassland, or when slurry and sludge is added to fields. Small amounts of loss by this route can occur also from senescing plant material. The ammonia enters the atmosphere and is deposited in rain or by dry deposition onto neighbouring farms or natural ecosystems.

Losses of nitrogen by erosion are very significant in some countries but not usually those in Western Europe. However, where peat occurs in upland high rainfall areas, the loss of organic nitrogen may attain significance especially under storm conditions and has been little quantified (Edwards *et al.*, 1985).

Ecological consequences

The appearance of nitrogen in water or in the atmosphere by any of the routes described above has been blamed for a number of deleterious environmental and health effects (Keeney, 1982).

Nitrates in water and food and human health

High amounts of NO_3 in water and food are of concern to humans mainly if the NO_3 is converted to NO_2. The nitrite can cause a problem called methemoglobinemia which results in the inability of the blood to transport O_2: this is of little significance in adults but can be fatal in infants. The risk to babies is not solely due to the level of nitrate in the water supply but relates also to the source of the water. Babies using bacterially contaminated water such as that from wells are much more likely to have the micro-organisms which can convert NO_3 to NO_2. Worldwide some 300 cases of 'Rocky Mountain Blue Water Disease' have been reported since 1946 (Bryson, 1987). The condition is rare in Western Europe and in England the last recorded case was in 1972 (Jenkinson, 1986).

Stomach cancer has been linked to high levels of nitrate in food and water although recent medical evidence (Forman *et al.*, 1985) suggests that this connection is very tenuous. These workers showed that the incidence of this cancer, thought to be triggered by nitrosamines and not directly by NO_3 or NO_2 was less in an area of England where daily nitrate intake was high than in areas where intake was low. Moreover on a world scale whilst use of fertiliser N has increased at 7% per annum (Fig. 1) since 1974 male gastric cancer deaths in Denmark, Britain and Holland have decreased since 1945 from about 55 to about 25 per 100,000 deaths in 1985 (Bryson, 1984; Bryson, 1987).

Nitrates and animal health

Nitrate poisoning has been found in some cattle grazing feeds or drinking water of high nitrate content and probably where micro-organisms in the rumen reduce NO_3 to NO_2 with consequent formation of nitrosamines or cause an effect equivalent to methemoglobinemia. However, as with methemoglobinemia in infants such occurrences are not widespread.

Nitrates and plant growth

Nitrate formed in soil can be toxic to plants and cause stunted growth and too much nitrate N in soil can cause excessive growth. The latter condition can give rise to problems of lodging and poor winter hardiness and in the case of vegetables can cause the accumulation of high amounts of nitrate in the leafy parts of the crops. This effect is enhanced by reduced light intensity, by soil moisture deficiency, and by other nutrient deficiencies in the plant leading to a failure to produce nitrate reductase

Nitrogen and environmental quality

Depletion of ozone in the stratosphere. Ozone (O_3) is formed high up in the atmosphere where it provides a protective shield against excessive ultraviolet radiation. Ozone is being formed and destroyed continuously in the stratosphere and a decrease in the average amount or the occurrence of 'holes' in the Ozone layer may cause problems for plant and animal life from increased exposure to ultraviolet radiation. The destruction of O_3 is catalysed by NO, halogens, hydroxyl and hydrogen. A possible source of NO is from N_2O the product of denitrification which can diffuse into the upper atmosphere. However, uncertainty remains as to the magnitude of this stimulant to O_3 depletion when compared to chlorofluorocarbons from spray can propellents or stratojets (Royal Commission on Environmental Pollution, 1979).

Eutrophication

The addition of nutrient to lakes and ponds can increase the growth of aquatic plants including algae giving rise to eutrophication. The enhanced growth of algae (blooms) can lead to a depletion of O_2 in the water so that fish and other animals cannot live. The odour and taste of the water are also affected. Recent work has indicated that phosphorus rather than nitrogen is the main nutrient limiting the growth of aquatic plants, and many algal blooms are caused by input of this element from industrial wastes, sewage treatment works and from detergents, rather than from agriculture.

Ecosystems damage

NH_3 volatilising from feed lots, heaps of farmyard manure or from intensively grazed pasture can be deposited on neighbouring natural ecosystems. The amounts of N being deposited by dry deposition and in rain can reach very high levels causing damage to the vegetation upon which they fall. Some of the NH_3 and some of the NO_2 entering the atmosphere from denitrification may be converted to nitric acid. This coupled with H_2SO_4 formed by emissions from the chimneys of factories or power stations where fossil fuels are burnt and with the waste gases emitted by motor cars builds up to make acid rain. Acid rain can affect plants directly and can acidify the water of lakes resulting in an increase in the amount of soluble aluminium which is toxic to fish and to plants.

Some dramatic examples of ecosystem damage from acid rain are evident in specific locations (Sasanow, 1985). What is not so clear is how small increases in the amount of N being added to closed stable natural ecosystems will affect them? For example the amount of N entering tundra ecosystems by rain and dry deposition has doubled from 2 to 4 kg N ha^{-1} yearly. There is little quantified information to describe how this will affect the growth of plants and the species balance in such systems. At a recent conference on the role of nitrogen in ecology Lee *et al.*, (1983) concluded that "N is an important factor limiting plant growth and development, but in ecological terms we are further than ever from generalisations. Too little attention has been paid to N as an ecological factor." At the same conference Pate (1983) stressed the role of nitrogen in providing protective mechanisms for plants by osmotic regulation in soil stress; defence against attack by animals; allelopathic substances, antifungal agents and defence against viruses. Thus extra though small inputs of nitrogen into ecosystems with little total N might result in major changes in the species composition and character of the vegetation. The changes which might result from such small 'sub-liminal' effects of nitrogen on normally unfertilised natural ecosystems merit further investigation.

N inputs to the oceans from rivers, rainfall, sewage sludge and biological N$_2$ fixation must be increasing as the quantity of fertiliser N used in agriculture increases. For example in the UK national nitrogen balance it is estimated 0.3 mt N

reaches the sea each year (Jenkinson 1986). Little is known of the nitrogen cycles of the oceans and in particular how widely nitrification to nitrous oxide occurs; it has been suggested that 4 to 10 mt N as N$_2$O may be lost annually from the oceans (Cohen and Gordon, 1979). Thus, the role of the oceans as a source-sink for nitrogen on a world scale is little understood and further study is required.

Effects on human health and the environment from the use of nitrogen in agriculture

Because of concern about the possible effects of nitrate in water the EEC following advice from the World Health Organisation (WHO) have specified a maximum admissable concentration in public water supplies of 50 mg nitrate or 11.3 mg NO$_3$-N l^{-1} (EEC, 1980). The limits are set to cover the worst case of infants exposed to contaminated well water and of necessity encompass much water that because of ameliorating factors do not expose people to potential hazard. Many sources of ground water are above or approaching this level and some river waters especially in late winter and early spring have levels approaching and in some cases exceeding these limits. The rising trend in annual nitrate levels of water is illustrated by data from Eastern England where cereals are grown intensively at the highest levels of nitrogen fertiliser input (Fig. 5).

Thus, the mean annual concentrations of nitrate in some rivers in the South of England used for

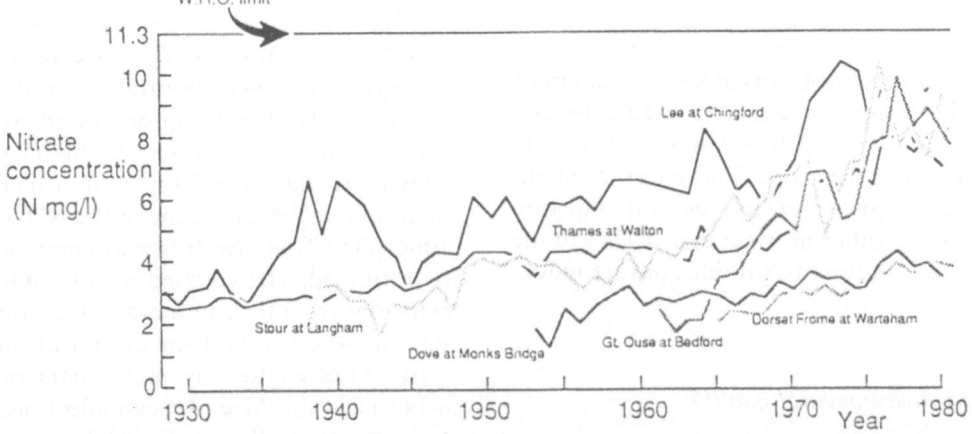

Fig. 5. Mean annual nitrate concentrations (mg NO$_3$-N l^{-1}) of rivers in Southern England. (Department of the Environment, 1984).

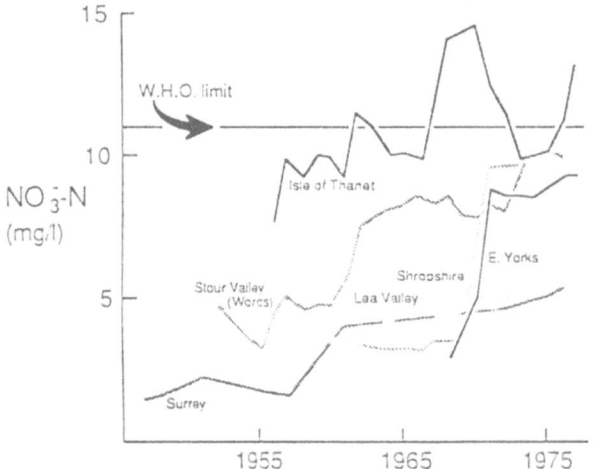

Fig. 6. Nitrate concentrations in some public water supply boreholes in chalk and sandstone aquifers in England. (Wilkinson and Green, 1982; Royal Society, 1983).

Table 3. Present and projected levels of nitrate in rivers in the UK (Department of the Environment 1984)

River	Mean concentration	Projected for AD2000
	μg NO$_3$–N l^{-1}	
Thurso	0.4	1.0
Findhorn	0.2	1.0
Tay	0.6	1.0
Tees	1.3	1.7
Yorkshire Ouse	3.5	5.2
Severn	5.6	7.6
Worcestershire Avon	10.3	14.5
Great Ouse	9.1	14.0
Thames	7.5	10.7
WHO limit	11.3	

public water supplies from 1928 to 1980 show a rising tend and although some approach the WHO limit none exceeds it. However data from deep boreholes on the chalk and triassic sandstone aquifers of the UK shown in Figure 6 also depict a rising trend with one source (Isle of Thanet) indicating levels well above the WHO recommended limit. Publications about water quality studies indicate an increasing number of cases of nitrate contamination in more or less extended areas *e.g.* in Northern France, in Hungary, The Netherlands, the Rhine-Western regions and other catchment areas in Germany and the Greater London catchment areas in England. In some cases current standards are permanently or periodically violated, in

others the observable trends of nitrate concentration point towards the same direction. (De Haen, 1982). Much of the public water supply in the UK comes from upland areas and it is encouraging to note that the mean concentrations of NO$_3$-N in water from rivers in Scotland and the North and West of the Country is low by comparison with that from the Worcestershire Avon, Great Ouse and Thames (Table 3). Model predictions for the latter rivers indicate that all will have water with nitrate levels close to or above the WHO limit in the year 2000.

Thus, the rising trend in the nitrate levels of water supplies in the UK cannot be doubted. The impact of these levels on health and the environment is less easy to discern (Cooke, 1986; Thompson, 1985). It must be remembered that water is only one of the sources of nitrate ingested by humans and data from the United States shown in Table 4 indicates that water provides only 0.7% of the input of nitrate to the average person. Vegetables provide 81.2% with major contributions coming from lettuce (112–205 ppm) and spinach (55–170 ppm) with lesser amounts from tomatoes (11–25) and potatoes (8–143 ppm). It should be noted that these concentrations of nitrate N were assessed in 1964 and that the upper levels of the range of values were considerably less than those determined for the same crops in 1907 (Sicilliano *et al.*, 1975).

Alternative views as to the seriousness of pollution with nitrate

In discussing the environmental problems of agriculture the Council of Advisors to the West Germany Government considered that the increasing danger to groundwater of pollution with nitrate

Table 4. Estimated average daily ingestion of nitrates for US residents (White, 1975)

Source of nitrate	mg	%
Vegetables	86.1	81.2
Fruits	1.4	1.3
Milk and products	0.2	0.2
Bread	2.0	1.9
Water	0.7	0.7
Cured meat	15.6	14.7
	106.0	100.0

was the second most significant impact of agriculture on the environment (Council of Environmental Advisors, 1985). They thought it was closely connected with the intensification of nitrogen usage on soils used for agricultural purposes many of which were water catchment areas. Nitrate not taken up by plants or superfluous to their needs and which stems from various sources including natural sources is not retained in the soil and is therefore transported in percolating water down to the groundwater. The use of slurry can cause particularly heavy nitrate leaching when nitrification has occurred.

De Wit *et al.* (1987) believe that "Damage to the natural environment is intrinsic to productive farming and even a reduction of the top rate of N fertiliser from 400 to 100 kg N ha^{-1} would be still too high to enable reconciliation with the demands of environmentalists".

Hauck (1984) recognised the inevitability of there being leaks of nitrogen from agricultural systems but asked "How can the amount of N taken up and used to produce a harvestable crop be increased, the amount lost be decreased, and the amount of N immobilised in the soil biomass be maintained in active forms of immediate or potential benefit to the plant?" Cooke (1986) concluded that "Scientific control of the nitrogen components of soil fertility, and of the use of fertilisers will provide drinking water that contains no more nitrate than would drainage from the same soil under unfertilised natural vegetation". Yet the Fertiliser Manufacturers Association (1985) believed "We should not delude ourselves that we do not have a problem. Clearly we do because nitrate levels in water sources are rising and fertilisers are part of the intensive agricultural system that contributes to these rising levels."

Statements of the latter type had been made much earlier for example by Parr (1973) who recognised also that non-agricultural sources of N will increase as nations become more urbanised. He said "The time has come for agriculturists to assume a less defensive and indeed more positive attitude to their approach to the plant nitrate-water quality issue. We have no alternative to accept the fact that in some cases agriculture is contributing to environmental pollution".

Despite some uncertainty about the severity and extent of the problems posed by the losses of nitrogen from agricultural systems researchers and farmers must acknowledge that there have been increases in the amount of NO_3 in river and ground water and of NO_x and NH_3 in the atmosphere. It must also be accepted that the use of N fertilisers, even if not entirely responsible for this situation, have made a major contribution and will do so in an ever increasing proportion if the trends of future fertiliser use shown in Figure 2 are fulfilled.

How to improve the efficiency of use of nitrogen fertiliser

From what was said earlier in relation to responses, a total ban of N fertiliser use on a world scale where between 400 and 800 m people are starving already is inconceivable. Even in a developed country such as the UK such a ban would completely upset the economy of the country since food would have to be imported or the diet of the population would have to change markedly. The amount of meat, milk and eggs in the diet would have to reduce. Reduction in the losses of N from farms and a marked increase in the use of plants that fix N biologically might help to alleviate the problem but such practice would require many years for wide acceptance. Moreover the use of legumes may increase the risk of leaching losses of N.

There are a multitude of ways to reduce losses of N from agricultural soils and its appearance in water supplies whilst still permitting its use by some if not all farmers. The spectrum of possibilities was well summarised by Aldrich (1980) and are shown in Table 5. All except numbers 9 and 11 depend on political action and would be unpopular and difficult to administer fairly, particularly since as already described the problem can vary widely across a country and between farms and fields on one farm. Some countries including Sweden have introduced taxes on nitrogen fertiliser but increases in price of 5–10% have had little impact. Economists have predicted that increases of 100 to 200% in the price of N fertiliser would be required to have any significant effect on levels used.

Farmers can preempt the need for draconian all-embracing taxes and laws by applying existing knowledge as to how to optimise use of nitrogen to the full. Researchers can contribute by developing alternative technologies and by research to expand

Table 5. How reduce nitrates in water? From Aldrich 1980

1. Limit amount of N added per ha.
2. Establish farm quotas or allotments.
3. Limit acreage of N-responsive crops.
4. Issue negotiable purchase rates for N.
5. Establish effluent standards.
6. Impose effluent charge or environmental user fee.
7. Impose a tax on N fertiliser
   = Nationwide
   = State or region
   = Watershed or drainage basin.
8. Subsidise use of lower application rates.
9. Establish a best management practice.
10. Solve N in water problems at point of use
    = Surface H_2O
    = Groundwater.
11. Increase research and education, rely on voluntary response.

the understanding of the N cycle in soils. In emergency situations nitrate can be removed from water by chemical and biological processes or water high in nitrate can be diluted with water low in nitrate but these procedures are expensive and would only be used widely as a last resort. It would be preferable to adapt strategies that make such practices unnecessary.

Apply existing knowledge to the full

The application by farmers of existing knowledge would help to lessen potential problems (Aldrich, 1984). For example the use of N could be concentrated in regions with a favourable climate and retentive soils. The rates of nitrogen should be adjusted to the length of the growing season, the temperature at the time of application and expected through the growing season and to the expected time, amount and distribution of annual rainfall. Responsive crops such as maize and grassland which can take up and utilise nitrate effectively should be used in sensitive areas. The losses of growth and hence nitrate uptake potential due to weeds, pests and disease should be minimised as far as possible. Most importantly the N fertiliser regime for each crop should be selected carefully to accord with the probable yield, the anticipated supply of N from the soil, the use of sewage sludge or farmyard manure, the best fit to economic optimum response curves (see Fig. 4) and with careful selection of the time, method and ideal form of N fertiliser. Less nitrate is lost from autumn than

from spring sown crops so this practice should be encouraged. Nitrate leaching can be reduced by avoiding autumn applications of fertiliser and by not applying nitrogen too early in spring. If a crop is harvested early any unused or freshly mineralised N in the soil should be retained by sowing a cover crop. Straw incorporation in autumn can reduce leaching of nitrate by absorbing nitrate at a time when it is vulnerable to leaching and then releasing it in the following spring when plants are in short supply.

Computer simulation models to assess the amount and pattern of nitrogen fertiliser application needed to achieve economically sound production levels were initiated for annual horticultural crops by Greenwood (1986). More recently other workers have developed models initially to describe leaching losses in winter wheat (Whitmore and Addiscott, 1986) and subsequently efficiency of utilisation of nitrogen fertiliser by winter wheat (Whitmore and Addiscott, 1987). Likewise Ingestad and Lund (1986) in Sweden have developed models to describe the quantities and patterns of nitrogen fertiliser needed to optimise growth of trees and other crops. Models of similar type are required for all arable crops and for grassland.

The application of these procedures, coupled with further development of ideas tested as yet in small scale studies should go some way to reducing the time nitrate is available in the soil for leaching or denitrification and would increase the overall efficiency of use of N fertiliser.

Utilise potentially applicable technology

Slow release fertilisers have potential to provide nitrogen at a rate more appropriate to plant requirements than mineral fertilisers. Many companies are working to develop these and the principle is akin to the accurate provision of nitrogen to meet plant requirements as postulated by Greenwood (1986) for horticultural crops and Ingestad and Lund (1986) for forest and coppice trees. Since it is unlikely that accurately metred irrigation systems with added nutrient can be provided for more than a fraction of the world's crops and forests the development of matched nitrogen supply patterns from slow release fertilisers holds considerable promise. Chemicals can be used to prevent nitrification that is to hold added fertiliser as the relatively non-leached NH_4^+ form. However, although some of these chemicals have worked well in small scale laboratory trials they have not yet succeeded in the field.

Improvements in the efficiency of biological nitrogen fixation by grain and pasture legumes and the wider use of legumes or non leguminous nitrogen fixing shrubs and trees have been suggested as the way to reduce use of nitrogen fertiliser. High N losses can occur from grazed grass/legume pastures and it might be necessary to keep livestock indoors to reduce losses with the added advantage that the N in the manure could be used more efficiently. However none of these changes will happen overnight without major breakthroughs in research. For example in the UK where only about 12.5% of the total annual N input to the national nitrogen cycle (*i.e.* 0.3 mt) comes from nitrogen fixation (Jenkinson, 1986) it is not practical to consider a total ban on the use of N fertiliser without catastropic effects on the national diet and on the rural economy.

The possibility that cereals and potatoes could effectively become legumes and fix their own nitrogen or that microbiological procedures could be used to manufacture nitrogen fertiliser has been described. These ideas remain extremely speculative and many years work will be required to research, develop and apply them.

Should the conservationist lobby and/or economics force a reduction in the use of nitrogen fertiliser, a much more likely scenario for the future, is the selective use of nitrogen fertiliser with enhanced dependence on legumes. In particular grassland has the most opportunities for the legume to make a larger contribution to the nitrogen budget, and the use of fertiliser on any grassland except strategically in emergency situations because of crop failures or peculiar weather patterns must be closely scrutinised. Such considerations are emphasised by the high losses of N that occur from grazed grassland.

The enhanced use of intercropping with more use of clovers and vetches intersown with cereals and other crops and ploughed in as green manures, looks an alternative way of increasing the role of biological N fixation but the present trend to winter cereals militates against this.

In summary, the future appears to lie in learning how to use fertiliser N more efficiently on cereals and root crops, on growing more grain legumes and on optimising the role of the legume in grassland farming. Strategies of this type should take us through any periods of possible nitrogen fertiliser deficiency before the advent of genetically-engineered plants and of novel bio-technological processes to manufacture fertilisers. It is apparent that biological nitrogen fixation alone could not sustain British agriculture in its present form and that it could only sustain a subsistence-type agriculture that could not support our present population even with a much changed diet. Moreover, reliance on biological nitrogen fixation alone does not eliminate the chances of environmental pollution. However, biological nitrogen fixation could and should make a larger contribution to present farming practices and particularly to grassland agriculture, by sparing use of mineral fertiliser for high value crops and by reducing the chances of environmental pollution.

Future research

As explained earlier there is a fund of knowledge which if applied carefully could help already to reduce the amount of nitrate-N reaching water supplies. In addition although much is known about the main processes taking place in the internal N cycle of the soil, research is needed to distinguish the amount of N in the biomass and the importance of this relatively small proportion of the total N in soils to the N cycle. Equally the organic and clay chemistry which distinguishes the active and stable fractions of humus as required in most attempts to model the behaviour of N in soil (Parton *et al.*, 1983) is not fully understood. Most measurements of N turnover in soil are net that is representing the sum of mineralisation and immobilisation but the precise nature of the individual processes and factors that affect them is little understood. The links between these processes and gaseous losses of N from soils and fertiliser N is incompletely known. When details of these and other biological processes are understood it may be possible to determine the long term fate of mineralised N and to devise management strategies for effective use of biologically and chemically fixed

N and to ensure the efficient use of N from plant and animal residues. The relationships between soil organic and inorganic N pools and soil type, crops, climate and residue management practice are needed to assist the process of devising appropriate management practices. Knowledge of all these factors should allow the amount of N taken up and used by crops to be increased while the amount lost is decreased and the amount of N immobilised in the soil biommas and/or humus is maintained in active forms of immediate or potential benefit to plants (Hauck, 1984).

Increased understanding of the processes of the internal N cycle will enable N fertiliser to be used more efficiently than at present on arable crops. For grassland, additional research strategies are needed to manage animal wastes so that losses of N are minimised because intensively grazed pastures are leaky systems. A greater role must be found for legumes but whilst this reduces the need for N fertiliser it does little to lessen leaks of N from the system. The late J.C. Ryden (1982) suggested that if ways could be found to denitrify surplus nitrogen from animal systems directly to N_2 and not to N_2O this would reduce the potentially deleterious effect of losses of N from grazing systems in the environment.

Further work is needed to minimise the appearance of nitrate in ground water (Table 6). This will be difficult since the residence time of NO_3 at depth in soil and underlying rock strata is thought to be about 25 years so that much remains stored in aquifers at the present time. Given knowledge of sensitive soils, catchments and management systems which contribute NO_3 to ground water it is

Table 6. Research needs to reduce nitrate leaching losses to groundwater

Define sensitive soils, farming systems and catchments.

Determine the profile of nitrate at depth under contrasted farming systems where these lie over aquifers used for public water supplies.

Monitor the effect of changes in farming practice on the profiles of nitrate in the soil and on the apperance of nitrate in ground water.

Breed crops including grasses capable of utilising all the nitrate available in a soil effectively.

Develop management systems and fertiliser strategies to suit both crops and soils in a range of climatic conditions so as to reduce losses of nitrate to ground water.

possible that farmers in these areas could be persuaded to alter their systems or lessen use of N fertiliser. Compensation payments might be needed to encourage this just as for farmers in Environmentally Sensitive Areas.

Conclusion

Nitrogen fertiliser will continue to be used in the foreseeable future. Should the main feed stock of natural gas run out and there be no replacement or no alternative way of chemically fixing nitrogen from the atmosphere its use may of necessity decline. However, at the present time farmers should be advised to apply all the available knowledge to increase the efficiency of use of N fertiliser (p. 291) and to minimise losses from the land and from pastures. If the levels of nitrate continue to rise in drinking water *etc.* politicians will act to control the use of nitrogen fertiliser. At the same time scientists must strive to find further technical solutions (see p. 292) so that farmers can continue to use N fertiliser without adding to the nitrogen pollution problem. Biological nitrogen fixation alone is unlikely to resolve the problem, moreover losses of nitrate can occur from leguminous grain crops and from pastures rich in legumes. More information is needed about the effect of nitrogen lost from agricultural systems on 'downstream' natural ecosystems.

In the present political climate applying in Western Europe and the United States, and with the widely held view that some farmers are feather bedded irresponsible polluters of the environment, it is important to stress that use of N fertiliser can be related directly to use of land (North, 1986). North showed (Table 7) for the UK that the more nitrogen fertiliser used per ha of wheat the less hectares should be needed to meet the national requirement for wheat. Provided policies can be devised to use the land so spared from agriculture to meet conservation and environmental objectives it should be possible for rational people to devise rural policies using all the knowledge briefly described in this paper to the benefit of the nation. Alternatively, all the land could continue to be used for wheat production but with lower overall rates of N fertiliser. Food and fibre could be produced in sufficient amounts using N fertiliser to best advantage in catchments unlikely to leak more than

Table 7. Wheat production in the UK — nitrogen fertiliser use and land area. (North, 1986). Production level 12 mt. Assume 1 kg N = 30 kg grain

Nitrogen kg/ha	Yield t/ha	Area m ha	Nitrogen 000 tonnes
0	4.0	3.00	nil
100	7.0	1.71	171
200	10.0	1.20	240
300	13.0	0.92	276

In 1985 N use on wheat in the UK was 380,000 t
Crop area: 2 mha
Output: 11.6 mt grain

modest amounts of nitrate whilst releasing other land for recreation and as nature reserves etc. However even the latter require management, and even they will leak some nitrogen to water and to the atmosphere.

The problem goes beyond national boundaries since escapes of nitrogen from one country are unwanted inputs to a neighbour and because of the closely intertwined nature of agricultural production and world economics. The imbalance of N use between poor countries with starving populations, and rich countries with many over fed people cries out for attention. The fact that the world has not been able to deal with this situation lends little confidence that rational policies to lower N use in the developed regions so making more available for use in the developing world will be established.

More people need to be educated about the internal N cycle in all soils, the facts of N use and misuse and of world agriculture and population issues in the hope that they will influence Governments to seek rational solutions. Should the predicted deleterious effects of NO_x releases on the ozone layer in the atmosphere be confirmed pressure for change in agricultural and industrial practices will increase. Scientists are generally poor communicators with lay people and with decision makers and when talking to themselves even in the presence of ministers they tend to talk in acronyms and jargon which soon makes the uninformed listener stop listening. Those people reading this volume have a duty to provide clear facts with the range of options for continued use of N fertiliser in the future to the general public. Only with full understanding of the processes of the N cycle by scientists and of the nature of nitrogen cycles in agricultural, regional and national ecosystems by all people will it be

possible to determine guidelines and policies for the sound management of nitrogen fertiliser in the world leading to effective, economically sound, environmentally kind and socially acceptable agricultural systems.

References

Aldrich S R 1980 Nitrogen in relation to food and energy. Sp. Publication 61, Agricultural Experiment Station, College of Agriculture, University of Illinois, USA.

Aldrich S R 1984 Nitrogen management to minimise adverse effects on the environment. *In* Nitrogen in Crop Production. pp 663–673. Ed. R D Hauck. American Society of Agronomy, Madison, Wisconsin, USA.

Bryson D D 1984 Nitrates and health. Proc. Fertil. Soc. 228.

Bryson D D 1987 Nitrates and our health. Fertiliser Review 1987, 8–10. Fertiliser Manufacturers Association Ltd, London.

Council of Environmental Advisors 1985 Summary of the report on environmental problems of the agriculture. Federal Republic of Germany (Bundestagsdrucksache 10/3613).

Cohen Y and Gordon L J 1979 Nitrous oxide production in the ocean. J. Geophys. Res. 84, 347–353.

Cooke G W 1986 Nitrates in surface and underground water. Span 29, 10–11.

De Haen H 1982 Economic aspects of policies to control nitrate contamination resulting from agricultural production. Euro. R. Agr. Eco. 9, 443–465

De Wit C T, Huisman H and Rabbinge R 1987 Agriculture and its environment: Are there other ways? Agric. Systems 23, 211–236.

Department of the Environment 1984 Standing Technical Advisory Committee on Water Quality: Fourth biennial report, February 1981 to March 1983, HMSO, London.

Edwards A C, Creasey J and Cresser M S 1985 Factors influencing nitrogen input and output in two Scottish upland catchments. Soil Use and Management 1, 83–87.

EEC 1980 Council directive on the quality of water for human consumption. Official Issue No. 80/778, EECL229, 111 European Economic Community, Brussels.

FAO 1981 Fertiliser Yearbook Vol 30, Food and Agriculture Organisation of the United Nations, Rome, Italy.

FAO 1982 Current fertiliser situation and outlook. Commission on Fertilisers, Seventh Session 7–10 September. Agriculture Organisation of the United Nations, Rome, Italy.

Fertiliser Manufacturers Association 1985 Environmental aspects of fertiliser use. Fertiliser Manufacturers Association Ltd, London.

Fertiliser Manufacturers Association 1987 Fertiliser Review. Fertiliser Manufacturers Association Ltd, London.

Forman D, Al-Dabbagh S and Doll R 1985 Gastric cancer and salivary nitrate and nitrite. Nature 315, 462.

Gasser J K R 1982 Agricultural productivity and the nitrogen cycle. Phil. Trans. R. Soc. Lond 296, 303–314.

Goh M and Haynes R J 1986 Nitrogen and agronomic practice. *In* Mineral Nitrogen in the Plant–Soil System. Ed. R J Haynes. pp 379–468. Academic Press, London.

Greenwood D J 1986 Advances in plant nutrition Vol. II. Eds. P B Tinker and A Lauchli. pp 1–61. Prediction of Nitrogen Fertiliser Needs of Arable Crops. Praeger, New York.

Haigh R A and White R E 1986 Nitrate leaching from a small underdrained, grassland clay catchment. Soil Use and Management 2, 65–70.

Hauck R D (Ed.) 1984 Nitrogen in Crop Production. American Society of Agronomy, Madison, Wisconsin.

Haynes R J 1986 Mineral Nitrogen in the Plant. Soil Systems, Academic Press, London.

Henzell E F and Ross P J 1973 The nitrogen cycle of pasture ecosystems. Eds. G W Butler and R W Bailey. *In* Chemistry and Biochemistry of herbage. Academic Press, London, pp 227–247.

Hood A E M 1976 Nitrogen, grassland and water quality in the United Kingdom. Outl. Agric. 8, 320–327.

Ingestad T and Lund A B 1986 Theory and techniques for steady state mineral nutrition and growth of plants. Scand. J. For Res 1, 439–453.

Jenkinson D S 1986? Nitrogen in UK arable agriculture. J. Roy. Agr. Soc. England 147, 178–189.

Jollans J L 1985 Fertiliser in UK farming. Centre for Agriculture Strategy Rpt 9, University of Reading, England.

Keeney D R 1982 Nitrogen management for maximum efficiency and minimum pollution. *In* Nitrogen in Agricultural Soils. Ed. F J Stevenson, pp 605–649. American Society of Agronomy.

Lastigzon J 1981 World fertiliser progress into the 1980's. Tech. Bull. T-22. Int. Fert. Development Centre, Muscle Shouls, Ala.

Lee J A *et al* 1983 Nitrogen as a limiting factor in plant communities Ch. 5 *In* Nitrogen as an Ecological Factor, Eds. J A Lee, S McNeill and I H Rorison, pp 95–112. Blackwell, Oxford.

Lidgate H J 1984 Benefits of fertiliser input to arable cropping. Chem. Indus. (17 September) 649–652.

North T J 1986 Use and Management of the Land: Current and future trends. Proc. Br. Crop Protection Counc. Conf – Pests and Diseases 1A-1, pp 3–14.

Parr J F 1973 Chemical and biochemical considerations for maximising the efficiency of fertiliser nitrogen. J. Environ. Qual. 2, 75–84.

Parton W J, Anderson D W, Cole C V and Stewart J W B 1983 Simulation of soil organic matter formations and mineralisation in semi-arid agro ecosystems. *In* Nutrient Cycling in Agricultural Ecosystems. Eds. R R Lowrance, R L Todd, L

E Asmussen and R A Leonard, pp 533–550. Sp Publ. 23, The University of Georgia, USA.

Pate J J 1983 Patterns of nitrogen metabolism in higher plants and their ecological significance. *In* Nitrogen as an Ecological Factor. Eds. J A Lee, S McNeill and I H Rorison. pp 225–255. Blackwell, Oxford.

Royal Commission on Environmental Pollution 1979. Agriculture and pollution 7th Rept. 87–125, HMSO, London.

Royal Society 1983 The nitrogen cycle of the UK: A study group report. The Royal Society, London.

Ryden J C 1982. The nitrogen cycle in grassland: A case for studies in grazed pasture. Animal and Grassland Research Institute, Ann. Rept. pp 150–166.

Ryden J C *et al* 1984 Nitrate leaching from grassland. Nature 311, 50–53.

Sasanow S 1985 Acid rain: A review of current controversy. Soil Use and Management 1, 34–36.

Sicilliano J, Krulick S, Heisler E G, Schwartz J H and White J W 1975 Nitrate and nitrite content of some fresh and processed market vegetables. J. Agric Food Chem 23, 461–464.

Stangel P J 1984 World nitrogen situation — trends outlook and requirement. *In* Nitrogen in Crop Production. Ed. R D Hauck. pp 23–54. American Society of Agronomy.

Stanford G and Legg J O 1984 Nitrogen and yield potential. *In* Nitrogen in Crop Production. Ed. R D Hauck pp 263–272. American Society of Agronomy.

Stevenson F J 1986 Cycles of Soil Carbon, Nitrogen, Phosphorus, Sulphur and micronutrients. J Wiley & Sons, New York.

Thompson R 1985 The nitrate issue: A review. Soil Use and Management 1, 102–103.

UNIDO 1980 World wide study of the fertiliser industry 1975–2000. Revised document 81. United Nations Industrial Development Organisation, Vienna, Austria.

White J N 1975 Relative significance of dietary sources of nitrate and nitrite. J. Agric. Food Chem 23, 886–891.

White R E 1984 Nitrogen Cycle — Nitrogen leached from grassland. Nature 322, 10.

Whitmore A P and Addiscott T M 1986 Computer simulation of winter leaching losses of nitrate from soils cropped with winter wheat. Soil Use and Management 2, 26–30.

Whitmore A P and Addiscott T M 1987 Applications of computer modelling to predict mineral nitrogen in soil and nitrogen in crops. Soil Use and Management 3, 38–43.

Wilkinson W B and Green L A 1982 The water industry and the nitrogen cycle. Phil. Trans. R. Soc. London B 296, 459–475.

Developments in Plant and Soil Sciences

1. J. Monteith and C. Webb, Eds., Soil water and nitrogen in Mediterranean-type environments. 1981. ISBN 90-247-2406-6
2. J. C. Brogan, Ed., Nitrogen losses and surface run-off from landspreading of manures. 1981. ISBN 90-247-2471-6
3. J. D. Bewley, Ed., Nitrogen and carbon metabolism. 1981. ISBN 90-247-2472-4
4. R. Brouwer, I. Gašparíková, J. Kolek and B. C. Loughman, Eds., Structure and function of plant roots. 1981. ISBN 90-247-2510-0
5. Y. R. Dommergues and H. G. Diem, Eds., Microbiology of tropical soils and plant productivity. 1982. ISBN 90-247-2719-7
6. G. P. Robertson, R. Herrara and T. Rosswall, Eds., Nitrogen cycling in ecosystems of Latin America and the Caribbean. 1982. ISBN 90-247-2719-7
7. D. Atkinson et al., Eds., Tree root systems and their mycorrhizas. 1983. ISBN 90-247-2821-5
8. M. R. Sarić and B. C. Loughman, Eds., Genetic aspects of plant nutrition. 1983. ISBN 90-247-2822-3
9. J. R. Freney and J. R. Simpson, Eds., Gaseous loss of nitrogen from plant-soil systems. 1983. ISBN 90-247-2820-7
10. United Nations Economic Commission for Europe. Efficient use of fertilizers in agriculture. 1983. ISBN 90-247-2866-5
11. J. Tinsley and J. F. Darbyshire, Eds., Biological processes and soil fertility. 1984. ISBN 90-247-2902-5
12. A. D. L. Akkermans, D. Baker, K. Huss-Danell and J. D. Tjepkema, Eds., *Frankia* symbioses. 1984. ISBN 90-247-2967-X
13. W. S. Silver and E. C. Schröder, Eds., Practical application of *Azolla* for rice production. 1984. ISBN 90-247-3068-6
14. P. G. L. Vlek, Ed., Micronutrients in tropical food crop production. 1985. ISBN 90-247-3085-6
15. T. P. Hignett, Ed., Fertilizer manual. 1985. ISBN 90-247-3122-4
16. D. Vaughan and R. E. Malcolm, Eds., Soil organic matter and biological activity. 1985. ISBN 90-247-3154-2
17. D. Pasternak and A. San Pietro, Eds., Biosalinity in action: Bioproduction with saline water. 1985. ISBN 90-247-3159-3
18. M. Lalonde, C. Camiré and J. O. Dawson, Eds., *Frankia* and actinorhizal plants. 1985. ISBN 90-247-3214-X
19. H. Lambers, J. J. Neeteson and I. Stulen, Eds., Fundamental, ecological and agricultural aspects of nitrogen metabolism in higher plants. 1986. ISBN 90-247-3258-1
20. M. B. Jackson, Ed., New root formation in plants and cuttings. 1986. ISBN 90-247-3260-3
21. F. A. Skinner and P. Uomala, Eds., Nitrogen fixation with non-legumes. 1986. ISBN 90-247-3283-2
22. A. Alexander, Ed., Foliar fertilization. 1986. ISBN 90-247-3288-3
23. H. G. v.d. Meer, J. C. Ryden and G. C. Ennik, Eds., Nitrogen fluxes in intensive grassland systems. 1986. ISBN 90-247-3309-X
24. A. U. Mokwunye and P. L. G. Vlek, Eds., Management of nitrogen and phosphorus fertilizers in sub-Saharan Africa. 1986. ISBN 90-247-3312-X
25. Y. Chen and Y. Avnimelech, Eds., The role of organic matter in modern agriculture. 1986. ISBN 90-247-3360-X
26. S. K. De Datta and W. H. Patrick Jr., Eds., Nitrogen economy of flooded rice soils. 1986. ISBN 90-247-3361-8
27. W. H. Gabelman and B. C. Loughman, Eds., Genetic aspects of plant mineral nutrition. 1987. ISBN 90-247-3494-0
28. A. van Diest, Ed., Plant and Soil: Interfaces and interactions. 1987. ISBN 90-247-3535-1
29. United Nations, Ed., The utilization of secondary and trace elements in agriculture. 1987. ISBN 90-247-3546-7
30. H. G. v.d. Meer, R. J. Unwin, G. C. Ennik and T. A. van Dijk, Eds., Animal manure on grassland and fodder crops: Fertilizer or waste? 1987. ISBN 90-247-3568-8

31. N. J. Barrow, Reactions with variable-charge soils. 1987. ISBN 90-247-3589-0
32. D. P. Beck and L. A. Materon, Eds., Nitrogen fixation by legumes in Mediterranean agriculture. 1988. ISBN 90-247-3624-2
33. R. D. Graham, R. J. Hannam and N. C. Uren, Eds., Manganese in soils and plants. 1988. ISBN 90-247-3758-3
34. J. G. Torrey and J. L. Winship, Eds., Applications of continuous and steady-state methods to root biology. 1989. ISBN 0-7923-0024-6
35. F. A. Skinner, R. M. Boddey and I. Fendrik, Eds., Nitrogen fixation with non-legumes. 1989. ISBN 0–7923–0059–9
36. B. C. Loughman, O. Gašparíková and J. Kolek, Eds., Structural and functional aspects of transport in roots. 1989. ISBN 0–7923–0060–2
37. P. Planquaert and R. Haggar, Eds., Legumes in farming systems. 1989. ISBN 0-7923-0134-X
38. A. E. Osman, M. M. Ibrahim and M. A. Jones, Eds., The role of legumes in the farming systems of the Mediterranean areas. 1989. ISBN 0-7923-0419-5
39. M. Clarholm and L. Bergström, Eds., Ecology of arable land – Perspectives and challenges. 1989. ISBN 0-7923-0424-1